RANDOM D

This book provides an exposition of discrete time dynamic modeling over an infinite horizon. Chapter 1 reviews some mathematical results from the theory of deterministic dynamical systems, with particular emphasis on applications to economics. The theory of irreducible Markov processes, especially Markov chains, is surveyed in Chapter 2. Equilibrium and long-run stability of a dynamical system in which the law of motion is subject to random perturbations are the central theme of Chapters 3–5. A unified account of relatively recent results, exploiting splitting and contractions, that have found applications in many contexts is presented in detail. Chapter 6 explains how a random dynamical system may emerge from a class of dynamic programming problems. With examples and exercises, readers are guided from basic theory to the frontier of applied mathematical research.

Rabi Bhattacharya is Professor of Mathematics at the University of Arizona. He has also taught at the University of California at Berkeley and Indiana University. Professor Bhattacharya has held visiting research professorships at the University of Goettingen, the University of Bielefeld, and the Indian Statistical Institute. He is a recipient of a Guggenheim Fellowship and an Alexander Von Humboldt Forschungspreis. He is a Fellow of the Institute of Mathematical Statistics and has served on the editorial boards of a number of international journals, including the *Annals of Probability*, *Annals of Applied Probability*, *Journal of Multivariate Analysis*, *Econometric Theory*, and *Statistica Sinica*. He has co-authored *Normal Approximations and Asymptotic Expansions* (with R. Ranga Rao), *Stochastic Processes with Applications* (with E. C. Waymire), and *Asymptotic Statistics* (with M. Denker).

Mukul Majumdar is H. T. and R. I. Warshow Professor of Economics at Cornell University. He has also taught at Stanford University and the London School of Economics. Professor Majumdar is a Fellow of the Econometric Society and has been a Guggenheim Fellow, a Ford Rotating Research Professor at the University of California at Berkeley, an Erskine Fellow at the University of Canterbury, an Oskar Morgenstern Visiting Professor at New York University, a Lecturer at the College de France, and an Overseas Fellow at Churchill College, Cambridge University. Professor Majumdar has served on the editorial boards of many leading journals, including *The Review of Economic Studies*, *Journal of Economic Theory*, *Journal of Mathematical Economics*, and *Economic Theory*, and he has edited the collection *Organizations with Incomplete Information* (Cambridge University Press, 1998).

To Urmi, Deepta, and Aveek

Random Dynamical Systems

Theory and Applications

RABI BHATTACHARYA

University of Arizona

MUKUL MAJUMDAR

Cornell University

CAMBRIDGE UNIVERSITY PRESS
Cambridge, New York, Melbourne, Madrid, Cape Town, Singapore, São Paulo

Cambridge University Press
32 Avenue of the Americas, New York, NY 10013-2473, USA

www.cambridge.org
Information on this title: www.cambridge.org/9780521825658

First published 2007

Printed in the United States of America

A catalog record for this publication is available from the British Library.

Library of Congress Cataloging in Publication Data

Bhattacharya, R. N. (Rabindra Nath), 1937–
Random dynamical systems : theory and applications / Rabi Bhattacharya, Mukul Majumdar.
p. cm.
Includes bibliographical references and index.
ISBN-13: 978-0-521-82565-8 (hardback)
ISBN-10: 0-521-82565-2 (hardback)
ISBN-13: 978-0-521-53272-3 (pbk.)
ISBN-10: 0-521-53272-8 (pbk.)

1. Random dynamical systems.   I. Majumdar, Mukul, 1944–   II. Title.

QA614.835.B53   2007
515′.39–dc22   2006024380

ISBN   978-0-521-82565-8 hardback
ISBN   978-0-521-53272-3 paperback

Contents

v

Preface

The scope of this book is limited to the study of discrete time dynamic processes evolving over an infinite horizon. Its primary focus is on models with a one-period lag: "tomorrow" is determined by "today" through an exogenously given rule that is itself stationary or time-independent. A finite lag of arbitrary length may sometimes be incorporated in this scheme. In the deterministic case, the models belong to the broad mathematical class, known as dynamical systems, discussed in Chapter 1, with particular emphasis on those arising in economics. In the presence of random perturbations, the processes are random dynamical systems whose long-term stability is our main quest. These occupy a central place in the theory of discrete time stochastic processes.

Aside from the appearance of many examples from economics, there is a significant distinction between the presentation in this book and that found in standard texts on Markov processes. Following the exposition in Chapter 2 of the basic theory of irreducible processes, especially Markov chains, much of Chapters 3–5 deals with the problem of stability of random dynamical systems which may not, in general, be irreducible. The latter models arise, for example, if the random perturbation is limited to a finite or countable number of choices. Quite a bit of this theory is of relatively recent origin and appears especially relevant to economics because of underlying structures of monotonicity or contraction. But it is useful in other contexts as well.

In view of our restriction to discrete time frameworks, we have not touched upon powerful techniques involving deterministic and stochastic differential equations or calculus of variations that have led to significant advances in many disciplines, including economics and finance.

It is not possible to rely on the economic data to sift through various possibilities and to compute estimates with the degrees of precision that natural or biological scientists can often achieve through controlled experiments. We duly recognize that there are obvious limits to the lessons that formal models with exogenously specified laws of motion can offer.

The first chapter of the book presents a treatment of deterministic dynamical systems. It has been used in a course on dynamic models in economics, addressed to advanced undergraduate students at Cornell. Supplemented by appropriate references, it can also be part of a graduate course on dynamic economics. It requires a good background in calculus and real analysis.

Chapters 2–6 have been used as the core material in a graduate course at Cornell on Markov processes and their applications to economics. An alternative is to use Chapters 1–3 and 5 to introduce models of intertemporal optimization/equilibrium and the role of uncertainty. Complements and Details make it easier for the researchers to follow up on some of the themes in the text.

In addition to numerous examples illustrating the theory, many exercises are included for pedagogic purposes. Some of the exercises are numbered and set aside in paragraphs, and a few appear at the end of some chapters. But quite a few exercises are simply marked as (Exercise), in the body of a proof or an argument, indicating that a relatively minor step in reasoning needs to be formally completed.

Given the extensive use of the techniques that we review, we are unable to provide a bibliography that can do justice to researchers in many disciplines. We have cited several well-known monographs, texts, and review articles which, in turn, have extended lists of references for curious readers.

The quote attributed to Toni Morrison in Chapter 1 is available on the Internet from Simpson's Contemporary Quotations, compiled by J. B. Simpson.

The quote from Shizuo Kakutani in Chapter 2 is available on the Internet at www.uml.edu/Dept/Math/alumni/tangents/tangents_Fall2004/ MathInTheNews.htm. Endnote 1 of the document describes it as "a joke by Shizuo Kakutani at a UCLA colloquium talk as attributed in Rick Durrett's book *Probability: Theory and Examples*." The other quote in this chapter is adapted from Bibhuti Bandyopadhyay's original masterpiece in Bengali.

The quote from Gerard Debreu in Chapter 4 appeared in his article in *American Economic Review* (Vol. 81, 1991, pp. 1–7).

The quote from Patrick Henry in Chapter 5 is from Bartlett's Quotations (no. 4598), available on the Internet.

The quote attributed to Freeman J. Dyson in the same chapter appeared in the circulated abstract of his Nordlander Lecture ("The Predictable and the Unpredictable: How to Tell the Difference") at Cornell University on October 21, 2004.

The quote from Kenneth Arrow at the beginning of Chapter 6 appears in Chapter 2 of his classic *Essays in the Theory of Risk-Bearing.*

Other quotes are from sources cited in the text.

Acknowledgment

We would like to thank Vidya Atal, Kuntal Banerjee, Seung Han Yoo, Benjarong Suwankiri, Jayant Ganguli, Souvik Ghosh, Chao Gu, and Wee Lee Loh for research assistance. In addition, for help in locating references, we would like to thank Vidhi Chhaochharia and Kameshwari Shankar.

For direct and indirect contributions we are thankful to many colleagues: Professors Krishna Athreya, Robert Becker, Venkatesh Bala, Jess Benhabib, William Brock, Partha Dasgupta, Richard Day, Prajit Dutta, David Easley, Ani Guerdjikova, Nigar Hashimzade, Ali Khan, Nicholas Kiefer, Kaushik Mitra, Tapan Mitra, Kazuo Nishimura, Manfred Nermuth, Yaw Nyarko, Bezalel Peleg, Uri Possen, Debraj Ray, Roy Radner, Rangarajan Sundaram, Edward Waymire, Makoto Yano, and Ithzak Zilcha. Professor Santanu Roy was always willing to help out, with comments on stylistic and substantive matters.

We are most appreciative of the efforts of Ms. Amy Moesch: her patience and skills transformed our poorly scribbled notes into a presentable manuscript.

Mukul Majumdar is grateful for the support from the Warshow endowment and the Department of Economics at Cornell, as well as from the Institute of Economic Research at Kyoto University. Rabi Bhattacharya gratefully acknowledges support from the National Science Foundation with grants DMS CO-73865, 04-06143.

Two collections of published articles have played an important role in our exposition: a symposium on *Chaotic Dynamical Systems* (edited by Mukul Majumdar) and a symposium on *Dynamical Systems Subject to Random Shocks* (edited by Rabi Bhattacharya and Mukul Majumdar) that appeared in *Economic Theory* (the first in Vol. 4, 1995, and the second in Vol. 23, 2004). We acknowledge the enthusiastic support of Professor C. D. Aliprantis in this context.

Finally, thanks are due to Scott Parris, who initiated the project.

Notation

$\mathbb{Z}$	set of *all* integers.
$\mathbb{Z}_+(\mathbb{Z}_{++})$	set of all *nonnegative* (*positive*) integers.
$\mathbb{R}$	set of *all* real numbers.
$\mathbb{R}_+(\mathbb{R}_{++})$	set of all *nonnegative* (*positive*) real numbers.
$\mathbb{R}^\ell$	set of all ℓ-vectors.
$\mathbf{x} = (x_i) = (x_1, \ldots, x_\ell)$	an element of $\mathbb{R}^\ell$.
$\mathbf{x} \geq 0$	$x_i \geq 0$ for $i = 1, 2, \ldots, \ell$; [$\mathbf{x}$ is *nonnegative*].
$\mathbf{x} > 0$	$x_i \geq 0$ for all i; $x_i > 0$ for some i; [$\mathbf{x}$ is *positive*].
$\mathbf{x} \gg 0$	$x_i > 0$ for all i; [$\mathbf{x}$ is *strictly positive*].
$(S, \mathcal{S})$	a measurable space [when S is a metric space, $\mathcal{S} = \mathcal{B}(S)$ is the Borel sigmafield unless otherwise specified].

1

Dynamical Systems

Not only in research, but also in the everyday world of politics and economics, we would all be better off if more people realized that simple nonlinear systems do not necessarily possess simple dynamical properties.

Robert M. May

There is nothing more to say – except why. But since why is difficult to handle, one must take refuge in how.

Toni Morrison

1.1 Introduction

There is a rich literature on discrete time models in many disciplines – including economics – in which dynamic processes are described formally by first-order difference equations (see (2.1)). Studies of dynamic properties of such equations usually involve an appropriate definition of a steady state (viewed as a dynamic equilibrium) and conditions that guarantee its existence and local or global stability. Also of importance, particularly in economics following the lead of Samuelson (1947), have been the problems of comparative statics and dynamics: a systematic analysis of how the steady states or trajectories respond to changes in some parameter that affects the law of motion. While the dynamic properties of linear systems (see (4.1)) have long been well understood, relatively recent studies have emphasized that "the very simplest" nonlinear difference equations can exhibit "a wide spectrum of qualitative behavior," from stable steady states, "through cascades of stable cycles, to a regime in which the behavior (although fully deterministic) is in many respects chaotic or indistinguishable from the sample functions of a random process" (May 1976, p. 459). This chapter is not intended to be a

1

comprehensive review of the properties of complex dynamical systems, the study of which has benefited from a collaboration between the more "abstract" qualitative analysis of difference and differential equations, and a careful exploration of "concrete" examples through increasingly sophisticated computer experiments. It does recall some of the basic results on dynamical systems, and draws upon a variety of examples from economics (see Complements and Details).

There is by now a plethora of definitions of "chaotic" or "complex" behavior, and we touch upon a few properties of chaotic systems in Sections 1.2 and 1.3. However, the map (2.3) and, more generally, the quadratic family discussed in Section 1.7 provide a convenient framework for understanding many of the definitions, developing intuition and achieving generalizations (see Complements and Details). It has been stressed that the qualitative behavior of the solution to Equation (2.5) depends crucially on the initial condition. Trajectories emanating from initial points that are very close may display radically different properties. This may mean that small changes in the initial condition "lead to predictions so different, after a while, that prediction becomes in effect useless" (Ruelle 1991, p. 47). Even within the quadratic family, complexities are not "knife-edge," "abnormal," or "rare" possibilities. These observations are particularly relevant for models in social sciences, in which there are obvious limits to gathering data to identify the initial condition, and avoiding computational errors at various stages.

In Section 1.2 we collect some basic results on the existence of fixed points and their stability properties. Of fundamental importance is the contraction mapping theorem (Theorem 2.1) used repeatedly in subsequent chapters. Section 1.3 introduces complex dynamical systems, and the central result is the Li–Yorke theorem (Theorem 3.1). In Section 1.4 we briefly touch upon linear difference equations. In Section 1.5 we explore in detail dynamical systems in which the state space is $\mathbb{R}_+$, the set of nonnegative reals, and the law of motion α is an increasing function. Proposition 5.1 is widely used in economics and biology: it identifies a class of dynamical systems in which all trajectories (emanating from initial x in $\mathbb{R}_{++}$) converge to a unique fixed point. In contrast, Section 1.6 provides examples in which the long-run behavior depends on initial conditions. In the development of complex dynamical systems, the "quadratic family" of laws of motion (see (7.11)) has played a distinguished role. After a review of some results on this family in Section 1.7, we turn to examples of dynamical systems from economics and biology.

We have selected some descriptive models, some models of optimization with a single decision maker, a dynamic game theoretic model, and an example of intertemporal equilibrium with overlapping generations. An interesting lesson that emerges is that variations of some well-known models that generate monotone behavior lead to dynamical systems exhibiting Li–Yorke chaos, or even to systems with the quadratic family as possible laws of motion.

✳1.2 Basic Definitions: Fixed and Periodic Points ✳

We begin with some formal definitions. A dynamical system is described by a pair (S, α) where S is a nonempty set (called the *state space*) and α is a function (called the *law of motion*) from S into S. Thus, if x_t is the state of the system in period t, then

$$x_{t+1} = \alpha(x_t) \tag{2.1}$$

is the state of the system in period $t + 1$.

In this chapter we always assume that the state space S is a *(nonempty) metric space (the metric is denoted by d)*. As examples of (2.1), take S to be the set $\mathbb{R}$ of real numbers, and define

$$\alpha(x) = ax + b, \tag{2.2}$$

where a and b are real numbers.

Another example is provided by $S = [0, 1]$ and

$$\alpha(x) = 4x(1 - x). \tag{2.3}$$

Here in (2.3), $d(x, y) \equiv |x - y|$.

The evolution of the dynamical system $(\mathbb{R}, \alpha)$ where α is defined by (2.2) is described by the difference equation

$$x_{t+1} = ax_t + b. \tag{2.4}$$

Similarly, the dynamical system $([0, 1], \alpha)$ where α is defined by (2.3) is described by the difference equation

$$x_{t+1} = 4x_t(1 - x_t). \tag{2.5}$$

Once the initial state x (i.e., the state in period 0) is specified, we write $\alpha^0(x) \equiv x$, $\alpha^1(x) = \alpha(x)$, and for every positive integer $j \geq 1$,

$$\alpha^{j+1}(x) = \alpha(\alpha^j(x)). \tag{2.6}$$

We refer to α^j as the *jth iterate* of α. For any initial x, the *trajectory* from x is the sequence $\tau(x) = \{\alpha^j(x)_{j=0}^{\infty}\}$. The *orbit* from x is the set $\gamma(x) = \{y: y = \alpha^j(x) \text{ for some } j \geq 0\}$. The limit set $w(x)$ of a trajectory $\tau(x)$ is defined as

$$w(x) = \bigcap_{j=1}^{\infty} \overline{[\tau(\alpha^j(x)]}, \qquad (2.7)$$

where $\bar{A}$ is the closure of A.

Fixed and periodic points formally capture the intuitive idea of a *stationary* state or an *equilibrium* of a dynamical system. In his *Foundations*, Samuelson (1947, p. 313) noted that "*Stationary* is a descriptive term characterizing the behavior of an economic variable over time; it usually implies constancy, but is occasionally generalized to include behavior periodically repetitive over time."

A point $x \in S$ is a *fixed point* if $x = \alpha(x)$. A point $x \in S$ is a *periodic point* of *period* $k \geq 2$ if $\alpha^k(x) = x$ and $\alpha^j(x) \neq x$ for $1 \leq j < k$. Thus, to prove that x is a periodic point of period, say, 3, one must prove that x is a fixed point of α^3 and that it is *not* a fixed point of α and α^2. *Some writers consider a fixed point as a periodic point of period 1.*

Denote the set of all periodic points of S by $\wp(S)$. We write $\aleph(S)$ to denote the set of nonperiodic points.

We now note some useful results on the existence of fixed points of α.

Proposition 2.1 *Let $S = \mathbb{R}$ and α be continuous. If there is a (nondegenerate) closed interval $I = [a, b]$ such that (i) $\alpha(I) \subset I$ or (ii) $\alpha(I) \supset I$, then there is a fixed point of α in I.*

Proof.

(i) If $\alpha(I) \subset I$, then $\alpha(a) \geq a$ and $\alpha(b) \leq b$. If $\alpha(a) = a$ or $\alpha(b) = b$, the conclusion is immediate. Otherwise, $\alpha(a) > a$ and $\alpha(b) < b$. This means that the function $\beta(x) = \alpha(x) - x$ is positive at a and negative at b. Using the intermediate value theorem, $\beta(x^*) = 0$ for some x^* in (a, b). Then $\alpha(x^*) = x^*$.

(ii) By the Weierstrass theorem, there are points x_m and x_M in I such that $\alpha(x_m) \leq \alpha(x) \leq \alpha(x_M)$ for all x in I. Write $\alpha(x_m) = m$ and $\alpha(x_M) = M$. Then, by the intermediate value theorem, $\alpha(I) = [m, M]$.

Since $\alpha(I) \supset I, m \le a \le b \le M$. In other words,

$$\alpha(x_m) = m \le a \le x_m,$$

and

$$\alpha(x_M) = M \ge b \ge x_M.$$

The proof can now be completed by an argument similar to that in case (i). ∎

Remark 2.1 Let $S = [a, b]$ and α be a continuous function from S into S. Suppose that for all x in (a, b) the derivative $\alpha'(x)$ exists and $|\alpha'(x)| < 1$. Then α has a unique fixed point in S.

Proposition 2.2 *Let S be a nonempty compact convex subset of $\mathbb{R}^\ell$, and α be continuous. Then there is a fixed point of α.*

A function $\alpha : S \to S$ is a *uniformly strict contraction* if there is some $C, 0 < C < 1$, such that for all $x, y \in X, x \ne y$, one has

$$d(\alpha(x), \, \alpha(y)) < Cd(x, y). \tag{2.8}$$

If $d(\alpha(x), \alpha(y)) < d(x, y)$ for $x \ne y$, we say that α is a *strict contraction*. If only

$$d(\alpha(x), \alpha(y)) \le d(x, y),$$

we say that α is a *contraction*.

If α is a contraction, α is continuous on S.

In this book, the following fundamental theorem is used many times:

Theorem 2.1 *Let (S, d) be a nonempty complete metric space and $\alpha : S \to S$ be a uniformly strict contraction. Then α has a unique fixed point $x^* \in S$. Moreover, for any x in S, the trajectory $\tau(x) = \{\alpha^j(x)\}_{j=0}^\infty\}$ converges to x^*.*

Proof. Choose an arbitrary $x \in S$. Consider the trajectory $\tau(x) = (x_t)$ from x, where

$$x_{t+1} = \alpha(x_t). \tag{2.9}$$

Note that $d(x_2, x_1) = d(\alpha(x_1), \alpha(x)) < Cd(x_1, x)$ for some $C \in (0, 1)$; hence, for any $t \geq 1$,

$$d(x_{t+1}, x_t) < C^t d(x_1, x). \tag{2.10}$$

We note that

$$\begin{aligned} d(x_{t+2}, x_t) &\leq d(x_{t+2}, x_{t+1}) + d(x_{t+1}, x_t) \\ &< C^{t+1} d(x_1, x) + C^t d(x_1, x) \\ &= C^t(1 + C)d(x_1, x). \end{aligned}$$

It follows that for any integer $k \geq 1$,

$$d(x_{t+k}, x_t) < [C^t/(1 - C)]d(x_1, x),$$

and this implies that (x_t) is a Cauchy sequence. Since S is assumed to be complete, $\lim_{t \to \infty} x_t = x^*$ exists. By continuity of α, and (2.9),

$$\alpha(x^*) = x^*.$$

If there are two distinct fixed points x^* and x^{**} of α, we see that there is a contradiction:

$$0 < d(x^*, x^{**}) = d(\alpha(x^*), \alpha(x^{**})) < Cd(x^*, x^{**}), \tag{2.11}$$

where $0 < C < 1$. ∎

Remark 2.2 For applications of this fundamental result, it is important to reflect upon the following:

(i) for any $x \in S$, $d(\alpha^n(x), x^*) \leq C^n(1 - C)^{-1}d(\alpha(x), x))$,
(ii) for any $x \in S$, $d(x, x^*) \leq (1 - C)^{-1}d(\alpha(x), x)$.

Theorem 2.2 *Let S be a nonempty complete metric space and $\alpha : S \to S$ be such that α^k is a uniformly strict contraction for some integer $k > 1$. Then α has a unique fixed point $x^* \in S$.*

Proof. Let x^* be the unique fixed point of α^k. Then

$$\alpha^k(\alpha(x^*)) = \alpha(\alpha^k(x^*)) = \alpha(x^*)$$

Hence $\alpha(x^*)$ is also a fixed point of α^k. By uniqueness, $\alpha(x^*) = x^*$. This means that x^* is a fixed point of α. But *any* fixed point of α is a fixed point of α^k. Hence x^* is the unique fixed point of α. ∎

Theorem 2.3 *Let S be a nonempty compact metric space and* $\alpha : S \to S$ *be a strict contraction. Then* α *has a unique fixed point.*

Proof. Since $d(\alpha(x), x)$ is continuous and S is compact, there is an $x^* \in S$ such that

$$d(\alpha(x^*), x^*) = \inf_{x \in S} d(\alpha(x), x). \qquad (2.12)$$

Then $\alpha(x^*) = x^*$, otherwise

$$d(\alpha^2(x^*), \alpha(x^*)) < d(\alpha(x^*), x^*),$$

contradicting (2.12). ∎

Exercise 2.1

(a) Let $S = [0, 1]$, and consider the map $\alpha : S \to S$ defined by

$$\alpha(x) = x - \frac{x^2}{2}.$$

Show that α is a strict contraction, but *not* a uniformly strict contraction. Analyze the behavior of trajectories $\tau(x)$ from $x \in S$.

(b) Let $S = \mathbb{R}$, and consider the map $\alpha : S \to S$ defined by

$$\alpha(x) = [x + (x^2 + 1)^{1/2}]/2.$$

Show that $\alpha(x)$ is a strict contraction, but does not have a fixed point. ∎

A fixed point x^* of α is (locally) *attracting* or (locally) *stable* if there is an open set U containing x^* such that for all $x \in U$, the trajectory $\tau(x)$ from x converges to x^*.

We shall often drop the caveat "local": note that *local attraction* or *local stability* is to be distinguished from the property of *global stability* of a dynamical system: (S, α) is *globally stable* if for all $x \in S$, the trajectory $\tau(x)$ converges to the unique fixed point x^*. Theorem 2.1 deals with *global stability*.

A fixed point x^* of α is *repelling* if there is an open set U containing x^* such that for any $x \in U, x \neq x^*$, there is some $k \geq 1, \alpha^k(x) \notin U$.

Consider a dynamical system (S, α) where S is a (nondegenerate) closed interval $[a, b]$ and α is continuous on $[a, b]$. Suppose that α is also continuously differentiable on (a, b). A fixed point $x^* \in (a, b)$ is *hyperbolic* if $|\alpha'(x^*)| \neq 1$.

Proposition 2.3 *Let $S = [a, b]$ and α be continuous on $[a, b]$ and continuously differentiable on (a, b). Let $x^* \in (a, b)$ be a hyperbolic fixed point of α.*

    (a) If $|\alpha'(x^*)| < 1$, then x^* is locally stable.
    (b) If $|\alpha'(x^*)| > 1$, then x^* is repelling.

Proof.

    (a) There is some $u > 0$ such that $|\alpha'(x)| < \mathbf{m} < 1$ for all x in $I = [x^* - u, x^* + u]$. By the mean value theorem, if $x \in I$,

$$|\alpha(x) - x^*| = |\alpha(x) - \alpha(x^*)| \leq \mathbf{m}|x - x^*| < \mathbf{m}u < u.$$

Hence, α maps I into I and, again, by the mean value theorem, is a uniformly strict contraction on I. The result follows from Theorem 2.1.
    (b) this is left as an exercise.   ■

We can define "a hyperbolic periodic point of period k" and define (locally) attracting and repelling periodic points accordingly.
    Let x_0 be a periodic point of period 2 and $x_1 = \alpha(x_0)$. By definition $x_0 = \alpha(x_1) = \alpha^2(x_0)$ and $x_1 = \alpha(x_0) = \alpha^2(x_1)$. Now if α is differentiable, by the chain rule,

$$[\alpha^2(x_0)]' = \alpha'(x_1)\alpha'(x_0).$$

More generally, suppose that x_0 is a periodic point of period k and its orbit is denoted by $\{x_0, x_1, \ldots, x_{k-1}\}$. Then,

$$[\alpha^k(x_0)]' = \alpha'(x_{k-1}) \cdots \alpha'(x_0).$$

It follows that

$$[\alpha^k(x_0)]' = [\alpha^k(x_1)]' \cdots [\alpha^k(x_{k-1})]'.$$

We can now extend Proposition 2.3 appropriately.
    While the contraction property of α ensures that, independent of the initial condition, the trajectories enter any neighborhood of the fixed point, there are examples of simple nonlinear dynamical systems in which trajectories "wander around" the state space. We shall examine this feature more formally in Section 1.3.

Example 2.1 Let $S = \mathbb{R}$, $\alpha(x) = x^2$. Clearly, the only fixed points of α are 0, 1. More generally, keeping $S = \mathbb{R}$, consider the family of dynamical systems $\alpha_\theta(x) = x^2 + \theta$, where θ is a real number. For $\theta > 1/4$, α_θ does not have any fixed point; for $\theta = 1/4$, α_θ has a unique fixed point $x = 1/2$; for $\theta < 1/4$, α_θ has a pair of fixed points.

When $\theta = -1$, the fixed points of the map $\alpha_{(-1)}(x) = x^2 - 1$ are $[1 + \sqrt{5}]/2$ and $[1 - \sqrt{5}]/2$. Now $\alpha_{(-1)}(0) = -1$; $\alpha_{(-1)}(-1) = 0$. Hence, both 0 and -1 are periodic points of period 2 of $\alpha_{(-1)}$. It follows that:

$$\tau(0) = (0, -1, 0, -1, \ldots), \quad \tau(-1) = (-1, 0, -1, 0, \ldots),$$
$$\gamma(-1) = \{-1, 0\}, \gamma(0) = \{0, -1\}.$$

Since

$$\alpha_{(-1)}^2(x) = x^4 - 2x^2,$$

we see that (i) $\alpha_{(-1)}^2$ has four fixed points: the fixed points of $\alpha_{(-1)}$, and $0, -1$; (ii) the derivative of $\alpha_{(-1)}^2$ with respect to x, denoted by $[\alpha_{(-1)}^2(x)]'$, is given by

$$[\alpha_{(-1)}^2(x)]' = 4x^3 - 4x.$$

Now, $[\alpha_{(-1)}^2(x)]'_{x=0} = [\alpha_{(-1)}^2(x)]'_{x=-1} = 0$. Hence, both 0 and -1 are attracting fixed points of α^2. ∎

Example 2.2 Let $S = [0, 1]$. Consider the "tent map" defined by

$$\alpha(x) = \begin{cases} 2x & \text{for } x \in [0, 1/2] \\ 2(1-x) & \text{for } x \in [1/2, 1]. \end{cases}$$

Note that α has two fixed points "0" and "2/3." It is tedious to write out the functional form of α^2:

$$\alpha^2(x) = \begin{cases} 4x & \text{for } x \in [0, 1/4] \quad [0, 0.25] \\ 2(1-2x) & \text{for } x \in [1/4, 1/2] \quad [0.25, 0.5] \\ 2(2x-1) & \text{for } x \in [1/2, 3/4] \quad [0.5, 0.75] \\ 4(1-x) & \text{for } x \in [3/4, 1]. \quad [0.75, 1] \checkmark \end{cases}$$

Verify the following:

(i) "2/5" and "4/5" are periodic points of period 2.

(ii) "2/9," "4/9," "8/9" are periodic points of period 3. It follows from a well-known result (see Theorem 3.1) that there are periodic points of *all* periods.

By using the graphs, if necessary, verify that the fixed and periodic points of the tent map are repelling.   ■

Example 2.3 In many applications to economics and biology, the state space S is the set of all nonnegative reals, $S = \mathbb{R}_+$. The law of motion $\alpha : S \to S$ has the special form

$$\alpha(x) = x\beta(x), \tag{2.11'}$$

where $\beta(0) \geq 0$, $\beta : \mathbb{R}_+ \to \mathbb{R}_+$ is continuous (and often has additional properties). Now, the fixed points $\hat{x}$ of α must satisfy

$$\alpha(\hat{x}) = \hat{x}$$

or

$$\hat{x}[1 - \beta(\hat{x})] = 0.$$

The fixed point $\hat{x} = 0$ may have a special significance in a particular context (e.g., extinction of a natural resource). Some examples of α satisfying (2.11') are

(Verhulst 1845)	$\alpha(x) = \dfrac{\theta_1 x}{x + \theta_2}$,	$\theta_1 > 0, \theta_2 > 0.$
(Hassell 1975)	$\alpha(x) = \theta_1 x (1 + x)^{-\theta_2}$,	$\theta_1 > 0, \theta_2 > 0.$
(Ricker 1954)	$\alpha(x) = \theta_1 x e^{-\theta_2 x}$,	$\theta_1 > 0, \theta_2 > 0.$

Here θ_1, θ_2 are interpreted as exogenous parameters that influence the law of motion α.

Assume that $\beta(x)$ is differentiable at $x \geq 0$. Then,

$$\alpha'(x) = \beta(x) + x\beta'(x).$$

Hence,

$$\alpha'(0) = \beta(0).$$

For each of the special maps, the existence of a fixed point $\hat{x} \neq 0$ and the local stability properties depend on the values of the parameters θ_1, θ_2. We shall now elaborate on this point.

For the Verhulst map $\alpha(x) = \theta_1 x / (x + \theta_2)$, where $x \geq 0, \theta_1 > 0$, and $\theta_2 > 0$, there are two cases:

Case I: $\theta_1 \leq \theta_2$. Here $x^* = 0$ is the unique fixed point;
Case II: $\theta_1 > \theta_2$. Here there are two fixed points $x_{(1)}^* = 0$ and $x_{(2)}^* = \theta_1 - \theta_2$.

Verify that $\alpha'(0) = (\theta_1/\theta_2)$. Hence, in Case I, $x^* = 0$ is locally attracting if $(\theta_1/\theta_2) < 1$. In Case II, however, $\alpha'(x_1^*) = \alpha'(0) > 1$, so $x_1^* = 0$ is repelling, whereas x_2^* is locally attracting, since

$$\alpha'(x_2^*) \equiv \alpha'(\theta_1 - \theta_2) = (\theta_2/\theta_1) < 1.$$

For the Hassell map, there are two cases:

Case I: $\theta_1 \leq 1$. Here $x^* = 0$ is the unique fixed point.
Case II: $\theta_1 > 1$. Here there are two fixed points $x_{(1)}^* = 0$, $x_{(2)}^* = (\theta_1)^{1/\theta_2} - 1$.

In Case I, if $\theta_1 < 1$, $x^* = 0$ is locally attracting. In Case II, $x_{(1)}^* = 0$ is repelling. Some calculations are needed to show that the fixed point $x_{(2)}^* = (\theta_1)^{1/\theta_2} - 1$ is locally stable if

$$\theta_1 < \left(\frac{\theta_2}{\theta_2 - 2} \right)^{\theta_2} \quad \text{and} \quad \theta_2 > 2.$$

For the Ricker map, there are two cases.

Case I: $\theta_1 \leq 1$. Here $x^* = 0$ is the unique fixed point.
Case II: $\theta_1 > 1$. Here $x_{(1)}^* = 0$ and $x_{(2)}^* = (\log \theta_1)/\theta_2$ both are fixed points. Note that for all $0 < \theta_1 < 1, x^* = 0$ is locally attracting. For $\theta_1 > 1, x_{(1)}^* = 0$ is repelling. The fixed point $x_{(2)}^* = (\log \theta_1)/\theta_2$ is locally attracting if

$$|1 - \log \theta_1| < 1$$

(which holds when $1 < \theta_1 < e^2$). ∎

1.3 Complexity

1.3.1 Li–Yorke Chaos and Sarkovskii Theorem

In this section we take the state space S to be a (nondegenerate) interval I in the real line, and α a continuous function from I into I.

A subinterval of an interval I is an interval contained in I. Since α is continuous, $\alpha(I)$ is an interval. If I is a compact interval, so is $\alpha(I)$.

Suppose that a dynamical system (S, α) has a periodic point of period k. Can we conclude that it also has a periodic point of some other period $k' \neq k$? It is useful to look at a simple example first.

Example 3.1 Suppose that (S, α) has a periodic point of period $k(\geq 2)$. Then it has a fixed point (i.e., a periodic point of period one). To see this, consider the orbit γ of the periodic point of period k, and let us write

$$\gamma = \{x^{(1)}, \ldots, x^{(k)}\},$$

where $x^{(1)} < x^{(2)} < \cdots < x^{(k)}$. Both $\alpha(x^{(1)})$ and $\alpha(x^{(k)})$ must be in γ. This means that

$$\alpha(x^{(1)}) = x^{(i)} \text{ for some } i > 1$$

and

$$\alpha(x^{(k)}) = x^{(j)} \text{ for some } j < k.$$

Hence, $\alpha(x^{(1)}) - x^{(1)} > 0$ and $\alpha(x^{(k)}) - x^{(k)} < 0$.

By the intermediate value theorem, there is some x in S such that $\alpha(x) = x$.   ∎

We shall now state the Li–Yorke theorem (Li and Yorke 1975) and provide a brief sketch of the proof of one of the conclusions.

Theorem 3.1 *Let I be an interval and $\alpha : I \to I$ be continuous. Assume that there is some point a in I for which there are points $b = \alpha(a)$, $c = \alpha(b)$, and $d = \alpha(c)$ satisfying*

$$d \leq a < b < c \text{ (or } d \geq a > b > c). \tag{3.1}$$

Then

[1] for every positive integer $k = 1, 2, \ldots$ there is a periodic point of period k, $x^{(k)}$, in I,

[2] there is an uncountable set $\aleph' \subset \aleph(I)$ such that

(i) for all x, y in $\aleph'$, $x \neq y$,

$$\limsup_{n \to \infty} |\alpha^n(x) - \alpha^n(y)| > 0; \tag{3.2}$$

$$\liminf_{n \to \infty} |\alpha^n(x) - \alpha^n(y)| = 0. \tag{3.3}$$

(ii) If $x \in \aleph'$ and $y \in \wp(I)$

$$\limsup_{n \to \infty} |\alpha^n(x) - \alpha^n(y)| > 0.$$

Proof of [1].

Step 1. Let G be a real-valued continuous function on an interval I. For any compact subinterval I_1 of $G(I)$ there is a compact subinterval Q of I such that $G(Q) = I_1$.

Proof of Step 1. One can figure out the subinterval Q directly as follows. Let $I_1 = [G(x), G(y)]$ where x, y are in I. Assume that $x < y$. Let r be the last point of $[x, y]$ such that $G(r) = G(x)$; let s be the first point after r such that $G(s) = G(y)$. Then $Q = [r, s]$ is mapped onto I_1 under G. The case $x > y$ is similar.

Step 2. Let I be an interval and $\alpha : I \to I$ be continuous. Suppose that $(I_n)_{n=0}^{\infty}$ is a sequence of compact subintervals of I, and for all n,

$$I_{n+1} \subset \alpha(I_n). \tag{3.4}$$

Then there is a sequence of compact subintervals (Q_n) of I such that for all n,

$$Q_{n+1} \subset Q_n \subset Q_0 = I_0 \tag{3.5}$$

and

$$\alpha^n(Q_n) = I_n. \tag{3.6}$$

Hence, there is

$$x \in \bigcap_n Q_n \text{ such that } \alpha^n(x) \in I_n \text{ for all } n. \tag{3.7}$$

Proof of Step 2. The construction of the sequence Q_n proceeds "inductively" as follows: Define $Q_0 = I_0$. Recall that α^0 is defined as the identity mapping, so $\alpha^0(Q_0) = I_0$ and $I_1 \subset \alpha(I_0)$. If Q_{n-1} is defined as a compact subinterval such that $\alpha^{n-1}(Q_{n-1}) = I_{n-1}$, then $I_n \subset \alpha(I_{n-1}) = \alpha^n(Q_{n-1})$. Use Step 1, with $G = \alpha^n$ on Q_{n-1}, in order to get a compact subinterval Q_n of Q_{n-1} such that $\alpha^n(Q_n) = I_n$. This completes the induction argument (establishing (3.5) and (3.6)). Compactness of Q_n leads to (3.7). ∎

Now we prove [1]. Assume that $d \leq a < b < c$ (the other case $d \geq a > b > c$ is treated similarly).

Write $K = [a, b]$ and $L = [b, c]$.

Let k be any positive integer.

For $k > 1$, define a sequence of intervals (I_n) as follows:
$$I_n = L \text{ for } n = 0, 1, 2, \ldots, k-2; \quad I_{k-1} = K; \text{ and } I_{n+k} = I_n \text{ for}$$
$n = 0, 1, 2, \ldots$.

For $k = 1$, let $I_n = L$ for all n.

Let Q_n be the intervals in Step 2. Note that $Q_k \subset Q_0 = I_0$ and $\alpha^k(Q_k) = I_k = I_0$. Hence, Proposition 2.1 applied to α^k gives us a fixed point x^k of α^k in Q_k. Now, x^k cannot have a period less than k; otherwise, we need to have $\alpha^{k-1}(x^k) = b$, contrary to $\alpha^{k+1}(x^k) \in L$. ∎

Proof of [2]. See Complements and Details. ∎

We shall now state *Sarkovskii's theorem* on periodic points. Consider the following Sarkovskii ordering of the positive integers:

$$3 \triangleright 5 \triangleright 7 \cdots \triangleright 2.3 \triangleright 2.5 \cdots \triangleright 2^2 3 \triangleright 2^2 5 \triangleright \cdots \quad \text{(SO)}$$
$$\triangleright 2^3.3 \triangleright 2^3.5 \cdots \triangleright 2^3 \triangleright 2^2 \triangleright 2 \triangleright 1$$

In other words, first list all the odd integers beginning with 3; next list 2 times the odds, 2^2 times the odds, etc. Finally, list all the powers of 2 in decreasing order.

Theorem 3.2 *Let $S = \mathbb{R}$ and α be a continuous function from S into S. Suppose that α has a periodic point of period k. If $k \triangleright k'$ in the Sarkovskii ordering (SO), then α has a periodic point of period k'.*

Proof. See Devaney (1986). ∎

It follows that if α has only finitely many periodic points, then they all necessarily have periods that are powers of two.

1.3.2 A Remark on Robustness of Li–Yorke Complexity

Let $S = [J, K]$, and suppose that a continuous function α satisfies the Li–Yorke condition (3.1) with strict inequality throughout; i.e., suppose that there are points a, b, c, d such that

$$d = \alpha(c) < a < b = \alpha(a) < c = \alpha(b) \quad (3.8)$$

or

$$d = \alpha(c) > a > b = \alpha(a) > c = \alpha(b). \qquad (3.9)$$

Consider the space $C(S)$ of all continuous (hence, bounded) real-valued functions on $S = [J, K]$. Let $\|\alpha\| = \max_{x \in S} \alpha(x)$. The conclusions of Theorem 3.1 hold with respect to the dynamical system (S, α). But the Li–Yorke complexity is now "robust" in a precise sense.

Proposition 3.1 *Let $S = [J, K]$, and let α satisfy (3.8). In addition, assume that*

$$J < m(S, \alpha) < M(S, \alpha) < K, \qquad (3.10)$$

where $m(S, \alpha)$ and $M(S, \alpha)$ are respectively the minimum and maximum of α on $[J, K]$. Then there is an open set N of $C(S)$ containing α such that $\beta \in N$ implies that [1] and [2] of Theorem 3.1 hold with β in place of α.

Proof. First, we show the following:
Fix $x \in [J, K]$. Given $k \geq 1$, $\varepsilon > 0$, there exists $\delta(k, \varepsilon) > 0$ such that "$\|\beta - \alpha\| < \delta(k, \varepsilon)$" implies $|\beta^j(x) - \alpha^j(x)| < \varepsilon$ for all $j = 1, \ldots, k$.

The proof is by induction on k. It is clearly true for $k = 1$, with $\delta(1, \varepsilon) \equiv \varepsilon$. Assume that the claim is true for $k = m$, but not for $k = m + 1$. Then there exist some $\varepsilon > 0$ and a sequence of functions $\{\beta_n\}$ satisfying $\|\beta_n - \alpha\| \to 0$ such that $|\beta_n^{m+1}(x) - \alpha^{m+1}(x)| \geq \varepsilon$. Let $\beta_n^m(x) = y_n$ and $\alpha^m(x) = y$. Then, by the induction hypothesis, $y_n \to y$. From Rudin (1976, Chapter 7) we conclude that $\beta_n(y_n) \to \alpha(y)$, which yields a contradiction.

Next, choose a real number ρ satisfying $0 < \rho < \min[1/2(a - d), 1/2(b - a), 1/2(c - b)]$ and a positive number r such that $\|\beta - \alpha\| < r$ implies $|\beta^j(a) - \alpha^j(a)| < \rho$ for $j = 1, 2, 3$, and also $0 < r < \min\{K - M(S, \alpha), m(S, \alpha) - J\}$.

Define the open set N as

$$N = \{\beta \in C(S) : \|\beta - \alpha\| < r\}.$$

It follows that any $\beta \in N$ maps S into S, since the maximum of β on $[J, K]$ is less than $M(S, \alpha) + r < K$. Similarly, the minimum of β on $[J, K]$ is likewise greater than J. It remains to show that the condition (3.8) also holds for any β in N. Recall that $\alpha(a) = b$, $\alpha(b) = c$, $\alpha(c) = d$.

Since $|\beta(a) - \alpha(a)| \equiv |\beta(a) - b| < \rho$, we have

(i) $\beta(a) > b - \rho > a + \rho > a$.
Likewise, since $\beta(a) < b + \rho$ and $|\beta^2(a) - c| < \rho$, we get
(ii) $\beta^2(a) > c - \rho > b + \rho > \beta(a)$.
Finally, since $|\beta^3(a) - d| < \rho$, we get
(iii) $\beta^3(a) < d + \rho < a - \rho < a$.   ∎

1.3.3 Complexity: Alternative Approaches

Attempts to capture the complexity of dynamical systems have led to
alternative definitions of chaos that capture particular properties. Here,
we briefly introduce two interesting properties: *topological transitivity*
and *sensitive dependence on initial condition*.

A dynamical system (S, α) is *topologically transitive* if for any pair of
nonempty open sets U and V, there exists $k \geq 1$ such that $\alpha^k(U) \cap V \neq
\phi$. Of interest are the following two results.

Proposition 3.2 *If there is some x such that $\gamma(x)$, the orbit from x, is
dense in S, then (S, α) is topologically transitive.*

Proof.   Left as an exercise.   ∎

Proposition 3.3 *Let S be a (nonempty) compact metric space. Assume
that (S, α) is topologically transitive. Then there is some $x \in S$ such that
the orbit $\gamma(x)$ from x is dense in S.*

Proof.   Since S is compact, it has a countable base of open sets; i.e.,
there is a family $\{V_n\}$ of open sets in S with the property that if M is any
open subset of S, there is some $V_n \subset M$.

Corresponding to each V_n, define the set O_n as follows:

$$O_n = \{x \in S : \alpha^j(x) \in V_n, \quad \text{for some } j \geq 0\}.$$

O_n is open, by continuity of α. By topological transitivity it is also dense
in S. By the Baire category theorem (see Appendix), the intersection
$O = \cap_n O_n$ is nonempty (in fact, dense in S). Take any $x \in O$, and consider the orbit $\gamma(x)$ from x. Take any y in S and any open M containing
y. Then M contains some V_n. Since x belongs to the corresponding O_n,
there is some element of $\gamma(x)$ in V_n. Hence, $\gamma(x)$ is dense in S.   ∎

It is important to reflect upon the behavior of a topologically transitive dynamical system and contrast it to one in which the law of motion satisfies the strict contraction property. We now turn to another concept that has profound implications for the long-run prediction of a dynamical system. A dynamical system (S, α) has *sensitive dependence on initial condition* if there is $\partial > 0$ such that for any $x \in S$ and any neighborhood N of x there exist $y \in N$ and an integer $j \geq 0$ with the property $|\alpha^j(x) - \alpha^j(y)| > \partial$.

Devaney (1986) asserted that if a dynamical system "possesses sensitive dependence on initial condition, then for all practical purposes, the dynamics defy numerical computation. Small errors in computation which are introduced by round-off may become magnified upon iteration. The results of numerical computation of an orbit, no matter how accurate, may bear no resemblance whatsoever with the real orbit."

Example 3.2 The map $\alpha(x) = 4x(1 - x)$ on $[0, 1]$ is topologically transitive and has sensitive dependence on initial condition (see Devaney 1986). ∎

1.4 Linear Difference Equations

Excellent coverage of this topic is available from many sources: we provide only a sketch. Consider

$$x_{t+1} = ax_t + b. \tag{4.1}$$

When $a = 1, x_t = x_0 + bt$. When $a = -1, x_t = -x_0 + b$ for $t = 1, 3, \ldots$, and $x_t = x_0$ for $t = 2, 4, \ldots$.

Now assume $a \neq 1$. The map $\alpha(x) = ax + b$ when $a \neq 1$, has a unique fixed point $x^* = \frac{b}{1-a}$. Given any initial $x_0 = x$, the solution to (4.1) can be verified as

$$x_t = (x - x^*)a^t + x^*. \tag{4.2}$$

The long-run behavior of trajectories from alternative initial x can be analyzed from (4.2). In this context, the important fact is that the sequence a^t converges to 0 if $|a| < 1$ and becomes unbounded if $|a| > 1$.

Example 4.1 Consider an economy where the output $y_t(\geq 0)$ in any period is divided between consumption $c_t(\geq 0)$ and investment $x_t(\geq 0)$. The return function is given by

$$y_{t+1} = rx_t, \quad r > 1, \, t \geq 0. \tag{4.3}$$

Given an initial stock $y > 0$, a *program* $\mathbf{x} = (x_t)$ (from y) is a nonnegative sequence satisfying $x_0 \leq y$, $x_{t+1} \leq rx_t$ (for $t \geq 1$). It generates a corresponding *consumption program* $\mathbf{c} = (c_t)$ defined by $c_0 = y - x_0$; $c_{t+1} = y_{t+1} - x_{t+1} = rx_t - x_{t+1}$ for $t \geq 0$.

(a) Show that a consumption program $\mathbf{c} = (c_t)$ must satisfy

$$\sum_{t=0}^{\infty} c_t / r^t \leq y.$$

(b) Call a program $\mathbf{x}$ from y [generating $\mathbf{c}$] *efficient* if there does not exist another program $\mathbf{x}'$ [generating $(\mathbf{c}')$] from y such that $c_t' \geq c_t$ for all $t \geq 0$ with strict inequality for some t. Show that a program $\mathbf{x}$ is efficient if and only if

$$\sum_{t=0}^{\infty} c_t / r^t = y.$$

(c) Suppose that for the economy to survive it must consume an amount $c > 0$ in every period. Informally, we say that a program $\mathbf{x}$ from y that satisfies $c_t \geq c$ for *all* $t \geq 0$ survives at (or above) c. Note that the law of motion of the economy that plans a consumption $c \geq 0$ in *every* period $t \geq 1$, with an investment x_0 initially, can be written as

$$x_{t+1} = rx_t - c. \tag{4.4}$$

This equation has a solution

$$x_t = r^t(x_0 - \xi) + \xi,$$

where

$$\xi = c/(r - 1). \tag{4.5}$$

Hence,

(1) x_t reaches 0 in finite time if $x_0 < \xi$;
(2) $x_t = \xi$ for all t if $x_0 = \xi$;
(3) x_t diverges to (plus) infinity if $x_0 > \xi$.

To summarize the implications for survival and sustainable development we state the following:

Proposition 4.1 *Survival at (or above) c > 0 is possible if and only if the initial stock y ≥ ξ + c, or equivalently,*

$$y \geq \left(\frac{r}{r-1} \right) c.$$

Of course if $r \leq 1$, there is no program that can guarantee that $c_t \geq c$ for any $c > 0$. ∎

In this chapter, our exposition deals with processes that are generated by an *autonomous* or *time invariant* law of motion [in (2.1), the function $\alpha : S \to S$ does *not* depend on time]. We shall digress briefly and study a simple example which explicitly recognizes that the law of motion may itself depend on time.

Example 4.2 Let $x_t (t \geq 0)$ be the stock of a natural resource at the *beginning* of period t, and assume that the evolution of x_t is described by the following (nonautonomous) difference equation

$$x_{t+1} = ax_t + b_{t+1}, \tag{4.6}$$

where $0 < a < 1$ and $(b_{t+1})_{t \geq 0}$ is a sequence of nonnegative numbers. During period t, $(1 - a)x_t$ is consumed or used up and b_{t+1} is the "new discovery" of the resource reported at the beginning of period $t + 1$.

Given an initial $x_0 > 0$, one can write

$$x_t = a^t x_0 + \sum_{i=0}^{t-1} a^i b_{t-i}. \tag{4.7}$$

Now, if the sequence $(b_{t+1})_{t \geq 0}$ is *bounded above*, i.e., if there is $B > 0$ such that $0 \leq b_{t+1} \leq B$ for all $t \geq 0$, then the sequence $\sum_{i=0}^{t-1} a^i b_{t-i}$ converges, so that the sequence (x_t) converges to a finite limit as well.

For other examples of such nonautonomous systems, see Azariadis (1993, Chapters 1–5). What happens if we want to introduce uncertainty in the new discovery of the resource? Perhaps the natural first step is to consider $(b_{t+1})_{t \geq 0}$ as a sequence of independent, identically distributed random variables (assuming values in $\mathbb{R}_+$). This leads us to the process (2.1) studied in Chapter 4. ∎

1.5 Increasing Laws of Motion

Consider first the dynamical system (S, α) where $S \equiv \mathbb{R}$ (the set of reals) and $\alpha : S \to S$ is continuous and nondecreasing (i.e., if $x, x' \in S$ and $x \geq x'$, then $\alpha(x) \geq \alpha(x')$). Consider the trajectory $\tau(x_0)$ from any initial x_0. Since

$$x_{t+1} = \alpha(x_t) \quad \text{for } t \geq 0, \tag{5.1}$$

there are three possibilities:

Case I: $x_1 > x_0$.
Case II: $x_1 = x_0$.
Case III: $x_1 < x_0$.

In Case I, $x_2 = \alpha(x_1) \geq \alpha(x_0) = x_1$. It follows that $\{x_t\}$ is a non-decreasing sequence.

In Case II, it is clear that $x_t = x_0$ for all $t \geq 0$.

In Case III, $x_2 = \alpha(x_1) \leq \alpha(x_0) = x_1$. It follows that $\{x_t\}$ is a non-increasing sequence.

In Case I, if $\{x_t\}$ is bounded above,

$$\lim_{t \to \infty} x_t = x^*$$

exists. From (5.1) by taking limits and using the continuity of α, we have

$$x^* = \alpha(x^*).$$

Similarly, in Case III, if $\{x_t\}$ is bounded below,

$$\lim_{t \to \infty} x_t = x^*$$

exists, and, again by the continuity of α, we have, by taking limits in (5.1),

$$x^* = \alpha(x^*).$$

We shall now identify some well-known conditions under which the long-run behavior of all the trajectories can be precisely characterized.

Example 5.1 *Let $S = \mathbb{R}_+$ and $\alpha : S \to S$ be a continuous, nondecreasing function that satisfies the following condition* (PI):
there is a unique $x^* > 0$ such that

$$\alpha(x) > x \quad \text{for all } 0 < x < x^*,$$
$$\alpha(x) < x \quad \text{for all } x > x^*. \tag{PI}$$

In this case, if the initial $x_0 \in (0, x^*)$, then

$$x_1 = \alpha(x_0) > x_0,$$

and we are in Case I. But note that
$$x_0 < x^*$$

implies that $x_1 = \alpha(x_0) \le \alpha(x^*) = x^*$. Repeating the argument, we get

$$x^* \ge x_{t+1} \ge x_t \ge x_1 > x_0 > 0. \tag{5.2}$$

Thus, the sequence $\{x_t\}$ is nondecreasing, bounded above by x^*. Hence, $\lim_{t \to \infty} x_t = \hat{x}$ exists and, using the continuity of α, we conclude that $\hat{x} = \alpha(\hat{x}) > 0$.

Now, the uniqueness of x^* implies that $\hat{x} = x^*$.

If $x_0 = x^*$, then $x_t = x^*$ for all $t \ge 0$ and we are in Case II.

If $x_0 > x^*$, then $x_1 = \alpha(x_0) < x_0$, and we are in Case III. Now,

$$x_1 \ge x^*,$$

and, repeating the argument,

$$x^* \le x_{t+1} \le x_t \cdots \le x_1 \le x_0.$$

Thus, the sequence $\{x_t\}$ is nonincreasing and bounded below by x^*. Hence, $\lim_{t \to \infty} x_t = \hat{x}$ exists. Again, by continuity of α, $\hat{x} = \alpha(\hat{x})$, so that the uniqueness of x^* implies that $x^* = \hat{x}$.

To summarize:

Proposition 5.1 *Let $S = \mathbb{R}_+$ and $\alpha : S \to S$ be a continuous, nondecreasing function that satisfies* (PI). *Then for any $x > 0$ the trajectory $\tau(x)$ from x converges to x^*. If $x < x^*$, $\tau(x)$ is a nondecreasing sequence. If $x > x^*$, $\tau(x)$ is a nonincreasing sequence.*

Here, the long-run behavior of the dynamical system is independent of the initial condition on $x > 0$.   ∎

Remark 5.1 Suppose α is continuous and *increasing*, i.e., "$x > x'$" implies "$\alpha(x) > \alpha(x')$."

Again, there are three possibilities:

Case I: $x_1 > x_0$.
Case II: $x_1 = x_0$.
Case III: $x < x_0$.

Now in Case I, $x_2 = \alpha(x_1) > \alpha(x_0) = x_1$. Hence $\{x_t\}$ is an increasing sequence. In Case II, $x_t = x_0$ for all $t \geq 0$. Finally, in Case III $\{x_t\}$ is a decreasing sequence. The appropriate rewording of Proposition 5.1 when α is a continuous, increasing function is left as an exercise.

Proposition 5.2 *(Uzawa–Inada condition)* $S = \mathbb{R}_+$, α *is continuous, nondecreasing, and satisfies the Uzawa–Inada condition* (UI):

$$A(x) \equiv [\alpha(x)/x] \text{ is decreasing in } x > 0;$$
$$\text{for some } \bar{x} > 0, \ A(\bar{x}) > 1,$$
$$\text{and for some } \bar{\bar{x}} > \bar{x} > 0, \ A(\bar{\bar{x}}) < 1. \tag{UI}$$

Then the condition (PI) *holds.*

Proof. Clearly, by the intermediate value theorem, there is some $x^* \in (\bar{x}, \bar{\bar{x}})$ such that $A(x^*) = 1$, i.e., $\alpha(x^*) = x^* > 0$. Since $A(x)$ is decreasing, $A(x) > 1$ for $x < x^*$ and $A(x) < 1$ for $x > x^*$. In other words, for all $x \in (0, x^*)$, $\alpha(x) > x$ and for all $x > x^*$, $\alpha(x) < x$, i.e., the property (PI) holds. ∎

In the literature on economic growth, the (UI) condition is implied by appropriate differentiability assumptions. We state a list of typical assumptions from this literature.

Proposition 5.3 *Let $S = \mathbb{R}_+$ and $\alpha : S \to S$ be a function that is*

(i) *continuous on S,*

(ii) *twice continuously differentiable at $x > 0$ satisfying:*

[E.1] $\lim\limits_{x \downarrow 0} \alpha'(x) = 1 + \theta_1, \ \theta_1 > 0,$

[E.2] $\lim\limits_{x \uparrow \infty} \alpha'(x) = 1 - \theta_2, \ \theta_2 > 0,$

[E.3] $\alpha'(x) > 0, \ \alpha''(x) < 0$ *at $x > 0$.*

Then the condition (UI) *holds.*

Proof. Take $\bar{\bar{x}} > \bar{x} \geq 0$; by the mean value theorem
$[\alpha(\bar{\bar{x}}) - \alpha(\bar{x})] = (\bar{\bar{x}} - \bar{x})\alpha'(z)$ where $\bar{x} < z < \bar{\bar{x}}$.
Since $\alpha'(z) > 0$, $\alpha(\bar{\bar{x}}) > \alpha(\bar{x})$.
Thus α is increasing. Also, $\alpha''(x) < 0$ at $x > 0$ means that α is strictly concave. Take $\bar{\bar{x}} > \bar{x} > 0$. Then $\bar{x} \equiv t\bar{\bar{x}} + (1 - t)0, \ 0 < t < 1$.

Now

$$\alpha(\bar{x}) = \alpha(t\bar{\bar{x}} + (1-t)0)$$
$$> t\alpha(\bar{\bar{x}}) + (1-t)\alpha(0)$$
$$\geq t\alpha(\bar{\bar{x}})$$

or

$$\frac{\alpha(\bar{x})}{\bar{x}} > \frac{t\alpha(\bar{\bar{x}})}{t\bar{\bar{x}}} = \frac{\alpha(\bar{\bar{x}})}{\bar{\bar{x}}}.$$

Hence $\frac{\alpha(x)}{x}$ is decreasing. Write $B(x) \equiv \alpha(x) - x$. By [E.1]–[E.3] there is some $\bar{x} > 0$ such that $\alpha'(x) > 1$ for all $x \in (0, \bar{x}]$. Hence, $B'(x) > 0$ for all $x \in (0, \bar{x}]$. By the mean value theorem

$$\alpha(\bar{x}) = \alpha(0) + \bar{x}\alpha'(z), \quad 0 < z < \bar{x} \qquad \frac{\alpha(\bar{x}) - \alpha(0)}{x - 0} = \alpha'(z)$$
$$\geq \bar{x}\alpha'(z)$$

or

$$\frac{\alpha(\bar{x})}{\bar{x}} \geq \alpha'(z) > \alpha'(\bar{x}) > 1.\checkmark$$

Now, if α is bounded, i.e., if there is some $N > 0$ such that $\alpha(x) \leq N$, then $[\alpha(x)/x] \leq [N/x]$. Hence there is $\bar{\bar{x}}$, sufficiently large, such that $\alpha(\bar{\bar{x}})/\bar{\bar{x}} \leq [N/\bar{\bar{x}}] < 1.\checkmark$

If α is not bounded, we can find a sequence of points (x_n) such that $\alpha(x_n)$ and x_n go to infinity as n tends to infinity. Then

$$\lim_{n\to\infty} \frac{\alpha(x_n)}{x_n} = \lim_{n\to\infty} \alpha'(x_n) = 1 - \theta_2 < 1.$$

Hence, we can find some point $\bar{\bar{x}}$ sufficiently large, such that $\alpha(\bar{\bar{x}})/\bar{\bar{x}} < 1$.

Thus, the Uzawa-Inada condition (UI) is satisfied. ∎

Exercise 5.1 In his *Economic Dynamics*, Baumol (1970) presented a simple model that captured some of the ideas of "classical" economists formally. Let P_t, the net total product in period t, be a function of the working population L_t:

$$P_t = F(L_t), \quad t \geq 0,$$

where $F : \mathbb{R}_+ \to \mathbb{R}_+$ is continuous and increasing, $F(0) = 0$.

At any time, the working population tends to grow to a size where output per worker is just enough to provide each worker with a "subsistence level" of minimal consumption $M > 0$. This is formally captured by the relation

$$L_{t+1} = \frac{P_t}{M}.$$

Hence,

$$L_{t+1} = \frac{F(L_t)}{M} \equiv \alpha(L_t).$$

Identify conditions on the average productivity function $\frac{F(L)}{L}$ $(L > 0)$ that guarantee the following:

(i) There is a unique $L^* > 0$ such that $(F(L^*)/L^*) = M$.
(ii) For any $0 < L < L^*$, the trajectory $\tau(L)$ is increasing and converges to L^*; for any $L > L^*$, the trajectory $\tau(L)$ decreases to L^*. ∎

Example 5.2 Consider an economy (or a fishery) which starts with an initial stock y (the metaphorical corn of one-sector growth theory or the stock of renewable resource, e.g., the stock of trouts, the population of dodos, ...). In each period t, the economy is required to *consume* or *harvest* a positive amount c out of the beginning of the period stock y_t. The remaining stock in that period $x_t = y_t - c$ is "invested" and the resulting output (principal plus return) is the beginning of the period stock y_{t+1}. The output y_{t+1} is related to the input x_t by a "production" function g. Assume that

[A.1] $g : \mathbb{R} \to \mathbb{R}$ *is continuous, and increasing on* $\mathbb{R}_+$, $g(x) = 0$ *for* $x \leq 0$;
[A.2] *there is some* $x^* > 0$ *such that* $g(x) > x$ *for* $0 < x < x^*$ *and* $g(x) < x$ *for* $x > x^*$;
[A.3] g *is concave.*

The evolution of the system (given the initial $y > 0$ and the planned harvesting $c > 0$) is described by

$$y_0 = y,$$
$$x_t = y_t - c, \qquad t \geq 0;$$
$$y_{t+1} = g(x_t) \quad \text{for } t \geq 0.$$

Let T be the first period t, if any, such that $x_t < 0$; if there is no such t, then $T = \infty$. If T is finite we say that the agent (or the resource) *survives up to* (but not including) period T. We say that the agent survives (forever) if $T = \infty$ (i.e., if $x_t \geq 0$ for all t).

Define the net return function $h(x) = g(x) - x$. It follows that h satisfies

$$h(x) \begin{Bmatrix} \geq \\ = \\ < \end{Bmatrix} 0 \quad \text{as} \quad \begin{cases} 0 < x < x^*; \\ x = 0, x^*; \\ x > x^*. \end{cases}$$

Since $g(x) < 0$ for all $x \leq 0$, all statements about g and h will be understood to be for nonnegative arguments unless something explicit is said to the contrary.

Actually, we are only interested in following the system up to the "failure" or "extinction" time T.

The *maximum sustainable harvest or consumption* is

$$H = \max_{[0, x^*]} h(x). \tag{5.3}$$

We write

$$x_{t+1} = x_t + h(x_t) - c.$$

If $c > H$, $x_{t+1} - x_t = h(x_t) - c < H - c < 0$. Hence, x_t will fall below (extinction) 0 after a finite number of periods. On the other hand, if $0 < c < H$, there will be two roots ξ' and ξ'' of the equation

$$c = h(x),$$

which have the properties

$$0 < \xi' < \xi'' < x^*.$$

and

$$h(x) - c \left.\begin{matrix} \geq \\ = \\ < \end{matrix}\right\} 0 \quad \text{as} \quad \begin{cases} \xi' < x < \xi''; \\ x = \xi', \xi''; \\ x < \xi', x > \xi''. \end{cases}$$

We can show that

(a) If $x_0 < \xi'$, then x_t reaches or falls below 0 in finite time.
(b) If $x_0 = \xi'$, then $x_t = \xi'$ for all t.
(c) If $x_0 > \xi'$, then x_t converges monotonically to ξ''.

Note that if $c = H$, there are two possibilities: either $\xi' = \xi''$ (i.e., $h(x)$ attains the maximum H at a unique period ξ') or for all x in a nondegenerate interval $[\xi', \xi'']$, $h(x)$ attains its maximum.

The implications of the foregoing discussion for survival and extinction are summarized as follows.

Proposition 5.4 *Let $c > 0$ be the planned consumption for every period and H be the maximum sustainable consumption.*

(1) If $c > H$, there is no initial y from which survival (forever) is possible.

(2) If $0 < c < H$, then there is ξ', with $h(\xi') = c$, $\xi' > 0$ such that survival is possible if and only if the initial stock $y \geq \xi' + c$

(3) $c = H$ implies $h(\xi') = H$, and ξ' tends to 0 as c tends to 0.

1.6 Thresholds and Critical Stocks

We now consider some examples of dynamical systems that have been of particular interest in various contexts in development economics and in the literature on the management of a renewable resource. It has been emphasized that the evolution of an economy may depend crucially on the initial condition (hence, on the "history" that leads to it). It has also been noted that if the stock of a renewable resource falls below a critical level, the biological reproduction law may lead to its eventual extinction. In sharp contrast with the dynamical systems identified in Proposition 5.1, in which trajectories from positive initial stocks all converge to a positive fixed point, we sketch some examples where the long-run behavior of trajectories changes remarkably as the initial condition goes above a threshold (see Complements and Details).

Consider the "no-harvesting" case where $S = \mathbb{R}_+$, and the biological reproduction law is described by a continuous function $\alpha : \mathbb{R}_+ \to \mathbb{R}_+$. Given an initial state $x \geq 0$, the state in period t is the stock of the resources at the beginning of that period and (2.1) is assumed to hold. When $x_t = 0$, the resource is *extinct*. We state a general result.

Proposition 6.1 *Let $S = \mathbb{R}_+$ and $\alpha : S \to S$ be a continuous, increasing function with the following properties [P2]:*

[P2.1] $\alpha(0) = 0$;
- *[P2.2] there are two positive fixed points $x_{(1)}^*$, $x_{(2)}^*$ $(0 < x_{(1)}^* < x_{(2)}^*)$ such that*

$$\text{(i)} \quad \alpha(x) < x \quad \text{for } x \in (0, x_{(1)}^*);$$
$$\text{(ii)} \quad \alpha(x) > x \quad \text{for } x \in (x_{(1)}^*, x_{(2)}^*);$$
$$\text{(iii)} \quad \alpha(x) < x \quad \text{for } x > x_{(2)}^*.$$

For any $x \in (0, x_{(1)}^*)$, the trajectory $\tau(x)$ from x is decreasing and converges to 0; for any $x \in (x_{(1)}^*, x_{(2)}^*)$, the trajectory $\tau(x)$ from x is increasing and converges to $x_{(2)}^*$; for any $x > x_{(2)}^*$, the trajectory $\tau(x)$ from x is decreasing and converges to $x_{(2)}^*$.

Proof. Left as an exercise. ∎

The striking feature of the trajectories from any $x > 0$ is that these are all convergent, but the limits depend on the initial condition. Also, these are all monotone, but again depending on the initial condition – some are increasing, others are decreasing. We interpret $x_{(1)}^* > 0$ as the *critical* level for survival of the resource.

Exercise 6.1 Consider $S = \mathbb{R}_+$ and $\alpha : S \to S$ defined by

$$\alpha(x) = \frac{\theta_1 x^2}{x^2 + \theta_2}$$

where θ_1, θ_2 are positive parameters. When $\theta_1 > 2\sqrt{\theta_2}$, compute the fixed points of α and verify [P2]. ∎

Exercise 6.2 Consider the problem of survival with constant harvesting discussed in Example 5.2. Work out the conditions for survival with a constant harvest $c > 0$ when the "production" function g is continuous, increasing on $\mathbb{R}_+$, and satisfies

[P2.1] $g(x) = 0$ for $x \leq 0$;
[P2.2] there are two positive fixed points $x^*_{(1)}, x^*_{(2)}$ $(0 < x^*_{(1)} < x^*_{(2)})$
such that

(i) $g(x) < x$ for $x \in (0, x^*_{(1)})$;
(ii) $g(x) < x$ for $x \in (x^*_{(1)}, x^*_{(2)})$;
(iii) $g(x) < x$ for $x > x^*_{(2)}$. ∎

Exercise 6.3 Let $S = \mathbb{R}_+$ and $\alpha : S \to S$ is defined by $\alpha(x) = x^m$, where $m \geq 1$. Show that (i) the fixed points of α are "0" and "1"; (ii) for all $x \in [0, 1)$, the trajectory $\tau(x)$ is monotonically decreasing, and converges to 0. For $x > 1$, the trajectory $\alpha(x)$ is monotonically increasing and unbounded. Thus, $x = 1$ is the "threshold" above which sustainable growth is possible; if the initial $x < 1$, the trajectory from x leads to extinction. ∎

Example 6.1 In parts of Section 1.9 we review dynamical systems that arise out of "classical" optimization models (in which "convexity" assumptions on the preferences and technology hold). Here we sketch an example of a "nonclassical" optimization model in which a critical level of initial stock has important policy implications.

Think of a competitive fishery (see Clark 1971, 1976, Chapter 7). Let $x_t (\geq 0)$ be the stock or "input" of fish in period t, and $f : \mathbb{R}_+ \to \mathbb{R}_+$ the biological reproduction relationship. The stock x in any period gives rise to output $y = f(x)$ in the *subsequent* period. The following assumptions on f are introduced:

[A.1] $f(0) = 0$;
[A.2] $f(x)$ *is twice continuously differentiable for $x \geq 0$; $f'(x) > 0$ for $x > 0$.*
[A.3] f *satisfies the following end-point conditions: $f'(\infty) < 1 < f'(0) < \infty$; $f'(x) > 0$ for $x > 0$.*
[A.4] *There is a (finite) $b_1 > 0$, such that (i) $f''(b_1) = 0$; (ii) $f''(x) > 0$ for $0 \leq x < b_1$; (iii) $f''(x) < 0$ for $x > b_1$.*

In contrast to the present ("nonclassical") model, the traditional (or "classical") framework would replace [A.4] by

[A.4′]. *f is strictly concave for $x \geq 0$ ($f''(x) < 0$ for $x > 0$), while preserving [A.1]–[A.3].*

In some versions, [A.2] and [A.3] are modified to allow $f'(0) = \infty$. In the discussion to follow, we find it convenient to refer to a model with assumptions [A.1]–[A.3] and [A.4′] as classical, and a model with [A.1]–[A.4] as nonclassical.

We define a function h (representing the *average product function*) as follows:

$$h(x) = [f(x)/x] \quad \text{for } x > 0; \quad h(0) = \lim_{x \to 0}[f(x)/x]. \qquad (6.1)$$

Under [A.1]–[A.4], it is easily checked that $h(0) = f'(0)$; furthermore, there exist positive numbers k^*, $\bar{k}$, b_2 satisfying (i) $0 < b_1 < b_2 < k^* < \bar{k} < \infty$; (ii) $f'(k^*) = 1$; (iii) $f(\bar{k}) = \bar{k}$; (iv) $f'(b_2) = h(b_2)$. Also, for $0 \leq x < k^*$, $f'(x) > 1$ and for $x > k^*$, $f'(x) < 1$; for $0 < x < \bar{k}$, $x < f(x) < \bar{k}$ and for $x > \bar{k}$, $\bar{k} < f(x) < x$; and for $0 < x < b_2$, $f'(x) > h(x)$ and for $x > b_2$, $f'(x) < h(x)$. Also note that for $0 \leq x < b_2$, $h(x)$ is increasing, and for $x > b_2$, $h(x)$ is decreasing; for $0 \leq x < b_1$, $f'(x)$ is increasing, and for $x > b_1$, $f'(x)$ is decreasing.

A *feasible production program* from $x > 0$ is a sequence $(\mathbf{x}, \mathbf{y}) = (x_t, y_{t+1})$ satisfying

$$x_0 = \underset{\sim}{x}; \quad 0 \leq x_t \leq y_t \quad \text{and} \quad y_t = f(x_{t-1}) \text{ for } t \geq 1. \qquad (6.2)$$

The sequence $\mathbf{x} = (x_t)_{t\geq0}$ is the *input* (or stock) *program*, while the corresponding $\mathbf{y} = (y_{t+1})_{t\geq0}$ satisfying (6.2) is the *output program*. The *harvest program* $\mathbf{c} = (c_t)$ generated by $(\mathbf{x}, \mathbf{y})$ is defined by $c_t \equiv y_t - x_t$ for $t \geq 1$. We will refer to $(\mathbf{x}, \mathbf{y}, \mathbf{c})$ briefly as *a program* from $\underset{\sim}{x}$, it being understood that $(\mathbf{x}, \mathbf{y})$ is a feasible production program, and $\mathbf{c}$ is the corresponding harvest program.

A slight abuse of notation: we shall often specify only the stock program $\mathbf{x} = (x_t)_{t\geq0}$ from $x > 0$ to describe a program $(\mathbf{x}, \mathbf{y}, \mathbf{c})$. It will be understood that $x_0 = \underset{\sim}{x}$; $0 \leq x_t \leq f(x_{t-1})$ for all $t \geq 1$, $c_t = y_t - x_t$ for $t \geq 1$.

Let the profit per unit of harvesting, denoted by $q > 0$, and the rate of interest $\gamma > 0$ remain constant over time. Consider a firm that has an objective of maximizing the discounted sum of profits from harvesting.

A program $\mathbf{x}^* = (x_t^*)$ of stocks from $\underset{\sim}{x} > 0$ is optimal if

$$\sum_{t=1}^{\infty} \left[\frac{q}{(1+\gamma)^{t-1}} \right] c_t^* \geq \sum_{t=1}^{\infty} \left[\frac{q}{(1+\gamma)^{(t-1)}} \right] c_t$$

for every program $\mathbf{x}$ from $\underset{\sim}{x}$. Write $\delta = 1/(1+\gamma)$. Models of this type have been used to discuss the possible conflict between profit maximization and conservation of natural resources.

The program $(\mathbf{x}, \mathbf{y}, \mathbf{c})$ from $\underset{\sim}{x} > 0$ defined as $x_0 = \underset{\sim}{x}$, $x_t = 0$ for $t \geq 1$ is the *extinction program*. Here the entire output $f(\underset{\sim}{x})$ is harvested in period 1, i.e., $c_1 = f(\underset{\sim}{x})$, $c_t = 0$ for $t \geq 2$.

In the qualitative analysis of optimal programs, the roots of the equation $\delta f'(x) = 1$ play an important role. This equation might not have a nonnegative real root at all; if it has a pair of unique nonnegative real roots, denote it by Z; if it has nonnegative real roots, the smaller one is denoted by z and the larger one by Z.

The qualitative behavior of optimal programs depends on the value of $\delta = 1/(1+\gamma)$. Three cases need to be distinguished. The first two were analyzed and interpreted by Clark (1971).

Case 1. Strong discounting: $\delta f'(b_2) \leq 1$

This is the case when δ is "sufficiently small," i.e., $1 + \gamma \geq f(x)/x$ for all $x > 0$.

Proposition 6.2 *The extinction program is optimal from any $\underset{\sim}{x} > 0$, and is the unique optimal program if $\delta f'(\hat{b}_2) < 1$.*

Remark 6.1 First, if $\delta f'(b_2) = 1$, there are many optimal programs (see Majumdar and Mitra 1983, p. 146). Second, if we consider the classical model (satisfying [A.1]–[A.3] and [A.4']), it is still true that if $\delta f'(0) \leq 1$, the extinction program is the unique optimal program from any $\underset{\sim}{x} > 0$.

Case 2. Mild discounting: $\delta f'(0) \geq 1$

This is the case where δ is "sufficiently close to 1" ($\delta > 1/f'(0)$) and $Z > b_2$ exists (if z exists, $z = 0$).

Now, given $\underset{\sim}{x} < Z$, let M be the smallest positive integer such that $x_M^1 \geq Z$; in other words, M is the first period in which the *pure accumulation* program from $\underset{\sim}{x}$ defined by $x_0^1 = \underset{\sim}{x}, x_{t+1}^1 = f(x_t^1)$ for $t \geq 0$ attains Z.

Proposition 6.3 *If* $\underset{\sim}{x} \geq Z$, *then the program* $\mathbf{x}^* = (x_t^*)_{t \geq 0}$ *from* $\underset{\sim}{x}$ *defined by* $x_0^* = \underset{\sim}{x}, x_t^* = Z$ *for* $t \geq 1$ *is the unique optimal program from* $\underset{\sim}{x}$.

Proposition 6.4 *If* $\underset{\sim}{x} < Z$, *the program* $\mathbf{x}^* = (x_t^*)_{t \geq 0}$ *defined by* $x_0^* = \underset{\sim}{x}$, $x_t^* = x_t^1$ *for* $t = 1, \ldots, M - 1$, $x_t^* = Z$ *for* $t \geq M$ *is the unique optimal program.*

In the corresponding classical model for $\delta f'(0) > 1$, there is a unique positive K_δ^*, solving $\delta f'(x) = 1$. Propositions 6.3 and 6.4 continue to hold with Z replaced by K_δ^* (also in the definition of M).

Case 3. Two turnpikes and the critical point of departure $[\delta f'(0) < 1 < \delta f'(k_2)]$

In case (i) the extinction program $(\mathbf{x}, \mathbf{y}, \mathbf{c})$ generated by $x_t = 0$ for all $t \geq 1$ and in case (ii) the stationary program generated by $x_t = Z$ for all $t \geq 0$ (which is also the optimal program from Z) serve as the "turnpikes" approached by the optimal programs. *Both* the classical and nonclassical models share the feature that the *long-run behavior* of optimal programs is *independent* of the positive *initial stock*. The "intermediate" case of discounting, namely when

$$1/f'(b_2) < \delta < 1/f'(0),$$

turned out to be difficult and to offer a sharp contrast between the classical and nonclassical models. In this case

$$0 < z < b_1 < b_2 < Z < k^*.$$

The qualitative properties of optimal programs are summarized in two steps.

Proposition 6.5 $\underset{\sim}{x} \geq Z$, *the program* $\mathbf{x}^* = (x_t^*)_{t \geq 0}$ *defined by* $x_0^* = \underset{\sim}{x}$, $x_t^* = Z$ *for* $t \geq 1$ *is optimal.*

A program $\mathbf{x} = (x_t)_{t \geq 0}$ from $\underset{\sim}{x} < Z$ is a *regeneration program* if there is some positive integer $N \geq 1$ such that $x_t > x_{t-1}$ for $1 \leq t \leq N$, and $x_t = Z$ for $t \geq N$. It should be stressed that a regeneration program may allow for positive consumption in *all* periods, and need not specify "pure accumulation" in the initial periods. For an interesting example of a regeneration program that allows for positive consumption and is optimal, the reader is referred to Clark (1971, p. 259).

Proposition 6.6 *Let $x < Z$. There is a critical stock $K_c > 0$ such that if $0 < x < K_c$, the extinction program from x is an optimal program. If $K_c < x < Z$, then any optimal program is a regeneration program.*

In the literature on renewable resources, K_c is naturally called the "minimum safe standard of conservation." It has been argued that a policy that prohibits harvesting of a fishery till the stock exceeds K_c will ensure that the fishery will not become extinct, even under pure "economic exploitation."

Some conditions on x can be identified under which there is a unique optimal program. But if $x = K_c$, then both the extinction program and a regeneration program are optimal. For further details and proofs, see Majumdar and Mitra (1982, 1983).   ∎

1.7 The Quadratic Family

Let $S = [0, 1]$ and $A = [0, 4]$. The *quadratic family of maps* is then defined by

$$\alpha_\theta(x) = \theta x(1 - x) \quad \text{for } (x, \theta) \in S \times A. \tag{7.1}$$

We interpret x as the *variable* and θ as the *parameter* generating the family.

We first describe some basic properties of this family of maps. First, note that $F_\theta(0) = 0(= F_\theta(1)) \ \forall \theta \in [0, 4]$, so that 0 is a fixed point of F_θ. By solving the quadratic equation $F_\theta(x) = x$, one notes that $p_\theta = 1 - 1/\theta$ is the only other fixed point that occurs if $\theta > 1$. For $\theta > 3$, one can show (e.g., by numerical calculations) that the fourth-degree polynomial equation $F_\theta^2(x) := F_\theta \circ F_\theta(x)$ has two other solutions (in addition to 0 and p_θ). This means that for $3 < \theta \leq 4$, F_θ has a period-two orbit. For $\theta > 1 + \sqrt{6}$, a new period-four orbit appears. We refer to the Li–Yorke and Sarkovskii theorems, stated in Section 1.3, for the successive appearance of periodic points of period 2^k ($k \geq 0$), as θ increases to a limit point of $\theta_c \approx 3.57$, which is followed by other cascades of periodic orbits according to the Sarkovskii ordering. A period-three orbit appears for the first time when θ exceeds a critical value $\theta^* \approx 3.8284$ so that, for $\theta > \theta^*$, F_θ has a periodic orbit of every period. If a new periodic orbit appears for the first time after $\theta = \theta_0$, say, then there exists an interval $(\theta_0, \theta_1]$ such that this new periodic orbit $\{x_1(\theta), x_2(\theta), \ldots, x_m(\theta)\}$, say, remains (locally) attracting or stable. For $\theta > \theta_1$ this orbit becomes repelling or unstable, as do all previously arising periodic orbits and fixed points. There are uncountably many points θ for which F_θ has no locally attracting periodic or fixed points. Indeed, while we do not develop the range of formal definitions capturing *chaos* or *complexity*, we note that the set of values of θ for which F_θ is chaotic has positive Lebesgue measure, according to the results of Misiurewicz (1983) and Jakobson (1981) (see Theorem 7.1 for a particular case and Complements and Details).

For each $\theta \in A$, α_θ has exactly one *critical point* (i.e., a point where $\alpha'_\theta(x) = 0$, and this critical point ($z^* = 0.5$) is independent of the parameter θ.

✒ 1.7.1 Stable Periodic Orbits

Even though there may be an infinite number of periodic orbits for a given dynamical system (as in the Li–Yorke theorem), a striking result due to Julia and Singer (see Singer 1978) informs us that there can be *at most one (locally) stable* periodic orbit.

Proposition 7.1 *Let $S = [0, 1]$, $A = [1, 4]$; given some $\hat{\theta} \in A$, define $\alpha_{\hat{\theta}}(x) = \hat{\theta}x(1 - x)$ for $x \in S$. Then there can be at most one stable periodic orbit. Furthermore, if there is a stable periodic orbit, then $w(0.5)$, the limit set of $z^* = 0.5$, must coincide with this orbit.*

Suppose, now, that we have a stable periodic orbit. This means that the asymptotic behavior (limit sets) of trajectories from all initial states "near" this periodic orbit must coincide with the periodic orbit. But what about the asymptotic behavior of trajectories from other initial states? If one is interested in the behavior of a "typical" trajectory, a remarkable result, due to Misiurewicz (1983), settles this question.

Proposition 7.2 *Let $S = [0, 1]$, $A = [1, 4]$; given some $\hat{\theta} \in A$, define $\alpha_{\hat{\theta}}(x) = \hat{\theta}x(1 - x)$ for $x \in S$. Suppose there is a stable periodic orbit. Then for (Lebesgue) almost every $x \in [0, 1]$, $w(x)$ coincides with this orbit.*

Combining these two results, we have the following scenario. Suppose we do have a stable periodic orbit. Then there are no other stable periodic orbits. Furthermore, the (unique) stable periodic orbit "attracts" the trajectories from almost every initial state. Thus we can make the qualitative prediction that the asymptotic behavior of the "typical" trajectory will be just like the given stable periodic orbit.

It is important to note that the above scenario (existence of a stable periodic orbit) is by no means inconsistent with condition (3.1) of the Li–Yorke theorem (and hence with its implications). Let us elaborate on this point following Devaney (1986) and Day and Pianigiani (1991). Consider $\theta = 3.839$, and for this θ, simply write $\alpha(x) = \theta x(1 - x)$ for $x \in S$. Choosing $x^* = 0.1498$, it can be checked then that there is $0 < \varepsilon < 0.0001$ such that $\alpha^3(x)$ maps the interval $U \equiv [x^* - \varepsilon, x^* + \varepsilon]$ into itself, and $|[\alpha^3(x)]'| < 1$ for all $x \in U$. Hence, there is $\widehat{x} \in U$ such that $\alpha^3(\widehat{x}) = \widehat{x}$, and $|[\alpha^3(\widehat{x})]'| < 1$. Thus, $\widehat{x}$ is a periodic point of period 3, and it can be checked (by choice of the range of ε) that $\alpha^3(\widehat{x}) = \widehat{x} < \alpha(\widehat{x}) < \alpha^2(\widehat{x})$ so that condition (3.1) of Theorem 3.1 is satisfied. Also, $\widehat{x}$ is a periodic point of period 3, which is *stable*, so that Proposition 7.2 is also applicable. Then we may conclude that the set $\aleph'$ of "chaotic" initial states in Theorem 3.1 must be of Lebesgue measure zero. In other words, Li–Yorke chaos exists but is *not* "observed" when $\theta = 3.839$: the mere existence of a set of initial conditions giving rise to complex trajectories does not quite mean that the typical trajectory is "unpredictable."

We note another formal definition of the concept of sensitive dependence on initial conditions due to Guckenheimer (1979) that is of interest in the context of the quadratic family. Let the state space S be a (non-degenerate) closed interval and α be a continuous map from S into S. The dynamical system (S, α) has *Guckenheimer dependence* if there are a set $T \subset S$ of positive Lebesgue measure and $\varepsilon > 0$ such that given any $x \in T$ and every neighborhood U of x, there is $y \in U$ and $n \geq 0$ such that $|\alpha^n(x) - \alpha^n(y)| > \varepsilon$. We note the following important result.

Proposition 7.3 *Let $S = [0, 1]$ and $I = [1, 4]$; given some $\widehat{\theta} \in I$ consider $\alpha_{\widehat{\theta}}(x) = \widehat{\theta} x(1 - x)$, $x \in S$. Suppose there is a stable periodic orbit. Then the dynamical system $(S, \alpha_{\widehat{\theta}})$ does not have Guckenheimer dependence on initial conditions.*

Finally, we state a special case of Jakobson's theorem, which is a landmark in the theory of chaos.

Theorem 7.1 *Let* $S = [0, 1]$, $J = [3, 4]$, *and* $\alpha_\theta(x) = \theta x(1 - x)$ *for* $(x, \theta) \in S \times J$. *Then the set* $\Delta = \{\theta \in J: (S, \alpha_\theta)$ *has Guckenheimer dependence*$\}$ *has positive Lebesgue measure.*

For our purposes in later chapters the following proposition suffices.

Proposition 7.4 *(a) Let* $0 \leq \theta \leq 1$. *Then* $F_\theta^n(x) \to 0$ *as* $n \to \infty$, $\forall x \in [0, 1]$. *(b) Let* $1 < \theta \leq 3$. *Then* $F_\theta^n(x) \to p_\theta \equiv 1 - 1/\theta$ *as* $n \to \infty$, $\forall x \in (0, 1)$. *(c) For* $3 < \theta \leq 1 + \sqrt{5}$, F_θ *has an attracting period-two orbit.*

Proof.

(a) Let $0 \leq \theta \leq 1$. Then, for every $x \in [0, 1]$, $0 \leq F_\theta(x) \leq x \leq 1$, so that $F_\theta^n(x)$, $n \geq 1$, is a nonincreasing sequence, which must converge to some x^*. This limit is a fixed point. But the only fixed point of F_θ is 0. Hence $x^* = 0$.

(b) First, consider the case $1 < \theta \leq 2$. Then $p_\theta \leq 1/2$, and $x < F_\theta(x) < p_\theta$ $\forall x \in (0, p_\theta)$. Since F_θ is *increasing* on $(0, p_\theta) \subset (0, 1/2]$, it follows that $F_\theta(x) < F_\theta^2(x) < F_\theta^3(x) < \cdots < p_\theta$. The increasing sequence $F_\theta^n(x)$ converges to a fixed point x^*, which then must be p_θ. Next let $x \in (p_\theta, 1/2]$. Here $1/2 \geq x \geq F_\theta(x) \geq p_\theta$ and F_θ is increasing. Hence, $F_\theta(x) \geq F_\theta^2(x) \geq F_\theta^3(x) \geq \cdots$, that is, as $n \uparrow \infty$, $F_\theta^n(x) \downarrow y^*$, say $y^* \geq p_\theta$. Again, by the uniqueness of the fixed point of F_θ on $[p_\theta, 1)$, $y^* = p_\theta$. For $x \in [1/2, 1)$, $y := F_\theta(x) \in (0, 1/2]$, so that $F_\theta^n(x) = F_\theta^{n-1}(y) \to p_\theta$.

Now consider the case $2 < \theta \leq 1 + \sqrt{5}$. Then $1/2 < p_\theta < \theta/4 \equiv F_\theta(1/2)$, and F_θ is *decreasing* on $[1/2, \theta/4]$. Therefore, F_θ^2 is *increasing* on $[1/2, \theta/4]$, mapping this interval onto $[F_\theta^2(1/2), F_\theta^3(1/2)]$. Observe that the function $\theta \to g(\theta) := F_\theta^2(1/2) \equiv (\theta^2/4)(1 - \theta/4)$ is strictly *concave* (i.e., $g''(\theta) < 0$) on $[2, 4]$ and $g(2) = g(1 + \sqrt{5}) = 1/2$, implying that $g(\theta) > 1/2$ for $\theta \in (2, 1 + \sqrt{5})$. Applying F_θ^2 repeatedly to the inequality $1/2 \leq F_\theta^2(1/2)$, one gets $1/2 \leq F_\theta^2(1/2) < F_\theta^4(1/2) < F_\theta^6(1/2) < \cdots$, that is, $F_\theta^{2n}(1/2) \uparrow q_0$, say as $n \uparrow \infty$, implying $q_0 = q_0(\theta)$ *is a fixed point of* F_θ^2 $\forall \theta \in (2, 1 + \sqrt{5}]$, with $q_0 \leq p_\theta$ (since $1/2 < p_\theta$ implies $F_\theta^{2n}(1/2) < p_\theta$ $\forall n$). Similarly, applying F_θ to the inequality $p_\theta \leq F_\theta(1/2)$ one gets $p_\theta \geq F_\theta(1/2) \geq F_\theta^3(1/2)$, and repeatedly applying F_θ^2 to the last inequality one arrives at $F_\theta(1/2) \geq F_\theta^3(1/2) \geq F_\theta^5(1/2) \geq \cdots$. Thus $F_\theta^{2n+1}(1/2) \downarrow q_1 = q_1(\theta) \geq p_\theta$ as $n \uparrow \infty$,

$\forall \theta \in (2, 1 + \sqrt{5})$ (note that $p_\theta < F_\theta(1/2)$ implies $p_\theta < F_\theta^{2n}(F_\theta(1/2))$ $\forall n$); in particular, q_1 *is a fixed point of* F_θ^2.

Let $1/2 \le x \le p_\theta$, *and* $2 < \theta \le 3$. Then, by the preceding arguments $F_\theta^{2n}(1/2) \le F_\theta^{2n}(x) \le p_\theta \le F_\theta^{2n+1}(x) \le F_\theta^{2n+1}(1/2)$, $\forall n = 1, 2, \ldots$. Since F_θ has no periodic orbit of (prime) period 2 (i.e., F_θ^2 has no fixed point other than p_θ on $(0, 1)$), one has $q_0 = q_1 = p_\theta$. Hence $F_\theta^n(x) \to p_\theta \ \forall x \in [1/2, p_\theta]$. Define $\hat{p}_\theta = 1 - \hat{p}_\theta \equiv 1/\theta$. Then $\hat{p}_\theta < 1/2$ and $F_\theta(\hat{p}_\theta) = p_\theta$. On $[\hat{p}_\theta, 1/2]$, F_θ is increasing and, therefore, if $x \in (\hat{p}_\theta, 1/2]$, then $F_\theta(x) \in (p_\theta, F_\theta(1/2)]$, $F_\theta^2(x) \in [F_\theta^2(1/2), p_\theta)$ (since F_θ is decreasing on $[p_\theta, F_\theta(1/2)]$). Hence, $\forall x \in [\hat{p}_\theta, 1/2]$, with $z := F_\theta^2(x)$, one has $F_\theta^n(x) \equiv F_\theta^{n-2}(z) \to p_\theta$. Next, let $x \in (0, \hat{p}_\theta)$. Since F_θ is strictly increasing on $(0, \hat{p}_\theta]$ and $x < F_\theta(x)$, it follows that there is a smallest integer k such that the sequence $F_\theta^n(x)$ increases as n increases, $n \le k - 1$, and $F_\theta^k(x) \ge \hat{p}_\theta$ (else, $F_\theta^n(x) \uparrow x^* \le \hat{p}_\theta$, implying the existence of a fixed point $x^* \in (0, \hat{p}_\theta]$). Also, since $F_\theta^{k-1}(x) < \hat{p}_\theta$, $F_\theta^k(x) \le p_\theta$, so that $F_\theta^k(x) \in [\hat{p}_\theta, p_\theta]$ and $F_\theta^n(x) \equiv F_\theta^{n-k}(F_\theta^k(x)) \to p_\theta$ as $n \to \infty$. Similarly, if $x \in (p_\theta, 1)$ there exists a smallest integer k such that $F_\theta^k(x) \ge p_\theta$ (noting that F_θ is decreasing on $(p_\theta, 1)$ and $F_\theta(x) < x$ on $(p_\theta, 1)$). Hence $F_\theta^n(x) \to p_\theta$. We have now shown that $F_\theta^n(x) \to p_\theta$ as $n \to \infty$, $\forall x \in (0, 1)$.

(c) Let $3 < \theta \le 1 + \sqrt{5}$. Since $F_\theta'(p_\theta) = 2 - \theta < 1$, p_θ is a *repelling fixed point*, as is 0. For $1/2 < x < F_\theta^2(1/2)(< p_\theta)$ one then gets $F_\theta^{2n}(1/2) < F_\theta^{2n}(x) < F_\theta^{2n+2}(1/2) < p_\theta \ \forall n$. Therefore, $F_\theta^{2n}(x) \to q_0$ as $n \to \infty$. Similarly, for $F_\theta^3(1/2) < x < F_\theta(1/2)$, $F_\theta^{2n+3}(1/2) < F_\theta^{2n}(x) < F_\theta^{2n+1}(1/2) \ \forall n$, so that $F_\theta^{2n}(x) \to q_1$ as $n \to \infty$. Since p_θ is repelling, it follows that $q_1 > p_\theta > q_0$, and $\{q_0, q_1\}$ is an attracting periodic orbit of F_θ. ∎

Example 7.1 *A Ricardian Model.* Consider the model of a "Ricardian system" discussed by Bhaduri and Harris (1987).

Suppose that the total product of "corn" (Y) as a function of "labor" (N) is given by

$$Y_t = \alpha N_t - bN_t^2/2.$$

Hence, the average product of labor (AP_t) is given by

$$AP_t = \frac{Y_t}{N_t} = \alpha - \frac{bN_t}{2}.$$

The "rent" R_t in period t emerges as

$$R_t = \left(\frac{Y_t}{N_t} - \frac{dY_t}{dN_t} \right) N_t = \frac{bN_t^2}{2}.$$

"Profit" P_t is the residual after payment of rent and replacement of the wage fund W_t advanced to employ labor. Thus,

$$P_t \equiv Y_t - R_t - W_t.$$

Bhaduri and Harris focus on the dynamics of the wage fund as representing the process of accumulation. The size of the wage fund governs the amount of labor that can be employed on the land. At a given wage rate $w > 0$ we have

$$W_t = wN_t.$$

Accumulation of the wage fund comes from the reinvestment of profits accruing to the capitalists. If there is no consumption out of profit income, we have

$$W_{t+1} - W_t = P_t.$$

This leads to

$$N_{t+1} - N_t = \frac{P_t}{w} = \frac{a}{w}N_t - \frac{b}{w}N_t^2 - N_t$$

or upon simplification

$$N_{t+1} = \frac{a}{w}N_t - \frac{b}{w}N_t^2.$$

Since the positivity of the marginal product of labor requires $a > bN_t$, corresponding to all meaningful employment levels, Bhaduri and Harris define $n_t \equiv bN_t/a < 1$ and obtain

$$n_{t+1} = \frac{a}{w}n_t - \frac{a}{w}\frac{b^2N_t^2}{a^2} = \frac{a}{w}n_t(1 - n_t).$$

Thus, the basic equation governing the dynamics of the Ricardian system is

$$n_{t+1} = \mu n_t(1 - n_t),$$

where $\mu = a/w$.

We can use the results on the quadratic family to study the emergence of chaos in the Ricardian system. ∎

1.8 Comparative Statics and Dynamics

Samuelson's influential monograph (Samuelson 1947, p. 351) emphasized the need to obtain results on *comparative statics* and *dynamics*: "the
usefulness of any theoretical structure lies in the light which it throws
upon the way economic variables will change when there is a change in
some datum or parameter." In comparative statics, one investigates how
a stationary state responds to a change in a parameter explicitly affecting the law of motion (see Proposition 8.1). Of particular interest is to
identify unambiguously the *direction of change*, and we shall see some
examples in the subsequent sections. The central notion of comparative
dynamics is to analyze the effects of a change in a parameter affecting
the law of motion on the behavior of the trajectories of the dynamical
system. Again, even for 'simple' nonlinear systems, a complete comparative dynamic analysis is a formidable challenge, and we work out in
some detail several examples to stress this point.

To proceed more formally, consider a *family* of dynamical systems
(all sharing the same state space S) in which the law of motion depends
explicitly on a parameter θ:

$$x_{t+1} = \alpha_\theta(x_t) \tag{8.1}$$

where, for each admissible value of θ, α_θ is a map from S into itself. As
an example, take the quadratic family of maps:

$$x_{t+1} = \theta x_t(1 - x_t), \tag{8.2}$$
$$\text{where} \quad S = [0, 1] \quad \text{and} \quad \theta \in [0, 4].$$

Now, with variations of θ we can pose the following questions:

(i) How do fixed and periodic points behave in response to changes
in θ?

(ii) How does a change in θ affect the qualitative properties of the
various trajectories?

The first question is at the core of *bifurcation theory*. We have seen that
even for simple families like (8.2), the *number* as well as *local stability
properties* of *periodic points may change abruptly as the parameter* θ
passes through threshold values. This point is elaborated in Section 1.8.1.

The second question leads to the issue of *robustness of chaotic behavior*, which can be precisely posed in alternative ways. First, one
can begin with a particular dynamical system (say, (2.3)) and show that

according to some formal definition, complicated behavior persists on some nonnegligible set of initial conditions.

Next, one can proceed to look at the family of dynamical systems, for example, (8.2) corresponding to different values of the parameter θ, and try to assert that the class of parameter values giving rise to complicated behavior from a non-negligible set of initial conditions is itself non-negligible. (An example of this approach is Jakobson's theorem mentioned in Section 1.7.)

One can also take a topological approach to robustness. Consider first a given law of motion α for which the dynamical system (2.1) is chaotic in some sense, and then vary α in an appropriate function space and see whether chaos persists on a neighborhood containing α (recall Proposition 3.1).

1.8.1 Bifurcation Theory

Bifurcation theory deals with the question of changes in the stationary state of a dynamical system with respect to variations of a parameter that affects the law of motion. We continue to introduce the main ideas through some explicit computations.

Example 8.1 (*continuation of Example 2.1*)   Consider a class of dynamical systems with a common state space $S = \mathbb{R}$ and with the laws of motion given by

$$\alpha_\theta(x) = x^2 + \theta, \tag{8.3}$$

where "θ" (a real number) is a parameter. Our task is to study the implications of changes in the parameter θ. The parallelism between the dynamics of the family (8.3) and our earlier quadratic family (7.1) is not an accident. But we do not take up the deeper issue of *topological conjugacy* here (see Devaney 1986, Chapter 1.7 and Exercise 1, p. 47). In order to understand stationary states, we have to explore the nature of fixed points of α_θ and its iterates α_θ^j. Consider first the dependence of the fixed points of α_θ on the value of θ. The fixed points of α_θ, for a given value of θ, are obtained by solving the equation

$$\alpha_\theta(x) = x \tag{8.4}$$

and are given by

$$p_+(\theta) = (1 + \sqrt{1 - 4\theta})/2$$
$$p_-(\theta) = (1 - \sqrt{1 - 4\theta})/2. \tag{8.5}$$

Observe that $p_+(\theta)$ and $p_-(\theta)$ are real if and only if $1 - 4\theta \geq 0$ or $\theta \leq 1/4$. Thus, when $\theta > 1/4$, we see that the dynamical system (with the state space $S = \mathbb{R}$) has no fixed point. When $\theta = 1/4$, we have

$$p_+ (1/4) = p_- (1/4) = 1/2. \tag{8.6}$$

Now when $\theta < 1/4$, both $p_+(\theta)$ and $p_-(\theta)$ are real and distinct (and $p_+(\theta) > p_-(\theta)$).

Thus, as the parameter θ decreases through $1/4$, the dynamical system undergoes a *bifurcation* from the situation of no fixed point to a unique fixed point at $\theta = 1/4$ and then to a pair of fixed points.

We turn to the "local" dynamics of the system from initial states close to the fixed point(s). Since $\alpha'_\theta(x) = 2x$ (does not depend on θ), we see that $\alpha'_{1/4}(1/2) = 1$ (so the fixed point $1/2$ is nonhyperbolic when $\theta = 1/4$). Now we see that

$$\alpha'_\theta(p_+(\theta)) = 1 + \sqrt{1 - 4\theta} \tag{8.7}$$
$$\alpha'_\theta(p_-(\theta)) = 1 - \sqrt{1 - 4\theta}. \tag{8.8}$$

For $\theta < 1/4$, $\alpha'_\theta(p_+(\theta)) > 1$ so that the fixed point $(p_+(\theta))$ is repelling.

Now consider $p_-(\theta)$; of course, $\alpha'_\theta(p_-(\theta)) = 1$ when $\theta = 1/4$. When $\theta < 1/4$ but is sufficiently close to $1/4$, $\alpha'_\theta(p_-(\theta)) < 1$, so that the fixed point $p_-(\theta)$ is locally attracting. It will continue to be attracting as long as $|\alpha'_\theta(p_-(\theta))| < 1$, i.e.,

$$-1 < \alpha'_\theta(p_-(\theta)) < 1$$

or

$$-1 < 1 - \sqrt{1 - 4\theta} < 1.$$

It follows that $p_-(\theta)$ is locally stable for all θ satisfying

$$-3/4 < \theta < 1/4$$

when $\theta = -3/4$, $p_-(\theta)$ is again non-hyperbolic (i.e., $|\alpha'_\theta(p_\theta)| = 1$) and for $\theta < -3/4$, $|\alpha'_\theta(p_-(\theta))| > 1$ so that $p_-(\theta)$, too, becomes repelling.

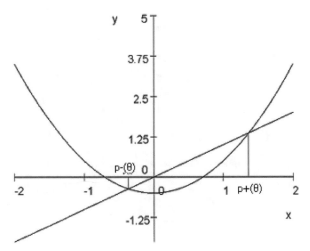

Figure 1.1a: $\theta = -\frac{1}{2}$

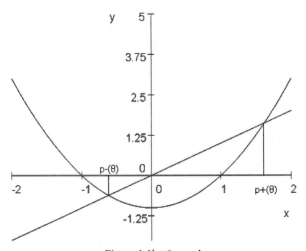

Figure 1.1b: $\theta = -1$

Using a graphical analysis (see Figure 1.1) one can show the following:

(i) For $\theta \leq 1/4$, if the initial state $x > p_+(\theta)$ or $x < -p_+(\theta)$, then the trajectory from x tends to infinity;

(ii) for $-3/4 < \theta < 1/4$, all the trajectories starting from $x \in (-p_+(\theta), p_+(\theta))$ tend to the attracting fixed point $p_-(\theta)$.

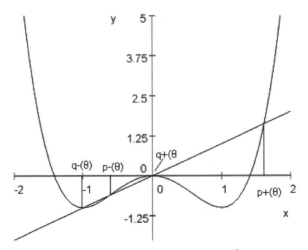

Figure 1.1c: Second iterate of $\alpha(x) = x^2 - 1$

As the parameter θ decreases through $-3/4$, the fixed point $p_-(\theta)$ loses its stability property; but more is true. We see a *period-doubling bifurcation*: a pair of periodic points is "born." To examine this, consider the equation $\alpha_\theta^2(x) = x$, i.e.,

$$x^4 + 2\theta x^2 - x + \theta^2 + \theta = 0 \tag{8.9}$$

using the fact that both p_+ and p_- are solutions to (8.2), we see that there are two other roots given by

$$q_+(\theta) = (-1 + \sqrt{-4\theta - 3})/2 \tag{8.10}$$

$$q_-(\theta) = (-1 - \sqrt{-4\theta - 3})/2. \tag{8.11}$$

Clearly, $q_+(\theta)$ and $q_-(\theta)$ are real if and only if $\theta \leq -3/4$.

Of course, when $\theta = -3/4$, $q_+(\theta) = q_-(\theta) = -1/2 = p_-(\theta)$. Furthermore, for $-5/4 < \theta < -3/4$, the periodic points are locally stable (see Figures 1.1a–1.1c).

To summarize, we have confirmed that

as the parameter passes through a "threshold" value, the number as well as the local stability properties of fixed and periodic points of a family of dynamical systems may change.   ∎

Example 8.2 (*Verhulst equation*) We return to the Verhulst equation discussed in Example 2.3. Consider the dynamical system (S, α) where $S = \mathbb{R}_+$ and $\alpha : S \to S$ is defined by

$$\alpha(x) = \frac{\theta_1 x}{x + \theta_2}, \tag{8.12}$$

where θ_1, θ_2 are two *positive* numbers, treated as parameters. Note that α is increasing:

$$\alpha'(x) = \frac{\theta_1 \theta_2}{(x + \theta_2)^2} > 0 \quad \text{for } x \geq 0.$$

Also,

$$\alpha''(x) = \frac{-2\theta_1\theta_2}{(x + \theta_2)^3} < 0 \quad \text{for } x \geq 0.$$

Verify that

$$\lim_{x \to 0} \alpha'(x) = (\theta_1/\theta_2)$$

and

$$\lim_{x \to \infty} \alpha'(x) = 0.$$

To study the long-run behavior of alternative trajectories, three cases are distinguished (of which one is left as an exercise):

Case I: $\theta_1 < \theta_2$
Case II: $\theta_1 = \theta_2$
Case III: $\theta_1 > \theta_2$

In Case I, the unique fixed point $x^* = 0$ is locally attracting. But in this case, $\alpha'(0) < 1$ and $\alpha''(x) < 0$ for $x \geq 0$ imply that $\alpha'(x) < 1$ for all $x > 0$ (the graph of α is "below" the 45°-line). By the mean value theorem, for $x > 0$

$$\alpha(x) = \alpha(0) + x\alpha'(z), \quad \text{where } 0 < z < x$$
$$= x a'(z) < x$$

Hence, from any $x > 0$,

$$x_{t+1} = \alpha(x_t) < x_t.$$

Thus, the trajectory $\tau(x) = \{x_t\}$ is decreasing. Since $x_t \geq 0$, the sequence $\tau(x)$ must converge to some fixed point $\hat{x}$ of α (by using the continuity of α). Since 0 is the unique fixed point, we conclude the following:

when $\theta_1 < \theta_2$, for any $x > 0$ the trajectory $\tau(x)$ is decreasing and converges to the unique fixed point of α, namely, $x^* = 0$.

(A graphical analysis might be useful to visualize the behavior of trajectories.)

We now turn to Case III. Here, $x^*_{(1)} = 0$ is repelling and $x^*_{(2)} = \theta_1 - \theta_2$ is locally attracting. But by the mean value theorem, for $x > 0$ and "sufficiently small," $\alpha(x) > x$. But this also means that for *all* x satisfying $0 < x < x^*_{(2)}$, it must be that $\alpha(x) > x$ (otherwise, there must be another fixed point between 0 and $x^*_{(2)}$, and this leads to a contradiction).

Consider $x > x^*_{(2)}$. Since $\alpha'(x^*_{(2)}) < 1$ and $\alpha''(x) < 0$ for *all* x, it follows that $\alpha'(x) < 1$ for *all* $x > x^*_{(2)}$. Write

$$\eta(x) = \alpha(x) - x.$$

Then $\eta'(x) < 0$ for *all* $x \geq x^*_{(2)}$ and $\eta(x^*_{(2)}) = 0$. This implies that $\eta(x) < 0$ for *all* $x > x^*_{(2)}$, i.e., $\alpha(x) < x$ for all $x > x^*_{(2)}$.

To summarize, we have

$$\alpha(x) > x \quad \text{for } x \text{ satisfying } 0 < x < x^*_{(2)}$$
$$\alpha(x) < x \quad \text{for } x > x^*_{(2)}.$$

We can now adapt the discussion of Example 5.1, to conclude

when $\theta_1 > \theta_2$, for any $x > 0$, $\tau(x)$ converges to $x^*_2 = \theta_1 - \theta_2$. If $0 < x < x^*_2$, $\tau(x)$ is increasing; if $x > x^*_2$, $\tau(x)$ is decreasing.

Finally, the case $\theta_1 = \theta_2$ is left as an exercise. ∎

Example 8.3 Here $S = \mathbb{R}_+$ and $\alpha : S \to S$ is defined by

$$\alpha(x) = \frac{\theta_1 x^2}{x^2 + \theta_2}, \tag{8.13}$$

where θ_1, θ_2 are two positive real numbers (parameters).

First, note that α is increasing:

$$\alpha'(x) = \frac{2\theta_1\theta_2 x}{(x^2 + \theta_2)^2} > 0 \quad \text{for } x > 0. \tag{8.14}$$

As to the fixed points of α, there are three cases:

Case I: If $\theta_1 < 2\sqrt{\theta_2}$, the only fixed point of α is $x^*_{(1)} = 0$.
Case II: If $\theta_1 = 2\sqrt{\theta_2}$, the fixed points of α are $x^*_{(1)} = 0$, $x^*_{(2)} = \frac{\theta_1}{2}$.

Case III: If $\theta_1 > 2\sqrt{\theta_2}$, the fixed points of α are

$$x_{(1)}^* = 0$$

$$x_{(2)}^* = [\theta_1 - \sqrt{\theta_1^2 - 4\theta_2}]/2$$

$$x_{(3)}^* = [\theta_1 + \sqrt{\theta_1^2 - 4\theta_2}]/2. \tag{8.15}$$

Since $\alpha'(0) = 0$, we see that the fixed point $x_{(1)}^* = 0$ is locally attracting (in all cases).

In Case II, the fixed point $x_{(2)}^* = \theta_1/2$ satisfies $\alpha'(\frac{\theta_1}{2}) = 1$, i.e., $x_{(2)}^*$ is non-hyperbolic.

In Case III, after some simplifications, we get

$$\alpha'(x_{(2)}^*) = \frac{4\theta_2}{\theta_1(\theta_1 - \sqrt{\theta_1^2 - 4\theta_2})}$$

$$\alpha'(x_{(3)}^*) = \frac{4\theta_2}{\theta_1(\theta_1 + \sqrt{\theta_1^2 - 4\theta_2})}. \tag{8.16}$$

Using the fact that $\theta_1 > 2\sqrt{\theta_2}$, in this case one can show that

$$\alpha'(x_{(2)}^*) > 1,$$
$$\alpha'(x_{(3)}^*) < 1.$$

Hence, $x_{(2)}^*$ is repelling while $x_{(3)}^*$ is locally attracting.

We now turn to the global behavior of trajectories and their sensitivity to changes in θ_1 and θ_2.

In Case I, by continuity of α', $\alpha'(x) < 1$ for all $x \in (0, \tilde{x})$ for some $\tilde{x} > 0$. Hence, $\alpha(x) < x$ for $x \in (0, \tilde{x})$. But this means that $\alpha(x) < x$ for *all* $x > 0$, otherwise there has to be a fixed point of α, say $z > 0$. But this would contradict the fact that in this case 0 is the unique fixed point.

Since $\alpha(x) < x$ for all $x > 0$, it follows from each $x > 0$, the trajectory $\tau(x)$ is decreasing and converges to 0.

In Case II we can similarly prove that for all $x \in (0, x_{(2)}^*)$, the trajectory $\tau(x)$ from x is decreasing and converges to 0. Next, one can verify that for x sufficiently large, $\alpha(x) < x$. But this implies that $\alpha(x) < x$ for *all* $x > x_{(2)}^*$. Otherwise there would have to be another fixed point z of α such that $z > x_{(2)}^*$, and this would contradict the fact that $x_{(2)}^*$ is the only positive fixed point of α. Hence, for each $x > x_{(2)}^*$, the trajectory $\tau(x)$ from x is decreasing and converges to $x_{(2)}^*$.

In Case III, arguments similar to those given above show that for all $x \in (0, x_{(2)}^*)$, $\alpha(x) < x$. Hence, the trajectory $\tau(x)$ is monotonically decreasing and converges to 0. Similarly, for $x > x_{(3)}^*$, $\alpha(x) < x$. Hence, for each $x > x_{(3)}^*$ the trajectory $\tau(x)$ is decreasing and converges to $x_{(3)}^*$.

Finally, $\alpha'(x_{(2)}^*) > 1$ implies that for $x > x_2^*$ and x sufficiently close to $x_{(2)}^*$, $\alpha'(x) > 1$. By using the mean value theorem we can show that $\alpha(x) > x$. But this means that for all $x \in (x_{(2)}^*, x_{(3)}^*)$, $\alpha(x) > x$ (otherwise there must be a fixed point z of α between $x_{(2)}^*$ and $x_{(3)}^*$, a contradiction). Also $\alpha'(x) > 0$ for $x > 0$ implies that $\alpha(x) \le \alpha(x_{(3)}^*) = x_{(3)}^*$ for all $x \in (x_{(2)}^*, x_{(3)}^*)$. We can use our analysis in Section 1.5 to show that for each $x \in (x_{(2)}^*, x_{(3)}^*)$, the trajectory $\tau(x)$ is increasing and converges to $x_{(3)}^*$.   ∎

The above discussion suggests that bifurcations occur near non-hyperbolic fixed and periodic points. This is indeed the only place where bifurcations of fixed points occur, as the following result on comparative statics demonstrates.

Proposition 8.1 *Let α_θ be a one-parameter family of functions, and suppose that $\alpha_{\theta_0}(x_0) = x_0$ and $\alpha'_{\theta_0}(x_0) \ne 1$. Then there are intervals I about x_0 and N about θ_0, and a smooth function $p : N \to I$ such that $p(\theta_0) = x_0$ and $\alpha_\theta(p(\theta)) = p(\theta)$. Moreover, α_θ has no other fixed points in I.*

Proof. Consider the function defined by $G(x, \theta) = \alpha_\theta(x) - x$. By hypothesis, $G(x_0, \theta_0) = 0$ and

$$\frac{\partial G}{\partial x}(x_0, \theta_0) = \alpha'_{\theta_0}(x_0) - 1 \ne 0.$$

By the implicit function theorem, there are intervals I about x_0 and N about θ_0, and a smooth function $p : N \to I$ such that $p(\theta_0) = x_0$ and $G(p(\theta), \theta) = 0$ for all $\theta \in N$. Moreover, $G(x, \theta) \ne 0$ unless $x = p(\theta)$. This concludes the proof.   ∎

1.9 Some Applications

1.9.1 The Harrod–Domar Model

We begin with the Harrod–Domar model, which has left an indelible impact on development planning and has figured prominently in policy debates on the appropriate strategy for raising the growth rate of an economy.

The "income" or output at the "end" of period t, Y_t, is determined by the amount of capital K_t at the "beginning" of the period according to a linear production function:

$$Y_t = \beta K_t, \tag{9.1}$$

where $\beta > 0$ is the productivity index (or the "output–capital" ratio).

At the end of period t, a fraction s ($0 < s < 1$) is saved and invested to augment the capital stock. Write S_t as the saving in period t. The stock of capital does not depreciate. Hence we have

$$S_t = s Y_t, \quad 0 < s < 1, \tag{9.2}$$

and the capital stock "at the beginning" of period $t + 1$ is given by

$$\begin{aligned} K_{t+1} &= K_t + S_t \\ &= K_t + s Y_t. \end{aligned} \tag{9.3}$$

From (9.1) and (9.3)

$$\beta K_{t+1} = \beta K_t + \beta s Y_t$$
$$\text{or,} \quad Y_{t+1} = Y_t(1 + \beta s). \tag{9.4}$$

This law of motion (9.4) describes the evolution of income (output) over time. Clearly,

$$\frac{Y_{t+1} - Y_t}{Y_t} = s\beta. \tag{9.5}$$

The last equation has figured prominently in the discussion of the role of saving and productivity in attaining a high rate of growth (it is assumed that there is an "unlimited" supply of labor needed as an input in the production process).

1.9.2 The Solow Model

1.9.2.1 Homogeneous Functions: A Digression

Let $F(K, L) : \mathbb{R}_+^2 \to \mathbb{R}_+$ be a homogeneous function of degree 1, i.e., $F(\lambda K, \lambda L) = \lambda F(K, L)$ for all $(K, L) \in \mathbb{R}_+^2$, $\lambda > 0$. Assume that F is twice continuously differentiable at $(K, L) >> 0$, strictly concave, and $F_K > 0$, $F_L > 0$, $F_{KK} < 0$, $F_{LL} < 0$ at $(K, L) >> 0$.

For $L > 0$, note $F(K, L) = F(K/L, 1)/L$. Write $k \equiv K/L$, and $f(k) \equiv F(K/L, 1)/L$. A standard example is the Cobb–Douglas form:

$$F(K, L) = K^a L^b, \quad a > 0, \quad b > 0, \quad a + b = 1.$$

Then $f(k) \equiv k^a$, and $0 < a < 1$.

One can show that

$$
\begin{aligned}
f'(k) &\equiv \frac{d}{dk} f(k) = F_K & \text{at } (K, L) >> 0 \\
F_L &= f(k) - k f'(k) & \text{at } (K, L) >> 0 \\
F_{KK} &= f''/L < 0 & \text{at } (K, L) >> 0 \\
F_{LL} &= [k^2 f''(k)]/L < 0 & \text{at } (K, L) >> 0.
\end{aligned}
$$

1.9.2.2 The Model

There is only one producible commodity that can be either consumed or used as an input along with labor to produce more of itself. When consumed, it simply disappears from the scene. Net output at the "end" of period t, denoted by $Y_t (\geq 0)$, is related to the input of the producible good K_t (called "capital") and labor L_t employed "at the beginning of" period t according to the following technological rule ("production function"):

$$Y_t = F(K_t, L_t), \tag{9.6}$$

where $K_t \geq 0$, $L_t \geq 0$.

The fraction of output saved (at the end of period t) is a constant s, so that total saving S_t in period t is given by

$$S_t = s Y_t, \quad 0 < s < 1. \tag{9.7}$$

Equilibrium of saving and investment plans requires

$$S_t = I_t, \tag{9.8}$$

where I_t is the net investment in period t. For the moment, assume that capital stock does not depreciate over time, so that at the beginning of period $t + 1$, the capital stock K_{t+1} is given by

$$K_{t+1} \equiv K_t + I_t. \tag{9.9}$$

Suppose that the total supply of labor in period t, denoted by $\hat{L}_t$, is determined completely exogenously, according to a "natural" law:

$$\hat{L}_t = \hat{L}_0(1 + \eta)^t, \quad \hat{L}_0 > 0, \quad \eta > 0. \tag{9.10}$$

Full employment of the labor force requires that

$$L_t = \hat{L}_t. \tag{9.10'}$$

Hence, from (9.7), (9.8), and (9.10'), we have

$$K_{t+1} = K_t + sF(K_t, \hat{L}_t).$$

Assume that F is homogeneous of degree 1. We then have

$$\frac{K_{t+1}}{\hat{L}_{t+1}} \frac{\hat{L}_{t+1}}{\hat{L}_t} = \frac{K_t}{\hat{L}_t} + sF\left(\frac{K_t}{\hat{L}_t}, 1\right).$$

Writing $k_t \equiv K_t/\hat{L}_t$, we get

$$k_{t+1}(1 + \eta) = k_t + s\, f(k_t), \tag{9.11}$$

where

$$f(k) \equiv F(K/L, 1).$$

From (9.11)

$$k_{t+1} = k_t/(1 + \eta) + s\, f(k_t)/(1 + \eta)$$

or

$$k_{t+1} = \alpha(k_t), \tag{9.12}$$

where

$$\alpha(k) \equiv [k/(1 + \eta)] + s[f(k)/(1 + \eta)]. \tag{9.12'}$$

Equation (9.12) is the fundamental dynamic equation describing the intertemporal behavior of k_t when *both* the full employment condition and the condition of short-run savings–investment equilibrium (see (9.10') and (9.8)) are satisfied. We shall refer to (9.12) as the law of motion of the Solow model in its *reduced form*.

Assume that

$$f(0) = 0, \quad f'(k) > 0, \quad f''(k) < 0 \quad \text{for } k > 0$$

and

$$\lim_{k\downarrow 0} f'(k) = \infty, \quad \lim_{k\uparrow\infty} f'(k) = 0.$$

Then, using (9.12′),

$$\alpha(0) = 0,$$

$$\alpha'(k) = (1 + \eta)^{-1}[1 + sf'(k)] > 0 \quad \text{at } k > 0,$$

$$\alpha''(k) = (1 + \eta)^{-1}sf''(k) < 0 \quad \text{at } k > 0.$$

Also verify the boundary conditions:

$$\lim_{k\downarrow 0}\alpha'(k) = \lim_{k\downarrow 0}[(1 + \eta)^{-1} + (1 + \eta)^{-1}sf'(k)] = \infty,$$

$$\lim_{k\uparrow\infty}\alpha'(k) = (1 + \eta)^{-1} < 1.$$

By using Propositions 5.1 and 5.3 we get

Proposition 9.1 *There is a unique $k^* > 0$ such that*

$$k^* = \alpha(k^*);$$

equivalently,

$$k^* = [k^*/(1 + \eta)] + s[f(k^*)/1 + \eta]. \tag{9.13}$$

If $k < k^*$, the trajectory $\tau(k)$ is increasing and converges to k^*.
If $k > k^*$, the trajectory $\tau(k)$ is decreasing and converges to k^*.

A few words on comparative statics are in order. Simplifying (9.13) we get

$$k^* = \frac{s}{\eta}f(k^*). \tag{9.14}$$

How do changes in the parameters s and η affect the steady state value k^*? Keep η fixed, and by using the implicit function theorem we get

$$\frac{\partial k^*}{\partial s} = \left[\frac{f(k^*)}{\eta}\right] \bigg/ \left[1 - \frac{f'(k^*)k^*}{f(k^*)}\right].$$

By our assumptions,

$$\frac{f(k)}{k} > f'(k) \quad \text{for all } k > 0.$$

Hence,

$$1 > \frac{f'(k)k}{f(k)} \quad \text{for all } k > 0.$$

This leads to

$$\frac{\partial k^*}{\partial s} > 0,$$

$$\frac{\partial y^*}{\partial s} = f'(k^*)\frac{\partial k^*}{\partial s} > 0.$$

We see that an increase in s increases the steady state per capita capital and output. A similar analysis of the impact of a change in η on k^* and y^* is left as an exercise.

Exercise 9.1 An extension of the model introduces the possibility of depreciation in the capital stock in the production process and replaces Equation (9.9) by

$$K_{t+1} = (1 - d)K_t + I_t, \tag{9.9'}$$

where $d \in [0, 1]$ (the depreciation factor) is the fraction of capital "used up" in production. In our exposition so far $d = 0$. The other polar case is $d = 1$, where all capital is used up as an input. In this case (9.9) is replaced by

$$K_{t+1} = I_t. \tag{9.9''}$$

It is useful to work out the details of the dynamic behavior of the more general case (9.9'). ∎

In his article Solow (1956) noted that the "strong stability" result that he established "is not inevitable" but depends on the structure of the law of motion α in (9.12) ("on the way I have drawn the productivity curve"). By considering simple variations of his model, Solow illustrated some alternative laws of motion that could generate qualitatively different trajectories.

This point was subsequently elaborated by Day (1982), who stressed the possibility that with alternative specifications of the basic ingredients of the Solow model, one could obtain laws of motion that are consistent with cyclical behavior and Li–Yorke chaos. Day's examples involved

(a) allowing the fraction of income saved to be a function of output (i.e., dispensing with the assumption that s is a constant in (9.7));

(b) allowing for a "pollution effect" of increasing per capita capital.

Day's examples provide the motivation behind one of the examples in Chapter 3.

Exercise 9.2 Consider the variation of the Solow model in which $\eta = 0$ (i.e., labor force is stationary) and (9.9″) holds (i.e., capital depreciates fully). Then (9.12) is reduced to

$$k_{t+1} = sf(k_t). \tag{9.12″}$$

In one of Day's examples (Day 1982), $f(k)$ is specified as

$$f(k) = \begin{cases} Bk^\beta(m-k)^\gamma & \text{for } 0 \le k \le m \\ 0 & \text{for } k \ge m \end{cases} \tag{9.D}$$

where $B, m > 0$ and $\beta, \gamma \in (0, 1)$. The term $(m-k)^\gamma$ on the right-hand side of (9.D) attempts to capture the pollution effect. The law of motion can be explicitly obtained as

$$k_{t+1} = \alpha(k_t) = sBk^\beta(m-k)^\gamma \quad \text{for } 0 \le k \le m.$$

(1) Fix the savings rate $s \in (0, 1)$, $B \in (0, \frac{2}{s}]$, and $\beta = \gamma = \frac{1}{2}$. Verify that under these assumptions $\alpha(k) = sBk^{\frac{1}{2}}(m-k)^{\frac{1}{2}}$ is a map from $[0, m]$ into itself. [Hint: Compute $k^* = \arg\max_{k \in [0,m]} \alpha(k)$, and let $k^m = \alpha(k^*) = \frac{Bsm}{2}$. Use this information along with the inequalities implied by the restriction on B to prove the result.]

(2) Now fix $B = B^* = \frac{2}{s}$. Verify that k^c, the smallest root of

$$sB^*k^{\frac{1}{2}}(m-k)^{\frac{1}{2}} = k^*$$

is given by $k^c = \frac{1}{4}m(2 - \sqrt{3})$, and the following string of inequalities hold: $0 < k^c < k^* < k^m$ (i.e., $\alpha^3(k^c) < k^c < \alpha(k^c) < \alpha^2(k^c)$).

(3) Using (1) and (2) above, apply the Li–Yorke sufficiency theorem. ∎

1.9.3 Balanced Growth and Multiplicative Processes

Notation For two square $(n \times n)$ matrices $A = (a_{ij})$ and $B = (b_{ij})$, write

$$A \geq B \quad \text{if } a_{ij} \geq b_{ij} \quad \text{for all } i, j;$$
$$A > B \quad \text{if } A \geq B \quad \text{and } A \neq B;$$
$$A >> B \quad \text{if } a_{ij} > b_{ij} \quad \text{for all } i, j.$$

A is *nonnegative* (resp. *positive, strictly positive*) if $A \geq$ (resp. $>, >>$) 0.

Primed letters denote transposes. When A is a square matrix, $A_T = TAT^{-1}$ denotes the transform of A by the nonsingular $n \times n$ matrix T.

Consider the dynamical system (S, α) where $S = \mathbb{R}^n_+$ and $\alpha : S \to S$ is defined by

$$\alpha(x) = Ax, \tag{9.15}$$

where $A \geq 0$ is a *nonnegative* $n \times n$ matrix. The dynamical system (9.15) leads to the study of

$$x_t = A^t x_0, \tag{9.16}$$

where $x_0 \geq 0$. We shall first prove a fundamental result on nonnegative indecomposable matrices. A *permutation matrix* is obtained by permuting the columns of an identity matrix.

An $n \times n$ matrix $A(n \geq 2)$ is said to be *indecomposable* if for no permutation matrix Π does $A_\pi = \Pi A \Pi^{-1} = \begin{pmatrix} A_{11} & A_{12} \\ 0 & A_{22} \end{pmatrix}$ where A_{11}, A_{22} are square. ($\Pi A \Pi^{-1}$ is obtained by performing the same permutation on the rows and on the columns of A.)

Proposition 9.2 *Let $A \geq 0$ be indecomposable. Then*

(1) A has a characteristic root $\lambda > 0$ such that
(2) to λ can be associated an eigenvector $x_0 >> 0$;
(3) if β is any characteristic root of A, $|\beta| \leq \lambda$;
(4) λ increases when any element of A increases;
(5) λ is a simple root of $\phi(t) \equiv \det(tI - A) = 0$

Proof.

(1) (a) If $x > 0$, then $Ax > 0$. For if $Ax = 0$, A would have a column of zeros, and so would not be indecomposable.

(1) (b) A has a characteristic root $\lambda > 0$.

Let $\Omega = \{x \in \mathbb{R}^n \mid x > 0, \sum x_i = 1\}$. If $x \in \Omega$, we define $T(x) = [1/\rho(x)]Ax$ where $\rho(x) > 0$ is so determined that $T(x) \in \Omega$ (by (a) such a ρ exists for every $x \in \Omega$). Clearly, $T(x)$ is a continuous transformation of S into itself, so, by Proposition 2.2, there is an $x_0 \in \Omega$ with $x_0 = T(x_0) = [1/\rho(x_0)]Ax_0$. Put $\lambda = \rho(x_0)$.

(2) $x_0 >> 0$. Suppose that after applying a proper Π,

$$\tilde{x}_0 = \begin{pmatrix} \xi \\ 0 \end{pmatrix}, \quad \xi >> 0.$$

Partition A_π accordingly. $A_\pi \tilde{x}_0 = \lambda \tilde{x}_0$ yields

$$\begin{pmatrix} A_{11} & A_{12} \\ A_{21} & A_{22} \end{pmatrix} \begin{pmatrix} \xi \\ 0 \end{pmatrix} = \begin{pmatrix} \lambda \xi \\ 0 \end{pmatrix};$$

thus $A_{21}\xi = 0$, so $A_{21} = 0$, violating the indecomposability of A.

If $M = (m_{ij})$ is a matrix, we henceforth denote by M^* the matrix $M^* = (|m_{ij}|)$.

(3–4) If $0 \le B \le A$, and if β is a characteristic root of B, then $|\beta| \le \lambda$. Moreover, $|\beta| = \lambda$ implies $B = A$.

A' is indecomposable and therefore has a characteristic root $\lambda_1 > 0$ with an eigenvector $x_1 >> 0$: $A'x_1 = \lambda_1 x_1$. Moreover, $\beta y = By$. Taking absolute values and using the triangle inequality, we obtain

(i) $|\beta| y^* \le By^* \le Ay^*$. So

(ii) $|\beta| x_1' y^* \le x_1' Ay^* = \lambda_1 x_1' y^*$.

Since $x_1 >> 0$, $x_1' y^* > 0$, thus $|\beta| \le \lambda_1$.

Putting $B = A$ one obtains $|\beta| \le \lambda_1$. In particular $\lambda \le \lambda_1$ and since, similarly, $\lambda_1 \le \lambda$, $\lambda_1 = \lambda$.

Going back to the comparison of B and A and assuming that $|\beta| = \lambda$, one gets from (i) and (ii)

$$\lambda y^* = By^* = Ay^*.$$

From $\lambda y^* = Ay^*$, application of (2) gives $y^* >> 0$. Thus $By^* = Ay^*$ together with $B \le A$ yields $B = A$.

(5) (a) If β is a principal submatrix of A, and β is any characteristic root of B, $|\beta| < \lambda$.

β is also a characteristic root of the matrix $\bar{B} = \begin{pmatrix} B & 0 \\ 0 & 0 \end{pmatrix}$. Since A is indecomposable, $\bar{B} < A_\pi$ for a proper Π. Hence $|\beta| < \lambda$ by (3–4).

(5) (b) λ is a simple root of $\phi(t) = \det(tI - A) = 0$. $\phi'(\lambda)$ is the sum of the principal $(n-1) \times (n-1)$ minors of $\det(\lambda I - A)$. Let A_i be one of the principal $(n-1) \times (n-1)$ submatrices of A. By (5) (a), $\det(tI - A)$ cannot vanish for $t \geq \lambda$. Hence, $\det(\lambda I - A_i) > 0$ and $\phi'(\lambda) > 0$. ∎

As an immediate consequence of (4), one obtains

$$\min_i \sum_j a_{ij} \leq \lambda \leq \max_i \sum_j a_{ij},$$

and one equality holds only if all row sums are equal (then they both hold).

This is proved by increasing (respectively decreasing) some elements of A so as to make all row sums equal to

$$\max_i \sum_j a_{ij} \left(\text{resp. } \min_i \sum_j a_{ij} \right).$$

A similar result naturally holds for column sums.

We write $\lambda(A)$ to denote the (Perron–Frobenius or *dominant*) characteristic root whose existence is asserted in Proposition 9.2.

Remark 9.1

(a) $\lambda(A) = \lambda(A'), \lambda(\theta A) = \theta \lambda(A)$ for $\theta > 0$; $\lambda(A^k) = (\lambda(A))^k$ for any positive integer k.

(b) Any eigenvector $y \geq 0$ associated with $\lambda(A)$ is, in fact, strictly positive, i.e., $y >> 0$ and $y = \theta x_0$ for some $\theta > 0$.

The definitive result on the long-run behavior of (9.16) is the following.

Proposition 9.3 *Let A be an indecomposable nonnegative matrix, and $\lambda = \lambda(A)$.*

(i) $\lim_{t \to +\infty} (A/\lambda)^t$ *exists if and only if there is a positive integer k such that $A^k >> 0$.*

*(ii) In the case of the convergence of $(A/\lambda)^t$, every column of the limit
matrix is a strictly positive column eigenvector of A and every row is a
strictly positive row eigenvector of A, both associated with λ.*

Proof. Proposition 9.2 ensures that $\lambda > 0$, so that (A/λ) makes sense.
Moreover, the dominant root of (A/λ) is unity. This enables us to assume,
without loss of generality, that $\lambda(A) = 1$.

First we shall prove (ii). To this end, let $B = \lim A^t$. Clearly, $B \geq 0$.
We also have $\lim A^{t+1} = B$, so that $AB = A \lim A^t = \lim A^{t+1} = B$
and, similarly, $BA = B$. $AB = B$ implies that $Ab = b$ for any column
of B. It follows, by Remark 9.1, that b is either a strictly positive column eigenvector of A associated with $\lambda = 1$ or the column zero vector.
Likewise $BA = B$ implies that any row of B is either a strictly positive
row eigenvector of A associated with $\lambda = 1$ or the row zero vector. In
fact, $B = (b_{ij})$ is a strictly positive matrix. For if some $b_{ij} = 0$, the above
results imply that both the ith row and the jth column of B vanish, and
hence, all the rows and columns vanish. But the possibility of $B = 0$ is
ruled out, because $Bx_0 = \lim A^t x_0 = x_0$ for a strictly positive column
eigenvector x_0 of A associated with $\lambda = 1$. This proves (ii).

The "only if" part of (i) immediately follows from (ii). In fact, if
$\lim A^t = B$ exists, B must be strictly positive by (ii). Hence A^t must also
be strictly positive for $t \geq k$, whenever k is large enough. For proof of the
"if" part of (i), first assume that the assertion is true for any strictly positive matrix. Then, since $A^k >> 0$ and $\lambda(A^k) = \lambda(A)^k = 1$ by Remark
9.1, $\lim(A^k)^t = B >> 0$ exists. Every positive integer t is uniquely ex-
pressible in the form $t = p(t)k + r(t)$, where $p(t), r(t)$ are nonnegative
integers with $0 \leq r(t) < k$, $\lim p(t) = +\infty$. Therefore, letting $(A^k)^{p(t)} =
B + R(t)$, we have $\lim R(t) = 0$. Hence $A^t = A^{r(t)}(A^k)^{p(t)} = A^{r(t)}(B +
R(t)) = A^{r(t)}B + A^{r(t)}R(t) \to B$ because $A^{r(t)}B = B$ and $A^{r(t)}$ is
bounded. The reason for $A^{r(t)}B = B$ is that since A and A^k have eigen-
vectors in common, the columns of B are strictly positive column eigen-
vectors of A associated with $\lambda = 1$; hence $A^{r(t)}B = AA\cdots AB = B$.
Thus it remains to prove the assertion for strictly positive matrices.

Now let A be a strictly positive matrix with $\lambda(A) = 1$. To prove the
convergence of A^t, we only have to show that the vector sequence $\{A^t y\}$
is convergent for any positive $y > 0$.

Set $y_t = A^t y$ for any given $y > 0$. y_t is a solution of the difference
equation

$$y_{t+1} = Ay_t, \quad y_0 = y.$$

Take a strictly positive column eigenvector $x = (x_i)$ of A associated with $\lambda = 1$, and let

$$\theta_{i,t} = y_{i,t}/x_i,$$
$$m_t = \min[\theta_{i,t}, \ldots, \theta_{n,t}],$$
$$M_t = \max[\theta_{i,t}, \ldots, \theta_{n,t}].$$

We let $l(t)$ be an integer for which $m_t = \theta_{l(t),t}$. Then, for any i

$$
\begin{aligned}
y_{i,t+1} &= \sum_{j=1}^{n} a_{ij} y_{j,t} \\
&= \sum_{j=1}^{n} a_{ij}(\theta_{j,t}) x_j \\
&= m_t a_{il(t)} x_{l(t)} + \sum_{j \neq l(t)} a_{ij} \theta_{j,t} x_j \\
&\leq m_t a_{il(t)} x_{l(t)} + M_t \sum_{j \neq l(t)} a_{ij} x_j \\
&= (m_t - M_t) a_{il(t)} x_{l(t)} + M_t \sum_{j=1}^{n} a_{ij} x_j \\
&= (m_t - M_t) a_{il(t)} x_{l(t)} + M_t x_i.
\end{aligned}
$$

Now let

$$\delta = \min_{i,j} a_{ij}, \quad \varepsilon = \min_{i,j} x_j/x_i.$$

Then $\delta > 0$, $\varepsilon > 0$, and the above relations, if divided by x_i, become

$$y_{i,t+1}/x_i \leq (m_t - M_t)\varepsilon\delta + M_t \quad (i = 1, \ldots, n),$$

from which we get

$$M_{t+1} \leq (m_t - M_t)\varepsilon\delta + M_t,$$

which can be rearranged to

$$\varepsilon\delta(M_t - m_t) \leq M_t - M_{t+1}.$$

Similarly, we have

$$\varepsilon\delta(M_t - m_t) \leq m_{t+1} - m_t.$$

Since $M_t \geq m_t$, the above two inequalities imply that $\{m_t\}$ is nondecreasing and bounded above, and $\{M_t\}$ is nonincreasing and bounded below. Hence $\lim m_t = m$ and $\lim M_t = M$ exist. Then, either of the above two inequalities entails $m = M$. In view of the definition of m_t, M_t, we have thereby proved $\lim y_t = mx$. ∎

Remark 9.2 An economic implication of the above result is the *relative stability* of a balanced-growth solution $x_t = \lambda^t x$, $\lambda = \lambda(A)$, $\lambda x = Ax$ of the self-sustained system:

$$x_{t+1} = Ax_t \tag{9.17}$$

where A is a nonnegative matrix such that $A^k >> 0$ for some k. For any solution $x_t = A^t x_0$ of the equation, x_t / λ^t converges to Bx_0, as t tends to infinity, where $B = \lim(A/\lambda)^t$. Stated differently, all the component ratios $x_{i,t}/\lambda^t x_i$ converge to a common limit. The latter statement can be justified in the following way: x_t/λ^t is a solution of $y_{t+1} = (A/\lambda)y_t$, with $\lambda(A/\lambda) = 1$, so that $x_{i,t}/\lambda^t x_i$ corresponds to $\theta_{i,t}$ in the proof of Proposition 9.3, which converges to a limit common to all i.

Remark 9.3 A nonlinear multiplicative process, first studied by Solow and Samuelson (1953) was explored in a number of contexts in economics and mathematical biology. We sketch the simplest case.

Take $S = \mathbb{R}_+^\ell$ and let $\alpha : S \to S$ satisfy:

[A.1] *α is continuous;*
[A.2] *α is homogeneous of degree 1, i.e., $\alpha(\mu x) = \mu\alpha(x)$ for $\mu > 0$ and $x \geq 0$;*
[A.3] *"$x' > x$" implies $\alpha(x') >> a(x)$.*

One can consider a large population consisting of ℓ types of individuals (e.g., the population can be divided into ℓ age groups) or a community consisting of ℓ interacting species (see Demetrius (1971) for a review of examples of this type). One can also think of $x_{i,t}$ as the stock of good i in period t. Then the interpretation of the mathematical model as a description of a dynamic multisector economy was that

we suppose all optimization and allocation problems to have been solved in one way or the other. We do not question what happens inside the sausage grinder: we simply observe that inputs flow into the economy and outputs appear. Since the economy is assumed to be closed..., all outputs become the inputs in the next period. Thus, it is as if we had a single production process

permitting continuous substitution of inputs (Solow and Samuelson 1953, p. 412).

One can then show that there are $\lambda > 0$ and a unique (in Ω) $v = (v_i)$, such that

$$\lambda v_i = \alpha_i(v) \quad \text{for } i = 1, \ldots, \ell.$$

Moreover, let x_t be the trajectory from $x_0 >> 0$. Then $\lim_{t \to \infty} x_{i,t}/(v_i \lambda^t) = c > 0$, where c is independent of i. Significant progress has been made to replace [A.3] with weaker assumptions (see Nikaido 1968).

1.9.4 Models of Intertemporal Optimization with a Single Decision Maker

We now turn to some infinite horizon economic models of dynamic optimization with a single decision maker. These give rise to dynamical systems that describe the evolution of *optimal* states or actions. In what follows, we use some of the concepts and insights from the literature on dynamic programming. In Chapter 6 we explore a more general problem of discounted dynamic programming under uncertainty, and prove a few deep results by appealing to the contraction mapping theorem. In contrast, in this section we rely on rather elementary methods to characterize optimal actions and to touch upon some fundamental issues on the role of prices in decentralization. We shall now informally sketch a model of deterministic dynamic programming (with discounting). It is a useful exercise to spell out formally how this framework encompasses the specific cases that we study in detail.

A deterministic stationary dynamic programming problem (SDP) is specified by a triple $< S, A, \varphi, f, u, \delta >$ where:

- S is the *state space* with generic element s; we assume that S is a nonempty complete separable metric space (typically, a closed subset of $\mathbb{R}^\ell$).
- A is the set of all conceivable *actions*, with generic element. We assume that A is a compact metric space.
- $\varphi: S \to A$ is a set valued mapping: for each $s \in S$, the nonempty set $\varphi(s) \subset A$ describes the set of all *feasible* actions to the decision

maker, given that the state he observes is (s) (some continuity properties of φ are usually needed, see Chapter 6, especially Section 6.3).

- $f : S \times A \to S$ is the *law of transition* for the system: for each current
  state–action pair (s, a) the state in the next period is given by $f(s) \in S$
  (again, f typically satisfies some continuity property).
- $u : S \times A \to R$ is the *immediate return* (or *one-period utility*) function
  that specifies a return $u(s, a)$ immediately accuring to the decision
  maker when the current state is s, and the action a is chosen (some
  continuity and boundedness properties are usually imposed on u).
- $\delta \in [0, 1)$ is the *discount factor*.

The sequential decision-making process is described as follows: An
initial state $s_0 \in S$ is given, and if the decision maker selects an action
a_0 in the set $\varphi(s_0)$ of actions feasible to him or her, two things happen.
First, an immediate return $u(s_0, a_0)$ is generated, and secondly, the system
moves to a state $s_1 = f(s_0, a_0)$, which will be observed tomorrow. The
story is repeated. Now, when the action a_t is chosen in $\varphi(s_t)$ in period
t, the *present value* of the immediate return $u(s_t, a_t)$ in period t is given
by $\delta^t u(s_t, a_t)$. Given $s_0 \in S$, the aim of the decision maker is to choose
a sequence (a_t) of action that solves

$$\text{maximize} \sum_{t=0}^{\infty} \delta^t u(s_t, a_t)$$
$$\text{subject to } s_{t+1} = f(s_t, a_t), \quad t = 0, 1 \ldots$$
$$a_t \in \phi(s_t), \quad t = 0, 1, 2, \ldots$$

1.9.4.1 Optimal Growth: The Aggregative Model

In this subsection we review several themes that have been central to
the literature on optimal economic development. We use an aggregative, discrete time, "discounted" model (in contrast with the pioneering
"undiscounted" continuous time model of Ramsey (1928)), and deal with
the following questions:

- the existence of an optimal program;
- the "duality" theory of dynamic competitive prices (attainment of
  the optimal program in an informationally decentralized economy by
  using competitive prices);
- optimal stationary programs ("golden rules");

– asymptotic or long-run properties of optimal programs ("turnpike theorems").

Consider an infinite horizon, one-good ("aggregative") model with a stationary *technology* described by a gross output function f, which satisfies the following assumptions [F]:

[F.1] $f : \mathbb{R}_+ \to \mathbb{R}_+$ *is continuous on* $\mathbb{R}_+$.
[F.2] $f(0) = 0$.
[F.3] $f(x)$ *is twice differentiable at* $x > 0$, *with* $f'(x) > 0$ *and* $f''(x) < 0$.
[F.4] (a) $\lim_{x \downarrow 0} f'(x) = \infty$; $\lim_{x \uparrow \infty} f'(x) = 0$;
     (b) $f(x) \to \infty$ *as* $x \to \infty$.

Note that the assumptions imply that f is strictly increasing and strictly concave on $\mathbb{R}_+$.

Corresponding to any f satisfying [F.1]–[F.4], there exists a unique number $\mathbf{K} > 0$, such that

(i) $f(\mathbf{K}) = \mathbf{K}$,
(ii) $f(x) > x$ for $0 < x < \mathbf{K}$,
(iii) $f(x) < x$ for $x > \mathbf{K}$.

We refer to $\mathbf{K}$ as the *maximum sustainable stock* for the technology described by f.

A *feasible production program* from $x > 0$ is a sequence $(\mathbf{x}, \mathbf{y}) = (x_t, y_{t+1})$ satisfying

$$x_0 = x; \quad 0 \le x_t \le y_t \quad \text{and} \quad y_t = f(x_{t-1}) \quad \text{for } t \ge 1. \qquad (9.18)$$

The sequence $\mathbf{x} = (x_t)_{t \ge 0}$ is the *input program*, while the corresponding $\mathbf{y} = (y_{t+1})_{t \ge 0}$ satisfying (9.18) is the *output program*. The *consumption program* $\mathbf{c} = (c_t)$, generated by $(\mathbf{x}, \mathbf{y})$, is defined by

$$c_t \equiv y_t - x_t \quad \text{for } t \ge 1. \qquad (9.19)$$

We refer to $(\mathbf{x}, \mathbf{y}, \mathbf{c})$ briefly as *a program* from x, it being understood that $(\mathbf{x}, \mathbf{y})$ is a feasible production program, and $\mathbf{c}$ is the corresponding consumption program.

A program $(\mathbf{x}, \mathbf{y}, \mathbf{c})$ from x is called *strictly positive* (written $(\mathbf{x}, \mathbf{y}, \mathbf{c}) >> 0$) if $x_t > 0$, $y_{t+1} > 0$, $c_{t+1} > 0$ for $t \ge 0$. It is a standard exercise to check that for any program $(\mathbf{x}, \mathbf{y}, \mathbf{c})$ from x we have $(x_t, y_{t+1}, c_{t+1}) \leq (\mathbf{B}, \mathbf{B}, \mathbf{B})$ for $t \geq 0$, where $\mathbf{B} = \max(x, \mathbf{K})$.

Again, a slight abuse of notation: we shall often specify only the input program $\mathbf{x} = (x_t)_{t \geq 0}$ from $x > 0$ to describe a program $(\mathbf{x}, \mathbf{y}, \mathbf{c})$. It will be understood that $x_0 = x$ and $0 \leq x_t \leq f(x_{t-1})$ for all $t \geq 1$, and (9.18) and (9.19) hold. Indeed we shall also refer to $(\mathbf{x}, \mathbf{y}, \mathbf{c})$ as a program from $y_1 > 0$ to mean that it is really a program from the unique $x > 0$ such that $y_1 = f(x)$. Note that decisions on "consumption today versus consumption tomorrow" begin in period 1.

A social planner evaluates alternative programs according to some welfare criterion. The criterion we examine is a "discounted sum of utilities generated by consumptions." More precisely, let $u : \mathbb{R}_+ \to \mathbb{R}_+$ be the *utility* (or *return*) *function* and δ be the *discount factor* satisfying $0 < \delta < 1$. In what follows, u satisfies the following assumptions [U]:

[U.1] *u is continuous on* $\mathbb{R}_+$.
[U.2] $u(c)$ *is twice continuously differentiable at* $c > 0$ *with* $u'(c) > 0$, $u''(c) < 0$, *and* $\lim_{c \downarrow 0} u'(c) = \infty$.

Clearly, [U.1] and [U.2] imply that u is strictly concave and increasing on $\mathbb{R}_+$. We normalize $u(0) = 0$.

For any program $(\mathbf{x}, \mathbf{y}, \mathbf{c})$ from $x > 0$, one has

$$0 \leq u(c_t) \leq u(\mathbf{B}) \quad \text{for all } t \geq 1.$$

A program $(\mathbf{x}^*, \mathbf{y}^*, \mathbf{c}^*)$ from $x > 0$ is *optimal* if

$$\sum_{t=1}^{\infty} \delta^{t-1} \left[u(c_t) - u(c_t^*) \right] \leq 0$$

for *all* programs $(\mathbf{x}, \mathbf{y}, \mathbf{c})$ from $x > 0$.

It is useful to recast the optimization problem just outlined in a dynamic programming framework. One can take $S = [0, \mathbf{B}]$, and $A = [0, 1] \equiv \varphi(s)$ for all $s \in S$.

We first deal with the question of *existence* of an optimal program.

Theorem 9.1 *There is a unique optimal program from any* $x > 0$.

Proof. Let $b \equiv u(\mathbf{B})$. If $(\mathbf{x}, \mathbf{y}, \mathbf{c})$ is any program from $\underset{\sim}{x}$, then $0 \leq u(c_t) \leq b$, for $t \geq 1$. So for every $T \geq 1$

$$\sum_{t=1}^{T} \delta^{t-1} u(c_t) \leq b/(1 - \delta).$$

Hence,

$$\sum_{t=1}^{\infty} \delta^{t-1} u(c_t) \equiv \lim_{T \to \infty} \sum_{t=1}^{T} \delta^{t-1} u(c_t) \leq b/(1 - \delta).$$

Let

$$a = \sup \left\{ \sum_{t=1}^{\infty} \delta^{t-1} u(c_t) : (\mathbf{x}, \mathbf{y}, \mathbf{c}) \text{ is a program from } \underset{\sim}{x} \right\}.$$

Choose a sequence of programs $(\mathbf{x}^n, \mathbf{y}^n, \mathbf{c}^n)$ from $\underset{\sim}{x}$ such that

$$\sum_{t=1}^{\infty} \delta^{t-1} u(c_t^n) \geq a - (1/n) \quad \text{for } n = 1, 2, 3, \ldots \qquad (9.20)$$

Using the continuity of f, and the diagonalization argument, there is a program $(\mathbf{x}^*, \mathbf{y}^*, \mathbf{c}^*)$ from $\underset{\sim}{x}$, and a subsequence of n (retain notation) such that for $t \geq 0$,

$$\left(x_t^n, y_{t+1}^n, c_{t+1}^n \right) \longrightarrow \left(x_t^*, y_{t+1}^*, c_{t+1}^* \right) \quad \text{as } n \to \infty. \qquad (9.21)$$

We claim that $(\mathbf{x}^*, \mathbf{y}^*, \mathbf{c}^*)$ is an optimal program from $\underset{\sim}{x}$. If the claim is false, then we can find $\varepsilon > 0$ such that

$$\sum_{t=1}^{\infty} \delta^{t-1} u(c_t^*) < a - \varepsilon.$$

Pick T such that $\delta^T [(b/1 - \delta)] < (\varepsilon/3)$. Using (9.20), and the continuity of u, we can find $\bar{n}$, such that for all $n \geq \bar{n}$,

$$\sum_{t=1}^{T} \delta^{t-1} u(c_t^n) < \sum_{t=1}^{T} \delta^{t-1} u(c_t^*) + (\varepsilon/3).$$

Thus, for all $n \geq \bar{n}$ we get

$$\sum_{t=1}^{\infty} \delta^{t-1} u(c_t^n) \leq \sum_{t=1}^{T} \delta^{t-1} u(c_t^n) + \delta^T [(b/1 - \delta)]$$

$$< \sum_{t=1}^{T} \delta^{t-1} u(c_t^n) + (\varepsilon/3)$$

$$< \sum_{t=1}^{T} \delta^{t-1} u(c_t^*) + (2\varepsilon/3) \leq \sum_{t=0}^{\infty} \delta^{t-1} u(c_t^*) + (2\varepsilon/3)$$

$$< a - (\varepsilon/3).$$

But this leads to a contradiction to (9.19) for $n > \max[\bar{n}, (3/\varepsilon)]$. This establishes our claim. Uniqueness of the optimal program follows from the strict concavity of u and f.   ∎

Note that for any $x > 0$, the unique optimal program $(\mathbf{x}^*, \mathbf{y}^*, \mathbf{c}^*)$ from x is *strictly positive*. For any $y > 0$, define the *value function* $V : \mathbb{R}_+ \to \mathbb{R}_+$ by

$$V(y_1) = \sum_{t=1}^{\infty} \delta^{t-1} u(c_t^*),$$

where $(\mathbf{x}^*, \mathbf{y}^*, \mathbf{c}^*)$ is the unique optimal program from $x = f^{-1}(y_1)$. We see $V(0) = 0$. For any $y_1 > 0$, define the *optimal consumption policy function* $c : \mathbb{R}_+ \to \mathbb{R}_+$ by

$$c(y_1) = c_1^*,$$

where $(\mathbf{x}^*, \mathbf{y}^*, \mathbf{c}^*)$ is the unique optimal program from y_1 (or, more formally, from $x = f^{-1}(y_1)$).

We now state some of the basic properties of the value and consumption policy functions.

Remark 9.4 Let $(\mathbf{x}^*, \mathbf{y}^*, \mathbf{c}^*)$ be the optimal program from x (or $y_1^* = f(x)$). Then for any $t \geq 2$, $(x_{t+1}^*, y_{t+1}^*, c_{t+1}^*)_{t \geq T}$ is optimal from x_T^* (or $y_{T+1}^* = f(x_T^*)$). It follows that

$$c(y_t^*) = c_t^* \quad \text{for all } t \geq 1,$$

and, for $T \geq 1$,

$$V\left(y_T^*\right) = \sum_{t=T}^{\infty} \delta^{t-T} u\left(c_t^*\right).$$

Lemma 9.1 *The value function V is increasing, strictly concave on $\mathbb{R}_+$, and differentiable at $y_1 > 0$. Moreover,*

$$V'(y_1) = u'(c(y_1)). \tag{9.22}$$

Proof. We shall only prove differentiability of V at $y > 0$. By concavity of V on $\mathbb{R}_+$, at *any* $y_1 > 0$ the right-hand derivative $V_+'(y_1)$ and the left-hand derivative $V_-'(y_1)$ both exist (see Nikaido 1968, Theorem 3.15 (ii), p. 47) and satisfy

$$V_-'(y_1) \geq V_+'(y_1). \tag{9.22a}$$

Let $(\mathbf{x}^*, \mathbf{y}^*, \mathbf{c}^*)$ be the unique optimal program from y_1. Recall that $(\mathbf{x}^*, \mathbf{y}^*, \mathbf{c}^*)$ is *strictly positive*. Take $h > 0$ and consider the program $(\mathbf{x}', \mathbf{y}', \mathbf{c}')$ from $y_1 + h$ defined as

$$x_1' = x_1^*, \quad c_1' = c_1^* + h, \quad y_1' = y_1 + h$$
$$x_t' = x_t^*, \quad c_t' = c_t^*, \quad y_t^* = y_t \quad \text{for } t \geq 2$$

Since $(\mathbf{x}', \mathbf{y}', \mathbf{c}')$ is clearly a *feasible program* from $y_1 + h$, one has

$$V(y_1 + h) \geq \sum_{t=1}^{\infty} \delta^{t-1} u(c_t').$$

Now, $V(y_1) \equiv \sum_{t=1}^{\infty} \delta^{t-1} u(c_t^*)$. Hence,

$$V(y_1 + h) - V(y_1) \geq u(c_1') - u\left(c_1^*\right)$$
$$= u\left(c_1^* + h\right) - u\left(c_1^*\right)$$
$$\text{or} \quad V_+'(y_1) \geq u'\left(c_1^*\right). \tag{9.22.b}$$

Take $\bar{h} > 0$, but sufficiently small, so that $c_1^* - \bar{h} > 0$. Consider the program $(\mathbf{x}'', \mathbf{y}'', \mathbf{c}'')$ from $y_1 - h$ defined as

$$x_1'' = x_1^*, \quad c_1'' = c_1^* - \bar{h}; \quad y_1'' = y_1^* - \bar{h};$$
$$x_t'' = x_t^*, \quad c_t'' = c_t^*, \quad y_t'' = y_t^* \quad \text{for } t \geq 2.$$

Since $(\mathbf{x}'', \mathbf{y}'', \mathbf{c}'')$ is clearly a *feasible program* from $y_1 - \bar{h}$, one has

$$V(y_1 - \bar{h}) \geq \sum_{t=1}^{\infty} \delta^{t-1} u\left(c_t''\right).$$

Then,

$$V(y_1) - V(y_1 - \bar{h}) \leq u\left(c_1^*\right) - u\left(c_1^* - \bar{h}\right)$$
$$\text{or} \quad V_-'(y_1) \leq u'\left(c_1^*\right) \tag{9.22c}$$

From (9.22b) and (9.22c)

$$V_+'(y_1) \geq u'\left(c_1^*\right) \geq V_-'(y_1). \tag{9.22d}$$

But (9.22a) and (9.22d) enable us to conclude that $V_-'(y_1) = V_+'(y_1) \equiv V'(y_1) = u'(c_1^*)$. ∎

Lemma 9.2 *The optimal consumption policy function c is increasing,* i.e.,

$$\text{``}y > y'(\geq 0)\text{''} \quad implies \quad \text{``}c(y) > c(y').\text{''} \tag{9.23}$$

Moreover, c is continuous on $\mathbb{R}_+$.

Proof.   This follows directly from the relation (9.22):

$$V'(y) = u'(c(y)) \quad \text{for all } y > 0.$$

Since V is strictly concave, "$y > y' > 0$" implies "$V'(y) < V'(y')$." Hence

$$c(y) > c(y'). \tag{9.24}$$

If $y' = 0$, $c(y') = 0$ and $y > 0$ implies $c(y) > 0$.

Continuity of c follows from a standard dynamic programming argument (see, e.g., Proposition 9.8 and, also Chapter 6, Remark 3.1). ∎

1.9.4.2 On the Optimality of Competitive Programs

The notion of a competitive program has been central to understanding the role of prices in *coordinating* economic decisions in a *decentralized* economy.

A program $(\mathbf{x}, \mathbf{y}, \mathbf{c})$ from $x > 0$ is intertemporal profit maximizing (IPM) if there is a positive sequence $\mathbf{p} = (p_t)$ of *prices* such that for $t \geq 0$,

$$p_{t+1}f(x_t) - p_t x_t \geq p_{t+1}f(x) - p_t x \quad \text{for } x \geq 0. \qquad (M)$$

A price sequence $\mathbf{p} > 0$ associated with an IPM program for which (M) holds is called a sequence of *support* or *Malinvaud* prices. A program $(\mathbf{x}, \mathbf{y}, \mathbf{c})$ from x is *competitive* if there is a positive sequence $\mathbf{p} = (p_t)$ of prices such that (M) holds for $t \geq 0$, *and* for $t \geq 1$

$$\delta^{t-1}u(c_t) - p_t c_t \geq \delta^{t-1}u(c) - p_t c \quad \text{for } c \geq 0. \qquad (G)$$

A price sequence $\mathbf{p} > 0$ associated with a competitive program $(\mathbf{x}, \mathbf{y}, \mathbf{c})$ for which both (M) and (G) hold will be called a sequence of *competitive* or *Gale* prices. The conditions (M) and (G) are referred to as competitive conditions. One can think of the conditions (G) and (M) in terms of a separation of the sequence of consumption and production decisions by means of a price system, the central theme of the masterly exposition of a static Robinson Crusoe economy by Koopmans (1957). Observe that a price sequence $\mathbf{p} = (p_t)$ associated with a competitive program $(\mathbf{x}, \mathbf{y}, \mathbf{c})$ is, in fact, *strictly positive*. Clearly, if for some $t \geq 1$, $p_t = 0$, then (G) implies

$$\delta^{t-1}u(c_t) \geq \delta^{t-1}u(c) \quad \text{for } c \geq 0$$

and this contradicts the assumed property that u is increasing. On the other hand if $p_0 = 0$, (M) implies

$$p_1 f(x) \geq p_1 f(x) \quad \text{for all } x \geq 0.$$

Since $p_1 > 0$, we get a contradiction from the assumed property that f is increasing. Here, we see that $\mathbf{p} >> 0$.

Our interest in competitive programs is partly due to the following basic result.

Theorem 9.2 *If* $(\mathbf{x}, \mathbf{y}, \mathbf{c}; \mathbf{p})$ *from* $x > 0$ *is a competitive program and*

$$\lim_{t \to \infty} p_t x_t = 0, \qquad (IF)$$

then $(\mathbf{x}, \mathbf{y}, \mathbf{c})$ *is the optimal program from* $x > 0$.

Proof. Let $(\bar{x}, \bar{y}, \bar{c})$ be *any* program from $x > 0$. By using the conditions (G) and (M) one gets

$$\sum_{t=1}^{T} \delta^{t-1}[u(\bar{c}_t) - u(c_t)] \leq \sum_{t=1}^{T} p_t[\bar{c}_t - c_t]$$

$$= \sum_{t=0}^{T-1}\{[p_{t+1}f(\bar{x}_t) - p_t\bar{x}_t] - [p_{t+1}f(x_t) - p_t x_t]\} + p_T(x_T - \bar{x}_T) \leq p_T x_T.$$

Hence,

$$\sum_{t=1}^{T} \delta^{t-1}u(\bar{c}_t) - \sum_{t=1}^{T} \delta^{t-1}u(c_t) \leq p_T x_T.$$

Now, as $T \to \infty$, the terms on the left-hand side have limits, and the condition (IF) ensures that the right-hand side converges to zero. Hence,

$$\sum_{t=1}^{\infty} \delta^{t-1}u(\bar{c}_t) \leq \sum_{t=1}^{\infty} \delta^{t-1}u(c_t).$$

This establishes the optimality of the program $(x, y, c; p)$. ∎

In his analysis of the problem of decentralization in a dynamic economy, Koopmans (1957) observed that "by giving sufficiently free rein to our imagination, we can visualize conditions like (G) and (M) as being satisfied through a decentralization of decisions among an infinite number of agents, each verifying a particular utility maximization (G) or profit maximization (M) condition. A further decentralization among many contemporaneous agents within each period can also be visualized through a more disaggregated model. But even at this level of abstraction, it is difficult to see how the task of meeting the condition (IF) can be "pinned down on any particular decision maker. This is a new condition to which there is no counterpart in the finite model." It should be stressed that if any agent in a particular period is allowed to observe only a *finite* number of prices and quantities (or, perhaps, an infinite subsequence of $p_t^* x_t$), it is not able to check whether the condition (IF) is satisfied.

The relationship between competitive and optimal programs can be explored further; specifically, it is of considerable interest to know that the converse of Theorem 9.2 is also true.

We begin by noting the following characterization of competitive programs.

Lemma 9.3 *Let* $(\mathbf{x}, \mathbf{y}, \mathbf{c})$ *be a program from* $x > 0$. *There is a strictly positive price sequence* $\mathbf{p} = (p_t)$ *satisfying* (G) *and* (M) *if and only if*

$$(i)\ x_t > 0, \quad y_t > 0, \quad c_t > 0 \quad for\ t \geq 0$$

$$(ii)\ u'(c_t) = \delta u'(c_{t+1}) f'(x_t) \quad for\ t \geq 0 \tag{RE}$$

Proof. (Necessity) Using [(U.2] and (G), $c_t > 0$ for $t \geq 1$. Since $y_t \geq c_t$, we get $y_t > 0$ for $t \geq 1$. Using [F.2], $x_t > 0$ for $t \geq 0$. This establishes (i).

Using (G) and $c_t > 0$, one obtains

$$p_t = \delta^{t-1} u'(c_t) \quad \text{for } t \geq 1.$$

Using (M) and $x_t > 0$, one obtains

$$p_{t+1} f'(x_t) = p_t \quad \text{for } t \geq 0.$$

Combining the above two equations establishes (ii).

(Sufficiency) Let $p_t = \delta^{t-1} u'(c_t)$ for $t \geq 1$ and $p_0 = p_1 f'(\underline{x})$. Now, using (ii), one obtains

$$p_{t+1} f'(x_t) = p_t \quad \text{for } t \geq 1. \tag{9.25}$$

Given any $t \geq 1$ and $c \geq 0$, concavity of u leads to

$$\delta^{t-1}[u(c) - u(c_t)] \leq \delta^{t-1} u'(c_t)(c - c_t) = p_t(c - c_t),$$

so that (G) follows by transposing terms. Given any $t \geq 0$ and $x \geq 0$, concavity of f leads to

$$p_{t+1}[f(x) - f(x_t)] \leq p_{t+1} f'(x_t)(x - x_t) = p_t(x - x_t),$$

using (9.25). Thus (M) follows by transposing terms. ∎

Remark 9.5 The condition (RE), known as the Ramsey–Euler condition, asserts the equality of the marginal product of input with the intertemporal marginal rate of substitution on the consumption side.

Now we are in a position to establish a converse of Theorem 9.2.

Theorem 9.3 *Suppose* $(\mathbf{x}, \mathbf{y}, \mathbf{c})$ *is the optimal program from* $x > 0$. *Then there is a strictly positive price sequence* $\mathbf{p} = (p_t)$ *satisfying (G) and (M), and*

$$\lim_{t \to \infty} p_t x_t = 0. \tag{IF}$$

Proof. Since $(\mathbf{x}, \mathbf{y}, \mathbf{c})$ is optimal from $x > 0$, it is strictly positive i.e., $(\mathbf{x}, \mathbf{y}, \mathbf{c}) >> 0$.

Since $(\mathbf{x}, \mathbf{y}, \mathbf{c})$ is optimal from $x > 0$, for each $t \geq 0$, x_t must maximize

$$W(x) \equiv \delta^{t-1} u(y_t - x) + \delta^t u(f(x) - x_{t+1})$$

among all x satisfying $0 \leq x \leq y_t$ and $f(x) \geq x_{t+1}$. Since $c_t > 0$ and $c_{t+1} > 0$, the maximum is attained at an interior point. Thus, $W'(x_t) = 0$, which can be written as

$$u'(c_t) = \delta u'(c_{t+1}) f'(x_t).$$

Hence, $(\mathbf{x}, \mathbf{y}, \mathbf{c})$ satisfies (i) and (ii) of Lemma 9.3. Consequently, there is a strictly positive price sequence $\mathbf{p} = (p_t)$ satisfying (G) and (M). In fact, using (G) and $c_t > 0$, we have

$$p_t = \delta^{t-1} u'(c_t) \quad \text{for } t \geq 1.$$

and

$$p_0 = p_1 f'(x).$$

Using Lemma 9.1 we also get

$$p_t = \delta^{t-1} V'(y_t) \quad \text{for } t \geq 1. \tag{9.25'}$$

Using the concavity of V and (9.25'), we obtain for any $y > 0$ and $t \geq 1$

$$\delta^{t-1}[V(y) - V(y_t)] \leq \delta^{t-1} V'(y_t)(y - y_t)$$
$$= p_t(y - y_t).$$

Thus for all $t \geq 1$ and any $y > 0$

$$\delta^{t-1} V(y) - p_t y \leq \delta^{t-1} V(y_t) - p_t y_t. \tag{9.26}$$

Choosing $y = (y_t/2)$ in (9.26), we get

$$0 \leq (1/2) p_t y_t \leq \delta^{t-1}[V(y_t) - V(y_t/2)] \leq \delta^{t-1} V(y_t) \tag{9.27}$$

For any y, $V(y) \leq [b/1 - \delta]$ (see the proof of Theorem 3.1). Also for $t \geq 1$, $0 \leq x_t \leq y_t$ and $0 < \delta < 1$. The condition (IF) follows from (9.27) as we let t tend to infinity. ∎

Lemma 9.4 *The optimal investment policy function* $i: \mathbb{R}_+ \rightarrow \mathbb{R}_+$ *defined as*

$$i(y) \equiv y - c(y) \tag{9.28}$$

is increasing, i.e.,

$$\text{"}y > y'(\geq 0)\text{"} \quad \textit{implies} \quad \text{"}i(y) > i(y').\text{"} \tag{9.29}$$

Moreover, i is continuous on $\mathbb{R}_+$.

Proof. Suppose, to the contrary that $i(y) \leq i(y')$, where $y > y' > 0$. This implies

$$f'(i(y)) \geq f'(i(y')) \tag{9.29'}$$

The Ramsey-Euler condition (RE) gives us

$$\delta u'[\mathcal{C}(f(i(y)))]f'(i(y)) = u'(c(y))$$
$$\delta u'[\mathcal{C}(f(i(y')))]f'(i(y')) = u'(c(y')).$$

Using (9.23) and [U.2], we get

$$\delta u'[\mathcal{C}(f(i(y)))]f'(i(y)) = u'(c(y))$$
$$< u'(c(y')) = \delta u'[c(f(i(y')))]f'(i(y')).$$

Hence, by using (9.29') and [U.2],
$u'[c(f(i(y)))] < u'[c(f(i(y')))]$, which leads to

$$i(y) > i(y'),$$

a contradiction. If $y' = 0$, $i(y') = 0$ and $y > 0$ implies $i(y) > 0$. Continuity of i follows from the continuity of c. ∎

1.9.4.3 Stationary Optimal Programs
In view of the assumption [F.4], we see that the equation

$$\delta f'(x) = 1,$$

(where $0 < \delta < 1$) has a *unique* solution, denoted by $x_\delta^* > 0$. Write $y_\delta^* = f(x_\delta^*)$ and $c_\delta^* = y_\delta^* - x_\delta^*$. It is not difficult to verify that $c_\delta^* > 0$. The triplet

$(x_\delta^*, y_\delta^*, c_\delta^*)$ is referred to as the (δ-modified) *golden rule input, stock*, and *consumption*, respectively. Define the *stationary* program $(\mathbf{x}^*, \mathbf{y}^*, \mathbf{c}^*)_\delta$ from $x_\delta^* > 0$ by (again $x_0 = x_\delta^*$)

$$x_t^* = x_\delta^*, \quad y_t^* = f(x_\delta^*), \quad c_t^* = c_\delta^* \quad \text{for } t \geq 1$$

Define a price system:

$$\mathbf{p}^* = (p_t^*) \quad \text{by } p_0^* = \frac{u'(c_\delta^*)}{\delta}, \quad p_t^* = \delta^{t-1} u'(c_\delta^*) \quad \text{for } t \geq 1.$$

It is an exercise to show that $\mathbf{p}^* = (p_t^*)$ is a strictly positive sequence of competitive prices, relative to which the stationary program $(\mathbf{x}^*, \mathbf{y}^*, \mathbf{c}^*)_\delta$ satisfies the conditions (G) and (M). Also, with $0 < \delta < 1$, the condition (IF) is satisfied (since $\lim_{t\to\infty} p_t^* x_\delta^* = 0$). Hence, the stationary program $(\mathbf{x}^*, \mathbf{y}^*, \mathbf{c}^*)_\delta$ is *optimal* (among *all* programs) from $x_\delta^* > 0$.

✳1.9.4.4 Turnpike Properties: Long-Run Stability

Let $0 < \underset{\sim}{x} < x_\delta^*$, and suppose that $(\mathbf{x}^*, \mathbf{y}^*, \mathbf{c}^*)$ is the optimal program from $\underset{\sim}{x}$. Again, recall that the sequences $(x_t^*, y_t^*, c_t^*)_{t\geq 1}$ are also optimal programs from y_t^*.

Since $(\mathbf{x}^*, \mathbf{y}^*, \mathbf{c}^*)_\delta$ is the optimal program from x_δ^*, it follows that $y_1^* < f(x_\delta^*)$. Hence, $x_1^* = i(y_1^*) < i(f(x_\delta^*)) = x_\delta^*$, and this leads to $y_2^* = f(x_1^*) < f(x_\delta^*)$. Thus,

$$
\begin{array}{ll}
y_t^* < f(x_\delta^*) & \text{for all } t \geq 1 \\
x_t^* < x_\delta^* & \text{for all } t \geq 1 \\
c_t^* < c_\delta^* = f(x_\delta^*) - x_\delta^* & \text{for all } t \geq 1.
\end{array}
\tag{9.30}
$$

By the Ramsey-Euler condition

$$\delta u'(c_2^*) f'(x_1^*) = u'(c_1^*)$$

Since $\delta f'(x_1^*) > 1$, it follows that $u'(c_2^*) < u'(c_1^*)$. Hence $c_2^* > c_1^*$. Since c is increasing, $y_2^* > y_1^*$. Repeating the argument, $y_{t+1}^* > y_t^*$ for all $t \geq 1$. This leads to $x_{t+1}^* > x_t^*$ for all $t \geq 0$, and $c_{t+1}^* > c_t^*$ for all $t \geq 1$. Using (9.30), we conclude that as t tends to infinity, the sequences (x_t^*), (y_{t+1}^*), (c_{t+1}^*) are all convergent to some $\hat{x}$, $\hat{y}$, $\hat{c}$, respectively. Clearly, $0 < \hat{x} \leq x_\delta^*$, $0 < \hat{y} \leq f(x_\delta^*)$, $0 < \hat{c} \leq c_\delta^*$. Again, using the Ramsey-Euler condition,

$$\delta u'(c_{t+1}^*) f'(x_t^*) = u'(c_t^*),$$

we get in the limit (as t tends to infinity) that

$$\delta f'(\hat{x}) = 1.$$

Hence $\hat{x} = x_\delta^*$, and this implies $\hat{c} = c_\delta^*$, $\hat{y} = y_\delta^*$. To summarize:

Theorem 9.4 *If $\underset{\sim}{x} < x_\delta^*$, the optimal program $(\mathbf{x}^*, \mathbf{y}^*, \mathbf{c}^*)$ from $\underset{\sim}{x}$ satisfies*

$$x_\delta^* > x_{t+1}^* > x_t^* \quad \text{for } t \geq 0, \quad \text{and } \lim_{t\to\infty} x_t^* = x_\delta^*$$
$$y_\delta^* > y_{t+1}^* > y_t^* \quad \text{for } t \geq 1, \quad \text{and } \lim_{t\to\infty} y_t^* = y_\delta^*$$
$$c_\delta^* > c_{t+1}^* > c_t^* \quad \text{for } t \geq 1, \quad \text{and } \lim_{t\to\infty} c_t^* = c_\delta^*.$$

Similarly, if $\underset{\sim}{x} > x_\delta^*$, the optimal program $(\mathbf{x}^*, \mathbf{y}^*, \mathbf{c}^*)$ from $\underset{\sim}{x}$ satisfies

$$x_\delta^* < x_{t+1}^* < x_t^* \quad \text{for } t \geq 0, \quad \text{and } \lim_{t\to\infty} x_t^* = x_\delta^*$$
$$y_\delta^* < y_{t+1}^* < y_t^* \quad \text{for } t \geq 1, \quad \text{and } \lim_{t\to\infty} y_t^* = y_\delta^*$$
$$c_\delta^* < c_{t+1}^* < c_t^* \quad \text{for } t \geq 1, \quad \text{and } \lim_{t\to\infty} c_t^* = c_\delta^*.$$

Going back to (9.28), we see that the optimal input program $\mathbf{x} = (x_t^*)_{t\geq 0}$ is described by

$$x_{t+1}^* = i(f(x_t^*))$$
$$\equiv \alpha(x_t^*) \tag{9.30'}$$

where $\alpha : \mathbb{R}_+ \to \mathbb{R}_+$ is the *composition map $i \cdot f$*. We shall often refer to α as the optimal transition function.

Exercise 9.3 Using [F.4] and Lemma 9.4 show that α is *increasing*. Show that for $x < x_\delta^*$, $\alpha(x) > x$ and for $x > x_\delta^*$, $\alpha(x) < x$ (recall Example 5.1). ∎

Exercise 9.4 Consider the dynamic optimization problem:

$$\text{maximize} \sum_{t=0}^{\infty} \delta^t \log c_t$$
$$\text{subject to} \quad c_{t+1} + x_{t+1} = Ax_t^\beta,$$
$$c_t \geq 0, \quad x_t \geq 0, \quad \text{for } t \geq 0,$$

where $A > 0, 0 < \beta < 1, 0 < \delta < 1$, and $x_0 = x$ is given.

(a) Show that there is a unique optimal program which is interior.

(b) Show that the *optimal investment policy* function $i(y)$ and the *optimal consumption policy* function $c(y)$ are both linear:

$$i(y) = \delta\beta y,$$
$$c(y) = (1 - \delta\beta)y.$$

Hence, the optimal input program is described by

$$x_{t+1} = (A\delta\beta)x_t^\beta, \quad x_0 > 0. \qquad \blacksquare$$

Exercise 9.5 (*Optimal growth in a productive economy*) Consider the one-sector aggregative model of optimal growth with discounting:

$$\text{maximize} \sum_{t=0}^{\infty} \delta^t u(c_t)$$

$$\text{subject to } c_0 + x_0 \le y_0,$$
$$c_t + x_t \le y_t = f(x_{t-1}) \quad \text{for } t \ge 1;$$
$$y_0 > 0 \text{ given.}$$

Assume that

(i) $u(c) = c^\alpha$ for all $c \ge 0$ and $\alpha \in (0, 1)$;
(ii) $f : \mathbb{R}_+ \longrightarrow \mathbb{R}_+$ is continuous, concave, and strictly increasing;
(iii) f is differentiable and $f'(x) > 0$ on $\mathbb{R}_{++}$;
(iv) $f(0) = 0,\ f'(0) > 1$.

Given the utility function, one easily checks that u is continuous, strictly concave and strictly increasing, and differentiable on $\mathbb{R}_{++}$. Furthermore, $\lim_{c \to 0} u'(c) = +\infty$. As usual a discount factor $\delta \in (0, 1)$ is given.

Let the *asymptotic productivity* be defined by

$$\gamma = \lim_{x \to \infty} \left[\frac{f(x)}{x} \right] = \lim_{x \to \infty} f'(x).$$

Lemma 9.5.1 *If $\delta < \gamma^{-a}$, then for any given $y_0 > 0$ there exists $B \in \mathbb{R}_{++}$ such that every feasible consumption program $(c_t)_{t=0}^{\infty}$ from initial stock y_0 satisfies $\sum_{t=0}^{\infty} \delta^t u(c_t) < B$.*

Proof. If $\gamma < 1$, then the proof is direct. So, confine attention to a situation where $\gamma \geq 1$ and $a > 0$. Fix $y_0 > 0$. Define the pure accumulation program $\overline{\mathbf{y}} = (\overline{y}_t)$ where $\overline{y}_t = f(\overline{y}_{t-1})$, $\forall t \geq 1, \overline{y}_0 = y_0$. It is easy to check that every feasible consumption program from initial stock y_0 satisfies $c_t \leq \overline{y}_t$. Let $B = \sum_{t=0}^{\infty} \delta^t u(\overline{y}_t)$. It is sufficient to show that $B < \infty$. Note that $\overline{\mathbf{y}} = (\overline{y}_t)$ is monotone increasing, so it must diverge to plus infinity if it is not bounded. If $(\overline{y}_t)$ is bounded, clearly $B < \infty$. Observe that

$$\frac{\overline{y}_{t+1}}{\overline{y}_t} = \frac{f(\overline{y}_t)}{\overline{y}_t} \longrightarrow \gamma \quad \text{as } t \longrightarrow \infty$$

Since $\delta < \gamma^{-a}$, there exists $\beta > 1, \epsilon > 0$ such that

$$\delta < (\beta\gamma)^{-(a+\epsilon)}$$

There exists T such that for $t \geq T$,

$$\overline{y}_t > 1$$
$$\overline{y}_{t+1} < \beta\gamma\overline{y}_t$$

so that, in particular,

$$\overline{y}_{T+k} < (\beta\gamma)^k\overline{y}_T, \quad k \geq 1,$$

and

$$u(\overline{y}_{T+k}) \leq (\overline{y}_{T+k})^{a+\epsilon} \leq [(\beta\gamma)^k\overline{y}_T]^{a+\epsilon}, \quad k \geq 0.$$

Thus,

$$\sum_{t=0}^{\infty} \delta^t u(\overline{y}_t) = \sum_{t=0}^{T-1} \delta^t u(\overline{y}_t) + \delta^T \left[\sum_{k=0}^{\infty} \delta^k u(\overline{y}_{T+k}) \right]$$

$$\leq \sum_{t=0}^{T-1} \delta^t u(\overline{y}_t) + \delta^T (\overline{y}_T)^{a+\epsilon} \sum_{k=0}^{\infty} [\delta(\beta\gamma)^{a+\epsilon}]^k$$

$$< \infty, \quad \text{since } \delta < (\beta\gamma)^{-(a+\epsilon)}.$$

The proof is complete. ∎

Remark 9.5.1 *There exists a unique optimal program. Every optimal program* $(\mathbf{x}^*, \mathbf{c}^*, \mathbf{y}^*)$ *is strictly positive. The value function* $V(y)$ *is strictly*

increasing, strictly concave, differentiable on $\mathbb{R}_{++}$ and $V'(y_t) = u'(c_t)$.
Further, the following Ramsey-Euler equation holds:

$$u'(c_t^*) = \delta u'(c_{t+1}^*) f'(x_t^*)$$
$$V'(y_t^*) = \delta V'(y_{t+1}^*) f'(x_t^*).$$

Remark 9.5.2 *The optimal consumption and optimal investment functions are increasing.*

Remark 9.5.3 *Every optimal program* $(\mathbf{x}^*, \mathbf{c}^*, \mathbf{y}^*)$ *is monotonic.*

Remark 9.5.4 *If $f'(0) < \frac{1}{\delta}$, every optimal program converges to zero.*

For the simplicity of notation, we drop the superscript $*$ in the next two results when we deal with optimal programs.

Remark 9.5.5 *If $f'(0) > \frac{1}{\delta}$, then every optimal program is bounded away from zero.*

Hint. It is sufficient to show that there exists $\varepsilon > 0$ such that for each initial stock $y_0 \in (0, \varepsilon)$, the optimal $y_1 > y_0$. To see this, suppose not. Then, there exists a sequence of strictly positive initial stocks $\{y_0^n\} \downarrow 0$ such that the next period's optimal stock $y_1^n = f(x_0^n) \leq y_0^n$. Using concavity of the value function, we have $V'(y_1^n) \geq V'(y_0^n)$. However, for n large enough, $\delta f'(x_0^n) > 1$ so that the Ramsey-Euler equation implies

$$V'(y_0^n) = \delta V'(y_1^n) f'(x_0^n) > V'(y_1^n),$$

a contradiction.

Remark 9.5.6 *If $\delta > \gamma^{-1}$, then every optimal program of consumption, input, and output diverges to infinity.*

Hint. Fix any $y_0 > 0$. As the optimal input program $\{x_t\}$ is monotonic, if it does not diverge to infinity, it must converge to either zero or a strictly positive stationary optimal stock. However, $\delta > \gamma^{-1}$ implies $\delta f'(x) > 1, \forall x \geq 0$, which in turn implies (using Proposition 9.8) that $\{x_t\}$ is bounded away from zero and, furthermore, there does not exist a strictly positive stationary optimal stock. Thus, $\{x_t\} \to \infty$ and, therefore, $\{y_t\} \to \infty$. Finally, to see that the optimal consumption $\{c_t\} \to \infty$, suppose not. Then, $\{c_t\}$ is bounded above. Using the strict concavity of u and the Ramsey-Euler equation,

$$u'(c_t) = \delta u'(c_{t+1}) f'(x_t).$$

Since $\delta f'(x) > 1$, $\forall x \geq 0$, we have that $\{c_t\}$ is an increasing sequence which must then converge to some $c' > 0$. Taking limits on both sides of the Ramsey-Euler equation, we obtain $u'(c') = \delta \gamma u'(c')$, i.e., $\delta = \gamma^{-1}$, a contradiction. ∎

1.9.5 Optimization with Wealth Effects: Periodicity and Chaos

In the optimal growth model of Section 1.9.4.1, the utility (or return) function u is defined on consumption only. The need for a more general framework in which the utility function depends on *both* consumption and input has been stressed in the theory of optimal growth as well as the theory of natural resource management. A step in this direction – even in the aggregative or one good model – opens up the possibility of periodic or chaotic behavior of the optimal program. In this section, we sketch some examples from Majumdar and Mitra (1994a,b) (see Complements and Details).

1.9.5.1 Periodic Optimal Programs

We consider an aggregative model, specified by a *production function* $f : \mathbb{R}_+ \to \mathbb{R}_+$, a *welfare function* $w : \mathbb{R}_+^2 \to \mathbb{R}_+$, and a *discount factor* $\delta \in (0, 1)$.

The following assumptions on f are used:

[F.1] $f(0) = 0$; f is continuous on $\mathbb{R}_+$.
[F.2] f is nondecreasing and concave on $\mathbb{R}_+$.
[F.3] There is some $\mathbf{K} > 0$ such that $f(x) > x$ for $0 < x < \mathbf{K}$, and $f(x) < x$ when $x > \mathbf{K}$.

We define a set $\Im \subset \mathbb{R}_+^2$ as follows:

$$\Im = \{(x, z) \in \mathbb{R}_+^2 : z \leq f(x)\}. \tag{9.31}$$

The following assumptions on w are used:

[W.1] $w(x, c)$ is continuous on $\mathbb{R}^2_+$.

[W.2] $w(x, c)$ is nondecreasing in x given c, and nondecreasing in c, given x on $\mathbb{R}^2_+$. Further, if $x > 0$, $w(x, c)$ is strictly increasing in c on $\Im$.

[W.3] $w(x, c)$ is concave on $\mathbb{R}^2_+$. Furthermore, if $x > 0$, $w(x, c)$ is strictly concave in c on $\Im$.

A *program* $\mathbf{x} = (x_t)_0^\infty$ from $\underset{\sim}{x} \geq 0$ is a sequence satisfying

$$x_0 = \underset{\sim}{x}, \quad 0 \leq x_{t+1} \leq f(x_t) \quad \text{for } t \geq 0.$$

The *consumption program* $\mathbf{c} = (c_{t+1})_0^\infty$ is given by

$$c_{t+1} = f(x_t) - x_{t+1} \quad \text{for } t \geq 0.$$

As in Section 1.9.5.1, $(x_t, c_{t+1})_{t\geq0} \leq B \equiv \max(K, \underset{\sim}{x})$ for any program $\mathbf{x}$ from $\underset{\sim}{x} \geq 0$, we have $(x_t, c_{t+1})_{t\geq0} \leq \mathbf{B} \equiv \max(K, \underset{\sim}{x})$ for $t \geq 0$. In particular, if $\underset{\sim}{x} \in [0, \mathbf{B}]$, then $x_t, c_{t+1} \leq \mathbf{B}$ for $t \geq 0$.

A program $\mathbf{x}^* = (x_t^*)_0^\infty$ from $\underset{\sim}{x} \geq 0$ is *optimal* if

$$\sum_{t=0}^\infty \delta^t w(x_t^*, c_{t+1}^*) \geq \sum_{t=0}^\infty \delta^t w(x_t, c_{t+1})$$

for every program $\mathbf{x}$ from $\underset{\sim}{x}$.

A program $\mathbf{x}$ from $\underset{\sim}{x}$ is *stationary* if $x_t = x$ for $t \geq 0$. It is a *stationary optimal program* if it is also an optimal program from $\underset{\sim}{x}$. In this case, $\underset{\sim}{x}$ is called a *stationary optimal stock*. (Note that 0 is a stationary optimal stock.) A stationary optimal stock $\underset{\sim}{x}$ is *nontrivial* if $x > 0$. We turn first to the possibility of a *periodic* optimal program.

Example 9.1 Define $f : \mathbb{R}_+ \to \mathbb{R}_+$ by

$$f(x) = \begin{cases} (32/3)x - 32x^2 + (256/3)x^4 & \text{for } 0 \leq x < 0.25 \\ 1 & \text{for } x \geq 0.25 \end{cases}.$$

Define $w : \mathbb{R}^2_+ \to \mathbb{R}_+$ by

$$w(x, c) = 2x^{1/2}c^{1/2} \quad \text{for all } (x, c) \in \mathbb{R}^2_+.$$

Finally, define $\hat{\delta} = (1/3)$.

It can be checked that f satisfies [F.1]–[F.3] with $\mathbf{B} = 1$ and w satisfies [W.1]–[W.3].

Now, let δ be any discount factor satisfying $\hat{\delta} < \delta < 1$. We fix this δ in what follows. Define an open interval

$$A(\delta) = \{x: 0.25 < x < 3\delta^2/(1 + 3\delta^2)\}$$

and also a point

$$x(\delta) = \delta/(1 + \delta)$$

It can be shown then that

(a) $x(\delta)$ is the unique non-trivial stationary optimal stock (note that $x(\delta) \in A(\delta)$);

(b) for every $\underset{\sim}{x} \in A(\delta)$, with $\underset{\sim}{x} \neq x(\delta)$, the optimal program $\mathbf{x}^* = (x_t^*)_0^\infty$ from $\underset{\sim}{x}$ is *periodic with period* 2. ∎

1.9.5.2 Chaotic Optimal Programs

We consider a family of economies indexed by a parameter θ (where $\theta \in I \equiv [1, 4]$). Each economy in this family has the same production function (satisfying [F.1]–[F.3]) and the same discount factor $\delta \in (0, 1)$. The economies in this family differ in the specification of their utility or one-period return function $w : \mathbb{R}_+^2 \times I \to \mathbb{R}_+$ (w depending explicitly on the parameter θ). For a *fixed* $\theta \in [1, 4]$, the utility function $w(\cdot, \theta)$ can be shown to satisfy [W.1]–[W.3].

Example 9.2 The numerical specifications are as follows:

$$f(x) = \begin{cases} (16/3)x - 8x^2 + (16/3)x^4 & \text{for } x \in [0, 0.5) \\ 1 & \text{for } x \geq 0.5 \end{cases}$$
$$\delta = 0.0025. \tag{9.32}$$

The function w is specified in a more involved fashion. To case the writing denote $L \equiv 98$, $a \equiv 425$. Also, denote by S the closed interval $[0, 1]$ and by I the closed interval $[1, 4]$, and define the function $m : S \times I \to S$ by

$$m(x, \theta) = \theta x(1 - x) \quad \text{for } x \in S, \ \theta \in I$$

and $u : S^2 \times I \to \mathbb{R}$ by

$$u(x, z, \theta) = ax - 0.5Lx^2 + zm(x, \theta) - 0.5z^2$$
$$- \delta\{az - 0.5Lz^2 + 0.5[m(z, \theta)]^2\}. \qquad (9.33)$$

Define a set $\mathbf{D} \subset S^2$ by

$$\mathbf{D} = \{(x, c) \in \mathbb{R}_+ \times S : c \le f(x)\}$$

and a function $w : \mathbf{D} \times I \to \mathbb{R}_+$ by

$$w(x, c, \theta) = u(x, f(x) - c, \theta) \quad \text{for } (x, c) \in \mathbf{D} \quad \text{and} \quad \theta \in I. \quad (9.34)$$

We now extend the definition of $w(\cdot, \theta)$ to the domain $\mathfrak{S}$. For $(x, c) \in \mathfrak{S}$ with $x > 1$ (so that $f(x) = 1$ and $c \le 1$), define

$$w(x, c, \theta) = w(1, c, \theta). \qquad (9.35)$$

Finally, we extend the definition of $w(\cdot, \theta)$ to the domain $\mathbb{R}_+^2$. For $(x, c) \in \mathbb{R}_+^2$ with $c > f(x)$, define

$$w(x, c, \theta) = w(x, f(x), \theta). \qquad (9.36)$$

One of the central results in Majumdar and Mitra (1994b) is the following:

Proposition 9.4 *The optimal transition functions for the family of economies* $(f, w(\cdot, \theta), \delta)$ *are given by*

$$\alpha_\theta(x) = \theta x(1 - x) \quad \text{for all } x \in S. \qquad (9.37)$$

In other words, for a fixed $\hat{\theta}$, the optimal input program is described by $x_{t+1}^* = \hat{\theta} x_t^*(1 - x_t^*)$. ∎

1.9.5.3 Robustness of Topological Chaos

In Example 9.2, we provide a specification of $(f, w, \bar{\delta})$ for which topological chaos is seen to occur. A natural question to ask is whether this is fortuitous (so that if the parameters f, w, or δ were perturbed ever so little, the property of topological chaos would disappear) or whether this is a "robust" phenomenon (so that small perturbations of f, w, or δ would preserve the property of topological chaos).

We proceed now to formalize the above question as follows. Define $\mathfrak{F} = \{f : \mathbb{R}_+ \to \mathbb{R}_+$ satisfying [F.1]–[F.3]$\}$; $\mathfrak{W} = \{w : \mathbb{R}_+^2 \to \mathbb{R}_+$ satisfying [W.1]–[W.3]$\}$; $\Delta = \{\delta : 0 < \delta < 1\}$. An *economy e* is defined by a triple $(f, w, \delta) \in \mathfrak{F} \times \mathfrak{W} \times \Delta$. The set of economies, $\mathfrak{F} \times \mathfrak{W} \times \Delta$ is defined by E.

Consider the economy $\mathbf{e} = (f, w, \bar{\delta})$ defined in Example 9.3 where $\theta = 4$. We would like to demonstrate that all economies $e \in E$ "near" the economy $\mathbf{e}$ will exhibit topological chaos. Thus, the property of topological chaos will be seen to persist for small perturbations of the original economy.

A convenient way to make the above idea precise is to define for $e \in E$, the "distance" between economies e and $\mathbf{e}$, by

$$d(e, \mathbf{e}) = \sup_{x \geq 0} |f(x) - \mathbf{f}(x)| + \sup_{(x,c) \geq 0} |w(x, c) - \mathbf{w}(x, c)| + |\delta - \bar{\delta}|.$$

Note that $d(e, \mathbf{e})$ may be infinite.

Before we proceed further, we have to clarify two preliminary points. First, if we perturb the original economy $\mathbf{e}$ (i.e., choose another economy $e \neq \mathbf{e}$), we will, in general, change the set of programs from any given initial stock, as also the optimal program from any given initial stock. That is, programs and optimal programs (and hence optimal policy functions) are economy specific. Thus, given an economy e, we use the expressions like "e-program," "e-optimal program," and "e-optimal transition function" with the obvious meanings.

Second, recall that for the original economy $\mathbf{e}$, $\mathbf{K} = 1$ (recall [F.3]), and so if the initial stock was in $[0, 1]$, then for any program from the initial stock, the input stock in every period is confined to $[0.1]$. Furthermore, for any initial stock not in $[0, 1]$, the input stock on any program belongs to $[0, 1]$ from the very next period. In this sense $([0, 1, \alpha_4))$ is the "natural" dynamical system for the economy $\mathbf{e}$. When we perturb the economy, we do not wish to restrict the *kind* of perturbation in any way, and so we would have to allow the new economy's production function to satisfy [F.3] with a $\mathbf{K} > 1$. This changes the "natural" state space of the dynamical system for the new economy. However, recalling that we are only interested in "small" perturbations, it is surely possible to ensue that $d(e, \mathbf{e}) \leq 1$, so that we can legitimately take the "natural" state space choice to be $J = [0, 2]$.

We can now describe our result (from Majumdar and Mitra 1994a) formally as follows.

Proposition 9.5 *Let* $\mathbf{e} = (f, w, \bar{\delta})$ *be the economy described in Example 9.2. There exists some* $\varepsilon > 0$ *such that for every economy* $e \in E$ *with*

$d(e, \mathbf{e}) < \varepsilon$, the dynamical system (J, α) exhibits Li–Yorke chaos, where a is the e-optimal transition function and $J = [0, 2]$.

Example 9.3 (*Optimal exploitation of a fishery*)   We return to the model of a fishery introduced in Example 6.1 and, following Dasgupta (1982), sketch a dynamic optimization problem in which the return function depends on both the "harvest" and the "stock."

The resource stock (biomass of the fish species) at the end of period t is denoted by x_t. It generates y_{t+1}, the stock in the next period, according to f, the biological reproduction function (also called the stock recruitment function):

$$y_{t+1} = f(x_t).$$

If c_{t+1} is harvested in period $(t + 1)$, then the stock at the end of time period $(t + 1)$ is x_{t+1}, given by

$$x_{t+1} = f(x_t) - c_{t+1}.$$

It is usual to assume that $f(0) = 0$, f is continuous, nondecreasing, and concave, with $\lim_{x \to 0} [f(x)/x] > 1$ and $\lim_{x \to \infty} [f(x)/x] = 0$ (although there are important variations, especially dealing with nonconcave functions, f).

We can find $B > 0$ such that whenever x_t is in $[0, B]$, x_{t+1} is also in $[0, B]$. Thus $[0, B]$ is a legitimate choice for the state space, S.

Harvesting the fishery is not costless. Specifically, the harvest, c_{t+1}, depends on the labor effort, e_{t+1}. We express the effort e_{t+1} required to harvest c_{t+1} when the biomass available for harvest is y_{t+1} as

$$e_{t+1} = H(c_{t+1}, y_{t+1}),$$

with H increasing in its first argument and decreasing in its second. This is the *cost* of exploiting the fishery. The *benefit* obtained depends on the harvest level c_{t+1}:

$$B(c_{t+1}) = pc_{t+1},$$

where $p > 0$ is the price (assumed constant over time) per unit of the fish. The *return* in period $(t + 1)$ is then the benefit minus the cost:

$$w(c_{t+1}, x_t) = B(c_{t+1}) - H(c_{t+1}, f(x_t))$$

The optimization problem can then be written as

$$(P') \begin{cases} \text{maximize} \quad \sum_{t=0}^{\infty} \delta^t w(c_{t+1}, x_t) \\ \text{subject to } c_{t+1} = f(x_t) - x_{t+1} \quad \text{for } t \in \mathbb{Z}_+ \\ c_{t+1} \geq 0, x_{t+1} \geq 0 \quad\quad\quad \text{for } t \in \mathbb{Z}_+ \\ x_0 = x > 0 \end{cases}$$

∎

1.9.6 Dynamic Programming

We now introduce a model of intertemporal optimization that is particularly amenable to the use of the techniques of dynamic programming and accommodates a variety of "reduced" models both in micro- and macroeconomics. There is a single decision maker (a firm, a consumer, a social planner . . .) facing an optimization problem that can be formally specified by $(\mathfrak{J}, u, \delta)$. Here, $\mathfrak{J}$ is a (nonempty) set in $\mathbb{R}_+ \times \mathbb{R}_+$, to be interpreted as a *technology* or a *transition possibility* set. Let x, y be (nonnegative) numbers (interpreted as *stocks* of a commodity). Then

$$\mathfrak{J} = \{(x, y): \mathbb{R}_+ \times \mathbb{R}_+ \in y \text{ can be reached from } x\}.$$

The *utility* (immediate return, felicity, profit, . . .) function u is a nonnegative real-valued function defined on $\mathfrak{J}$, i.e., $u : \mathfrak{J} \to \mathbb{R}_+$ and δ is the *discount factor* $0 < \delta < 1$.

A *program* $\mathbf{x} = (x_t)_0^\infty$ from $\underset{\sim}{x}$ is a sequence satisfying

$$x_0 = \underset{\sim}{x}$$

$$(x_t, x_{t+1}) \in \mathfrak{J} \quad \text{for } t = 0, 1, 2, \ldots. \tag{9.38}$$

Given a program $\mathbf{x}$ from an initial $\underset{\sim}{x} \in S$, we write

$$R(\mathbf{x}) \equiv \sum_{t=0}^{\infty} \delta^t u(x_t, x_{t+1})$$

$R(\mathbf{x})$ is the discounted sum of one-period returns generated by the program $\mathbf{x}$.

A program $\mathbf{x}^* = (x_t^*)_0^\infty$ from $\underset{\sim}{x}$ is *optimal* if

$$\sum_{t=0}^{\infty} \delta^t u(x_t^*, x_{t+1}^*) \geq \sum_{t=0}^{\infty} \delta^t u(x_t, x_{t+1}) \tag{9.39}$$

for *all* programs from $\underset{\sim}{x}$.

The following basic assumptions will be maintained throughout:

[A.1] $\Im \subset \mathbb{R}_+ \times \mathbb{R}_+$ *is a closed and convex set containing* $(0, 0)$, "$(0, y) \in \Im$" *implies* "$y = 0$"; *for any* $x \geq 0$, *there is some* $y \geq 0$ *such that* $(x, y) \in \Im$.

[A.2] $u : \Im \to \mathbb{R}_+$ *is a bounded, concave, and upper semicontinuous function.*

[A.3] *There is some* $\mathbf{K}_1 > 0$ *such that* "$(x, y) \in \Im$, $x > \mathbf{K}_1$" *implies* "$y < x$."

We now prove a useful boundedness property.

Lemma 9.5 *There is some* $B' > 0$ *such that*

$$\text{``}(x, y) \in \Im, \quad x \leq \mathbf{K}_1\text{''} \quad \text{implies} \quad \text{``}0 \leq y \leq B'\text{.''} \qquad (9.40)$$

Proof. If the claim is false, there is a sequence $(x^n, y^n) \in \Im$ such that $0 \leq x^n \leq \mathbf{K}_1$ and $y^n \geq n$ for all $n \geq 1$. Since $(0, 0) \in \Im$ and $\Im$ is convex, it follows that $(\frac{x^n}{y^n}, 1) \in \Im$ for all $n = 1, 2, \dots$. Since $\Im$ is closed, by taking the limit we get $(0, 1) \in \Im$, a contradiction. ∎

Now choose $B = \max(\mathbf{K}_1, B')$. Then it is not difficult to verify that if $\mathbf{x} = (x_t)_0^\infty$ is *any* program from $\underset{\sim}{x} \in [0, B]$

$$0 \leq x_t \leq B \quad \text{for all } t \geq 0.$$

We shall consider $S = [0, B]$ as the state space, and assume hereafter that $\underset{\sim}{x} \in S$. Now, let b be any constant such that

$$0 \leq u(x, y) \leq b, \quad \text{where } 0 \leq x \leq B, \quad 0 \leq y \leq B.$$

Then for *any* program $\mathbf{x} = (x_t)_{t=0}^\infty$ from $\underset{\sim}{x} \in S$, one has

$$R(\mathbf{x}) \leq \frac{b}{1 - \delta}.$$

Proposition 9.6 *If* $(\Im, u, \delta)$ *satisfies [A.1]–[A.3], then there exists an optimal program from every initial* $\underset{\sim}{x} \in S$.

Proof. The proof is left as an exercise (consult the proof of Theorem 9.1). ∎

In this section, we provide some basic results by using the dynamic programming approach. We can associate with each dynamic optimization problem two functions, called the (optimal) *value function* and the (optimal) *transition function*. The value function defines the optimal return (i.e., the left-hand side of (9.39)) corresponding to each initial state x. If there is a unique optimal solution $\mathbf{x}^* = (x_t^*)_0^\infty$ for each initial state $x \in S$, the transition function defines the first period optimal state x_1^*, corresponding to each initial state x.

Given the stationary recursive nature of the optimization problem, the optimal transition function, in fact, generates the entire optimal solution, starting from any initial state. The value function helps us to study the properties of the transition function, given the connection between the two through the *functional equation of dynamic programming* (see (9.40)).

1.9.6.1 The Value Function

Given the existence of an optimal program from every $x \in S$ (Proposition 9.6), we can define the *value function*, $V : S \to \mathbb{R}_+$, as follows:

$$V(x) = \sum_{t=0}^{\infty} \delta^t u\left(x_t^*, x_{t+1}^*\right), \tag{9.40}$$

where $\mathbf{x}^* = (x_t^*)_0^\infty$ is an optimal program from the initial state $x \in S$.

The following result summarizes the basic properties of the value function. Define, for any $x \geq 0$

$$\Im_x = \{y \in S \colon (x, y) \in \Im\}.$$

Clearly, $\Im_x$ is a (nonempty) compact, convex subset of S.

Proposition 9.7

(i) *The value function V is a concave and continuous function on S.*

(ii) *V satisfies the following functional equation of dynamic programming*

$$V(x) = \max_{y \in \Im_x}\{u(x, y) + \delta V(y)\} \tag{9.41}$$

for all $x \in S$.

(iii) $\mathbf{x} = (x_t)_0^\infty$ *is an optimal program if and only if*

$$V(x_t) = u(x_t, x_{t+1}) + \delta V(x_{t+1}) \quad \text{for } t \in \mathbb{Z}_+ \qquad (9.42)$$

Proof.

(i) Concavity of V follows from a direct verification.

In order to establish the continuity of V on S, we first establish its upper semicontinuity.

If V were not upper semicontinuous on S, we could find $x^n \in S$ for $n = 1, 2, 3, \ldots$, with $x^n \to x^0$ and $V(x^n) \to V$ as $n \to \infty$, with $V > V(x^0)$. Let us denote by (x_t^n) an optimal program from x^n for $n = 0, 1, 2, \ldots$.

Denoting $[V - V(x^0)]$ by ε, we can find T large enough so that $b\delta^{T+1}/(1-\delta) \le (\varepsilon/4)$.

Clearly, we can find a subsequence n' (of n) such that for $t \in \{0, 1, \ldots, T\}$,

$$x_t^{n'} \to x_t^0 \quad \text{as } n' \to \infty.$$

Then, using the upper semicontinuity of u, we can find N such that for $t \in \{0, 1, \ldots, T\}$,

$$u\left(x_t^{n'}, x_{t+1}^{n'}\right) \le u\left(x_t^0, x_{t+1}^0\right) + [\varepsilon(1-\delta)/4]$$

whenever $n' \ge N$. Thus, for $n' \ge N$, we obtain the following string of inequalities:

$$V(x^0) = \sum_{t=0}^\infty \delta^t u\left(x_t^0, x_{t+1}^0\right)$$

$$\ge \sum_{t=0}^T \delta^t u\left(x_t^0, x_{t+1}^0\right) - (\varepsilon/4)$$

$$\ge \sum_{t=0}^T \delta^t u\left(x_t^{n'}, x_{t+1}^{n'}\right) - (\varepsilon/2)$$

$$\ge \sum_{t=0}^\infty \delta^t u\left(x_t^{n'}, x_{t+1}^{n'}\right) - 3(\varepsilon/4)$$

$$= V(x^{n'}) - 3(\varepsilon/4).$$

Since $V(x^{n'}) \to V$, we have $V(x^0) \ge V - 3(\varepsilon/4) > V - \varepsilon = V(x^0)$, a contradiction. Thus, V is upper semicontinuous on S.

V is concave on S, and hence continuous on int $S = (0, B)$. To show that V is continuous at 0, let x^n be a sequence of points in S ($n = 1, 2, 3, \ldots$) converging to 0. Then $V(x^n) = V((1 - (x^n/B)) 0 + (x^n/B)B) \geq [1 - (x^n/B)]V(0) + (x^n/B)V(B)$. Letting $n \to \infty$, $\liminf_{n\to\infty} V(x^n) \geq V(0)$. On the other hand, since V is upper semi-continuous on X, $\limsup_{n\to\infty} V(x^n) \leq V(0)$. Thus, $\lim_{n\to\infty} V(x^n)$ exists and equals $V(0)$. The continuity of V at B is established similarly.

(ii) Let $y \in \Im_x$, and let $(y_t)_0^\infty$ be an optimal program from y. Then $(x, y_0, y_1, \ldots.)$ is a program from x, and hence, by definition of V,

$$V(x) \geq u(x, y_0) + \sum_{t=1}^{\infty} \delta^t u(y_{t-1}, y_t)$$

$$= u(x, y_0) + \delta \sum_{t=1}^{\infty} \delta^{t-1} u(y_{t-1}, y_t)$$

$$= u(x, y_0) + \delta \sum_{t=0}^{\infty} \delta^t u(y_t, y_{t+1})$$

$$= u(x, y_0) + \delta V(y).$$

So we have established that

$$V(x) \geq u(x, y) + \delta V(y) \quad \text{for all } y \in \Im_x. \tag{9.43}$$

Next, let $\mathbf{x} = (x_t)_0^\infty$ be an optimal program from x, and note that

$$V(x) = u(x_0, x_1) + \delta \left[\sum_{t=1}^{\infty} \delta^{t-1} u(x_t, x_{t+1}) \right]$$

$$= u(x_0, x_1) + \delta \left[\sum_{t=0}^{\infty} \delta^t u(x_{t+1}, x_{t+2}) \right]$$

$$\leq u(x_0, x_1) + \delta V(x_1).$$

Using (9.43), we then have

$$V(x) = u(x_0, x_1) + \delta V(x_1). \tag{9.44}$$

Now, (9.43) and (9.44) establish (9.41).

(iii) If $(x_t)_0^\infty$ satisfies (9.41), then we get for any $T \geq 1$,

$$V(x) = \sum_{t=0}^{T} \delta^t u(x_t, x_{t+1}) + \delta^{T+1} V(x_{T+1}). \tag{9.45}$$

Since $0 \le V(x) \le b/(1 - \delta)$ for all $x \in S$ and $\delta^{T+1} \to 0$ as $T \to \infty$, we have

$$8V(x) = \sum_{t=0}^{\infty} \delta^t u(x_t, x_{t+1}).$$

Then, by the definition of V, $(x_t)_0^{\infty}$ is an optimal program (from x).

To establish the converse implication, let $(x_t^*)_0^{\infty}$ be an optimal program from $x = x_0^*$. Then, for each $T \ge 1$, $(x_T^*, x_{T+1}^*, \ldots)$ is an optimal program from x_T^*. For, if $(y_t)_0^{\infty}$ is a program from $y_0 = x_T^*$ such that

$$\sum_{t=0}^{\infty} \delta^t u(y_t, y_{t+1}) > \sum_{t=T}^{\infty} \delta^{t-T} u\left(x_t^*, x_{t+1}^*\right),$$

then the sequence $(x_0^*, \ldots, x_T^*, y_1, y_2, \ldots)$ has a discounted sum of utilities

$$\sum_{t=0}^{T-1} \delta^t u\left(x_t^*, x_{t+1}^*\right) + \sum_{t=T}^{\infty} \delta^t u(y_t, y_{t+1}) > \sum_{t=0}^{\infty} \delta^t u\left(x_t^*, x_{t+1}^*\right),$$

which contradicts the optimality of $(x_t^*)_0^{\infty}$ from $x = x_0^*$. Now, using the result (9.43) in (ii) above, we have $V(x_t^*) = u(x_t^*, x_{t+1}^*) + \delta V(x_{t+1}^*)$ for $t \in \mathbb{Z}_+$. ∎

1.9.6.2 The Optimal Transition Function

Recall that Proposition 9.6 guarantees existence of an optimal program, but *not its uniqueness*. However, a strict concavity assumption on u is sufficient to guarantee uniqueness.

[A.4] *u is strictly concave in its second argument.*

If [A.4] holds, then given any $x \in S$, there is a *unique solution* to the maximization problem on the right-hand side of (9.40). For if y and $y'(y' \ne y)$ in $\Im_x$ both solve this problem, then $V(x) = u(x, y) + \delta V(y)$ and $V(x) = u(x, y') + \delta V(y')$. However, $(x, 0.5y + 0.5y') \in \Im$, and $u(x, 0.5y + 0.5y') + \delta V(0.5y + 0.5y') > 0.5u(x, y) + 0.5u(x, y') + \delta[0.5V(y) + 0.5V(y')] = V(x)$, a contradiction to (9.40).

For each x, denote the unique state $y \in \Im_x$, which solves the maximization problem on the right-hand side of (9.40) by $h(x)$. We will call $h : S \to S$ the *optimal transition function*.

The following result summarizes the basic properties of h.

Proposition 9.8

(i) *The optimal transition function $h : S \to S$ is continuous on S.*
(ii) *For all $(x, y) \in \Im$ with $y \neq h(x)$ we have*

$$u(x, y) + \delta V(y) < V(x) = u(x, h(x)) + \delta V(h(x)). \tag{9.46}$$

(iii) *$(x_t)_0^\infty$ is an optimal program from x if and only if*

$$x_{t+1} = h(x_t) \quad for\ t \in \mathbb{Z}_+. \tag{9.47}$$

Proof.

(i) Let (x^n) $(n = 1, 2, 3, \ldots)$ be a sequence of points in S converging to x^0. Then for $n = 1, 2, 3, \ldots,$

$$V(x^n) = u(x^n, h(x^n)) + \delta V(h(x^n)). \tag{9.48}$$

Let $(y^{n'})$ be an arbitrary convergent subsequence of $(y^n) \equiv (h(x^n))$ converging to y^0. We claim that $y^0 = h(x^0)$. For the subsequence n', using (9.48), the upper semicontinuity of u and the continuity of V,

$$V(x^0) = \limsup_{n' \to \infty} u(x^{n'}, y^{n'}) + \delta V(y^0)$$
$$\leq u(x^0, y^0) + \delta V(y^0).$$

But, by (9.41), $V(x^0) \geq u(x^0, y^0) + \delta V(y^0)$, so that $V(x^0) = u(x^0, y^0) + \delta V(y^0)$. This means that $y^0 = h(x^0)$, establishing our claim. Thus, $h(x^n) \to h(x^0)$ as $n \to \infty$, establishing continuity of h.

(ii) We have $V(x) = u(x, h(x)) + \delta V(h(x))$ from (9.41) and the definition of h. For all $y \in \Im_x$, we have $u(x, y) + \delta V(y) \leq V(x)$ by (9.41). Further, if equality holds in the previous weak inequality, then $y = h(x)$. Thus if $y \neq h(x)$, the strict inequality must hold. This establishes (9.46).

(iii) If $(x_t)_0^\infty$ is an optimal program from x, then by Proposition 9.7 (iii), for every $t \in \mathbb{Z}_+$,

$$V(x_t) = u(x_t, x_{t+1}) + \delta V(x_{t+1})$$

so that by the just established part (ii) of the proposition, $x_{t+1} = h(x_t)$.

Conversely, if (9.48) holds, then by part (ii) of this proposition, we have for $t \in \mathbb{Z}_+$

$$V(x_t) = u(x_t, x_{t+1}) + \delta V(x_{t+1})$$

so that $(x_t)_0^\infty$ is an optimal program by Proposition 9.7 (iii). ∎

1.9.6.3 Monotonicity With Respect to the Initial State

The concept that is crucial to establishing monotonicity properties of the policy function is known as *supermodularity*, and was introduced into the optimization theory literature by Topkis (1978).

Let A be a subset of $\mathbb{R}^2$ and f be a function from A to $\mathbb{R}$. Then f is *supermodular* on A if whenever (a, b) and (a', b') belong to A with $(a', b') \geq (a, b)$, we have

$$f(a, b) + f(a', b') \geq f(a, b') + f(a', b), \qquad (9.49)$$

provided (a, b') and (a', b) belong to A.

If A is a rectangular region, then whenever (a, b) and (a', b') belong to A, we have (a, b') and (a', b) also in A. Further, for such a region if f is continuous on A and C^2 on int A, then

$$f_{12}(a, b) \geq 0 \quad \text{for all } (a, b) \in \text{int } A \qquad (9.50)$$

is equivalent to the condition that f is supermodular on A. (For this result, see Ross (1983); Benhabib and Nishimura (1985).)

Let $(\Im, u, \delta)$ be a dynamic optimization model. The principal result on the montonicity of its optimal transition function h with respect to the initial condition x is that if u is supermodular on $\Im$, then h is monotone nondecreasing in x. (This result is based on the analysis in Topkis (1978), Benhabib and Nishimura (1985).) We need an additional assumption on $\Im$:

[A.5]   *If* $(x, z) \in \Im$ *and* $x' \in S, x' \geq x, 0 \leq z' \leq z,$ *then* $(x', z') \in \Im.$

Proposition 9.9 *Let* $(\Im, u, \delta)$ *be a dynamic optimization model, with value function V and optimal transition function h. If u is supermodular on $\Im$, then h is monotone nondecreasing on S.*

Proof.   Let x, x' belong to S with $x' > x$. Denote $h(x)$ by z and $h(x')$ by z'. We want to show that $z' \geq z$. Suppose, on the contrary, that $z' < z$. Since $(x, z) \in \Im$ and $z' \in S$ with $z' \leq z$, we have $(x, z') \in \Im$. Since $(x, z) \in \Im$ and $x' \in S$ with $x' \geq x$, we have $(x', z) \in \Im$. Using the definition of supermodularity, and $(x', z) \geq (x, z')$, we have

$$u(x, z') + u(x', z) \geq u(x, z) + u(x', z'). \qquad (9.51)$$

Since $z = h(x)$ and $z' = h(x')$, we have

$$V(x) = u(x, z) + \delta V(z) \qquad (9.52)$$

and

$$V(x') = u(x', z') + \delta V(z'). \tag{9.53}$$

Since $(x, z') \in \Im$ and $(x', z) \in \Im$, and $z' \neq z$, we have $z' \neq h(x)$, $z \neq h(x')$, so

$$V(x) > u(x, z') + \delta V(z') \tag{9.54}$$

and

$$V(x') > u(x', z) + \delta V(z). \tag{9.55}$$

Adding (9.52) and (9.53),

$$V(x) + V(x') = u(x, z) + u(x', z') + \delta V(z) + \delta V(z'). \tag{9.56}$$

Adding (9.54) and (9.55)

$$V(x) + V(x') > u(x, z') + u(x', z) + \delta V(z') + \delta V(z). \tag{9.57}$$

Using (9.56) and (9.57),

$$u(x, z) + u(x', z') > u(x, z') + u(x', z),$$

which contradicts (9.51) and establishes the result. ∎

Exercise 9.6 Consider the optimization problem 1.9.4.1. Write down the reduced form of this problem. [Hint: think of $u(c_t) = u(f(x_{t-1}) - x_t)$.] ∎

Remark 9.6 It follows from the above result that, in the framework of Proposition 5.2, if (x_t) is an optimal program starting from $x \in S$, then either (i) $x_{t+1} \geq x_t$ for all $t \in \mathbb{Z}_+$ or (ii) $x_{t+1} \leq x_t$ for all $t \in \mathbb{Z}_+$.

Remark 9.7 Some definitive results illuminating the link between the discount factor δ and Li–Yorke chaos have been obtained. A comprehensive treatment of this problem is available in Chapters 9–13 (written by Mitra, Nishimura, Sorger, and Yano) of Majumdar, Mitra, and Nishimura (2000). By introducing appropriate strict concavity and monotonicity assumptions on $\Im$ and u, Mitra (Chapter 11) proves that if the optimal transition function h of a model $(\Im, u, \delta)$ exhibits a three-period cycle, then $\delta < [(\sqrt{5} - 1)/2]^2$. Furthermore, this restriction on δ is the best possible: whenever $\delta < [(\sqrt{5} - 1)/2]^2$, Mitra constructs a technology $\Im$

and a utility function u such that the optimal transition function h of $(\mathfrak{S}, u, \delta)$ exhibits a three-period cycle. This construction is based on the example of Nishimura, Sorger and Yano (Chapter 9).

Here is a special case of Sorger (1992) (the proof of which is due to T. Mitra). Choose units of measurement so that $S = [0, 1]$, and assume $u(0, 0) = 0$, strengthen [A.4] and assume that $u : \mathfrak{S} \to \mathbb{R}_+$ is *strictly concave*: $u(\theta(x, y) + (1 - \theta)(x', y')) > \theta u(x, y) + (1 - \theta)u(x', y')$, for $0 < \theta < 1$ (where $(x, y) \in \mathfrak{S}$, $(x', y') \in \mathfrak{S}$). Suppose that in this model $(\mathfrak{S}, u, \delta)$ the optimal transition function h is of the form

$$h(x) = 4x(1 - x)\ 0 \le x \le 1. \tag{9.57.1}$$

Then the discount factor δ must satisfy

$$\delta < \frac{1}{2}, \tag{R}$$

To see how the restriction (R) on δ emerges, note that if the optimal transition function h is given by (9.57.1) then the sequence

$$\mathbf{x}^* = \left\{x_t^*\right\}_0^\infty = \{(1/2),\ 1, 0, 0, \ldots\} \tag{9.57.2}$$

is an optimal program from $(1/2)$.

Now, suppose, to the contrary, that $\delta \ge (1/2)$. Define $\theta = 1/(1 + 2\delta)$, and note that $0 < \theta \le (1/2)$. One establishes a contradiction by showing that a program that starts at $(1/2)$ then goes to θ in the next period and thereafter follows an optimal program from θ, provides higher discounted sum of utilities than the program (9.57.2) from $1/2$; that is, $u[(1/2), \theta] + \delta V(\theta) > V(1/2)$.

Since $u(0, 0) = 0$ we have $V(0) = 0$. Now $[(1/2), \theta] = [\theta(1/2) + \theta\delta(1) + \theta\delta(0), \theta(1) + \theta\delta(0) + \theta\delta(0)]$. Since $((1/2), 1) \in \mathfrak{S}$ and $(1, 0) \in \mathfrak{S}$ and $(0, 0) \in \mathfrak{S}$, and $\theta + \theta\delta + \theta\delta = 1$, we have $[(1/2), \theta] \in \mathfrak{S}$ by convexity of $\mathfrak{S}$. Further, strict concavity of u and $u(0, 0) = 0$ imply that

$$u((1/2), \theta) > \theta u((1/2), 1) + \theta\delta u(1, 0). \tag{9.57.3}$$

Now, since $1 = h(1/2)$ and $0 = h(1)$, we have by the optimality equation

$$u((1/2), 1) = V(1/2) - \delta V(1) \tag{9.57.4}$$

$$u(1, 0) = V(1) - \delta V(0) = V(1). \tag{9.57.5}$$

Combining (9.57.4) and (9.57.5),

$$\theta u((1/2), 1) + \theta \delta u(1, 0) = \theta V(1/2). \qquad (9.57.6)$$

Using (9.57.6) in (9.57.3), we finally get

$$
\begin{aligned}
u[(1/2), \theta] + \delta V(\theta) &> \theta V(1/2) + \delta V(\theta) \\
&= \theta V(1/2) + \delta V(2\theta(1/2) + (1 - 2\theta)(0)) \\
&\geq \theta V(1/2) + 2\delta\theta V(1/2) \\
&= V(1/2),
\end{aligned}
$$

a contradiction to the optimality equation. This establishes the restriction (R). Note that this result, and the proof, remain valid if the optimal transition function h satisfies the equation of the tent map:

$$
h(x) = \begin{cases} 2x & \text{for } 0 \leq x \leq 1/2 \\ 2 - 2x & \text{for } 1/2 \leq x \leq 1 \end{cases}.
$$

Example 9.4 (*The two-sector optimal growth model*) The two-sector model of optimal economic growth, originally developed by Uzawa (1964) and Srinivasan (1964), is a generalization of the one-sector model, discussed in Section 1.9.4.1. The restriction imposed in the one-sector model, that consumption can be traded off against investment on a one-to-one basis, is relaxed in the two-sector model, and it is principally this aspect that makes the two-sector model a considerably richer framework in studying economic growth problems than its one-sector predecessor.

Production uses two inputs, capital and labor. Given the total amounts of capital and labor available to the economy, the inputs are allocated to two "sectors" of production, the consumption good sector and the investment good sector. Output of the former sector is consumed (and cannot be used for investment purposes), and output of the latter sector is used to augment the capital stock of the economy (and cannot be consumed). Thus, the consumption–investment decision amounts to a decision regarding allocation of capital and labor between the two production sectors.

The discounted sum of outputs of the consumption good sector is to be maximized to arrive at the appropriate sectoral allocation of inputs in each period. (A welfare function on the output of the consumption good sector can be used, instead of the output of the consumption good sector itself, in the objective function, but it is usual to assimilate this welfare function, when it is increasing and concave, in the "production function" of the consumption good sector.)

Formally, the model is specified by (F, G, ρ, δ), where

(a) the production function in the consumption good sector, $F : \mathbb{R}_+^2 \to \mathbb{R}_+$, satisfies the following:

[F.1] *F is continuous and homogeneous of degree 1 on $\mathbb{R}_+^2$.*
[F.2] *F is nondecreasing on $\mathbb{R}_+^2$.*
[F.3] *F is concave on $\mathbb{R}_+^2$.*

(b) the production function in the investment good sector, $G : \mathbb{R}_+^2 \to \mathbb{R}_+$, satisfies the following:

[G.1] *G is continuous and homogeneous of degree 1 on $\mathbb{R}_+^2$.*
[G.2] *G is nondecreasing on $\mathbb{R}_+^2$.*
[G.3] *G is concave on $\mathbb{R}_+^2$.*
[G.4] $\lim_{K \to \infty}[G(K, 1)/K] = 0.$

(c) the depreciation factor ρ satisfies $0 < \rho \leq 1$.
(d) the discount factor δ satisfies $0 < \delta < 1$.

The optimal growth problem can be written as

$$(P) \begin{cases} \text{maximize } \sum_{t=0}^{\infty} \delta^t c_{t+1} \\ \text{subject to } c_{t+1} = F(k_t, n_t) & \text{for } t \in \mathbb{Z}_+ \\ x_{t+1} = G(x_t - k_t, 1 - n_t) + (1 - \rho)x_t & \text{for } t \in \mathbb{Z}_+ \\ 0 \leq k_t \leq x_t, \ 0 \leq n_t \leq 1, & \text{for } t \in \mathbb{Z}_+ \\ x_0 = x > 0. \end{cases}$$

Here x_t is the total capital available at date t, which is allocated between the consumption good sector (k_t) and the investment good sector $(x_t - k_t)$. Labor is exogenously available at a constant amount (normalized to unity), which is allocated between the consumption good sector (n_t) and the investment good sector $(1 - n_t)$. Note that an exogenously growing labor force (at a constant growth rate) can be accommodated easily by interpreting k_t and x_t as per worker capital stocks, and reinterpreting the depreciation factor, ρ.

As in the one-sector model, we can find $B > 0$, such that for $x \in [0, B]$, we have $G(x, 1) + (1 - \rho)x$ in $[0, B]$, and furthermore for any solution to (P), $x_t \in [0, B]$ for some t. Thus, it is appropriate to define $S = [0, B]$ as the state space.

To convert the problem to its reduced form, we can define the transition possibility set $\Im$ as

$$\Im = \{(x, z) \text{ in } \mathbb{R}_+^2 : z \leq G(x, 1) + (1 - \rho)x\}$$

and the utility function u as

$$u(x, z) = \text{maximize } F(k, n)$$
$$\text{subject to } 0 \leq k \leq x, \ 0 \leq n \leq 1$$
$$z \leq G(x - k, 1 - n) + (1 - d)x.$$

Then a solution to (P) corresponds exactly to an optimal program for $(\Im, u, \delta)$, and vice versa. ∎

1.9.7 Dynamic Games

Dynamical systems arise naturally from dynamic games that capture problems of intertemporal optimization faced by an agent ("player") who has a conflict of interest with other agents.

Think of a renewable resource that grows according to a biological law described by a function $f : \mathbb{R}_+ \to \mathbb{R}_+$ satisfying the following properties:

[F.1] $f(0) = 0$;
[F.2] *f is continuous and increasing*;
[F.3] *there is $\bar{y} > 0$ such that $f(y) < y$ for all $y > \bar{y}$ and $f(y) > y$ for $0 < y < \bar{y}$.*

Thus, if the stock of the resource in period t is y_t and no extraction (harvesting/consumption) takes place, the stock in the next period is given by

$$y_{t+1} = f(y_t).$$

There are two players involved in extraction with *identical preferences*. The (common) *utility function* $u : \mathbb{R}_+ \to \mathbb{R}_+$ satisfies:

[U.1] *u is continuous, increasing, and strictly concave*;
[U.2] *u(c) is continuously differentiable at c > 0 and*

$$\lim_{c \downarrow 0} u'(c) = \infty.$$

The (common) discount factor δ satisfies $0 < \delta < 1$.

Let $K = \max[y_0, \bar{y}]$ and $S = [0, K]$. We refer to S as the *state space* and interpret y as the stock at the beginning of each period. In the Cournot tradition, each player chooses a consumption sequence over time that maximizes the discounted sum of utilities, taking the actions of the other player *as given*. The structure of the model makes it convenient to state the dynamic optimization in a generalized game form in the sense of Debreu (1952). (Here we are sidestepping some finer points on modeling independent choices with conflicting interests, and suggest that the reader should carefully study the more complete elaboration of Sundaram (1989).)

Let $h_t = \{y_0, (c_{1,0}, c_{2,0}), \ldots, y_{t-1}, (c_{1,t-1}, c_{2,t-1}), y_t\}$ be a history of the game up to period t; the stocks y_t and the consumption extraction choices (actions) up to period t. h_t is the set of all possible histories up to period t. A *strategy* $g^{(i)} = (g_t^{(i)})$ for player i is a sequence of functions $g_t^{(i)}$, where $g_t^{(i)}$ specifies the action c_{it} (consumption/extraction) in period t as a (Borel measurable) function of h_t (subject to the feasibility constraint faced by i at h_t, to be formally spelled out below). A (Borel measurable) function $\hat{g}^{(i)} : S \to S$ (satisfying $g(y) \le y$) defines a *stationary strategy* $\hat{g}^{(i)} = (\hat{g}_{(\infty)}^{(i)})$ for player i: irrespective of the history up to period t, and independently of t, it specifies a consumption $c = \hat{g}^{(i)}(y)$ (satisfying $c \le y$). We shall refer to $\hat{g}^i$ as a *policy function*, and often with an abuse of notation, denote $(\hat{g}_{(\infty)}^{(i)})$ simply by $\hat{g}^{(i)}$.

Now, think of the dynamic optimization problem of player i when player $j \ne i$ [$i, j = 1, 2$] uses a stationary strategy defined by a policy function $\hat{g}^j : S \to S$, assumed to satisfy $0 \le \hat{g}^j(y) \le y$ for $y \in S$. For simplicity of notation we write $g \equiv \hat{g}^j$. Given the initial $y_0 \in S$, the problem of player i can be written as

$$\underset{\{c_{it}\}}{\text{maximize}} \sum_{t=0}^{\infty} \delta^t u(c_{i,t}) \tag{9.58}$$

subject to

$$y_{t+1} = f[y_t - g(y_t) - c_{i,t}] \quad \text{for } t \ge 0, \tag{9.59}$$

$$c_{i,t} \in [0, \ y_t - g(y_t)] \quad \text{for } t \ge 0. \tag{9.60}$$

Equations (9.58)–(9.60) define an SDP for player i. Let G_i be the set of all strategies (including history dependent strategies) for player i that satisfy the constraints (9.59) and (9.60) for each t. For $g^{(i)} \in G_i$ denote by $W_g(g^{(i)})$ the *payoff* to player i from employing strategy $g^{(i)}$ given g, and $y_0 = y$. A strategy $g^{*(i)} \in G$ is said to be *optimal* if $W_g(g^{*(i)})(y) \geq W_g(g^{(i)})(y)$ for all $y > 0$ and all $g^{(i)} \in G_i$. If an optimal strategy $g^{*(i)}$ exists, the corresponding payoff function $W_g(g^{*(i)}) : S \to \mathbb{R}$ is called the *value function* of player i, arising from responding optimally to the policy function (more precisely, the stationary strategy defined by the policy function) g. Note that if $g^{*(i)}$ and $\bar{g}^{*(i)}$ are both optimal strategies (given $(g^j_{(\infty)})$ for j) then $W_g[g^{*(i)}] = W_g[\bar{g}^{*(i)}]$.

The following result is derived from a suitable adaptation of results in Maitra (1968) (see also Chapter 6).

Lemma 9.6 *Suppose* $g : S \to S$ *is a lower semicontinuous function satisfying*

(i) $0 \leq g(y) \leq y$ *for* $y \in S$ *and*
(ii) $[(g(y) - g(y'))/y - y'] \leq 1$ *for* $y, y' \in S, y \neq y'$.
Then
(a) *there exists an optimal strategy* $(\hat{g}^{*(i)}_{(\infty)})$ *for player* i *that is stationary (the policy function being* $\hat{g}^{*(i)}$),
(b) *the value function* $W_g(\hat{g}^{*(i)})$ *is a nondecreasing, upper semicontinuous function on* S.

For notational simplicity write $W_g(\bar{g}^{*(i)}) \equiv V_g$. It should be stressed that Lemma 9.5 identifies a class of stationary strategies of player j (defined by policy function g) such that the best response of player i is also a *stationary strategy* (optimal in the class of *all* strategies G_i).

A *symmetric* Nash equilibrium *in stationary strategies* is a pair of policy functions $(\bar{g}^{*(1)}, \bar{g}^{*(2)})$ such that

(a) $\bar{g}^{*(1)} = \bar{g}^{*(2)} = g^*$; and
(b) given $\bar{g}^{*(j)}$ ($j = 1, 2$), $\bar{g}^{*(i)}$ ($i \neq j$) solves the problem (9.58)–(9.60).

Going back to the optimization problem (9.58)–(9.60), we note that a symmetric Nash equilibrium results when player i, accepting the strategy of j defined by a policy function g, obtains g *itself* as a possible response, i.e., a solution to (9.58)–(9.60). A symmetric Nash equilibrium (g^*, g^*) is *interior* if $0 < g^*(y) < y$.

One of the fundamental results in the context of the resource allocation game is the following.

Theorem 9.5 *There is a symmetric Nash equilibrium in stationary strategies* (g^*, g^*). *The policy function* g^* *can be chosen to be lower semi-continuous, and for distinct* y_1', y_2'' *in S,*

$$\frac{g^*(y_1') - g^*(y_2'')}{y_1' - y_2''} \leq \frac{1}{2}.$$

For a proof of this theorem, the reader is referred to Sundaram (1989). ∎

Remark 9.8 The symmetric Nash equilibrium in stationary strategies is *subgame perfect* (Dutta 1999, Chapter 13). Consider the function

$$\hat{\psi}(y) = y - g(y) - g(y) \equiv y - 2g(y),$$

$\hat{\psi}$ is the savings function of the entire economy after the consumption of both players. Then, for $y' > y''$

$$\hat{\psi}(y') - \hat{\psi}(y'') = (y' - y'') - 2[g(y') - g(y')]$$

or

$$\frac{\hat{\psi}(y_1) - \hat{\psi}(y_2)}{y_1 - y_2} = 1 - 2\frac{[g(y_1') - g(y_2)]}{y_1 - y_2} \geq 0$$

or

$$\hat{\psi}(y_1) - \hat{\psi}(y_2) \geq y_1 - y_2 > 0.$$

It follows that in the equilibrium proved above, the generated sequence of the stock levels (y_t) is monotone (see Complements and Details).

1.9.8 Intertemporal Equilibrium

Consider an infinite horizon economy with only one producible good that can be either consumed or used as an input along with labor (a nonproducible, "primary" factor of production) to produce more of itself. Net output at the "end" of period t, Y_t, is determined by the production function $F(K_t, L_t)$ where $K_t \geq 0$ is the quantity of producible good used as input (called "capital") and $L_t > 0$ is the quantity of labor, both available at the "beginning" of period $t \geq 0$:

$$Y_t = F(K_t, L_t). \tag{9.61}$$

Here we assume that F is homogeneous of degree 1. We shall switch to the "per capita" representation for $L_t > 0$. Write $y_t \equiv \frac{Y_t}{L_t}$ and $k_t \equiv \frac{K_t}{L_t}$ (when $L_t > 0$) and using homogeneity, we get

$$y_t \equiv \frac{Y_t}{L_t} = \frac{F(K_t, L_t)}{L_t} = F\left(\frac{K_t}{L_t}, 1\right) \equiv f(k_t).$$

We shall also assume the Uzawa-Inada condition:

$$\lim_{k \to 0} f'(k) = \infty \quad \text{and} \quad \lim_{k \to \infty} f'(k) = 0. \tag{9.61'}$$

Individuals in the economy live for two periods: in the first period each "supplies" one unit of labor and is "retired" in the second period. The total endowment or supply of labor L_t in period t is determined exogenously

$$L_t = L_0(1 + \eta)^t, \quad \eta \geq 0, \ L_0 > 0. \tag{9.61''}$$

We assume that when used as an input in the production process, capital does not depreciate. At the end of the production process in period t, the total stock of the producible commodity is $K_t + Y_t$ and is available for consumption $C_t (\geq 0)$ in period t. The amount not consumed (saved) is available as the capital stock in period $t + 1$

$$K_{t+1} = (K_t + Y_t) - C_t \geq 0. \tag{9.62}$$

The division of the producible good between consumption and saving is determined through a market process that we shall now describe.

A typical individual born in period t supplies the unit of labor in period t for which he receives a wage $w_t (>0)$. Equilibrium in the labor market must satisfy the condition

$$w_t = F_L(K_t, L_t). \tag{9.63}$$

The individual consumes a part $c_t^{(1)}$ of the wage income in the first (working) period of his life and supplies the difference

$$s_t \equiv w_t - c_t^{(1)}$$

as capital ("lends" his saving to the capital market). Thus, S_t, the total saving of the economy in period t, is simply

$$S_t = L_t s_t, \tag{9.64}$$

and this is available as K_{t+1}, the capital stock to be used as input in period $t+1$

$$K_{t+1} = S_t. \tag{9.65}$$

In the second period of his life the retired individual consumes $c_{t+1}^{(2)} = (1 + r_{t+1})s_t$, where $1 + r_{t+1}$ is the return to capital (r_{t+1} is the 'interest' rate). Equilibrium in capital market requires

$$r_{t+1} = F_K(K_{t+1}, L_{t+1}). \tag{9.66}$$

Thus, in period t the aggregate consumption C_t can be written as

$$
\begin{aligned}
C_t &= [w_t L_t - s_t L_t] + (1 + r_t)S_{t-1} \\
&= F_L L_t - K_{t+1} + [1 + F_K]K_t \\
&= K_t + F_L L_t + F_K K_t - K_{t+1}.
\end{aligned}
$$

Since F is homogeneous of degree 1, $Y \equiv F_K K + F_L L$ by Euler's theorem, so that

$$C_t = K_t + Y_t - K_{t+1}$$

and we get back relation (9.62).

The individual in period t has a utility function $u(c_t^{(1)}, c_{t+1}^{(2)}) = (c_t^{(1)})^{\rho}(c_{t+1}^{(2)})^{1-\rho}$, where $0 < \rho < 1$.
Its problem is to maximize (for any $w_t > 0$)

$$u\left(c_t^{(1)}, c_{t+1}^{(2)}\right) = u(w_t - s_t, (1 + r_{t+1})s_t)$$

subject to

$$0 \le s_t \le w_t.$$

The solution $\hat{s}_t$ to this maximization problem is easily found:

$$\hat{s}_t \equiv (1 - \rho)w_t. \tag{9.67}$$

Thus, the evolution of the economy in equilibrium is described by the following (using (9.61'), (9.64), (9.65), (9.67) for the first and (9.63) for the second relation):

$$k_{t+1} = [(1 - \rho)w_t]/1 + \eta$$
$$w_t = f(k_t) - k_t f'(k_t). \tag{9.68}$$

Combining these we have

$$k_{t+1} = \frac{1 - \rho}{1 + \eta}[f(k_t) - k_t f'(k_t)]$$
$$= \alpha(k_t).$$

Note that $\alpha'(k) = -kf''(k) > 0$. The Uzawa-Inada condition on f' (see (9.61')) is *not* sufficient to derive the condition (PI) in Example 5.1.

1.9.9 Chaos in Cobb–Douglas Economies

A description of markets in which prices adjust when the forces of demand and supply are *not* balanced was provided by Walras, and was formally studied by Samuelson (1947). Subsequently the analysis of a Walras–Samuelson process challenged the very best researchers in economic theory, and some conceptual as well as analytical difficulties involving the model were exposed. Here we give an example of chaotic behavior of such a price adjustment process.

Consider an exchange economy with two (price taking) agents and two commodities. Denote the two agents by $i = 1, 2$ and the goods by x and y. Agent 1 has an initial endowment $w^1 = (4, 0)$ while agent 2 has $w^2 = (0, 2)$. Let α, β be two real numbers in the interval $(0, 1)$. The utility function of agent 1 is $u^1(x, y) = x^\alpha y^{1-\alpha}$ and that of agent 2 is $u^2(x, y) = x^\beta y^{1-\beta}$ for (x, y) in the consumption space R_+^2; a typical economy e can thus be described by the pair (α, β) which refers to the exponents of the utility functions of the agents. We shall denote the space of Cobb–Douglas economies defined in this manner by $\mathcal{E}_{\mathrm{CD}} \equiv (0, 1) \times (0, 1)$. Of particular interest for our purposes is the economy $\bar{e} = (3/4, 1/2)$.

Fixing the price of good y as 1, and that of x as $p > 0$, the excess demand function for good x is denoted by $z(p; e)$ and is given by

$$z(p; e) = 2\beta/p - 4(1 - \alpha), \quad e = (\alpha, \beta). \tag{9.69}$$

This is derived from the optimization problem solved by each agent independently, taking the price p (rate of exchange between x and y) as given. Each agent i maximizes its utility u^i subject to 'budget constraint' determined by p. More explicitly, the *income* m^1 of agent 1 at prices $(p, 1)$ is the value of its endowment w^1:

$$m^1 = 4p.$$

The income m^2 of agent 2 at prices $(p, 1)$ is the value of its endowment w^2:

$$m^2 = 2.$$

Then the optimization problem of agent $i(= 1, 2)$ is given by

$$\text{maximize } u^i(x, y)$$
$$\text{subject to } px + y \le m^i$$
$$x \ge 0, \quad y \ge 0. \tag{Pi}$$

The solution $(\bar{x}^i, \bar{y}^i)$ to (Pi) gives us the *demand* for x and y, coming from agent i. The *excess demand* for commodity x in the economy at $(1, p)$ is defined as

$$z(p; e) = (\bar{x}^1 + \bar{x}^2) - 4$$

(4 being the *total* endowment or supply of x), and can be easily verified as that in (9.69).

For every $e \in \mathcal{E}_{CD}$ there is a unique competitive equilibrium $p^*(e)$ given by

$$p^*(e) = \beta / [2(1 - \alpha)]. \tag{9.70}$$

In particular, for $\bar{e} = (3/4, 1/2)$, $p^*(\bar{e}) = 1$.

The Walras–Samuelson ("tatonnement") price adjustment process is defined as

$$p_{t+1} = T(p_t, e, \lambda) = p_t + \lambda z (p_t; e) \tag{9.71}$$

where

$$T(p_t, e, \lambda) = p_t + \lambda z(p_t; e), \tag{9.71'}$$

and $\lambda > 0$ is a "speed of adjustment" parameter. Of particular interest for our purpose is the value $\bar{\lambda} = 361/100$.

Substituting Equation (9.69) in Equation (9.71) we get

$$p_{t+1} = T(p_t, e, \lambda) = p_t + \lambda[2\beta/p_t - 4(1-\alpha)]. \qquad (9.72)$$

Equation (9.72) is the law of motion of dynamical system parameterized by $e = (\alpha, \beta)$ and $\lambda \in \mathbb{R}_{++}$. We then have

Proposition 9.10 *There exist open sets* $N \subset E_{CD}$ *and* $V \subset \mathbb{R}_{++}$ *such that if* $e = (\alpha, \beta) \in N$ *and* $\lambda \in V$ *then the process (9.72) exhibits Li–Yorke chaos.*

Proof. Let $K \subset \mathcal{E}_{CD}$ and $\Lambda \subset \mathbb{R}_{++}$ be compact sets containing $\bar{e} = (3/4, 1/2)$ and $\bar{\lambda} = 361/100$ in their respective interiors. For $e = \bar{e}$ and $\lambda = \bar{\lambda}$, Equation (9.72) becomes

$$p_{t+1} = T(p_t, \bar{e}, \bar{\lambda}) = p_t + 361/(100p_t) - 361/100. \qquad (9.73)$$

For convenience, we shall write $T(p, \bar{e}, \bar{\lambda})$ simply as $T(p)$. Let S denote the interval $[0.18, 17]$. If $p_t \in S$ then $p_{t+1} \in S$ as well. This follows because the global minimum of T is at 1.9 (from the first and second-order conditions) where T takes the value $0.19 > 0.18$. Likewise the maximum value of T is at 0.18 where it takes the value $16.63 < 17$. Hence T maps the interval S strictly into the interior of S.

It is then easy to verify that T satisfies condition (3.1) on the interval S, i.e., that there are points a, b, c, d in S such that $c < b < a < d$ and $T(a) = b$, $T(b) = c$, $T(c) = d$. Specifically, let $a = 3.75$. Then $b = T(a) = 1.9$, $c = T(b) = 0.19$ and $d = T(c) = 15.58 > a$. Thus the process (9.73) exhibits Li–Yorke chaos on S.

Clearly, the conditions of Proposition 3.1 are satisfied by the map T (regarded as a function from S to S). Hence, there exists some $\varepsilon > 0$ such that if $g \in C(S)$ satisfies

$$\sup_{p \in S} |T(p) - g(p)| < \varepsilon, \qquad (9.74)$$

then g also exhibits Li–Yorke chaos on S.

Recall that $T(p) \equiv T(p, \bar{e}, \bar{\lambda})$, where $\bar{e}$ lies in the interior of the compact set $K \subset \mathcal{E}_{CD}$ and $\bar{\lambda}$ is in the interior of the compact set Λ. Now, for $e = (\alpha, \beta) \in K$ and $\lambda \in \Lambda$ the map $T(p, e, \lambda)$ given by Equation (9.72) is clearly jointly continuous on the compact set $S \times K \times \Lambda$. Let e' be another economy in K and $\lambda' \in \Lambda$. By uniform continuity, there exists $\delta > 0$ such that if the (four-dimensional) Euclidean distance

$$d((p, e, \lambda), (p', e', \lambda')) < \delta \qquad (9.75)$$

then

$$|T(p, e, \lambda) - T(p', e', \lambda')| < \varepsilon. \qquad (9.76)$$

Let $N \times V$ be an open set in the interior of $K \times \Lambda$ containing $(\bar{e}, \bar{\lambda})$ and having radius less than δ. If $\varepsilon \in N$ and $\lambda \in V$ then for each $p \in S$,

$$d((p, \bar{e}, \bar{\lambda}), (p, e, \lambda)) < \delta$$

and, hence,

$$|T(p, \bar{e}, \bar{\lambda}) - T(p, e, \lambda)| < \varepsilon. \qquad (9.77)$$

Taking the supremum over all $p \in S$ we note that $(e, \lambda) \in N \times V$ implies $T(\cdot, e, \lambda)$ is within ε in the sup norm distance of $T(\cdot, \bar{e}, \bar{\lambda})$. Since relation (9.77) is satisfied by every mapping $T(\cdot, e, \lambda)$ with $(e, \lambda) \in N \times V$, the tatonnement process exhibits Li–Yorke chaos on an open set of Cobb–Douglas economies, for an open set of "rate of adjustment" parameters λ.   ∎

1.10 Complements and Details

Section 1.1. Simple rules of compounding interest and population growth are early examples of dynamic analysis in economics. Professor Samuelson in his *Foundations* talked about "a time when pure economic theory has undergone a revolution of thought – from statical to dynamical models. While many earlier foreshadowings can be found in the literature, we may date this upheaval from the publication of Ragnar Frisch's Cassel Volume essay." Our quote at the beginning of Chapter 3 is from this Frisch's (1933) article. Frisch, Goodwin, Harrod, Hicks, Kalecki, Lundberg, Metzler, Samuelson, and others made definitive contributions to studying endogenous (short run) cycles in an economy.

A useful compendium of formal trade cycle models of Hicks, Samuelson , Goodwin, and Kalecki was already available in Allen (1956). Some of the pioneering papers were collected in Gordon and Klein (1965) and Cass and McKenzie (1974). We should stress that the developments went significantly beyond the scope of our presentation in this chapter in many respects (in particular, in the treatment of lagged variables). We should also recall that difference, differential, and "mixed" systems (see Frisch and Holme 1935) were all experimented with.

The models of Samuelson (1939) and Metzler (1941) are celebrated examples of "second-order" systems. In the Samuelson model, one has

$$Y_t = C_t + I_t, \tag{C1.1}$$

(where Y, C, I denote income, consumption, and investment in period t).

$$C_t = cY_{t-1}, \quad 0 < c < 1 \tag{C1.2}$$

(where c is the fraction of I_{t-1} consumed in period t).

$$I_t = B[Y_{t-1} - Y_{t-2}], \tag{C1.3}$$

where $B > 0$ is the "accelerator coefficient."

Upon substitution we get

$$Y_t = (C + B)Y_{t-1} - BY_{t-2}. \tag{C1.4}$$

Subsequent research in dynamic economics continued in several directions: descriptive models of growth and market adjustments, models of intertemporal optimization with the duals interpreted as competitive equilibria, models of intertemporal equilibria with overlapping generations of utility maximizing agents, and dynamic game theoretic models capturing conflicts of interests. The emergence of chaos in simple frameworks was noted by Day (1982, 1994) and Benhabib and Day (1981, 1982). Goodwin, with his life-long commitment to exploring nonlinearity in economics (Goodwin 1990), remarked that if one were looking for "a system capable of endogenous, irregular, wave-like growth," the "discovery and elaboration of chaotic attractors seemed to me to provide the kind of conceptualization that we economists need."

By now, we have a relative abundance of examples (that are variations of the standard models) which generate complex or chaotic behavior. The striking feature that has been duly emphasized is that complexities arise out of the essential nonlinearity of the models even in their very simple formulations. Moreover, the examples do not arise out of any "knife-edge matching" of the parameters, making them atypical or accidental. For useful review articles, see Baumol and Benhabib (1989), Day and Pianigiani (1991), Grandmont (1986), Zarnowitz (1985). Some collections of influential articles are Benhabib (1992), Grandmont (1987),

Majumdar et al. (2000). The book by Day (1994) is an excellent introduction to chaotic dynamics for students in economics.

Section 1.2. For additional material, see Brauer and Castillo-Chavez
(2001), Kaplan and Glass (1995), Day (1994), Devaney (1986).

Section 1.3. See Collet and Eckmann (1980), Devaney (1986), Majumdar et al. (2000, Chapter 13 by Mitra). Proposition 3.1 is contained in
Bala and Majumdar (1992) who dealt with the question of robustness
of Li–Yorke chaos in a discrete time version of the Walras–Samuelson
price adjustment process.

Theorem 3.1 (*Proof of [2]*)   Let $\mathcal{M}$ be the set of sequences $M = \{M_n\}_{n=1}^{\infty}$ of intervals with

  [A.1] $M_n = K$ *or* $M_n \subset L$, *and* $\alpha(M_n) \supset M_{n+1}$.
    If $M_n = K$, *then*
  [A.2] n *is the square of an integer and* $M_{n+1}, M_{n+2} \subset L$.
    Of course if n is the square of an integer, then $n + 1$ and $n + 2$ are
  not, so the last requirement in [A.2] is redundant. For $M \in \mathcal{M}$ let
  $P(M, n)$ denote the number of i's in $\{1, \ldots, n\}$ for which $M_i = K$.
  For each $r \in (3/4, 1)$ choose $M^r = \{M_n^r\}_{n=1}^{\infty}$ to be a sequence in $\mathcal{M}$
  such that
  [A.3] $\lim_{n \to \infty} P(M^r, n^2)/n = r$.
    Let $\mathcal{M}_0 = \{M^r : r \in (3/4, 1)\} \subset \mathcal{M}$. Then $\mathcal{M}_0$ is uncountable
  since $M^{r_1} \neq M^{r_2}$ for $r_1 \neq r_2$. For each $M^r \in \mathcal{M}_0$, by Step 2, there
  exists a point x_r, with $\alpha^n(x_r) \in M_n^r$ for all n.
    Let $T = \{x_r : r \in (3/4, 1)\}$. Then T is also uncountable. For
  $x \in T$, let $P(x, n)$ denote the number of i's in $\{1, \ldots, n\}$ for
  which $\alpha^i(x) \in K$. We can never have $\alpha^k(x_r) = b$, because then x_r
  would eventually have period 3, contrary to [A.2]. Consequently
  $P(x_r, n) = P(M^r, n)$ for all n, and so

$$\rho(x_r) = \lim_{n \to \infty} P(M_r, n^2) = r$$

    for all r. We claim that
  [A.4] *for* $p, q \in T$, *with* $p \neq q$, *there exist finitely many* n's *such that*
    $\alpha^n(p) \in K$ *and* $\alpha^n(q) \in L$ *or vice versa.*

We may assume $\rho(p) > \rho(q)$. Then $P(p, n) - P(q, n) \to x$, and so there must be infinitely many n's such that $\alpha^n(p) \in K$ and $\alpha^n(q) \in L$.

Since $\alpha^2(b) = d \leq a$ and α^2 is continuous, there exists $\delta > 0$ such that $\alpha^2(x) < (b + d)/2$ for all $x \in [b - \delta, b] \subset K$. If $p \in T$ and $\alpha^n(p) \in K$, then [A.2] implies $\alpha^{n+1}(p) \in L$ and $\alpha^{n+2}(p) \in L$. Therefore $\alpha^n(p) < b - \delta$. If $\alpha^n(q) \in L$, then $\alpha^n(q) \geq b$, so

$$|\alpha^n(p) - \alpha^n(q)| > \delta.$$

By claim [A.4], for any $p, q \in T$, $p \neq q$, it follows that

$$\limsup_{n \to \infty} |\alpha^n(p) - \alpha^n(q)| \geq \delta > 0.$$

Hence (3.2) is proved. This technique may be similarly used to prove that (2(ii)) is satisfied.

Proof of (3.3) Since $\alpha(b) = c$, $\alpha(c) = d$, $d \leq a$, we may choose intervals $[b^n, c^n]$, $n = 0, 1, 2, \ldots$, such that

(a) $[b, c] = [b^0, c^0] \supset [b^1, c^1] \supset \cdots \supset [b^n, c^n] \supset \cdots$,
(b) $\alpha(x) \in (b^n, c^n)$ for all $x \in (b^{n+1}, c^{n+1})$,
(c) $\alpha(b^{n+1}) = c^n$, $\alpha(c^{n+1}) = b^n$.

Let $A = \bigcap_{n=0}^{\infty} [b^n, c^n]$, $b^* = \inf A$ and $c^* = \sup A$; then $\alpha(b^*) = c^*$ and $\alpha(c^*) = b^*$, because of (c).

In order to prove (3.3) we must be more specific in our choice of the sequences M^r. In addition to our previous requirements on $M \in \mathcal{M}$, we assume that if $M_k = K$ for both $k = n^2$ and $(n + 1)^2$, then $M_k = [b^{2n-(2j-1)}, b^*]$ for $k = n^2 + (2j - 1)$, $M_k = [c^*, c^{2n-2j}]$ for $k = n^2 + 2j$ where $j = 1, \ldots, n$. For the remaining k's that are not squares of integers, we assume $M_k = L$.

It is easy to check that these requirements are consistent with [A.1] and [A.2], and that we can still choose M^r so as to satisfy [A.3]. From the fact that $\rho(x)$ may be thought of as the limit of the fraction of n's for which $\alpha^{n^2}(x) \in K$, it follows that for any r^*, $r \in (3/4, 1)$ there exist infinitely many n such that $M_k^r = M_k^{r^*} = K$ for both $k = n^2$ and $(n + 1)^2$. To show (3.3), let $x_r \in S$ and $x_{r^*} \in S$. Since $b^n \to b^*$, $c^n \to c^*$ as $n \to \infty$, for any $\varepsilon > 0$ there exists N with $|b^n - b^*| < \varepsilon/2$, $|c^n - c^*| < \varepsilon/2$ for all $n > N$. Then, for any n with $n > N$ and $M_k^r = M_k^{r^*} = K$ for both $k = n^2$ and $(n + 1)^2$, we have

$$\alpha^{n^2+1}(x_r) \in M_k^r = [b^{2n-1}, b^*]$$

with $k = n^2 + 1$ and $\alpha^{n^2+1}(x_r)$ and $\alpha^{n^2+1}(x_{r*})$ both belong to $[b^{2n-1}, b^*]$. Therefore, $|\alpha^{n^2+1}(x_r) - \alpha^{n^2+1}(x_{r*})| \leq \varepsilon$. Since there are infinitely many n with this property, $\liminf_{n\to\infty} |\alpha^n(x_r) - \alpha^n(x_{r*})| = 0$.   ∎

Section 1.4. See Allen (1956), Baumol (1970), Devaney (1986, Chapter 2), Goldberg (1958), Sandefur (1990), Elaydi (2000). Parts (a)–(c) of Example 4.1 are contained in Majumdar (1974, Theorem 4.1). For an elaboration of the survival problem, see Majumdar and Radner (1992).

Consider the difference equation

$$x_{t+1} = Ax_t, \tag{C1.5}$$

where A is an $n \times n$ matrix and x_t is a column vector in $\mathbb{R}^n$. Given x_0, iteration of (C1.5) gives us

$$x_t = A^t x_0. \tag{C1.6}$$

The long-run behavior of x_t is clearly linked to the behavior of A^t. Here is a definitive result:

Proposition C1.1 *Let A be an $n \times n$ complex matrix. The sequence $A, \ldots, A^t, \ldots$ of its powers converges if and only if*

(1) each characteristic root r of A satisfies $|r| < 1$ or $r = 1$;
(2) when the second case occurs the order of multiplicity of the root 1 equals the dimension of the eigenvector space associated with that root.

For a proof see Debreu and Herstein (1953, Theorem 5).

Section 1.5. For possible generalizations of Proposition 5.1 one is led to fixed point theorems in partially ordered sets. See the review in Dugundji and Granas (1982, Chapter 1.4). A detailed analysis of the survival problem treated in Example 5.2 is in Majumdar and Radner (1992).

Section 1.6. The monographs by Clark (1976), Dasgupta and Heal (1979), and Dasgupta (1982) deal with the optimal management of exhaustible and renewable resources. For extensions of the basic qualitative results summarized in Example 6.1, see a Majumdar and Nermuth (1982), Dechert and Nishimura (1983), Mitra and Ray (1984), Kamihigashi and Roy (2005), and the review by Majumdar (2005).

Section 1.7. Consider $S = [a, b] \subset \mathbb{R}$. Let $\mathcal{S}$ be the Borel sigmafield of S, and P a probability measure on $\mathcal{S}$. If $\alpha : S \to S$ is measurable, then P is *invariant under* α if $P(E) = P(\alpha^{-1}(E))$ for all E in $\mathcal{S}$. P is *ergodic* if "$E \in \mathcal{S}, \alpha^{-1}(E) = E$" implies "$P(E) = 0$ or 1." The dynamical system (S, α) is said to exhibit *ergodic chaos* if there is an ergodic, invariant measure ν that is absolutely continuous with respect to the Lebesgue measure $\mathbf{m}$ on $\mathcal{S}$ (i.e., if $S \in \mathcal{S}$, and $\mathbf{m}(E) = 0$, then $\nu(E) = 0$). In this case, for brevity, call β *an ergodic measure* of α.

There are alternative sufficient conditions for ensuring that a dynamical system exhibits ergodic chaos. We (due to Misiurewicz 1981) recall the one that is most readily applicable to the quadratic family.

Proposition C7.1 *Let $S = [0, 1]$ and $J = [3, 4]$ given some $\hat{\theta}$ in J, define $\alpha_{\hat{\theta}}(x) = \hat{\theta}x(1 - x)$ for $x \in S$. Suppose that there is $k \geq 2$ such that $y = \alpha_{\hat{\theta}}^k(0, 5)$ satisfies $\alpha_{\hat{\theta}}(y) = y$ and $|a'_{\hat{\theta}}(y)| > 1$. Then $(S, \alpha_{\hat{\theta}})$ exhibits ergodic chaos.*

As an example, consider $S = [0, 1]$ and $\hat{\theta} = 4$. Here, $\alpha_4^2(0.5) = 0$, $\alpha_4(0) = 0$ and $\alpha'_4(0) = 4 > 1$, so the above proposition applies directly. For this example, the density ρ of the ergodic measure ν has been calculated to be

$$\rho(x) = \frac{1}{\pi[x(1 - x)]^{1/2}} \quad \text{for } 0 < x < 1.$$

The outstanding result due to Jakobson can now be stated:

Theorem C7.1 *Let $S = [0, 1]$, $J = [3, 4]$ and $\alpha_\theta(x) = \theta x(1 - x)$ for $(x, 0) \in S \times J$. Then the set $\Delta = \{\theta \in J : (S, \alpha_\theta) \text{ exhibits ergodic chaos}\}$ has positive Lebesgue measure.*

In fact, more is known. Consider the interval $J_\eta = [4 - \eta, 4]$ for $0 < \eta \leq 1$. Denote by $\mathbf{m}_\eta$ the Lebesgue measure of the set of θ in J_η for which the dynamical system (S, α_θ) exhibits ergodic chaos (formally, $\mathbf{m}_\eta = \mathbf{m}(J_\eta \cap \Delta)$). Then, given $\varepsilon > 0$, there is $0 < \eta < 1$ such that

$$\frac{\mathbf{m}_\eta}{\eta} > 1 - \varepsilon.$$

Thus, as η gets smaller (so that we are looking at θ "close to" 4), *most*
of the values of θ exhibit ergodic chaos (see Day (1994) and Day and
Pianigiani (1991) for more details).

Section 1.8. For a more complete discussion of bifurcation theory see
Devaney (1986, Section 1.12). By now there are many examples of
macroeconomic models with "endogenous" periodic behavior and bifurcations (in discrete and continuous time. See Azariadis (1993) and
Grandmont (1985) for very accessible presentations and references. Also
of interest are Torre (1977), Benhabib and Nishimura (1979), Becker and
Foias (1994), and Sorger (1994)).

Section 1.9.1. Following Chakravarty (1959) we outline the two-sector
Feldman-Mahalanobis model which was also prominent in the literature
on national planning.

C.1.9.1. The Feldman–Mahalanobis model portrays an economy (with
unlimited supply of labor) in which there are two sectors: the investment
good sector and the consumption good sector. Total investment in any
period t can be broken up into two components, one part going to augment
productive capacity in the *investment* good sector, the other part to the
consumption good sector.

Thus, investment I_t is divided into two parts $\lambda_k I_t$ and $\lambda_c I_t$ where
$\lambda_k > 0$ and $\lambda_c > 0$ are the fractions of investment allocated to the two
sectors. It follows, of course, that $\lambda_k + \lambda_c \equiv 1$. If β_k and β_c are the
respective output–capital ratios of the two sectors, it follows that

$$I_t - I_{t-1} = \beta_k \lambda_k I_{t-1} \tag{C1.7}$$

$$C_t - C_{t-1} = \beta_c \lambda_c I_{t-1}. \tag{C1.8}$$

Equation (C1.7) can be solved to get

$$I_t = I_0 (1 + \lambda_k \beta_k)^t, \tag{C1.9}$$

where I_0 is the initial investment (i.e., investment in period $t = 0$); hence,

$$I_t - I_0 = I_0\{(1 + \lambda_k \beta_k)^t - 1\}. \tag{C1.10}$$

From Equation (C1.8) it can be deduced that

$$C_t - C_0 = \beta_c \lambda_c I_0[\{(1 + \lambda_k \beta_k)^t - 1\}/\lambda_k \beta_k]. \tag{C1.11}$$

Adding (C1.8) and (C1.9) we get

$$Y_t - Y_0 = I_0\{(1 + \beta_k\lambda_k)^t - 1\}([\beta_c\lambda_c/\beta_k\lambda_k] + 1). \qquad (C1.12)$$

We can replace I_0 in (C1.12) by $\alpha_0 Y_0$, where α_0 is the fraction of income that is assigned to investment initially. We then have

$$Y_t = Y_0[1 + \alpha_0 \left[\frac{\beta_c\lambda_c + \beta_k\lambda_k}{\beta_k\lambda_k}\right] \{(1 + \beta_k\lambda_k)^t - 1\}]. \qquad (C1.13)$$

What happens to the growth of income if we choose a higher value of λ_k? Note that (since β_c, β_k are technologically given) a higher value of λ_k *increases* the terms $(1 + \beta_k\lambda_k)^t$ and *lowers* the term $(\beta_c\lambda_c + \beta_k\lambda_k)/\beta_k\lambda_k$. While for small values of t the overall impact is unclear, the term $(1 + \beta_k\lambda_k)^t$ will dominate in the long run and will imply a higher Y_t in future.

Section 1.9.3. For detailed expositions of the theory and applications of linear and nonlinear multiplicative processes, see Debreu and Herstein (1953), Gantmacher (1960), Gale (1960), Morishima (1964), and Nikaido (1968).

Section 1.9.4.1. The question of existence of a program that is optimal (according to the Ramsey–Weizsacker criterion) in the undiscounted model is a subtle one. Gale (1967), and McFadden (1967) treated this problem in discrete time, multisector frameworks. A complete outline of the existence proof for the one-good model in which the technology is allowed to have an initial phase of increasing returns is in Majumdar and Mitra (1982).

The problem of attaining an optimal program in a decentralized, price-guided economy with no terminal date has also challenged economic theorists. A particular problem is to replace the "insignificant future" (IF) condition in Theorems 9.2 and 9.3 by a sequence of verifications each of which can be achieved through the observation of a finite number of prices and quantities. A symposium in *Journal of Economic Theory* (volume 45, number 2, August 1988) explored this theme in depth: see the introduction to this symposium by Majumdar (1988) and the collection edited by Majumdar (1992).

On comparative dynamic problems, the reader is referred to Brock (1971), Nermuth (1978), Mitra (1983), Becker (1985) Dutta (1991),

Dutta, Majumdar, and Sundaram (1994) and Mitra (Chapter 2 in Majumdar et al. (2000)).

Exercise 9.5 was sketched out by Santanu Roy (who should be responsible for the blemishes!). In this context see Brock and Gale (1969), Jones and Manuelli (1990), Rebelo (1991), K. Mitra (1998) and T. Mitra (1998).

Section 1.9.5. In his evaluation of the progress of research on optimal intertemporal allocation of resources, Koopmans (1967, p. 2) observed:

In all of the models considered it is assumed that the objective of economic growth depends exclusively on the path of consumption as foreseen for the future. That is, the capital stock is not regarded as an end in itself, or as a means to other end other than consumption. We have already taken a step away from reality by making this assumption. A large and flexible capital stock has considerable importance for what is somewhat inadequately called "defense." The capital stock also helps to meet the cost of retaining all aspects of national sovereignty and power in a highly interdependent world.

See Kurz (1968), Arrow and Kurz (1970), and Roskamp (1979) for models with "wealth effects." In the literature on renewable resources and environmental economics, incorporating the direct "stock effects" is acknowledged to be a significant ingredient in formulating appropriate policies. Dasgupta (1982, p. 1070) summarized the importance of such effects as follows:

As a flow DDT is useful in agriculture as an input; as a stock it is hazardous for health. Likewise, fisheries and aquifers are useful not only for the harvest they provide: as a stock they are directly useful, since harvesting and extraction costs are low if stocks are large. Likewise, forests are beneficial not only for the flow of timber they can supply: as a stock they prevent soil erosion and maintain a genetic pool.

Benhabib and Nishimura (1985) identified sufficient conditions for the existence of periodic optimal programs in two-sector models.

Section 1.9.6. This exposition is based entirely on parts of Mitra (2000). Boldrin and Montrucchio (1986) and Deneckere and Pelikan (1986) used the reduced form (two-sector) model to provide examples of chaotic behavior generated by optimization problems. Of particular importance also is the paper by Nishimura, Sorger and Yano (1994) on ergodic chaos. Radner (1966) used a dynamic programming approach to provide an example of a multisector model (a linear-logarithmic economy) of economic growth in which optimal programs can be explicitly calculated for alternative optimality criteria. Radner (1967) dealt with a more general framework. The reduced form model was used by Gale (1967) in his landmark paper on optimal development in a multisector economy, and subsequently proved convenient for mathematical analysis as well as construction of examples.

Section 1.9.7. Indeed, Sundaram (1989, pp. 164–165) has a stronger monotonicity result. However, if the stock effect is introduced in the utility function, the evolution of stocks may display complex behavior (see the example of Dutta and Sundaram (1993). See also the paper by Dana and Montrucchio (1987)).

Section 1.9.8. Our exposition of the overlapping generations model (Samuelson 1958) is based on Diamond (1965) and Galor and Ryder (1989); see also Wendner (2003). Gale (1973) considered a pure exchange model. The possibility of chaotic behavior in this context was confirmed by Benhabib and Day (1982).

Section 1.9.9. Some of the basic papers on the stability of the price adjustment processes are by Arrow and Hurwicz (1958), Arrow, Block, and Hurwicz (1959), Arrow (1958), Scarf (1960), Saari (1985), and Bala and Majumdar (1992).

1.11 Supplementary Exercises

(1) Let A be a *strictly positive* $(n \times n)$ matrix, and consider the following "backward" system:

$$x_t = A x_{t+1},$$
$$x_t \geq 0 \quad (t = 0, 1, \ldots). \tag{11.1}$$

Of course $x_t = 0$ is a *trivial* solution. Show that

(i) there is a *balanced growth* solution to (11.1): $x_t = (1/\eta)^t x$, where $x > 0$ and $\eta > 0$ (and this solution is unique up to a multiplication by a positive scalar).

(ii) the balanced growth solution is the sole possible nontrivial solution.

(2) Again, let $A \gg 0$, and $\lambda \equiv \lambda(A)$ be its dominant root. Consider

$$x_t = Ax_{t+1} + c_t,$$
$$x_t \geq 0, \quad c_t \geq 0 \quad (t = 0, 1, \ldots) \tag{11.2}$$

where (c_t) is a given sequence of vectors, $c_t \geq 0$ for all $t \geq 0$.

(i) Show that the convergence of either one of the following two series for a given sequence (c_t) $(c_t \geq 0,$ for $t \geq 0)$

$$\sum_{m=0}^{\infty} A^m c_m \tag{α}$$

$$\sum_{m=0}^{\infty} \lambda^m c_m, \quad \lambda \equiv \lambda(A) \tag{β}$$

entails the convergence of the other. Moreover, (11.2) has a solution if and only if (α) is convergent. If this condition is met, the series

$$\sum_{m=0}^{\infty} A^m c_{m+v}$$

is also convergent for any integer $v = 0, 1, \ldots$, and the solution of (11.2) is given by

$$x_t = u_t + \sum_{m=0}^{\infty} A^m c_{m+t}$$

where u_t is any solution of (11.1).

(For further discussions and applications of *backward* equations, see Nikaido (1968) and Azariadis (1993).)

(3) Consider the difference equation

$$x_t = px_{t+1} + qx_{t-1}, \quad t = 1, 2, \ldots \tag{11.3}$$

where $0 < p < 1$ and $q = 1 - p$.

Show that the general solution has the form

$$x_t = \begin{cases} A + B(q/p)^t & \text{if } p \neq q \\ A + Bt & \text{if } p = q \end{cases}.$$

Specifying x_t for any two distinct values of t suffices to determine A and B completely.

(4) In the study of (local) stability of a fixed point of a nonlinear dynamical system, we are led to the question of identifying conditions under which

$$\lim_{t \to \infty} A^t = 0, \tag{11.4}$$

where A is a square matrix ($n \times n$). If all the roots λ_i of the characteristic equation

$$\det(A - \lambda I) = 0 \tag{11.5}$$

satisfy $|\lambda_i| < 1$, then (11.4) holds. For a quadratic equation

$$f(\lambda) = \lambda^2 + a_1 \lambda + a_2 = 0, \tag{11.6}$$

(where $a_1 \neq 0, a_2 \neq 0$), both roots λ_1, λ_2 of (11.6) satisfy $|\lambda_i| < 1$ if and only if

$$1 + a_1 + a_2 > 0, \quad 1 - a_1 + a_2 > 0, \quad 1 - a_2 > 0$$

(see Brauer and Castillo-Chavez (2001, Chapter 2.7) and Samuelson (1941)).

(5) Period-doubling bifurcation to chaos in an overlapping generations model

(5.1) The Framework. Consider an economy described by $(a, b; u_1, u_2, \delta, v)$, where $(a, b) \in \mathbb{R}_+^2$ and $b > 0$; u_1 and u_2 are nondecreasing functions from $\mathbb{R}_+$ to $\mathbb{R}$, which are increasing on $[0, a + b]$; $\delta \in (0, 1)$; and $v \in \mathbb{R}_{++}$.

A *program* is a sequence $\{c_t, d_t\}_0^\infty$ such that

$$(c_t, d_t) \in \mathbb{R}_+^2 \quad \text{and} \quad c_t + d_t = a + b \quad \text{for } t \geq 0$$

A *competitive equilibrium* is a sequence $\{c_t, d_t, p_t, q\}_0^\infty$ such that
(i) $(c_t, d_t, p_t, q) \in \mathbb{R}_+^4$ for $t \geq 0$.
(ii) For each $t \geq 0, (c_t, d_{t+1})$ is in $B_{t+1} = \{c, d \in \mathbb{R}_+^2 : p_t c + p_{t+1} d \leq p_t a + p_{t+1} b\}$, and $u_1(c_t) + \delta u_2(d_{t+1}) \geq u_1(c) + \delta u_2(d)$ for all $(c, d) \in B_{t+1}$. Further, d_0 is in $B_0 = \{d \in \mathbb{R}_+ : p_0 d \leq p_0 b - q v\}$ and $u_2(d_0) \geq u_2(d)$ for all $d \in B_0$.
(iii) $c_t + d_t = a + b$ for $t \geq 0$.

Here, in each period $t \geq 0$, an agent is born who lives for two periods $(t, t + 1)$. The agent has an endowment (a, b) over its life time. There is one 'old' agent in period 0 with an endowment b (who disappears at the end of period 0). In a competitive equilibrium, each agent born in period $t \geq 0$ chooses (c_t, d_{t+1}) from its budget set determined by prices p_t, p_{t+1} so as to maximize the discounted sum of utilities $u_2(c) + \delta u_2(d)$ from consumption (c, d) over its lifetime. The old agent leaves part of his income to the 'clearing house.' The choices of the agent must meet the feasibility condition (iii) period after period.

For an interpretation of this competitive equilibrium in terms of the institution of a clearing house and IOUs, see Gale (1973). Our example can be paraphrased from Gale as follows.

"We suppose that people never trade directly with each other but work through a clearing house." The old agent in period 0 consumes d_0 less than its endowment b, and leaves an 'income' (vq) with the clearing house. The new generation wishes to consume this excess, and must buy it from the clearing house. It is not able to pay for it until its old age. Accordingly it leaves with the clearing house an IOU for the 'income.' In its old age it pays its debt with interest and gets the IOU back, which it can destroy. But, in the meantime, a new "generation" of young agents has come to the clearing house with a new IOU and so on.

(5.2) Properties of competitive equilibrium. Verify that if $\{c_t, d_t, p_t, q\}_0^\infty$ is a competitive equilibrium,

(i) $p_t > 0$ for $t \geq 0$;
(ii) $p_t c_t + p_{t+1} d_{t+1} = p_t a + p_{t+1} b$ for $t \geq 0$, $p_0 d_0 = p_0 b - qv$.

(5.3) Dynamics of monetary equilibrium. A *monetary equilibrium* $\{c_t, d_t, p_t, q\}_0^\infty$ is a competitive equilibrium that satisfies $q > 0$. It is *interior* if $c_t > 0$ and $d_t > 0$.

Given a monetary equilibrium, we can conclude that for each $t \geq 0$, (c_t, d_{t+1}) must solve the problem

$$\left. \begin{array}{r} \text{maximize } u_1(c) + \delta u_2(d) \\ \text{subject to } p_t c + p_{t+1} d = p_t a + p_{t+1} b \\ (c, d) \geq 0 \end{array} \right\} . \qquad \text{(P)}$$

Also, d_0 must solve the problem

$$\left. \begin{array}{r} \text{maximize } \delta u_2(d) \\ \text{subject to } p_0 d = p_0 b - qv \\ d \geq 0 \end{array} \right\} . \qquad \text{(Q)}$$

We now look at the dynamics that must be satisfied by an interior monetary equilibrium. For this purpose, we assume in what follows that u_1 and u_2 are concave and continuously differentiable on $\mathbb{R}_+$, and $u_1'(c) > 0$ for $c \in [0, a + b)$, and $u_2'(d) > 0$ for $d \in [0, a + b)$.

We note that for an interior monetary equilibrium, $d_t < a + b$ and $c_t < a + b$ all $t \geq 0$. Thus, there exists a sequence $\{\lambda_t\}_0^\infty$ such that

$$\left. \begin{array}{l} u_1'(c_t) = \lambda_t p_t \\ \delta u_2'(d_{t+1}) = \lambda_t p_{t+1} \\ \text{and } \delta u_2'(d_0) = \lambda_0 p_0 \end{array} \right\} \quad \text{for } t \geq 0$$

and $\lambda_t > 0$ for $t \geq 0$. This yields

$$p_{t+1} u_1'(c_t) = \delta p_t u_2'(d_{t+1}) \quad \text{for } t \geq 0. \tag{11.7}$$

Also, using the constraint equation in (P),

$$p_{t+1}(b - d_{t+1}) = p_t(c_t - a) \quad \text{for } t \geq 0. \tag{11.8}$$

Finally, check the feasibility constraint

$$c_t - a = b - d_t \quad \text{for } t \geq 0. \tag{11.9}$$

[Hint: Using the constraint in (Q), we know that $b > d_0$ (since $q > 0$, $v > 0$), so that by repeatedly using (11.9) and (11.8), $b > d_t$ for $t \geq 0$.]

To summarize, if $\{c_t, d_t, p_t, q\}_0^\infty$ is an interior monetary equilibrium, then

(i) $(c_t, d_t, p_t) >> 0$ and $b > d_t$ for $t \geq 0$;
(ii) $c_t + d_t = a + b$ for $t \geq 0$;
(iii) $p_t(c_t - a) = p_{t+1}(b - d_{t+1})$ for $t \geq 0$;
(iv) $[u_1'(c_t)/\delta u_2'(d_{t+1})] = [p_t/p_{t+1}]$ for $t \geq 0$.

Conversely, verify the following: suppose there is a sequence $\{c_t, d_t, p_t\}_0^\infty$ satisfying

(i) $(c_t, d_t, p_t) >> 0$ for $t \geq 0$ and $b > d_0$;
(ii) $c_t + d_t = a + b$ for $t \geq 0$;
(iii) $p_t(c_t - a) = p_{t+1}(b - d_{t+1})$ for $t \geq 0$;
(iv) $[u_1'(c_t)/\delta u_2'(d_{t+1})] = [p_t/p_{t+1}]$ for $t \geq 0$;
then $\{c_t, d_t, p_t, q\}_0^\infty$ is an interior monetary equilibrium, where $q = p_0[b - d_0]/v$.

Thus for any sequence $\{c_t, d_t, p_t\}_0^\infty$, conditions (i)–(iv) *characterize* interior monetary equilibria.

Consider, now, any sequence $\{c_t, d_t, p_t\}_0^\infty$ satisfying conditions (i)–(iv) above. Defining $x_t = b - d_t$ for $t \geq 0$, and using (ii) and (iii), we then have

$$x_{t+1} = [p_t/p_{t+1}]x_t \quad \text{for } t \geq 0. \tag{11.10}$$

Using (iv) in (11.10), we finally get

$$x_{t+1} = \frac{u_1'(c_t)}{\delta u_2'(d_{t+1})}x_t \quad \text{for } t \geq 0. \tag{11.11}$$

(5.4) An example of period doubling. We consider now a specific example (based on the one discussed in Benhabib and Day (1982)) in which

$$(a, b) = (0, 1)$$
$$(1/4) < \delta < 1$$
$$u_1(c) = \begin{cases} c - (1/2)c^2 & \text{for } 0 \leq c \leq 1 \\ 1/2 & \text{for } c > 1 \end{cases}$$
$$u_2(d) = d \quad \text{for } d \geq 0.$$

Note that (a, b), u_1, and u_2 satisfy all the assumptions imposed above. Now (11.11) translates to

$$x_{t+1} = (1/\delta)[1 - c_t]x_t \quad \text{for } t \geq 0.$$

And since $x_t = b - d_t = 1 - d_t = a + b - d_t = c_t$, we finally obtain the basic difference equation

$$x_{t+1} = (1/\delta)x_t[1 - x_t] \quad \text{for } t \geq 0$$
with $x_0 \in (0, 1)$.

Applying this well-known theory, we can observe how the *nature* of the dynamics in the overlapping generations model changes as the discount factor δ decreases from values close to 1 to values close to 0.25.

2

Markov Processes

A drunk man will find his way home, but a drunk bird may get lost forever.

Shizuo Kakutani

Ranu looked at Kajal and thought suddenly, "His father used to look exactly like this when he was up to some mischief." Kajal's father had made a small mistake in thinking that his state of childhood at Nischindipur was lost forever. After an absence of twenty four years the innocent child had returned to Nischindipur. The unvanquished mystery of life, travelling through the ages, how gloriously it had revealed itself again.

Bibhuti Bandyopadhyay

2.1 Introduction

In theory and applications Markov processes constitute one of the most important classes of stochastic processes. A sequence of random variables $\{X_n: n = 0, 1, 2, \ldots\}$ with values in a state space S (with a sigmafield S) is said to be a (discrete parameter) *Markov process* if, for each $n \geq 0$, the conditional distribution of X_{n+1}, given $\{X_0, X_1, \ldots, X_n\}$ depends only on X_n. If this conditional distribution does not depend on n (i.e., if it is the same for all n), then the Markov process is said to be *time-homogeneous*. All Markov processes will be considered time-homogeneous unless specified otherwise. The above conditional probability $p(x, B) := \text{Prob}(X_{n+1} \in B \mid X_n = x) \equiv \text{Prob}(X_{n+1} \in B \mid \{X_0, X_1, \ldots, X_n\})$ on $\{X_n = x\}$, $x \in S$, $B \in S$, is called the *transition probability* of the Markov process. Conversely, given a function $(x, B) \to p(x, B)$ on $S \times S$ such that (i) $x \to p(x, B)$ is measurable on S into $\mathbb{R}$ for every $B \in S$ and (ii) $B \to p(x, B)$ is a probability measure on $(S, \mathcal{S})$ for every $x \in S$, one may construct a Markov process with transition probability $p(.,.)$ for any given initial state $X_0 = x_0$ (or *initial distribution* μ of X_0).

The present chapter is devoted to the general theory of Markov processes. A broad understanding of this vast and fascinating field is facilitated by an analysis of these processes on countable state spaces (*Markov chains*). *Irreducible chains*, where every state can be reached from every other state with positive probability, are either *transient* or *recurrent* (Theorem 7.1). In a transient chain, every state eventually disappears from the view, i.e., after a while, the chain will not return to it. A recurrent chain, on the other hand, will reappear again and again at any given state. Among the most celebrated chains are *simple random walks* on the integer lattice $\mathbb{Z}^k$, and a famous result of Polya (Theorem 7.2) says that these walks are recurrent in dimensions $k = 1, 2$ and are transient in all higher dimensions $k \geq 3$.

Recurrent chains may be further classified as *null recurrent* or *positive recurrent* (Section 2.8). In a null recurrent chain, it takes a long time for any given state to reappear – the mean return time to any state being infinite. If the mean return time to some state is finite, then so is the case for every state, and in this case the chain is called positive recurrent. In the latter case the Markov process has a unique *invariant distribution*, or *steady state*, to which it approaches as time goes on, no matter where it starts (Theorem 8.1). Simple random walks in dimensions $k = 1, 2$ are null recurrent. A rich class of examples is provided by the *birth-and-death chains* in which the process moves a unit step to the left or to the right at each period of time (Examples 8.1–8.3), but the probabilities of moving in one direction or another may depend on the current state. Depending on these probabilities, the chain may be transient, null recurrent, or positive recurrent (Propositions 8.3 and 8.4). Two significant applications are to physics (*Ehrenfest model of heat exchange*) and economics (*a thermostat model of managerial behavior*, due to Radner).

Moving on to general state spaces in Section 2.9, two proofs are given of an important result on the *exponential convergence* to a unique steady state under the so-called *Doeblin minorization* condition (Theorem 9.1). Simple examples that satisfy this condition are irreducible and aperiodic Markov chains on finite state spaces, and Markov processes with continuous and positive transition probability densities on a compact rectangle in $\mathbb{R}^k$ ($k \geq 1$). A third proof of Theorem 9.1 is given in Chapter 3, using notions of random dynamical systems. Theorem 9.2 provides an extension

to general state spaces of the criterion of Theorem 8.1 for the existence
of a *unique steady state* and its *stability in distribution*.

Section 2.10 extends two important classical limit theorems to functions of positive recurrent Markov chains – the SLLN, the *strong law of
large numbers* (Theorem 10.1), and the CLT, the *central limit theorem*
(Theorem 10.2). They are important for the statistical estimation of invariant (or steady state) probabilities, and of steady state means or other
quantities of interest. Appropriate extensions of the CLT to general state
spaces are provided later in Chapter 5. Finally, Sections 2.11 and 2.12
define notions of *convergence in distribution* in metric spaces and that
of *asymptotic stationarity* and provide criteria for them, which are used
in later chapters.

A special significance of discrete time Markov processes in our
scheme is due to a rather remarkable fact: such processes may be realized or represented as *random dynamical systems*. To understand the
latter notion, consider a set $\Gamma = \{f_1, \ldots, f_k\}$ of measurable functions
on S into S, with f_i assigned a positive probability $q_i (1 \leq i \leq k)$,
$q_1 + \cdots + q_k = 1$. Given an initial state $x_0 \in S$, one picks a function
at random from Γ: f_i being picked with probability q_i, $1 \leq i \leq k$.
Let α_1 denote this randomly chosen function. Then the state at time
$n = 1$ is $X_1 = \alpha_1 x_0 \equiv \alpha_1(x_0)$. At time $n = 2$, one again chooses a
function, say α_2 from Γ according to the assigned probabilities and
independently of the first choice α_1. The state at time $n = 2$ is then
$X_2 = \alpha_2 \alpha_1 x_0 = \alpha_2(X_1)$. In this manner one constructs a random dynamical system $X_{n+1} = \alpha_{n+1}(X_n) = \alpha_{n+1}\alpha_n \cdots \alpha_1 x_0 (n \geq 1)$, $X_0 = x_0$,
where $\{\alpha_n : n \geq 1\}$ is a sequence of independent maps on S into S with
the common distribution $Q = \{q_1, q_2, \ldots, q_k\}$ on Γ. It is simple to
check that $\{X_n : n = 0, 1, \ldots\}$ is a Markov process having the transition
probability

$$p(x, B) = \mathrm{Prob}(\alpha_1 x \in B) = Q(\{\gamma \in \Gamma : \gamma x \in B\}), \quad (x \in S, B \in \mathcal{S}).$$
$$(1.1)$$

One may take X_0 to be random, *independent of* $\{\alpha_n : n \geq 1\}$ and having
a given (initial) distribution μ. It is an important fact that, conversely,
every Markov process on a state space S may be represented in a similar
manner, provided S is a *standard state space*, i.e., a Borel subset of a
Polish space (see Chapter 3, Complements and Details for a more formal
elaboration of this point). Recall that a topological space M is *Polish* if
it is homeomorphic to a complete separable metric space.

Exercise 1.1 Show that $(0, 1)^k$, $[0, 1]^k$, $[0, 1)^k$, $(0, \infty)^k$, $[0, \infty)^k$ with their usual Euclidean distances are Polish, $k \geq 1$. [Hint: $(0, 1)^k$ and $(0, \infty)^k$ are homeomorphic to $\mathbb{R}^k$; $[0, 1)^k$ is homeomorphic to $[0, \infty)^k$]. ∎

Although the present chapter focuses primarily on the asymptotic, or long-run, behavior of a Markov process based on its transition probability, in many applications the Markov process is specified as a random dynamical system. One may of course compute the transition probability in such cases using Equation (1.1). However, the asymptotic analysis is facilitated by the evolution of the process as directly depicted by the random dynamical system. A systematic account of this approach begins in Chapter 3, which also presents several important results that are not derivable from the standard theory presented in this chapter.

2.2 Construction of Stochastic Processes

A stochastic process represents the random evolution with time of the state of an object, e.g., the price of a stock, population or unemployment rate of a country, length of a queue, number of people infected with a contagious disease, a person's cholesterol levels, the position of a gas molecule. In discrete time a stochastic process may be denoted as $\{X_n: n = 0, 1, 2, \ldots, \}$, where X_n is the state of the object at time n defined on a probability space $(\Omega, \mathcal{F}, P)$. Sometimes the index n starts at 1, instead of 0. The state space can be quite arbitrary, e.g., $\mathbb{R}_+$ (for the price of a stock), $\mathbb{Z}_+$ (for the length of a queue), $\mathbb{R}^3$ (for the position of a molecule). In the case the state space S is a *metric space* with some metric d, the sigmafield $\mathcal{S}$ is taken to be the Borel sigmafield. Note that for every n, the distribution $\mu_{0,1,\ldots,n}$, say, of $(X_0, X_1, \ldots, X_n)$ can be obtained from the distribution $\mu_{0,1,\ldots,n+1}$ of $(X_0, X_1, \ldots, X_n, X_{n+1})$ by integrating out the last coordinate (e.g., from the joint density or joint probability function of the latter). We will refer to this as the *consistency condition* for the finite-dimensional distributions $\mu_{0,1,\ldots,n}(n = 0, 1, \ldots)$.

An important question is: Given a sequence of finite-dimensional distributions $\mu_{0,1,\ldots,n}(n = 0, 1, \ldots)$ satisfying the consistency condition, does there exist a stochastic process $\{X_n: n = 0, 1, 2, \ldots\}$ on some probability space with these as its finite-dimensional distributions? The answer to this is "yes" if S is a *Polish space* and $\mathcal{S}$ its Borel sigmafield, according to *Kolmogorov's existence theorem*. One may take $\Omega = S^\infty$ (or $S^{\mathbb{Z}_+}$) – the space of all sequences $\omega = (x_0, x_1, \ldots, x_n, \ldots)$ with x_n in S, $\mathcal{F} =$ the smallest sigmafield containing all sets of the form $\{\omega \in \Omega \colon x_i \in B_i$ for $i = 0, 1, \ldots, n\}$ where $B_i \in S$ for $i = 0, 1, \ldots, n$ and n is an arbitrary nonnegative integer. We will denote this (*infinite*) *product sigmafield* as $S^{\otimes \infty}$ or $S^{\otimes \mathbb{Z}_+}$. It is sometimes called the *Kolmogorov sigmafield*. The random variable X_n is then the nth *coordinate function* on Ω, i.e., $X_n(\omega) = x_n$ the nth coordinate of ω. Such a construction of a stochastic process (on $\Omega = S^\infty$) is often called a *canonical construction*, and in this case the sigma-field $\mathcal{F}$ is called the *Kolmogorov sigmafield*. Unless specified otherwise we assume in this chapter that the Markov processes discussed are canonically constructed. The use of this basic theorem for the construction of stochastic processes is illustrated by the three important examples below.

Example 2.1 (*Independent sequence of random variables*) Let μ_n, $n = 0, 1, \ldots$, be an arbitrary sequence of probability measures on $(S, \mathcal{S})$. Then the sequence of joint distributions defined by the product probability measures $\mu_{0,1,\ldots,n} = \mu_0 \times \mu_1 \times \cdots \times \mu_n (n = 0, 1, \ldots)$ (e.g., given by the products of individual marginal densities or probability functions) is consistent. Hence there exists a sequence of independent random variables $\{X_n \colon n = 0, 1, 2, \ldots\}$ on some probability space $(\Omega, \mathcal{F}, P)$ such that the distribution of X_n is $\mu_n (n = 0, 1, \ldots)$. The sequence is *i.i.d.* (independent and identically distributed) if the marginals μ_n are the same for all n. The hypothesis of S being Polish may be completely dispensed with in this case, and $(S, \mathcal{S})$ may be an arbitrary measurable space. ∎

Example 2.2 (*Gaussian processes*) Suppose there exists on some probability space a sequence of random variables $\{X_n \colon n = 0, 1, \ldots\}$ such that for every $n(X_0, \ldots, X_{n+1})$ has an $(n+2)$-dimensional normal (Gaussian) distribution, say $\mu_{0,1,\ldots,n+1}$, with $E(X_i) = 0$ for all $i = 0, 1, \ldots, n + 1$, and covariances $\sigma_{ij} = E(X_i X_j)$ $(i, j = 0, 1, \ldots, n + 1)$. Then the distribution $\mu_{0,1,\ldots,n}$ of $(X_0, \ldots, X_n)$ is $(n+1)$-dimensional normal with mean vector zero and covariances $\sigma_{ij}(i, j = 0, \ldots, n)$. Note that the $(n + 1) \times (n + 1)$ matrix $((\sigma_{ij}))_{i,j=0,\ldots,n}$ is *symmetric* and *nonnegative definite* for every n, i.e., $\sigma_{ij} = \sigma_{ji}$ for all i, j and $\sum_{i,j=0,\ldots,n} \sigma_{ij} c_i c_j \geq 0$ for all $(n+1)$-tuples $(c_0, c_1, \ldots, c_n) \in \mathbb{R}^n$. Note that the last sum is just the variance of $\sum_{i=0,\ldots,n} c_i X_i$. We will say that the sequence of random variables above is a mean-zero *Gaussian process* with covariances σ_{ij}.

Now, conversely, given any double sequence σ_{ij} of real numbers $(i, j = 0, \ldots)$ with the properties of symmetry and nonnegative definiteness above, one can construct for each n the $(n + 1)$-dimensional normal distribution $\mu_{0,\ldots,n}$ with zero mean vector and covariances $\sigma_{ij}(i, j = 0, \ldots, n)$. Indeed, as in Example 2.1, one may first construct an independent sequence $\{Z_n: n = 0, 1, \ldots\}$ of normal random variables each with mean zero and variance one. Then writing $Z^{(n)}$ for the vector $(Z_0, \ldots, Z_n)$, the random vector $X^{(n)} = A(Z^{(n)})'$ is normal with mean vector zero and covariance matrix AA', where A is any $(n + 1) \times (n + 1)$ matrix. By a well-known result from linear algebra, there exists an A such that $AA' = ((\sigma_{ij}))_{i,j=0,\ldots,n}$. Thus $X^{(n)}$ has the desired distribution $\mu_{0,\ldots,n}$. This being true for all n, one has a sequence of distributions satisfying Kolmogorov's consistency condition. Hence, by Kolmogorov's existence theorem, one can find a probability space $(\Omega, \mathcal{F}, P)$ on which is defined a zero-mean Gaussian process $\{X_n: n = 0, 1, \ldots\}$ with covariances σ_{ij}. Finally, given any sequence of numbers $\{b_n: n = 0, 1, \ldots\}$, the sequence $Y_n = X_n + b_n$ $(n = 0, 1, \ldots)$ is a *Gaussian process with the mean sequence* $\{b_n: n = 0, \ldots\}$ *and covariances* σ_{ij}. ∎

Example 2.3 (*Markov processes*)   First, consider a countable state space S whose elements are labeled i, j, k, or x, y, z, etc., with or without subscripts. A *transition probability matrix* or a *stochastic matrix* $\mathbf{p} = ((p_{ij}))_{i,j \in S}$ has the defining properties: (1) $p_{ij} \geq 0$ and (2) $\sum_{k \in S} p_{ik} = 1$, for all $i, j \in S$.

Given such a matrix and a probability distribution $\mu = (\mu(i), i \in S)$, one may construct, for each $n \geq 0$, the distribution $\mu_{0,1,\ldots,n}$ on S^{n+1} (with a sigmafield $\mathcal{S}^{\otimes(n+1)}$ comprising all subsets of S^{n+1}), the probability assigned to each singleton $\{(i_0, i_1, \ldots, i_n)\}$ being given by

$$\mu_{0,1,\ldots,n}(\{(i_0, i_1, \ldots, i_n)\}) = \mu(i_0)p_{i_0 i_1} p_{i_1 i_2} \cdots p_{i_{n-1} i_n}. \qquad (2.1)$$

To check that these finite-dimensional distributions form a consistent sequence, note that

$$\mu_{0,1,\ldots,n}(\{(i_0, i_1, \ldots, i_n)\}) = \sum_{i_{n+1} \in S} \mu_{0,1,\ldots,n+1}(\{(i_0, i_1, \ldots, i_{n+1})\}),$$

since $\sum_{i_{n+1} \in S} p_{i_n i_{n+1}} = 1$. By Kolmogorov's existence theorem, there exists a probability space $(\Omega, \mathcal{F}, P)$ on which is defined a sequence of random variables $\{X_n: n = 0, 1, \ldots\}$ such that X_0 has distribution μ and $(X_0, X_1, \ldots, X_n)$ has distribution $\mu_{0,1,\ldots,n}$ defined by (2.1), i.e., $P(X_0 = i_0, X_1 = i_1, \ldots, X_n = i_n)$ is given by the right-hand side of (2.1). This process is called a *Markov chain having the transition probability matrix* $\mathbf{p}$ *and initial distribution* μ. In the special case in which μ is a *point mass*, $\mu(\{i_0\}) = 1$ for some $i_0 \in S$, one says that the Markov chain has *initial state* i_0. Observe that S is a complete separable metric space with the metric $d(i, i) = 0, d(i, j) = 1 (i \neq j)$, having the class of all subsets of S as its Borel sigmafield.

Next, on a complete separable metric state space (S, S) one may similarly define a Markov process $\{X_n: n = 0, 1, \ldots\}$ with a *transition probability* (function) $p(x, B)$ $(x \in S, B \in S)$ and initial distribution μ. Here

(1) $x \to p(x, B)$ is a Borel-measurable function for each $B \in S$;
(2) $B \to p(x, B)$ is a probability measure on (S, S) for each $x \in S$.

For example, let $S = \mathbb{R}$ and suppose we are given a transition *probability density function* $p(x, y)$, i.e., (i) $p(x, y) \geq 0$, (ii) $(x, y) \to p(x, y)$ continuous on $S \times S$, and (iii)$\int_{\mathbb{R}} p(x, y) \, dy = 1$ for every x. Then, $p(x, B) := \int_B p(x, y) \, dy$ $(B \in S)$ satisfies the conditions (1)–(2) above.

Let us define for each $n \geq 0$ the distribution $\mu_{0,1,\ldots,n}$ on S^{n+1} (with Borel sigma-field $S^{\otimes(n+1)}$), by

$$\mu_{0,1,\ldots,n}(B_0 \times B_1 \times \cdots \times B_n)$$
$$= \int_{B_0} \int_{B_1} \cdots \int_{B_{n-1}} \int_{B_n} p(x_{n-1}, x_n) p(x_{n-2}, x_{n-1}) \cdots$$
$$p(x_0, x_1) \, dx_n \, dx_{n-1} \cdots dx_1 \mu (dx_0)$$
$$(B_i \in S; i = 0, 1, \ldots, n), \quad n \geq 1, \quad (2.2)$$

where μ is a given probability on (S, S). By integrating out the last coordinate from $\mu_{0,1,\ldots,n+1}$ one gets $\mu_{0,1,\ldots,n}$, proving consistency. Hence there exists a stochastic process $\{X_n: n = 0, 1, 2, \ldots\}$ on some probability space $(\Omega, \mathcal{F}, P)$ such that X_0 has a distribution μ and $(X_0, X_1, \ldots, X_n)$ has distribution $\mu_{0,1,\ldots,n}(n \geq 1)$. This process is called a *Markov process with transition probability density* $p(x, y)$ *and initial distribution* μ. ∎

As in the case of Example 2.1, Markov processes with given transition probabilities and initial distributions can be constructed on arbitrary measurable state spaces (S, S). (See Complements and Details.)

2.3 Markov Processes with a Countable Number of States

In this section we consider Markov processes on a countable (finite or denumerable) state space S. In this case $\mathcal{S}$ is the class of *all* subsets of S.

Definition 3.1 A stochastic process $\{X_0, \ldots, X_n, \ldots\}$ has the *Markov property* if, for each n and m, the conditional distribution of $X_{n+1}, \ldots, X_{n+m}$ given $X_0, \ldots, X_n$ is the same as the conditional distribution given X_n alone. A process having the Markov property is called a Markov process. If the state space is countable, then we call it a *Markov chain*.

The next proposition establishes that it is enough to check for $m = 1$ in the above definition.

Proposition 3.1 *A stochastic process $\{X_0, \ldots X_n, \ldots\}$ has the Markov property iff for each n the conditional distribution of X_{n+1} given $X_0, X_1, \ldots X_n$ is a function of X_n alone.*

Proof. Let us consider a countable state space. The necessity is obvious. For sufficiency observe that

- $P(X_{n+1} = j_1, \ldots, X_{n+m} = j_m \mid X_0 = i_0, \ldots, X_n = i_n) =$
  $P(X_{n+1} = j_1 \mid X_0 = i_0, \ldots, X_n = i_n)$
- $P(X_{n+2} = j_2 \mid X_0 = i_0, \ldots, X_n = i_n, X_{n+1} = j_1) \cdots$
- $P(X_{n+m} = j_m \mid X_0 = i_0, \ldots, X_{n+m-1} = j_{m-1}) =$
  $P(X_{n+1} = j_1 \mid X_n = i_n) P(X_{n+2} = j_2 \mid X_{n+1} = j_1) \cdots$
- $P(X_{n+m} = j_m \mid X_{n+m-1} = j_{m-1}).$

The last equality follows from the hypothesis of the proposition. But, again by a similar argument, the last product equals $P(X_{n+1} = j_1, \ldots, X_{n+m} = j_m \mid X_n = i_n)$. ∎

Recall that a *transition probability matrix* or a *stochastic matrix* is a square matrix $\mathbf{p} = ((p_{ij}))$, where i and j vary over a finite or denumerable set S, satisfying (i) $p_{ij} \geq 0$ for all i and j and (ii) $\sum_{j \in S} p_{ij} = 1$ for all $i \in S$. It is a simple exercise to check that the process constructed in Example 2.3 on a countable state space S, having a transition probability matrix $\mathbf{p}$ and some initial distribution, or initial state, has the homogeneous Markov property:

$$P(X_{n+1} = j \mid X_0 = i_0, \ldots, X_{n-1} = i_{n-1}, X_n = i) = p_{ij}. \quad (3.1)$$

Notation Occasionally, we will write $p_{i,j}$ for p_{ij} as, for example, $p_{i-1,i}$ instead of p_{i-1i}.

An example may help clarify matters discussed so far. In this example we explicitly compute the distributions of X_n and the *invariant distribution* of the Markov chain.

Example 3.1 Following Proposition 3.1 let us consider a simple example of a Markov chain with state space $S = \{0, 1\}$ and transition probabilities $P(X_{n+1} = 1 \mid X_n = 0) = p$ and $P(X_{n+1} = 0 \mid X_n = 1) = q$. Then the transition probability matrix is

$$\mathbf{p} = \begin{bmatrix} 1 - p & p \\ q & 1 - q \end{bmatrix}.$$

Assume that $p \in (0, 1)$ and $q \in (0, 1)$. The initial distribution is $\mu = \{\mu(0), 1 - \mu(0)\}$. Note that given this information we can compute the distribution of X_n, i.e., $P(X_n = 0)$ and $P(X_n = 1)$ for $n \geq 1$. Observe that

$$
\begin{aligned}
P(X_{n+1} &= 0) \\
&= P(X_{n+1} = 0 \text{ and } X_n = 0) + P(X_{n+1} = 0 \text{ and } X_n = 1) \\
&= P(X_{n+1} = 0 \mid X_n = 0)P(X_n = 0) \\
&\quad + P(X_{n+1} = 0 \mid X_n = 1)P(X_n = 1) \\
&= (1 - p)P(X_n = 0) + q P(X_n = 1) \\
&= (1 - p - q)P(X_n = 0) + q.
\end{aligned}
$$

Now, $P(X_0 = 0) = \mu(0)$, so $P(X_1 = 0) = (1 - p - q)\mu(0) + q$ and $P(X_2 = 0) = (1 - p - q)^2 \mu(0) + q[1 + (1 - p - q)]$. Repeating this procedure n times, we get for $n \geq 1$:

$$P(X_n = 0) = (1 - p - q)^n \mu(0) + q \left[\sum_{j=0}^{n-1} (1 - p - q)^j \right]. \quad (3.2)$$

Since $p + q > 0$,

$$\sum_{j=0}^{n-1}(1 - p - q)^j = \frac{1 - (1 - p - q)^n}{p + q}.$$

From (3.2) we have

$$P(X_n = 0) = \frac{q}{p + q} + (1 - p - q)^n \left[\mu(0) - \frac{q}{p + q}\right]. \quad (3.3)$$

Then, using $P(X_n = 1) = 1 - P[X_n = 0]$, we get

$$P(X_n = 1) = \left(1 - \frac{q}{p + q}\right) - (1 - p - q)^n \left[\mu(0) - \frac{q}{p + q}\right]$$

$$= \frac{p}{p + q} + (1 - p - q)^n \left[-\mu(0) + \frac{q}{p + q}\right]$$

$$= \frac{p}{p + q} + (1 - p - q)^n \left[\mu(1) - 1 + \frac{q}{p + q}\right]$$

$$= \frac{p}{p + q} + (1 - p - q)^n \left[\mu(1) - \frac{p}{p + q}\right]. \quad (3.4)$$

Given our assumptions about p and q, we have $0 < p + q < 2$. So, $|1 - p - q| < 1$. In this case we can let $n \to \infty$ and get (irrespective of the values $\mu(0)$ and $\mu(1)$)

$$\lim_{n \to \infty} P(X_n = 0) = \frac{q}{p + q}$$

$$\lim_{n \to \infty} P(X_n = 1) = \frac{p}{p + q}.$$

The probabilities $\frac{q}{p+q}$ and $\frac{p}{p+q}$ are significant from another viewpoint. Suppose we want to choose $\mu(0)$ and $\mu(1)$ so that $P(X_n = 0)$ and $P(X_n = 1)$ are *independent* of n; in other words, we are looking for a *stochastic steady state* or a *time invariant distribution* for the Markov chain. If we set $\mu(0) = \frac{q}{p+q}$ and $\mu(1) = \frac{p}{p+q}$, then from (3.3) and (3.4) we get that for all $n = 0, 1, 2 \ldots$, $P(X_n = 0) = \frac{q}{p+q}$ and $P(X_n = 1) = \frac{p}{p+q}$. ∎

Let us introduce some more notions on Markov chains. Recall that the Markov property is defined by $P(X_{n+1} = i_{n+1} \mid X_0 = i_0, \ldots, X_n = i_n) = P(X_{n+1} = i_{n+1} \mid X_n = i_n)$ for every choice of n and of the states

$i_0, \ldots, i_{n+1}$ in S. In what follows in the rest of this section we state some useful working formulae for the transition probabilities of a Markov process, beginning with

$$P(X_{n+1} = i_{n+1}, \ldots, X_{n+m} = i_{n+m} \mid X_0 = i_0, \ldots, X_n = i_n)$$
$$= p_{i_n i_{n+1}} \cdots p_{i_{n+m-1} i_{n+m}}. \tag{3.5}$$

Now let $B_1, \ldots, B_m$ be subsets of S. Then by (3.5),

$$P(X_{n+1} \in B_1, \ldots, X_{n+m} \in B_m \mid X_0 = i_0, \ldots, X_{n-1} = i_{n-1}, X_n = i)$$
$$= \sum_{j_1 \in B_1} \cdots \sum_{j_m \in B_m} p_{i j_1} p_{j_1 j_2} \cdots p_{j_{m-1} j_m}. \tag{3.6}$$

Let us denote the probabilities of various events defined in terms of the Markov chain starting at i by P_i. Then we can write (3.6) as

$$P(X_{n+1} \in B_1, \ldots, X_{n+m} \in B_m \mid X_0 = i_0, \ldots, X_{n-1} = i_{n-1}, X_n = i)$$
$$= P_i(X_1 \in B_1, \ldots, X_m \in B_m). \tag{3.7}$$

The relation (3.7) extends, with the same proof, to conditional probabilities of sets $\{(X_{n+1}, \ldots, X_{n+m}) \in G\}$ where G is an arbitrary subset of S^m. Also, by summing (3.6) over all $j_1 \in B_1 = S, \ldots, j_{m-1} \in B_{m-1} = S$, and letting $B_m = \{j\}$, one has

$$P(X_{n+m} = j \mid X_0 = i_0, X_1 = i_1, \ldots, X_{n-1} = i_{n-1}, X_n = i)$$
$$= P(X_{n+m} = j \mid X_n = i) = p_{ij}^{(m)}, \tag{3.8}$$

where $p_{ij}^{(m)} = (\mathbf{p}^m)_{ij}$ is the (i, j) element of the matrix $\mathbf{p}^m$, called the *m-step transition probability matrix*. For all $m \geq 0, n \geq 1$, one may now express the joint distribution of $(X_m, X_{m+1}, \ldots, X_{m+n})$ as

$$P(X_m = j_0, X_{m+1} = j_1, \ldots, X_{m+n} = j_n)$$
$$= \sum_{i_0 \in S} P(X_0 = i_0, X_m = j_0, X_{m+1} = j_1, \ldots, X_{m+n} = j_n)$$
$$= \sum_{i_0 \in S} P(X_0 = i_0) P(X_m = j_0, \ldots, X_{m+n} = j_n \mid X_0 = i_0)$$
$$= \sum_{i_0 \in S} \mu(i_0) p_{i_0 j_0}^{(m)} p_{j_0 j_1}, \ldots, p_{j_{n-1} j_n}. \tag{3.9}$$

We now introduce the fundamental notion of a stationary or invariant distribution of a Markov chain with transition probability **p**.

Definition 3.2 A probability measure π is said to be an *invariant distribution* (or *invariant probability*) for **p** if

$$\pi' \mathbf{p} = \pi'$$

$$\text{or,} \quad \sum_i \pi_i p_{ij} = \pi_j \quad \text{for all } j \in S. \tag{3.10}$$

We shall also refer to an invariant distribution as a *stationary distribution* or, simply, as a *steady state* of the Markov process.

Here π is viewed as a column vector with π' as its transpose. By iterating in the defining relation (3.10) for an invariant distribution π one obtains

$$\pi' \mathbf{p}^m = \pi' (m \geq 1). \tag{3.11}$$

Note that if the initial distribution happens to be π then, for any $j \in S$, the jth element of the left side of (3.11) is the probability that $X_m = j$, and thus (3.11) implies that the probability of the event $X_m = j$ is the same for all m. But more is true.

Suppose that π is an invariant distribution for **p**, and let $\{X_n\}$ be the Markov chain with transition law **p** starting with initial distribution π. It follows from the Markov property, using (3.9), that

$$
\begin{aligned}
P_\pi(X_m = i_0, X_{m+1} &= i_1, \ldots, X_{m+n} = i_n) \\
&= (\pi' \mathbf{p}^m)_{i_0} p_{i_0 i_1} p_{i_1 i_2} \cdots p_{i_{n-1} i_n} \\
&= \pi_{i_0} p_{i_0 i_1} p_{i_1 i_2} \cdots p_{i_{n-1} i_n} \\
&= P_\pi(X_0 = i_0, X_1 = i_1, \ldots, X_n = i_n), \tag{3.12}
\end{aligned}
$$

for any given positive integer n and arbitrary states $i_0, i_1, \ldots, i_n \in S$. In other words, the distribution of the process is invariant under time translation; i.e., $\{X_n\}$ is a *stationary* Markov process according to the following definition.

Definition 3.3 A stochastic process $\{Y_n\}$ is called a *stationary* stochastic process if for all n, m, the distribution of $(Y_0, \ldots, Y_n)$ is the same as that of $(Y_m, \ldots, Y_{m+n})$.

2.4 Essential, Inessential, and Periodic States of a Markov Chain

We now introduce a useful classification of the states of a Markov chain. We continue to label the states $i, j, k, \ldots$ or $x, y, z, \ldots$ with or without subscripts. We say that i *leads to* j (written as $i \to j$) if $p_{ij}^{(n)} > 0$ for some $n \geq 1$. Surely $i \to j$ if and only if there is a chain of transitions $(i, i_1, \ldots, i_{n-1}, j)$ such that $p_{ii_1}, p_{i_1 i_2}, \ldots, p_{i_{n-1} j}$ are all strictly positive. In what follows we also write $\rho_{ij} = P(X_n = j$ for some $n \geq 1 \mid X_0 = i)$ (see (7.2)–(7.4) below). A state i is *essential* if $i \to j$ implies $j \to i$; otherwise, i is *inessential* which means that there is some state j such that $i \to j$ but $j \not\to i$ (i.e., $\rho_{ij} > 0$ but $\rho_{ji} = 0$). The class $\mathcal{E}$ of all essential states is *closed*, i.e., if $i \in \mathcal{E}$ and $j \in \mathcal{E}^c$, then $i \not\to j$ (or $\rho_{ij} = 0$) (Exercise).

Define the relation "$i \leftrightarrow j$" if $i \to j$ and $j \to i$. One may express $i \leftrightarrow j$ as i and j *communicate*. It is easy to check that $\leftrightarrow$ is an equivalence relation on $\mathcal{E}$ i.e., it is (i) *transitive*: $i \leftrightarrow j$, $j \leftrightarrow k \Rightarrow i \leftrightarrow k$, (ii) *symmetric*: $i \leftrightarrow j \Rightarrow j \leftrightarrow i$, and (iii) *reflexive*: $i \leftrightarrow i$. It follows that $\mathcal{E}$ may be decomposed into *disjoint communicating classes* $\mathcal{E}_1, \mathcal{E}_2, \ldots$. This is a standard algebraic property of an equivalence relation. To see this, choose $x_1 \in \mathcal{E}$, and find the set $\mathcal{E}(x_1)$ of all y such that $x_1 \leftrightarrow y$. Then $\mathcal{E}(x_1)$ is a closed set. For if $z \in \mathcal{E}(x_1)$ and $z \leftrightarrow w$, it follows by transitivity that $x_1 \leftrightarrow w$, so that $w \in \mathcal{E}(x_1)$. Denote $\mathcal{E}(x_1) = \mathcal{E}_1$. Now pick $x_2 \in \mathcal{E} \backslash \mathcal{E}_1$ and label $\mathcal{E}(x_2)$ as $\mathcal{E}_2$. Next pick $x_3 \in \mathcal{E} \backslash (\mathcal{E}_1 \cup \mathcal{E}_2)$ and write $\mathcal{E}(x_3) = \mathcal{E}_3$ and so on. We have thus proved the following:

Proposition 4.1 *The class $\mathcal{E}$ of all essential states of a Markov chain is closed and is the union of disjoint closed sets of communicating classes $\mathcal{E}_1, \mathcal{E}_2, \ldots$.*

We shall see that in our analysis of the existence, uniqueness, and stability properties of invariant distributions, inessential states play no significant role. This motivates the following:

Definition 4.1 A Markov chain with a transition probability matrix **p** is *irreducible* if it has one class of essential states and no inessential state.

In other words, for an irreducible Markov chain, for every pair of states (i, j) there exists a positive integer $n = n(i, j)$ such that $p_{ij}^{(n)} > 0$.

We turn to periodic behavior within an essential class. If $i \to i$, then the *period* of i is the greatest common divisor of the integers in the set $\{n \geq 1 : p_{ii}^{(n)} > 0\}$. If d_i is the period of i, then $p_{ii}^{(n)} = 0$ whenever n is *not* a multiple of d_i. It is important to note that if $i \leftrightarrow j$, then i and j possess the same period (i.e., "period" is constant on each $\mathcal{E}_i$) (see Complements and Details). Also, let $i \in \mathcal{E}$ have a period $d = d_i$, then for each $j \in \mathcal{E}_i$ (in Proposition 4.1), there exists a unique integer r_j, $0 \leq r_j \leq d - 1$ such that $p_{ij}^{(n)} > 0$ implies $n = r_j$ (mod d), i.e., $n = r_j$ or $n = sd + r_j$ for some integer $s \geq 1$.

It is certainly true that if $p_{ii} > 0$ for an essential state, then the state i has period 1. However, as shown below by Example 4.1, it is *not* true that the period of an essential state i can be computed as $\min\{n \geq 1 : p_{ii}^{(n)} > 0\}$.

Definition 4.2 An irreducible Markov chain is *aperiodic* if, for some state i, $\{n \geq 1 : p_{ii}^{(n)} > 0\}$ has 1 as the greatest common divisor (g.c.d.). If the g.c.d. is $d > 1$, then the chain is *periodic with period d*.

Example 4.1 Consider a Markov chain with four states and a transition probability matrix

$$\mathbf{p} = \begin{bmatrix} 0 & 1 & 0 & 0 \\ 0 & 0 & 1 & 0 \\ 0 & 1/2 & 0 & 1/2 \\ 1 & 0 & 0 & 0 \end{bmatrix}.$$

Note that all states communicate with each other, and their common period is 2. Although $\min\{n \geq 1 : p_{11}^{(n)} > 0\} = 4$. ∎

A set C of a Markov chain is said to be *closed*, if "$x \in C$, $x \to y$" implies $y \in C$. Observe that the restriction of a Markov chain on a closed subset C (i.e., with $p(x, y)$ restricted to $x, y \in C$) is a Markov chain. If, in addition, C is a class of communicating states then the Markov chain on C is irreducible. For a further decomposition of states, if the irreducible chain on S is periodic of period $d > 1$, then the state space consists of a single communicating class of essential states, which may be divided into d disjoint *cyclic classes* $C_0, C_1, \ldots, C_{d-1}$ such that if the initial state X_0 is in C_j then the chain cyclically moves from $C_j \to C_{j+1} \to \cdots \to C_{j+d-1} \to C_j$ ($X_1 \in C_{j+1}$, $X_2 \in C_{j+2}, \ldots, X_{d-1} \in C_{j+d-1}, X_d \in C_j$) and, in general, $X_r \in C_{j+r}$, where

$C_m = C_{m \bmod(d)}$ for all $m = 0, 1, 2, \ldots$. In particular, if the initial state is in the class C_j, then the chain cyclically moves back to C_j in d steps. Thus if one views $\mathbf{p}^d$ as a one-step transition probability matrix on S, then the sets C_j become (disjoint) closed sets ($j = 0, 1, \ldots, d-1$) and the Markov chain restricted to C_j is irreducible and aperiodic (i.e., has period one).

2.5 Convergence to Steady States for Markov Processes on Finite State Spaces

As will be demonstrated in this section, if the state space is finite, a complete analysis of the limiting behavior of $\mathbf{p}^n$, as $n \to \infty$, may be carried out by elementary methods that also provide rates of convergence to the unique invariant distribution. Although, later, the asymptotic behavior of general Markov chains is analyzed in detail, including the *law of large numbers* and the *central limit theorem* for Markov chains that admit unique invariant distributions, the methods of the present section are also suited for applications to certain more general state spaces (e.g., closed and bounded intervals).

First, we consider what happens to the n-step transition law if all states communicate with each other in one step. In what follows, for a finite set A, $(\#A)$ denotes the number of elements in A.

Theorem 5.1 *Suppose S is finite and $p_{ij} > 0$ for all i, j. Then there exists a unique probability distribution $\pi = \{\pi_j : j \in S\}$ such that*

$$\sum_i \pi_i p_{ij} = \pi_j \quad \text{for all } j \in S \tag{5.1}$$

and

$$\left| p_{ij}^{(n)} - \pi_j \right| \leq (1 - (\#S)\chi)^n \quad \text{for all } i, j \in S, \quad n \geq 1, \tag{5.2}$$

where $\chi = \min\{p_{ij} : i, j \in S\}$ and $(\#S)$ is the number of elements in S. Also, $\pi_j \geq \chi$ for all $j \in S$ and $\chi \leq 1/(\#S)$.

Proof. Let $M_j^{(n)}$, $m_j^{(n)}$ denote the maximum and the minimum, respectively, of the elements $\{p_{ij}^{(n)} : i \in S\}$ of the jth column of $\mathbf{p}^n$. Since

$p_{ij} \geqslant \chi$ and $p_{ij} = 1 - \sum_{k \neq j} p_{ik} \leqslant 1 - [(\#S) - 1]\chi$ for all i, one has

$$M_j^{(1)} - m_j^{(1)} \leqslant [1 - ((\#S) - 1)\chi] - \chi = 1 - (\#S)\chi. \qquad (5.3)$$

Fix two states i, i' arbitrarily. Let $J = \{j \in S: p_{ij} > p_{i'j}\}$, $J' = \{j \in S: p_{ij} \leqslant p_{i'j}\}$. Then,

$$0 = 1 - 1 = \sum_j (p_{ij} - p_{i'j}) = \sum_{j \in J'} (p_{ij} - p_{i'j}) + \sum_{j \in J} (p_{ij} - p_{i'j}),$$

so that

$$\sum_{j \in J'} (p_{ij} - p_{i'j}) = - \sum_{j \in J} (p_{ij} - p_{i'j}), \qquad (5.4)$$

$$\begin{aligned}
\sum_{j \in J} (p_{ij} - p_{i'j}) &= \sum_{j \in J} p_{ij} - \sum_{j \in J} p_{i'j} \\
&= 1 - \sum_{j \in J'} p_{ij} - \sum_{j \in J} p_{i'j} \leqslant 1 - (\#J')\chi - (\#J)\chi \\
&= 1 - (\#S)\chi. \qquad (5.5)
\end{aligned}$$

Therefore,

$$\begin{aligned}
p_{ij}^{(n+1)} - p_{i'j}^{(n+1)} &= \sum_k p_{ik} p_{kj}^{(n)} - \sum_k p_{i'k} p_{kj}^{(n)} = \sum_k (p_{ik} - p_{i'k}) p_{kj}^{(n)} \\
&= \sum_{k \in J} (p_{ik} - p_{i'k}) p_{kj}^{(n)} + \sum_{k \in J'} (p_{ik} - p_{i'k}) p_{kj}^{(n)} \\
&\leqslant \sum_{k \in J} (p_{ik} - p_{i'k}) M_j^{(n)} + \sum_{k \in J'} (p_{ik} - p_{i'k}) m_j^{(n)} \\
&= \sum_{k \in J} (p_{ik} - p_{i'k}) M_j^{(n)} - \sum_{k \in J} (p_{ik} - p_{i'k}) m_j^{(n)} \\
&= \left(M_j^{(n)} - m_j^{(n)} \right) \left(\sum_{k \in J} (p_{ik} - p_{i'k}) \right) \\
&\leqslant (1 - (\#S)\chi)\left(M_j^{(n)} - m_j^{(n)} \right). \qquad (5.6)
\end{aligned}$$

Letting i, i' be such that $p_{ij}^{(n+1)} = M_j^{(n+1)}$, $p_{i'j}^{(n+1)} = m_j^{(n+1)}$, one gets from (5.6),

$$M_j^{(n+1)} - m_j^{(n+1)} \leqslant (1 - (\#S)\chi)\left(M_j^{(n)} - m_j^{(n)} \right). \qquad (5.7)$$

Iteration now yields, using (5.3) as well as (5.7),

$$M_j^{(n)} - m_j^{(n)} \leq (1 - (\#S)\chi)^n \quad \text{for } n \geq 1. \tag{5.8}$$

Now

$$M_j^{(n+1)} = \max_i p_{ij}^{(n+1)} = \max_i \left(\sum_k p_{ik} p_{kj}^{(n)} \right)$$

$$\leq \max_i \left(\sum_k p_{ik} M_j^{(n)} \right) = M_j^{(n)},$$

$$m_j^{(n+1)} = \min_i p_{ij}^{(n+1)} = \min_i \left(\sum_k p_{ik} p_{kj}^{(n)} \right)$$

$$\geq \min_i \left(\sum_k p_{ik} m_j^{(n)} \right) = m_j^{(n)},$$

i.e., $M_j^{(n)}$ is nonincreasing and $m_j^{(n)}$ is nondecreasing in n. Since $M_j^{(n)}, m_j^{(n)}$ are bounded above by 1, (5.7) now implies that both sequences have the same limit, say π_j. Also $\chi \leq m_j^{(1)} \leq m_j^{(n)} \leq \pi_j \leq M_j^{(n)}$ for all n, so that $\pi_j \geq \chi$ for all j and

$$m_j^{(n)} - M_j^{(n)} \leq p_{ij}^{(n)} - \pi_j \leq M_j^{(n)} - m_j^{(n)},$$

which, together with (5.8), implies the desired inequality (5.2).

Finally, taking limits on both sides of the identity

$$p_{ij}^{(n+1)} = \sum_k p_{ik}^{(n)} p_{kj} \tag{5.9}$$

one gets $\pi_j = \sum_k \pi_k p_{kj}$, proving (5.1). Since $\sum_j p_{ij}^{(n)} = 1$, taking limits, as $n \to \infty$, it follows that $\sum_j \pi_j = 1$. To prove uniqueness of the probability distribution π satisfying (5.1), let $\bar{\pi} = \{\bar{\pi}_j : j \in S\}$ be a probability distribution satisfying $\bar{\pi}'\mathbf{p} = \bar{\pi}'$. Then by iteration it follows that

$$\bar{\pi}_j = \sum_i \bar{\pi}_i p_{ij} = (\bar{\pi}'\mathbf{pp})_j = (\bar{\pi}'\mathbf{p}^2)_j = \cdots = (\bar{\pi}'\mathbf{p}^n)_j = \sum_i \bar{\pi}_i p_{ij}^{(n)}. \tag{5.10}$$

Taking limits as $n \to \infty$, one gets $\bar{\pi}_j = \sum_i \bar{\pi}_i \pi_j = \pi_j$.

Thus, $\bar{\pi} = \pi$. ∎

It is also known (see Complements and Details) that for an *irreducible,
aperiodic* Markov chain on a finite state space S there is a positive integer
v such that

$$\chi' := \min_{i,j} p_{ij}^{(v)} > 0. \tag{5.11}$$

Applying Theorem 5.1 with $\mathbf{p}$ replaced by $\mathbf{p}^v$ one gets a probability
$\pi = \{\pi_j : j \in S\}$ such that

$$\max_{i,j} \left| p_{ij}^{(nv)} - \pi_j \right| \leqslant (1 - (\#S)\chi')^n, \quad n = 1, 2, \ldots. \tag{5.12}$$

Now use the relations

$$\left| p_{ij}^{(nv+m)} - \pi_j \right| = \left| \sum_k p_{ik}^{(m)} (p_{kj}^{(nv)} - \pi_j) \right|$$

$$\leqslant \sum_k p_{ik}^{(m)} (1 - (\#S)\chi')^n = (1 - (\#S)\chi')^n$$

$$m \geqslant 1 \tag{5.13}$$

to obtain

$$\left| p_{ij}^{(n)} - \pi_j \right| \leqslant (1 - (\#S)\chi')^{[n/v]}, \quad n = 1, 2, \ldots, \tag{5.14}$$

where $[x]$ is the integer part of x. From here one obtains the following
corollary to Theorem 5.1:

Corollary 5.1 *Let $\mathbf{p}$ be an irreducible and aperiodic transition matrix
on a finite state space S. Then there is a unique probability distribution
π on S such that,*

$$\pi'\mathbf{p} = \pi'. \tag{5.15}$$

Also,

$$\left| p_{ij}^{(n)} - \pi_j \right| \leqslant (1 - \#S\chi')^{[n/v]} \text{ for all } i, j \in S, \ n \geq 1, \tag{5.16}$$

where $\chi' > 0$ and v are as in (5.11).

Notation To indicate the initial distribution, say, μ, we will often write
P_μ for P, when X_0 has a distribution μ. In the case when $\mu = \delta_i$, a point
mass at i, i.e., $\mu(\{i\}) = 1$, we will write P_i instead of P_{δ_i}. Expectations
under P_μ and P_i will be often denoted by E_μ and E_i, respectively.

Example 5.1 (*Bias in the estimation of Markov models under hetero-geneity*)   Consider the problem of estimating the transition probabilities of a two-state Markov model, e.g., with states "employed" (labeled 0) and "unemployed" (labeled 1). If the population is homogeneous, with all individuals having the same transition probabilities of moving from one state to the other in one period, and the states of different individuals evolve independently (or, now in a manner such that the law of large numbers holds for sample observations), then the proportion in a random sample of those individuals moving to state 1 from state 0 in one period among those in state 0 is an *unbiased* estimate of the corresponding true transition probability, and as the sample size increases the estimate converges to the true value (*consistency*).

Suppose now that the population comprises two distinct groups of people, group I and group II, in proportions θ and $1 - \theta$, respectively ($0 < \theta < 1$). Assume that the transition probability matrices for the two groups are

$$\begin{bmatrix} 1-p-\varepsilon & p+\varepsilon \\ q-\varepsilon & 1-q+\varepsilon \end{bmatrix}, \quad \begin{bmatrix} 1-p+\varepsilon & p-\varepsilon \\ q+\varepsilon & 1-q-\varepsilon \end{bmatrix}, \quad \begin{matrix} 0 < p-\varepsilon < p+\varepsilon < 1, \\ 0 - q-\varepsilon < q+\varepsilon < 1. \end{matrix}$$

The steady state distributions of the two groups are

$$\left(\frac{q-\varepsilon}{p+q}, \frac{p+\varepsilon}{p+q} \right) \quad \text{and} \quad \left(\frac{q+\varepsilon}{p+q}, \frac{p-\varepsilon}{p+q} \right).$$

Assume that both groups are in their respective steady states. Then the expected proportion of individuals in the overall population who are in state 0 (employed) at any point of time is

$$\theta \left(\frac{q-\varepsilon}{p+q} \right) + (1-\theta) \left(\frac{q+\varepsilon}{p+q} \right) = \frac{q+\varepsilon(1-2\theta)}{p+q}. \tag{5.17}$$

Now, the expected proportion of those individuals in the population who are in state 0 at a given time *and* who move to state 1 in the next period of time is

$$\theta \left(\frac{q-\varepsilon}{p+q} \right) (p+\varepsilon) + (1-\theta) \left(\frac{q+\varepsilon}{p+q} \right) (p-\varepsilon)$$
$$= \frac{pq - (1-2\theta)(q-p)\varepsilon - \varepsilon^2}{p+q}. \tag{5.18}$$

Dividing (5.18) by (5.17), we get the overall (transition) probability of moving from state 0 to state 1 in one period as

$$\frac{pq - (1 - 2\theta)(q - \varepsilon)\varepsilon}{q + \varepsilon(1 - 2\theta)} = p + \frac{2\theta q\varepsilon - q\varepsilon - \varepsilon^2}{q + \varepsilon(1 - 2\theta)}. \qquad (5.19)$$

Similarly, the overall (transition) probability of moving from state 1 to state 0 in one period is given by

$$q + \frac{-2\theta p\varepsilon + p\varepsilon - \varepsilon^2}{p - \varepsilon(1 - 2\theta)}. \qquad (5.20)$$

If one takes a simple random sample from this population and computes the sample proportion of those who move from state 0 to state 1 in one period, among those who are in state 0 initially, then it would be an unbiased and consistent estimate of

$$\theta(p + \varepsilon) + (1 - \theta)(p - \varepsilon) = p + 2\theta\varepsilon - \varepsilon = p - \varepsilon(1 - 2\theta). \quad (5.21)$$

This follows from the facts that (1) the proportion of those in the sample who belong to group I (respectively, group II) is an unbiased and consistent estimate of θ (respectively, $1 - \theta$) and (2) the proportion of those moving to state 1 from state 0 in one period, among individuals in a sample from group I (respectively, group II), is an unbiased and consistent estimate of the corresponding group probability, namely, $p + \varepsilon$ (respectively, $p - \varepsilon$).

It is straightforward to check (Exercise) that (5.21) *is strictly larger than the transition probability* (5.19) in the overall population, unless $\theta = 0$ or 1 (i.e., unless the population is homogeneous). In other words, the usual sample estimates would *overestimate* the actual transition probability (5.19) in the population. Consequently, the sample estimate would *underestimate* the population transition probability of remaining in state 0 in the next period, among those who are in state 0 initially at a given time.

Similarly, the sample proportion in a simple random sample of those who move to state 0 from state 1 in one period, among those who are in state 1 initially in the sample, is an unbiased and consistent estimate of

$$\theta(q - \varepsilon) + (1 - \theta)(q + \varepsilon) = q + \varepsilon(1 - 2\theta), \qquad (5.22)$$

but it is strictly larger than the population transition probability (5.20) it aims to estimate (unless $\theta = 0$ or 1). Consequently, the population transition probability of remaining in state 1 in the next period, starting in state 1, is *underestimated* by the usual sample estimate, if the population is heterogeneous.

Thus, in a heterogeneous population, one may expect (1) a longer period of staying in the same state (employed or unemployed) and (2) a shorter period of moving from one state to another, than the corresponding estimates based on a simple random sample may indicate.

This example is due to Akerlof and Main (1981). ∎

Let us now turn to *periodicity*. Consider an irreducible, periodic Markov chain on a finite state space with period $d > 1$. Then $\mathbf{p}^d$ viewed as a one-step transition probability on S has d (disjoint) closed sets C_j ($j = 0, \ldots, d-1$), and the chain *restricted to C_j is irreducible and aperiodic*. Applying Corollary 5.1 there exists a probability $\pi^{(j)}$ on C_j, which is the unique invariant distribution on C_j (with respect to the (one-step) transition probability $\mathbf{p}^d$), and there exists $\lambda_j > 0$ such that

$$\left| p_{ik}^{(nd)} - \pi_k^{(j)} \right| \le e^{-n\lambda_j} \quad \text{for all } i, \ k \in C_j, \ n \ge 1,$$

$$(\pi^{(j)})' \mathbf{p}^d = (\pi^{(j)})'; \quad (j = 0, 1, \ldots, d-1). \tag{5.23}$$

More generally, identifying $j + r$ with $j + r$ (mod d),

$$p_{ik}^{(nd+r)} = \sum_{k' \in C_{j+r}} p_{ik'}^{(r)} p_{k'k}^{(nd)} = \sum_{k' \in C_{j+r}} p_{ik'}^{(r)} \left(\pi_k^{(j+r)} + O(e^{-n\lambda_k}) \right)$$

$$= \pi_k^{(j+r)} + O(e^{-n\lambda_k})(i \in C_j, k \in C_{j+r})$$

$$p_{ik}^{(nd+r)} = 0 \quad \text{if } i \in C_j \quad \text{and} \quad k \notin C_{j+r}. \tag{5.24}$$

Now extend $\pi^{(j)}$ to a probability on S by setting $\pi_i^{(j)} = 0$ for all $i \notin C_j$. Then define a probability (vector) π on S by

$$\pi = \frac{1}{d} \sum_{j=0}^{d-1} \pi^{(j)}. \tag{5.25}$$

Theorem 5.2 *Consider an irreducible periodic Markov chain on a finite state space S, with period $d > 1$. Then π defined by (5.25) is the unique invariant probability on S, i.e., $\pi' \mathbf{p} = \pi'$, and one has*

$$p_{ik}^{(nd+r)} = \begin{cases} d\pi_k + O(e^{-\lambda n}) & \text{if } i \in C_j, \ k \in C_{j+r}, \\ 0 & i \in C_j, \ k \notin C_{j+r}, \end{cases} \tag{5.26}$$

where $\lambda = \min\{\lambda_j : j = 0, 1, \ldots, d - 1\}$. *Also, one has*

$$\lim_{N \to \infty} \frac{1}{N} \sum_{m=1}^{N} p_{ik}^{(m)} = \pi_k \quad (i, k \in S). \tag{5.27}$$

Proof. Relations (5.26) follow from (5.23), and the definition of π in (5.25). Summing (5.26) over $r = 0, 1, \ldots, d - 1$, one obtains

$$\sum_{r=0}^{d-1} p_{ik}^{(nd+r)} = d\pi_k + O(e^{-\lambda n}) \quad (i, k \in S), \tag{5.28}$$

so that

$$\frac{1}{nd} \sum_{m=1}^{nd} p_{ik}^{(m)} = \frac{1}{nd} \left(\sum_{r=1}^{d-1} p_{ik}^{(r)} + p_{ik}^{(nd)} + \sum_{m=1}^{n-1} \sum_{r=0}^{d-1} p_{ik}^{(md+r)} \right)$$

$$= O\left(\frac{1}{n}\right) + \frac{1}{nd} \sum_{m=1}^{n-1} \left(d\pi_k + O(e^{-\lambda m}) \right)$$

$$= \pi_k + O\left(\frac{1}{n}\right) \quad (i, k \in S). \tag{5.29}$$

Since every positive integer N can be expressed uniquely as $N = nd + r$, where n is a nonnegative integer and $r = 0, 1, \ldots,$ or $d - 1$, it follows from (5.29) that

$$\frac{1}{N} \sum_{m=1}^{N} p_{ik}^{(m)} = \frac{1}{nd + r} \left[\sum_{m=1}^{nd} p_{ik}^{(m)} + \sum_{1 \le r' \le r} p_{ik}^{(nd+r')} \right]$$

$$= \frac{nd}{nd + r} \left[\pi_k + O\left(\frac{1}{n}\right) \right] = \pi_k + O\left(\frac{1}{N}\right),$$

yielding (5.27).

To prove the invariance of π, multiply both sides of (5.27) by p_{kj} and sum over $k \in S$ to get

$$\lim_{N \to \infty} \frac{1}{N} \sum_{m=1}^{N} p_{ij}^{(m+1)} = \sum_{k \in S} \pi_k p_{kj}. \tag{5.30}$$

But, again by (5.27), the left-hand side also equals π_j. Hence

$$\pi_j = \sum_{k \in S} \pi_k p_{kj} \quad (j \in S), \tag{5.31}$$

proving invariance of π: $\pi' = \pi'\mathbf{p}$. To prove uniqueness of the invariant distribution, assume μ is an invariant distribution. Multiply the left-hand side of (5.27) by μ_i and sum over i to get μ_k. The corresponding sum on the right-hand side is π_k, i.e., $\mu_k = \pi_k (k \in S)$. ∎

From the proof of Theorem 5.2, the following useful general result may be derived (Exercise):

Theorem 5.3 *(a) Suppose there exists a state i of a Markov chain and a probability $\pi = (\pi_j : j \in S)$ such that*

$$\lim_{n \to \infty} \frac{1}{n} \sum_{m=1}^{n} p_{ij}^{(m)} = \pi_j \quad \text{for all } j \in S. \tag{5.32}$$

Then π is an invariant distribution of the Markov chain.
(b) If π is a probability such that (5.32) holds for all i, then π is the unique invariant distribution.

Example 5.2 Consider a Markov chain on $S = \{1, 2, \ldots, 7\}$ having the transition probability matrix

$$\mathbf{p} = \begin{bmatrix} \frac{1}{3} & 0 & 0 & 0 & \frac{2}{3} & 0 & 0 \\ 0 & 0 & 0 & \frac{1}{3} & 0 & 0 & \frac{2}{3} \\ \frac{1}{6} & \frac{1}{6} & \frac{1}{6} & \frac{1}{6} & \frac{1}{6} & \frac{1}{6} & 0 \\ 0 & \frac{1}{2} & 0 & 0 & 0 & \frac{1}{2} & 0 \\ \frac{2}{5} & 0 & 0 & 0 & \frac{3}{5} & 0 & 0 \\ 0 & 0 & 0 & \frac{5}{6} & 0 & 0 & \frac{1}{6} \\ 0 & \frac{1}{4} & 0 & 0 & 0 & \frac{3}{4} & 0 \end{bmatrix}$$

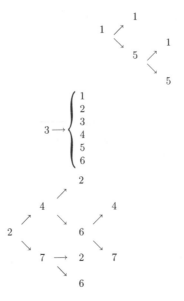

$1 \nrightarrow 3$.

The schematic diagram above shows that 3 is the only inessential state, so that $\mathcal{E} = \{1, 2, 4, 5, 6, 7\}$. There are two closed sets of communicating states, $\mathcal{E}_1 = \{1, 5\}$ and $\mathcal{E}_2 = \{2, 4, 6, 7\}$ of which $\mathcal{E}_1$ is aperiodic and $\mathcal{E}_2$ is periodic of period 2. When the chain is restricted to $\mathcal{E}_1$, the Markov chain on $\mathcal{E}_1$ has the transition probability matrix

$$
\mathbf{p}^{(1)} = \begin{array}{c} \\ 1 \\ 5 \end{array} \begin{array}{cc} 1 & 5 \\ \left[\begin{array}{cc} \frac{1}{3} & \frac{2}{3} \\ \frac{2}{5} & \frac{3}{5} \end{array} \right], \end{array}
$$

which has the unique invariant probability $\tilde{\boldsymbol{\pi}} = (\tilde{\pi}_1 = \frac{3}{8}, \tilde{\pi}_5 = \frac{5}{8})'$. For the Markov chain on $\mathcal{E}_2$ one has the transition probability matrix $\mathbf{p}^{(2)}$ given by

$$
\mathbf{p}^{(2)} = \begin{array}{c} \\ 2 \\ 4 \\ 6 \\ 7 \end{array} \begin{array}{c} 2 \quad 4 \quad 6 \quad 7 \\ \left[\begin{array}{cccc} 0 & \frac{1}{3} & 0 & \frac{2}{3} \\ \frac{1}{2} & 0 & \frac{1}{2} & 0 \\ 0 & \frac{5}{6} & 0 & \frac{1}{6} \\ \frac{1}{4} & 0 & \frac{3}{4} & 0 \end{array} \right] \end{array} \quad (\mathbf{p}^{(2)})^2 = \begin{array}{c} \\ 2 \\ 4 \\ 6 \\ 7 \end{array} \begin{array}{c} 2 \quad 4 \quad 6 \quad 7 \\ \left[\begin{array}{cccc} \frac{1}{3} & 0 & \frac{2}{3} & 0 \\ 0 & \frac{7}{12} & 0 & \frac{5}{12} \\ \frac{11}{24} & 0 & \frac{13}{24} & 0 \\ 0 & \frac{17}{24} & 0 & \frac{7}{24} \end{array} \right]. \end{array}
$$

The two cyclic classes are $C_0 = \{2, 6\}$ and $C_1 = \{4, 7\}$,

$$
\left(\mathbf{p}_{C_0}^{(2)}\right)^2 = \begin{array}{c} \\ 2 \\ 6 \end{array} \overset{\displaystyle 2 \qquad 6}{\begin{bmatrix} \frac{1}{3} & \frac{2}{3} \\ \frac{11}{24} & \frac{13}{24} \end{bmatrix}}, \quad \left(\mathbf{p}_{C_1}^{(2)}\right)^2 = \begin{array}{c} \\ 4 \\ 7 \end{array} \overset{\displaystyle 4 \qquad 7}{\begin{bmatrix} \frac{7}{12} & \frac{5}{12} \\ \frac{17}{24} & \frac{7}{24} \end{bmatrix}}.
$$

It follows by a simple calculation that the invariant distributions for $(\mathbf{p}^{(2)})^2$ on C_0 and C_1 are, respectively,

$$
\boldsymbol{\pi}^{(0)} = \left(\pi_2^{(0)} = \frac{11}{27}, \pi_6^{(0)} = \frac{16}{27} \right)', \quad \boldsymbol{\pi}^{(1)} = \left(\pi_4^{(1)} = \frac{17}{27}, \pi_7^{(1)} = \frac{10}{27} \right)'.
$$

Therefore, the unique invariant distribution for the Markov chain on $\mathcal{E}_2 = \{2, 4, 6, 7\}$ is given by

$$
\begin{aligned}
\boldsymbol{\pi} &= \frac{1}{2} \left(\frac{11}{27}, 0, \frac{16}{27}, 0 \right)' + \frac{1}{2} \left(0, \frac{17}{27}, 0, \frac{10}{27} \right)' \\
&= \left(\frac{11}{54}, \frac{17}{54}, \frac{16}{54}, \frac{10}{54} \right)' = (\pi_2, \pi_4, \pi_6, \pi_7)'. \qquad \blacksquare
\end{aligned}
$$

2.6 Stopping Times and the Strong Markov Property of Markov Chains

The Markov property defined in Section 2.2 has two important and stronger versions for Markov chains. The Markov property implies that the conditional distribution of $(X_n, X_{n+1}, \ldots, X_{n+m})$, given $\{X_0, X_1, \ldots, X_n\}$, depends only on X_n. Here we have taken $(X_n, X_{n+1}, \ldots, X_{n+m})$ in place of $(X_{n+1}, \ldots, X_{n+m})$ in the original Definition 3.1.

$$
\begin{aligned}
P(X_n = j_0, &X_{n+1} = j_1, \ldots, X_{n+m} = j_m \mid X_0 = i_0, X_1 = i_1, \ldots, X_n = i_n) \\
&= P(X_n = j_0, X_{n+1} = j_1, \ldots, X_{n+m} = j_m \mid X_n = i_n) \\
&= P_{i_n}(X_0 = j_0, X_1 = j_1, \ldots, X_m = j_m), \qquad (6.1)
\end{aligned}
$$

all three terms being zero if $j_0 \neq i_n$, and equal to $P_{i_n}(X_1 = j_1, \ldots, X_m = j_m)$ if $j_0 = i_n$.

To state the first stronger version of the Markov property, we need a definition.

Definition 6.1 For each $n \geq 0$, the process $X_n^+ = \{X_{n+m} : m = 0, 1, \ldots\}$ is called the *after-n process*.

The next result says that the conditional distributions of X_n^+, given $\{X_0, X_1, \ldots, X_n\}$, depend only on X_n and is in fact P_{X_n}, i.e., the distribution of a Markov chain with transition probability matrix **p** and initial state X_n.

Theorem 6.1 *The conditional distribution of X_n^+, given $\{X_0, X_1, \ldots, X_n\}$, is P_{X_n}, i.e., for every (measurable) set F in the infinite product space S^∞, one has*

$$P(X_n^+ \in F \mid X_0 = i_0, X_1 = i_1, \ldots, X_n = i_n) = P_{i_n}((X_0, X_1, \ldots,) \in F). \tag{6.2}$$

Proof. When F is finite-dimensional, i.e., of the form $F = \{\omega = (j_0, j_1, \ldots) \in S^\infty : (j_0, j_1, \ldots, j_m) \in G\} = \{(X_0, X_1, \ldots, X_m) \in G\}$ for some $G \subset S^{m+1}$, (6.2) is just the Markov property in the form (6.1). Since this is true for all m, and since the Kolmogorov sigmafield on the product space S^∞ is generated by the class of all finite-dimensional sets, (6.2) holds (see Complements and Details). ∎

To state the second and the most important strengthening of the Markov property, we need to define a class of random times τ such that the Markov property holds, given the past up to these random times.

Definition 6.2 A random variable τ with values in $\mathbb{Z}_+ \cup \{\infty\} = \{0, 1, 2, \ldots, \} \cup \{\infty\}$ is a *stopping time* if the event $\{\tau = m\}$ is determined by $\{X_0, X_1, \ldots, X_m\}$ for every $m \in \mathbb{Z}_+$.

If τ is a stopping time and $m' < m$, then the event $\{\tau = m'\}$ is determined by $\{X_0, X_1, \ldots, X_{m'}\}$ and, therefore, by $\{X_0, X_1, \ldots, X_m\}$. Hence an equivalent definition of a stopping time τ is $\{\tau \leq m\}$ is determined by $\{X_0, X_1, .., X_m\}$ for every $m \in \mathbb{Z}_+$.

Informally, whether or not the Markov chain stops at a stopping time τ depends only on $\{X_0, X_1, \ldots, X_\tau\}$, provided $\tau < \infty$.

Example 6.1 Let A be an arbitrary nonempty proper subset of the (countable) state space S. The *hitting time* η_A of a chain $\{X_n : n = 0, 1, \ldots\}$ is defined by

$$\eta_A = \inf\{n \geq 1 : X_n \in A\}, \tag{6.3}$$

the infinum being infinity (∞) if the set within $\{\ \ \}$ above is empty, i.e., if the process never hits A after time 0. This random time takes values in

$\{1, 2, \ldots\} \cup \{\infty\}$, and

$$\{\eta_A = 1\} = \{X_1 \in A\}$$

$$\{\eta_A = m\} = \left(\bigcap_{n=1}^{m-1}\{X_n \notin A\}\right) \cap \{X_m \in A\} \quad (m \geq 2), \quad (6.4)$$

proving that it is a stopping time. On the other hand, the *last time* γ_A, say, that the process hits A, is, in general, not a stopping time, since the event $\{\gamma_A = m\}$ equals $\{X_m \in A\} \cap_{n=m+1}^{\infty}\{X_n \notin A\}$. ∎

If, in definition (6.3) one allows $n = 0$ then the stopping time is called the *first passage time to A*, denoted by T_A:

$$T_A = \inf\{n \geq 0 : X_n \in A\}. \quad (6.5)$$

Thus if $X_0 \in A$, $T_A = 0$. As in the case η_A, it is easy to check that T_A is a stopping time.

For singleton sets $A = \{i\}$ one writes η_i and T_i instead of $\eta_{\{i\}}$ and $T_{\{i\}}$.

Definition 6.3 The *rth return time to i* of a Markov chain is defined by

$$\eta_i^{(0)} = 0, \quad \eta_i^{(r)} := \inf\{n > \eta_i^{(r-1)} : X_n = i\} \quad (r \geq 1). \quad (6.6)$$

Exercise 6.1 Check that $\eta_i^{(r)}$ $(r \geq 1)$ are stopping times. ∎

For a stopping time τ denote by X_τ^+ the *after-τ process* $\{X_{\tau+n} : n = 0, 1, \ldots\}$ on the set $\{\tau < \infty\}$ (if $\tau = \infty$, X_τ is not defined). Also write $(X_\tau^+)_n = X_{\tau+n}$ for the state of the after - τ process at time n. The strong Markov property defined below is an extension of (6.2) to stopping times.

Definition 6.4 A Markov chain $\{X_n : n = 0, 1, \ldots\}$ is to have the *strong Markov property* if, for every stopping time τ, the conditional distribution of the after - τ process X_τ^+, given the past up to time τ, is P_{X_τ} on the set $\{\tau < \infty\}$:

$$P(X_\tau^+ \in F, \tau = m \mid X_0, X_1, \ldots, X_m) = P_{X_m}(F)\mathbf{1}_{\{\tau=m\}}, \quad (6.7)$$

for every F belonging to the Kolmogorov sigmafield $\mathcal{F} = \mathcal{S}^{*\infty}$ on S^∞ and every $m \in \mathbb{Z}_+$.

Theorem 6.2 *Every Markov chain has the strong Markov property.*

Proof. If τ is a stopping time, then for all $F \in \mathcal{F}$ and every $m \geq 0$,

$$P(X_\tau^+ \in F, \tau = m \mid X_0 = i_0, X_i = i_1, \ldots, X_m = i_m)$$
$$= P(X_m^+ \in F, \tau = m \mid X_0 = i_0, X_1 = i_1, \ldots, X_m = i_m)$$
$$= P(X_m^+ \in F \mid X_0 = i_0, X_1 = i_1, \ldots, X_m = i_m)I_{\{\tau = m\}}.$$

Since the event $\{\tau = m\}$ is determined by $\{X_0, X_1, \ldots, X_m\}$, the last equality trivially holds if $\{X_0 = i_0, X_1 = i_1, \ldots, X_m = i_m\}$ implies $\{\tau \neq m\}$, both sides being zero. If $\{X_0 = i_0, X_1 = i_1, \ldots, X_m = i_m\}$ implies $\{\tau = m\}$, then both sides of the last equality are $P(X_m^+ \in F \mid X_0 = i_0, \ldots, X_m = i_m) = P_{i_m}((X_0, X_1, \ldots,) \in F) \equiv P_{i_m}(F)$. Hence (6.7) holds. ∎

Example 6.2 (*First passage times for the simple random walk*) Let $\{Z_n : n = 1, 2, \ldots\}$ be i.i.d. Bernoulli with $P(Z_n = +1) = p$, $P(Z_n = -1) = q$ where $q = 1 - p$ $(0 < p < 1)$. A *simple random walk starting at x* is the process on $S = \mathbb{Z}$ defined by $X_0 = x$, $X_n = x + Z_1 + \cdots + Z_n (n \geq 1)$, with $x \in \mathbb{Z}$. Since $\{X_n : n \geq 0\}$ is a *process with independent increments* it has the Markov property. Indeed its transition probabilities are given by $p_{i,i+1} = p$, $p_{i,i-1} = q$, $p_{ij} = 0$ if $|i - j| > 1$ or $i = j$. We first compute the probability $\varphi(x)$ that *the random walk starting at x reaches d before c*, where $c < d$ are given integers, and $c \leq x \leq d$, i.e., $\varphi(x) = P(T_d < T_c \mid X_0 = x)$. Now if $c < x < d$, then

$$P(\{X_n : n \geq 0\} \text{ reaches } d \text{ before } c \text{ and } Z_1 = +1 \mid X_0 = x)$$
$$= P(\{X_n' := x + 1 + Z_2 + \cdots + Z_{n+1}(n \geq 1), X_0' = x + 1\}$$
$$\text{reaches } d \text{ before } c \text{ and } Z_1 = +1) = p\varphi(x + 1),$$

since, $\{X_n' : n \geq 0\}$ is a simple random walk starting at $x + 1$, and it is independent of Z_1. Similarly

$$P(\{X_n : n \geq 0\} \text{ reaches } d \text{ before } c \text{ and } Z_1 = -1 \mid X_0 = x)$$
$$= q\varphi(x - 1).$$

Adding the two relations we get

$$\varphi(x) = p\varphi(x + 1) + q\varphi(x - 1),$$

or

$$\varphi(x + 1) - \varphi(x) = \frac{q}{p}(\varphi(x) - \varphi(x - 1)), \quad c < x < d. \quad (6.8)$$

In the case $p = \frac{1}{2} = q$, the increments of φ are constant, so that φ is linear. Since $\varphi(c) = 0$, $\varphi(d) = 1$, one has in this case

$$\varphi(x) = \frac{x - c}{d - c}, \quad c \le x \le d. \qquad (6.9)$$

If $p \neq \frac{1}{2}$, then an iteration of (6.8) yields

$$\varphi(x + 1) - \varphi(x) = \left(\frac{q}{p}\right)^{x-c} (\varphi(c + 1) - \varphi(c)) = \left(\frac{q}{p}\right)^{x-c} \varphi(c + 1). \qquad (6.10)$$

Summing both sides over $x = c, c + 1, \ldots, y - 1$, one gets

$$\varphi(y) = \varphi(c + 1) \sum_{x=c}^{y-1} \left(\frac{q}{p}\right)^{x-c} = \varphi(c + 1) \left[\frac{1 - (q/p)^{y-c}}{1 - (q/p)}\right]. \qquad (6.11)$$

To compute $\varphi(c + 1)$, let $y = d$ in (6.11) and use $\varphi(d) = 1$, to get

$$\varphi(c + 1) = \frac{1 - (q/p)}{1 - (q/p)^{d-c}}.$$

Substituting this in (6.11) we have

$$\varphi(y) = \frac{1 - (q/p)^{y-c}}{1 - (q/p)^{d-c}}, \quad c \le y \le d. \qquad (6.12)$$

To compute the probability $\psi(x) = P(T_c < T_d \mid X_0 = x)$, one similarity solves (6.8) (with ψ in place of φ), but with boundary conditions $\psi(c) = 1$, $\psi(d) = 0$, to get

$$\psi(y) = \frac{d - y}{d - c} \ (c \le y \le d), \qquad\qquad \text{if } p = \frac{1}{2},$$

$$\psi(y) = \frac{1 - (p/q)^{d-y}}{1 - (p/q)^{d-c}} \ (c \ge y \le d), \quad \text{if } p \neq \frac{1}{2}. \qquad (6.13)$$

Exercise 6.2 Alternatively, one may derive (6.13) from (6.9), (6.12) by "symmetry," noting that $\{-X_n : n = 0, 1, \ldots\}$ is a simple random walk starting at $-x$, and with the increment process $\{-Z_n : n \ge 1\}$. ∎

Note that $\varphi(y) + \psi(y) = 1$ for $c \le y \le d$. This implies that the simple random walk will reach the boundary $\{c, d\}$ with probability 1, whatever be $c < x < d$. Letting $c \downarrow -\infty$ in (6.9) and (6.12), one arrives at

$$\rho_{yd} := P(\eta_d < \infty \mid X_0 = y) = P(T_d < \infty \mid X_0 = y)$$
$$= \begin{cases} 1 & \text{if } p \geq \frac{1}{2}, \\ (p/q)^{d-y} & \text{if } p < \frac{1}{2}; \ (y < d) \end{cases} \tag{6.14}$$

Similarly, by letting $d \uparrow \infty$ in (6.12), one gets

$$\rho_{yc} := P(\eta_c < \infty \mid X_0 = y) = P(T_c < \infty \mid X_0 = y)$$
$$= \begin{cases} 1 & \text{if } p \leq \frac{1}{2}, \\ (q/p)^{y-c} & \text{if } p > \frac{1}{2}; \ (y > c) \end{cases} \tag{6.15}$$

Note that $T_y = \eta_y$ if $X_0 \equiv x \neq y$ (see (6.3) and (6.5)). It now follows from (6.14) and (6.15), with $d = y + 1$ and $c = y - 1$, respectively, and intersecting $\{\eta_y < \infty\}$ with $Z_1 = +1$ and $Z_1 = -1$, that

$$\rho_{yy} := P(\eta_y < \infty \mid X_0 = y) = p\rho_{y+1,y} + q\rho_{y-1,y} = \begin{cases} 1 & \text{if } p = \frac{1}{2}, \\ 2q & \text{if } p > \frac{1}{2}, \\ 2p & \text{if } p < \frac{1}{2}, \end{cases}$$

or in all cases

$$\rho_{yy} = 2\min\{p, q\}. \tag{6.16}$$

Next, use the strong Markov property with the stopping time $\eta_y^{(r)}$ (see (6.6)) to get

$$P\left(\eta_y^{(r+1)} < \infty \mid X_0 = x\right)$$
$$= E_x\left[P\left(\eta_y^{(r)} < \infty \text{ and } \left(X_{\eta_y^{(r)}}^+\right)_n\right.\right.$$
$$= y \text{ for some } n \geq 1 \mid \text{given the past up to } \eta_y^{(r)}\right]$$
$$= E_x\left[P_{X_{\eta_y^{(r)}}}(X_n = y \text{ for some } n \geq 1)\mathbf{1}\left[\eta_y^{(r)} < \infty\right]\right]$$
$$= E_x\left[P_y(X_n = y \text{ for some } n \geq 1)\mathbf{1}\left[\eta_y^{(r)} < \infty\right]\right]$$
$$= \rho_{yy}E_x\left(I\left[\eta_y^{(r)} < \infty\right]\right) = \rho_{yy}P\left(\eta_y^{(r)} < \infty \mid X_0 = x\right). \tag{6.17}$$

Iterating (6.17) one arrives at

$$P\left(\eta_y^{(r+1)} < \infty \mid X_0 = x\right) = \rho_{yy}^r P(\eta_y < \infty \mid X_0 = x)$$
$$= \rho_{yy}^r \rho_{xy} \quad (x, y \in S). \tag{6.18}$$

Letting $r \uparrow \infty$ and using (6.16) we get

$$P(X_n = y \text{ i.o.} \mid X_0 = x) = \begin{cases} 1 & \text{if } p = \frac{1}{2} \\ 0 & \text{if } p \neq \frac{1}{2} \end{cases}, \qquad (6.19)$$

where "*i.o.*" stands for *infinitely often*. One expresses (6.19) as follows *the simple random walk on $\mathbb{Z}$ is recurrent* (i.e., $\rho_{xx} = 1$ for all x) *if it is symmetric and transient* (i.e., $\rho_{xx} < 1$ for all x) *otherwise*. Note that transience, in case $p \neq \frac{1}{2}$, also follows from the SLLN (see Appendix):

$$\frac{X_n}{n} = \frac{x}{n} + \frac{Z_1 + \cdots + Z_n}{n} \to E Z_1 \neq 0, \quad \text{as } n \to \infty, \qquad (6.20)$$

with probability 1. ∎

Example 6.3 One may modify the simple random walk of the preceding example slightly by letting the step sizes Z_j have the distribution $P(Z_j = -1) = q$, $P(Z_j = 0) = r$, $P(Z_j = +1) = p$, with $p + q + r = 1$, $p > 0$, $q > 0$. Then the transition probabilities are

$$p_{i,i-1} = q, \quad p_{i,i} = r, \quad p_{i,i+1} = p \, (i \in \mathbb{Z}). \qquad (6.21)$$

Then $\varphi(x) := P(T_d < T_c \mid X_0 = x)$, $c \leq x \leq d$, satisfies

$$\varphi(x) = p\varphi(x + 1) + q\varphi(x - 1) + r\varphi(x), \quad c < x < d,$$

or

$$(p + q)\varphi(x) = p\varphi(x + 1) + q\varphi(x - 1).$$

From this we get the same relation as (6.8) and, in the symmetric case $p = q$, (6.9) holds. As a consequence all the relations (6.10)–(6.19) hold, provided one writes $p = q$ instead of $p = \frac{1}{2}$, and $p \neq q$ instead of $p \neq \frac{1}{2}$. Thus one has

$$\varphi(x) = \frac{x - c}{d - x}, \qquad\qquad c \leq x \leq d, \text{ if } p = q;$$

$$\varphi(x) = \frac{1 - (q/p)^{x-c}}{1 - (q/p)^{d-c}}, \qquad c \leq x \leq d, \text{ if } p \neq q. \qquad (6.22)$$

Then $\psi(x) := P(T_c < T_d \mid X_0 = x)$, $c \leq x \leq d$, is given by $\psi(x) = 1 - \varphi(x)$, as in (6.13). Based on these expressions for $\varphi(x)$ and $\psi(x)$, one derives (see (6.14) and (6.15))

$$\rho_{yd} = \begin{cases} 1 & \text{if } p \geq q \\ (p/q)^{d-y} & \text{if } p < q \end{cases} \quad (y < d);$$

$$\rho_{yc} = \begin{cases} 1 & \text{if } p \leq q \\ (q/p)^{y-c} & \text{if } p > q \end{cases} \quad (y > c). \tag{6.23}$$

From (6.23) one gets

$$\rho_{yy} = p\rho_{y+1,y} + q\rho_{y-1,y} + r,$$

or

$$= \begin{cases} 1 & \text{if } p = q, \\ 2\min\{p,q\} + r & \text{if } p \neq q \end{cases}. \tag{6.24}$$

Thus the chain is transient in the asymmetric case ($p \neq q$) and recurrent in the symmetric case ($p = q$).

We now turn to the computation of $m(x) := E(T_{\{c,d\}} \mid X_0 = x)$ for $c \leq x \leq d$, where $T_{\{c,d\}} = \inf\{n \geq 0 : X_n = c \text{ or } d\}$. By conditioning on $Z_1 = +1, -1,$ or 0, one obtains

$$m(x) = 1 + pm(x+1) + qm(x-1) + rm(x), c < x < d,$$
$$m(c) = 0, m(d) = 0, \tag{6.25}$$

whose solution is given by (see (C8.8) in Complements and Details)

$$m(x) = \begin{cases} \dfrac{(x-c)(d-x)}{p+q}, & \text{if } p = q, \\ \dfrac{d-c}{p-q}\left\{\dfrac{1-(q/p)^{x-c}}{1-(q/p)^{d-c}}\right\} - \dfrac{x-c}{p-q}, & \text{if } p \neq q. \end{cases} \tag{6.26}$$

Note that by letting $d \uparrow \infty$ in (6.26), one has $E_x T_c \equiv E_x \eta_c = \frac{x-c}{q-p} < \infty$ for $x > c$, if $p < q$, and $E_x T_d \equiv E_x \eta_d = \frac{d-x}{p-q} < \infty$ for $x < d$, if $p > q$. $E_x \eta_y = \infty$ in all other cases.   ∎

2.7 Transient and Recurrent Chains

We now introduce some concepts that are critical in understanding the asymptotic behavior of a Markov chain $\{X_n : n = 0, 1, \ldots\}$ with a (*countable*) state space S, and a transition probability matrix **p**. Write, as before,

$$\rho_{xy} = P_x(\eta_y < \infty). \tag{7.1}$$

Definition 7.1 A state $y \in S$ is *recurrent* if $\rho_{yy} = 1$. A state $y \in S$ is *transient* if $\rho_{yy} < 1$.

If $P_y(\eta_y = 1) = 1$ then $p_{yy} = 1$ and y is an *absorbing* state and y is necessarily recurrent.

Let us denote by $\mathbf{1}_y(z)$ the indicator function of the set $\{y\}$ defined by

$$\mathbf{1}_y(z) = \begin{Bmatrix} 1 & \text{if } z = y \\ 0 & \text{if } z \neq y \end{Bmatrix}. \tag{7.2}$$

Denote by $N(y)$ the number of times after time 0 that the chain is in state y. We see that

$$N(y) = \sum_{n=1}^{\infty} \mathbf{1}_y(X_n). \tag{7.3}$$

The event $\{N(y) \geq 1\}$ is the same as the event $\{\eta_y < \infty\}$. Thus,

$$P_x(N(y) \geq 1) = P_x(\eta_y < \infty) = \rho_{xy}. \tag{7.4}$$

One has

$$P_x(N(y) \geq 2) = \rho_{xy}\rho_{yy}. \tag{7.5}$$

For this note that

$$\{N(y) \geq 2\} = \{\eta_y < \infty \text{ and } X_{\eta_y}^+ \text{ returns to } y\}.$$

Hence, using the strong Markov property (see Theorem 6.2)

$$P_x(N(y) \geq 2) = P_x(\eta_y < \infty \text{ and } X_{\eta_y+n} = y \text{ for some } n \geq 1)$$
$$= E_x\left(\mathbf{1}_{[\eta_y<\infty]}P_{X_{\eta_y}}(X_n = y \text{ for some } n \geq 1)\right)$$
$$= E_x\left(\mathbf{1}_{[\eta_y<\infty]}P_y(X_\eta = y \text{ for some } n \geq 1)\right)$$
$$= E_x\left(\mathbf{1}_{[\eta_y<\infty]}\rho_{yy}\right)$$
$$= \rho_{yy}P_x(\eta_y < \infty) = \rho_{yy}\rho_{xy}.$$

Similarly we conclude that

$$P_x(N(y) \geq m) = \rho_{xy}\rho_{yy}^{m-1} \quad \text{for } m \geq 1. \tag{7.6}$$

Since

$$P_x(N(y) = m) = P_x(N(y) \geq m) - P_x(N(y) \geq m + 1), \tag{7.7}$$

it follows that

$$P_x(N(y) = m) = \rho_{xy}\rho_{yy}^{m-1}(1 - \rho_{yy}) \quad \text{for } m \geq 1. \tag{7.8}$$

Also,

$$P_x(N(y) = 0) = 1 - P_x(N(y) \geq 1) = 1 - \rho_{xy}. \tag{7.9}$$

Let us denote by $E_x(\cdot)$ the expectation of random variables defined in terms of a Markov chain starting at x. Then

$$E_x(N(y)) = E_x\left(\sum_{n=1}^{\infty} \mathbf{1}_y(X_n)\right)$$

$$= \sum_{n=1}^{\infty} \left[E_x \mathbf{1}_y(X_n)\right]$$

$$= \sum_{n=1}^{\infty} p_{xy}^{(n)}. \tag{7.10}$$

Write $G(x, y) = E_x(N(y)) = \sum_{n=1}^{\infty} p_{xy}^{(n)}$. Then $G(x, y)$ is the *expected number of visits to y for a Markov chain starting at x, after time 0*.

Theorem 7.1

   (i) *Let y be a transient state. Then, $P_x(N(y) < \infty) = 1$ and $G(x, y) = \frac{\rho_{xy}}{1-\rho_{yy}} < \infty \ \forall x \in S$.*

   (ii) *Let y be a recurrent state. Then, $P_y(N(y) = \infty) = 1$ and $G(y, y) = \infty$. Also $P_x(N(y) = \infty) = P_x(\eta_y < \infty) = \rho_{xy} \ \forall x \in S$. If $\rho_{xy} = 0$, then $G(x, y) = 0$. If $\rho_{xy} > 0$ then $G(x, y) = \infty$.*

Proof.

   (i) Let y be a transient state. Since $0 < \rho_{yy} < 1$, we see by (7.6) that

$$P_x(N(y) = \infty) = \lim_{m \to \infty} P_x(N(y) \ge m) = \lim_{m \to \infty} \rho_{xy}\rho_{yy}^{m-1} = 0.$$

Now by (7.8),

$$G(x, y) = E_x(N(y))$$

$$= \sum_{m=1}^{\infty} P_x(N(y) = m)$$

$$= \sum_{m=1}^{\infty} \rho_{xy}\rho_{yy}^{m-1}(1 - \rho_{yy}) = \rho_{xy}(1 - \rho_{yy})\sum_{m=1}^{\infty} m\rho_{yy}^{m-1}.$$

Now, differentiating both sides of the identity

$$(1 - t)^{-1} \equiv \sum_{m=0}^{\infty} t^m \quad (|t| < 1),$$

one arrives at

$$\sum_{m=1}^{\infty} m t^{m-1} = \frac{1}{(1-t)^2}, \quad G(x, y) = \frac{\rho_{xy}}{1 - \rho_{yy}} < \infty, \quad (\rho_{yy} < 1).$$

(ii) Let y be a recurrent state. Then $\rho_{yy} = 1$ and from (7.6) one gets

$$P_x(N(y) = \infty) = \lim_{m \to \infty} P_x(N(y) \geq m) = \rho_{xy}.$$

In particular, $P_y(N(y) = \infty) = 1$. If a nonnegative random variable has a positive probability of being infinite, then its expectation is infinite. Thus

$$G(y, y) = E_y(N(y)) = \infty.$$

If $\rho_{xy} = 0$, then $p_{xy}^{(n)} = 0$ for all $n \geq 1$. Thus $G(x, y) = 0$. If $\rho_{xy} > 0$, then again, $P_x(N(y) = \infty) = \rho_{xy} > 0$. Hence, $G(x, y) = E_x(N(y)) = \infty$. ∎

Next, we show that recurrence is a *class property*:

Proposition 7.1 *Let x be a recurrent state and suppose x leads to y, i.e., $x \to y$, then y is recurrent and $\rho_{xy} = \rho_{yx} = 1$.*

Proof. Assume that $x \neq y$, otherwise there is nothing to prove. Since $P_x(\eta_y < \infty) = \rho_{xy} > 0$ it must be that $P_x(\eta_y = n) > 0$ for some integer $n \geq 1$. Let n_0 be the smallest such positive integer, i.e., $n_0 = \min(n \geq 1: P_x(\eta_y = n) > 0)$. Then $p_{xy}^{(n_0)} > 0$ and $p_{xy}^{(m)} = 0$ for any m such that $1 \leq m < n_0$. Therefore, we can find states $y_1, \ldots, y_{n_0-1}$ none of them equal to either x or y, such that

$$P_x(X_1 = y_1, \ldots, X_{n_0-1} = y_{n_0-1}, X_{n_0} = y) = p_{xy_1} \cdots p_{y_{n_0-1}y} > 0.$$

We want to prove that $\rho_{yx} = 1$. Suppose that $\rho_{yx} < 1$. Then a Markov chain starting at y has a probability $1 - \rho_{yx}$ of never hitting x. This means that the chain starting at x has the positive probability $\delta = (p_{xy_1} \cdots p_{y_{n_0-1}y})(1 - \rho_{yx})$ of visiting $y_1, \ldots, y_{n_0-1}, y$ successively in the first n_0 periods and never hitting x after time n_0. Hence a Markov chain starting from x never returns to x with probability at least $\delta > 0$, contradicting the assumption that x is recurrent.

Since $\rho_{yx} = 1$, there is a positive integer n_1 such that $p_{yx}^{(n_1)} > 0$. Now,

$$
\begin{aligned}
p_{yy}^{(n_0+n+n_1)} &= P_y(X_{n_0+n+n_1} = y) \\
&\geq P_y(X_{n_1} = x, X_{n_1+n} = x, X_{n_1+n+n_0} = y) \\
&= p_{yx}^{(n_1)} p_{xx}^{(n)} p_{xy}^{(n_0)}.
\end{aligned}
$$

It follows that $G(y, y) \geq p_{yx}^{(n_1)} p_{xy}^{(n_0)} \sum_{n=1}^{\infty} p_{xx}^{(n)} = p_{yx}^{(n_1)} p_{xy}^{(n_0)} G(x, x) = \infty$. Hence y is a recurrent state. Since y is recurrent and $y \to x$, we can similarly prove that $\rho_{xy} = 1$. ∎

In view of Theorem 7.1 and Proposition 7.1, for an *irreducible* Markov chain either *all* states are transient or *all* states are recurrent. In the former case the *chain* is said to be *transient* and in the latter case the *chain* is called *recurrent*.

For studying the long-run behavior of a Markov chain, we note the following:

Proposition 7.2 *Every inessential state of a Markov chain is transient.*

Proof. Let y be inessential. Then there exists $x \neq y$ such that $\rho_{yx} > 0$, but $\rho_{xy} = 0$. Suppose, if possible, y is recurrent. Then, by Proposition 7.1, x is recurrent and $\rho_{yx} = 1$. Consider the Markov chain starting at y. Since $\rho_{yx} = 1, \eta_x := \inf\{n \geq 1: X_n = x\} < \infty$, with probability 1. Since $\rho_{xy} = 0$, the probability that y occurs after time η_x is zero. But this implies that, with probability 1, y occurs only finitely often, contradicting the supposition that y is recurrent. ∎

It follows that *every recurrent state is essential*. On the other hand, essential states need not be recurrent. For example, the simple asymmetric random walk on $\mathbb{Z}$ (with $0 < p < 1$) and the simple symmetric random walk $\mathbb{Z}^k, k \geq 3$, are irreducible (implying that all states are essential) and transient.

Example 7.1 (*Simple symmetric random walk on $\mathbb{Z}^k$*) Let $\mathbf{Z}_n, n \geq 1$, be i.i.d. with $P(\mathbf{Z}_1 = \mathbf{e}_i) = 1/2k = P(\mathbf{Z}_1 = -\mathbf{e}_i)$ for $i = 1, 2, \ldots, k$, where $\mathbf{e}_i (\in \mathbb{Z}^k)$ has 1 as its ith coordinate and 0's elsewhere. For any given $\mathbf{x} \in \mathbb{Z}^k$, the stochastic process $\mathbf{X}_n := \mathbf{x} + \mathbf{Z}_1 + \cdots + \mathbf{Z}_n (n \geq 1)$, $\mathbf{X}_0 = \mathbf{x}$, is called *the simple symmetric random walk on $\mathbb{Z}^k$, starting at $\mathbf{x}$.*

In view of independent increments $\mathbf{Z}_n (n \geq 1)$, this process is a Markov chain on $S = \mathbb{Z}^k$.   ∎

Theorem 7.2 (*Polya*)   *The simple symmetric random walk on $\mathbb{Z}^k$ is recurrent if $k = 1, 2$ and transient if $k \geq 3$.*

Proof.   The case $k = 1$ is proved in Section 2.6. Consider the case $k = 2$. Since every state $\mathbf{y} \in \mathbb{Z}^k$ leads to every other state $\mathbf{z}$ (i.e., the chain is irreducible), it is enough to prove that $\mathbf{0} = (0, 0)$ is a recurrent state, or $G(\mathbf{0}, \mathbf{0}) = \infty$. Since the return times to each state $\mathbf{x}$ are even integers $n = 2, 4, \ldots$ (i.e., the chain is periodic with period 2), one has

$$
\begin{aligned}
G(\mathbf{0}, \mathbf{0}) &= \sum_{m=1}^{\infty} p_{00}^{(2m)} = \sum_{m=1}^{\infty} P(\mathbf{Z}_1 + \mathbf{Z}_2 + \cdots + \mathbf{Z}_{2m} = (0, 0)) \\
&= \sum_{m=1}^{\infty} \sum_{r=0}^{m} \binom{2m}{r} \binom{2m-r}{r} \binom{2m-2r}{m-r} \binom{m-r}{m-r} \left(\frac{1}{4}\right)^{2m} \\
&= \sum_{m=1}^{\infty} \left(\frac{1}{4}\right)^{2m} \binom{2m}{m} \sum_{r=0}^{m} \binom{m}{r}^2,
\end{aligned}
\tag{7.11}
$$

since $\mathbf{Z}_1 + \cdots + \mathbf{Z}_{2m} = (0, 0)$ if and only if the number of $\mathbf{Z}_i's$, which equal $\mathbf{e}_1$, say r, is the same as the number which equal $-\mathbf{e}_1$, and the number of $\mathbf{Z}_i's$ which equal $\mathbf{e}_2$ is the same as the number which equal $-\mathbf{e}_2$, namely, $m - r$. Now

$$
\sum_{r=0}^{m} \binom{m}{r}^2 = \sum_{r=0}^{m} \binom{m}{r} \binom{m}{m-r} = \binom{2m}{m},
\tag{7.12}
$$

since the second sum equals the number of ways m balls can be selected from a box containing m distinct red balls and m distinct white balls. Thus

$$
G(\mathbf{0}, \mathbf{0}) = \sum_{m=1}^{\infty} \left(\frac{1}{4}\right)^{2m} \binom{2m}{m}^2.
\tag{7.13}
$$

By Stirling's formula,

$$
r! = (2\pi)^{1/2} r^{r+1/2} e^{-r} (1 + o(1)),
\tag{7.14}
$$

where $o(1) \to 0$ as $r \to \infty$. Hence

$$\binom{2m}{m} \underset{\cap}{\cup} \frac{(2m)^{2m+1/2}}{m^{2m+1}} = \frac{2^{2m+1/2}}{m^{1/2}}, \quad \binom{2m}{m}^2 \underset{\cap}{\cup} \frac{4^{2m+1/2}}{m}, \quad (7.15)$$

where $c(m) \underset{\cap}{\cup} d(m)$ means there exist numbers $0 < A < B$ independent of m, and a positive integer m_0, such that $A < c(m)/d(m) < B$ for all $m \geq m_0$. Using (7.15) in (7.13), one obtains $G(0, 0) = \infty$, since $\sum_{m=1}^{\infty} 1/m = \infty$. This proves recurrence for the case $k = 2$. For $k \geq 3$, one gets, analogous to (7.11),

$$G(\mathbf{0}, \mathbf{0}) = \sum_{m=1}^{\infty} \sum (2k)^{-2m} \frac{2m!}{(r_1!)^2 \cdots (r_k!)^2}, \quad (7.16)$$

where the inner sum is over all k-tuples of nonnegative integers $r_1, r_2, \ldots, r_k$, whose sum is equal to m.

Write the inner sum as

$$2^{-2m} \binom{2m}{m} \sum p_{k,m}^2(r_1, \ldots, r_k), \quad (7.17)$$

where $p_{k,m}(r_1, \ldots, r_k) := (k^{-m}) m!/(r_1!, \ldots, r_k!)$ is the probability of the ith box receiving r_i balls ($1 \leq i \leq k$) out of m balls distributed at random over k boxes. Let $a_{k,m}$ denote the maximum of these probabilities over all assignments r_i, $\sum_{i=1}^{k} r_i = m$. Then (7.17) is no more than

$$2^{-2m} \binom{2m}{m} a_{k,m}, \quad (7.18)$$

since $p_{k,m}^2 \leq p_{k,m} a_{k,m}$. In the case m is an integer multiple of k, one can show (see Exercise 7.1) that

$$a_{k,m} = (k^{-m}) m!/ \left(\frac{m}{k}!\right)^k. \quad (7.19)$$

For the general case one can show that (Exercise 7.1)

$$a_{k,rk+s} \leq a_{k,rk} \quad \text{for all } s = 1, 2, \ldots, k-1. \quad (7.20)$$

Grouping m in blocks of k consecutive values $rk \leq m < (r+1)k$ ($r = 1, 2, \ldots$), one then has

$$G(\mathbf{0}, \mathbf{0}) \leq \sum_{r=1}^{\infty} 2^{-2rk} \binom{2(r+1)k}{(r+1)k} k a_{k,rk} + \sum_{m=1}^{k-1} \binom{2m}{m} 2^{-2m}. \quad (7.21)$$

Using Stirling's formula (7.14), the infinite sum in (7.21) is no more than

$$A \sum_{r=1}^{\infty} 2^{-2rk} \frac{(2(r+1)k)^{2(r+1)k+\frac{1}{2}}}{((r+1)k)^{2(r+1)k+1}} k^{-rk+1} \frac{(rk)^{rk+\frac{1}{2}}}{r^{k(r+\frac{1}{2})}}$$

$$= A 2^{2k+\frac{1}{2}} k^{3/2} \sum_{r=1}^{\infty} \frac{1}{((r+1)k)^{1/2} r^{(k-1)/2}}$$

$$\leq A' \sum_{r=1}^{\infty} \frac{1}{r^{k/2}} < \infty \quad (k \geq 3), \tag{7.22}$$

where A and A' depend only on k. Thus, by Theorem 7.1, the simple symmetric random walk is transient in dimensions $k \geq 3$. ∎

Exercise 7.1

(a) Show that $r!(r+s)! \geq (r+1)!(r+s-1)!$ for all $s \geq 1$, with strict inequality for $s \geq 2$, and equality for $s = 1$. Use this to prove that, for fixed k and m, the maximum of $p_{k,m}(r_1, \ldots, r_k)$ occurs when r_i's are as close to each other as possible, differing by no more than one. [Hint: Take r to be the smallest among the r_i's and $r + s$ the largest, and alter them to $r + 1$ and $r + 1 - s$ (unless $s \leq 1$). Repeat the procedure with the altered set of r_i's until they are equal (if m is a multiple of k) or differ by at most one.]

(b) Show that

$$a_{k,rk+s} = k^{-(rk+s)} \frac{(rk+s)!}{(r!)^{k-s}((r+1)!)^s} \quad \text{for } 0 \leq s < k. \tag{7.23}$$

(c) Prove (7.20). ∎

One may give a proof of the recurrence of a one-dimensional simple symmetric random walk along the same lines as above, instead of the derivation given in Section 2.6. For this, write, using the first estimate in (7.15),

$$G(0, 0) = \sum_{m=1}^{\infty} p_{00}^{(2m)} = \sum_{m=1}^{\infty} \left(\frac{1}{2}\right)^{2m} \frac{2m!}{(m!)^2}$$

$$> A \sum_{m=1}^{\infty} m^{-1/2} = \infty, \tag{7.24}$$

for some positive number A.

It may be pointed out that an *asymmetric random walk* on $\mathbb{Z}^k (k \geq 1)$ defined by $\mathbf{X}_0 = \mathbf{x}, \mathbf{X}_n = \mathbf{x} + \mathbf{Z}_1 + \cdots + \mathbf{Z}_n (n \geq 1)$, where $\mathbf{x} \in \mathbb{Z}^k$ and $\mathbf{Z}_n (n \geq 1)$ are i.i.d. with values in $\mathbb{Z}^k$ and $E(\mathbf{Z}_1) \neq \mathbf{0}$, is *transient* for all $k \geq 1$. This easily follows from the SLLN: with probability 1, $\frac{\mathbf{X}_n}{n} = \frac{\mathbf{x}}{n} + \frac{\mathbf{Z}_1 + \cdots + \mathbf{Z}_n}{n} \to E\mathbf{Z}_1 \neq \mathbf{0}$ as $n \to \infty$. If $\mathbf{x}$ were to be a recurrent state then (with probability 1) there would exist a random sequence $n_1 < n_2 < \cdots$ such that $\mathbf{X}_{n_m} = \mathbf{x}$ for all $m = 1, 2, \ldots$, implying $\mathbf{X}_{n_m}/n_m = \frac{\mathbf{x}}{n_m} \to \mathbf{0}$ as $m \to \infty$, which is a contradiction.

For general recurrent Markov chains, the following consequence of the strong Markov property (Theorem 6.2) is an important tool in analyzing large-time behavior.

Recall (from (6.6)) that $\eta_x^{(r)}$ is the rth return time to state x. For simplicity of notation we sometimes drop the subscript x, i.e., $\eta^{(0)} = 0$ and $\eta^{(1)} \equiv \eta_x^{(1)} := \inf\{n \geq 1: X_n = x\}, \eta^{(r+1)} \equiv \eta_x^{(r+1)} = \inf\{n > \eta^{(r)}: X_n = x\} (r \geq 1)$.

Theorem 7.3 *Let x be a recurrent state of a Markov chain. Then the random cycles $\mathbf{W}_r = (X_{\eta^{(r)}}, X_{\eta^{(r)}+1}, \ldots, X_{\eta^{(r+1)}}), r \geq 1$, are i.i.d., and are independent of $\mathbf{W}_0 = (X_0, X_1, \ldots, X_{\eta_x^{(1)}})$. If $X_0 \equiv x$, then the random cycles are i.i.d. for $r \geq 0$.*

Proof. By Theorem 7.1 (ii), $\eta^{(r)} < \infty$ for all $r \geq 1$, outside a set of 0 probability. By the strong Markov property (Theorem 6.2), the conditional distribution of the after-$\eta^{(r)}$ process $X_{\eta^{(r)}}^+ = (X_{\eta^{(r)}}, X_{\eta^{(r)}+1}, \ldots,)$, given the past up to time $\eta^{(r)}$ is $P_{X_{\eta^{(r)}}} = P_x$. Since the latter is nonrandom, it is independent of the past up to time $\eta^{(r)}$. In particular, the cycle $\mathbf{W}_r = (X_{\eta^{(r)}}, X_{\eta^{(r)}+1}, \ldots, X_{\eta^{(r+1)}})$ is independent of all the preceding cycles. Also, the conditional distribution of $(X_{\eta^{(r)}}, \ldots, X_{\eta^{(r+1)}})$, given the past up to time $\eta^{(r)}$ is the P_x distribution of $\mathbf{W}_0 = (X_0 = x, X_1, \ldots, X_{\eta^{(1)}})$, which is therefore also the unconditional distribution. ∎

Finally, the following result may be derived in several different ways (see Corollary 11.1).

Proposition 7.3 *A Markov chain on a finite state space has at least one invariant probability.*

Proof. First we show that the class $\mathcal{E}$ of essential states is nonempty. If all states are inessential and, therefore, transient (Proposition 7.2), then $p_{ij}^{(n)} \to 0$ as $n \to \infty$ for all $i, j \in S$ (Theorem 7.1). Considering the fact that $\sum_{j \in S} p_{ij}^{(n)} = 1$ $\forall n$, since S is finite, one can take term-by-term limit of the sum to get $0 = 1$, a contradiction. Now if $\mathcal{E}$ comprises a single (equivalence) class of communicating states, then the Markov chain restricted to $\mathcal{E}$ is irreducible and one may use either Corollary 5.1 or Theorem 5.2 to prove the existence of a unique invariant probability $\bar{\pi}$ on $\mathcal{E}$. By letting $\bar{\pi}$ assign mass zero to all inessential states, one obtains an invariant probability on S. If $\mathcal{E}$ comprises several equivalence classes of communicating essential states $\mathcal{E}_1, \mathcal{E}_2, \ldots, \mathcal{E}_m$, then, by the above argument, there exists an invariant probability $\pi^{(j)}$ on $\mathcal{E}_j (1 \le j \le m)$. By assigning mass zero to $S \backslash \mathcal{E}_j$, one extends $\pi^{(j)}$ to an invariant probability on S. ∎

2.8 Positive Recurrence and Steady State Distributions of Markov Chains

By Theorem 5.2, every irreducible Markov chain on a finite state space has a unique invariant probability π and that chain approaches this steady state distribution over time, in the sense that $1/n \sum_{m=1}^{n} p_{ij}^{(m)} \to \pi_j$ as $n \to \infty$ (for all $i, j \in S$). Such a result does not extend to the case when S is denumerable (i.e., countable, but infinite). As we have seen in Section 2.7, many interesting irreducible Markov chains are transient and, therefore, cannot have an invariant distribution.

Exercise 8.1 Prove that a transient Markov chain cannot have an invariant distribution. [Hint: By Theorem 7.1 $p_{ij}^{(n)} \to 0$ as $n \to \infty$.] ∎

It is shown in this section that an irreducible Markov chain has a unique invariant distribution if and only if it is *positive recurrent* in the following sense:

Definition 8.1 A recurrent state x is *positive recurrent* if $E_x \eta_x < \infty$, where $\eta_x = \inf\{n \ge 1, X_n = x\}$ is the first return time to x. A recurrent state x is *null recurrent* if $E_x \eta_x = \infty$. An *irreducible* Markov chain *all* whose states are positive recurrent is said to be a *positive recurrent* Markov chain.

To prove our main result (Theorem 8.1) of this section we will need the following, which is of independent interest.

Proposition 8.1

(a) *If a Markov chain has a positive recurrent state i, then it has an invariant distribution* $\pi = (\pi_j \colon j \in S)$ *where*

$$\pi_j = \lim_{n \to \infty} \frac{1}{n} \sum_{m=1}^n p_{ij}^{(m)} (j \in S), \quad \pi_i = \frac{1}{E_i \eta_i}. \tag{8.1}$$

(b) *If i is positive recurrent and* $i \to j$, *then j is positive recurrent.*

Proof.

(a) Let i be a positive recurrent state, and consider the Markov chain $\{X_n \colon n = 0, 1, \ldots\}$ with initial state i. By Theorem 7.3, $\eta^{(r+1)} - \eta^{(r)}$ $(r = 0, 1, \ldots)$ are i.i.d., where $\eta^{(0)} = 0$, $\eta^{(r+1)} \equiv \eta_i^{(r+1)} = \inf\{n > \eta_i^{(r)} \colon X_n = i\}$ $(r \geq 0)$. In particular $\eta^{(1)} = \eta_i$. By the SLLN one has, with probability 1,

$$\frac{\eta^{(N)}}{N} = \frac{\sum_{r=0}^{N-1} (\eta^{(r+1)} - \eta^{(r)})}{N} \to E_i \eta^{(1)} \equiv E_i \eta_i \quad \text{as } N \to \infty. \tag{8.2}$$

Define $N_n = N_n(i) = \sup\{r \geq 0 \colon \eta^{(r)} \leq n\}$, namely, the number of returns to i by time $n(n \geq 1)$. Then, with probability 1, $N_n \to \infty$ and

$$0 \leq \frac{n - \eta^{(N_n)}}{N_n} \leq \frac{\eta^{(N_n+1)} - \eta^{(N_n)}}{N_n} \to 0 \quad \text{as } n \to \infty. \tag{8.3}$$

The last convergence follows from the fact that $Z_r / r \to 0$ (almost surely) as $r \to \infty$, where we write $Z_r = \eta^{(r+1)} - \eta^{(r)}$ (see Exercise 8.2). Using (8.3) in (8.2) we get

$$\lim_{n \to \infty} \frac{n}{N_n} = E_i \eta_i, \quad \lim_{n \to \infty} \frac{N_n}{n} = \frac{1}{E_i \eta_i} \quad \text{a.s.} \tag{8.4}$$

Next write $T_j^{(r)} = \#\{m \in [\eta^{(r)}, \eta^{(r+1)}) \colon X_m = j\}$, the number of visits to j during the rth cycle. Then $T_j^{(r)} (r \geq 0)$ are i.i.d. with a (common) expectation $E_i T_j^{(0)} \leq E_i \eta_i < \infty$, so that by the SLLN (and the fact that $N_n \to \infty$ as $n \to \infty$),

$$\lim_{N \to \infty} \frac{\sum_{r=0}^{N-1} T_j^{(r)}}{N} = E_i T_j^{(0)}, \quad \lim_{n \to \infty} \frac{\sum_{r=0}^{N_n-1} T_j^{(r)}}{N_n} = E_i T_j^{(0)}, \tag{8.5}$$

almost surely. Now $\sum_{r=0}^{N_n-1} T_j^{(r)}$ is the number of $m \in [0, \eta^{(N_n)})$ such that $X_m = j$, and differs from $\sum_{m=1}^n \mathbf{1}_{[X_m=j]}$ by no more than $\eta^{(N_n+1)} - \eta^{(N_n)}$.

Hence, again using (8.3), (8.4), and by (8.5),

$$\lim_{n \to \infty} \frac{\sum_{m=1}^{n} \mathbf{1}_{[X_m=j]}}{n} = \lim_{n \to \infty} \frac{\sum_{r=0}^{N_n-1} T_j^{(r)}}{N_n} \cdot \frac{N_n}{n} = \frac{E_i T_j^{(0)}}{E_i \eta_i}, \quad (8.6)$$

with probability 1. Since the sequence under the limit on the left is bounded by 1, one may take expectation to get

$$\lim_{n \to \infty} \frac{1}{n} \sum_{m=1}^{n} p_{ij}^{(m)} = \frac{E_i T_j^{(0)}}{E_i \eta_i} = \pi_j, \quad \text{say } (j \in S). \quad (8.7)$$

Note that $\sum_{j \in S} T_j^{(0)} = \eta^{(1)} \equiv \eta_i$, so

$$\sum_{j \in S} \pi_j = \frac{\sum_{j \in S} E_i T_j^{(0)}}{E_i \eta_i} = \frac{E_i \sum_{j \in S} T_j^{(0)}}{E_i \eta_i} = \frac{E_i \eta_i}{E_i \eta_i} = 1, \quad (8.8)$$

proving that $\pi = (\pi_j : j \in S)$ is a probability measure on S. As in the proof of Theorem 5.2 [see (5.27), (5.30), and (5.31)], (8.7) implies the invariance of π (also see Theorem 5.3). Since $T_i^{(0)} = 1$ a.s., the second relation in (8.1) also follows.

(b) First note that, under the given hypothesis, j is recurrent (Proposition 7.1). In (8.6), replace n by $\eta_j^{(r)}$ – the time of rth visit to $j (r \geq 1)$. Then one obtains

$$\lim_{r \to \infty} \frac{r}{\eta_j^{(r)}} = \frac{E_i T_j^{(0)}}{E_i \eta_i} \quad \text{a.s.} \quad (8.9)$$

Note that $E_i T_j^{(0)} = E_i T_j^{(r)} (r \geq 1)$ must be positive. For otherwise $T_j^{(r)} = 0$ for all r (with probability 1), implying $\rho_{ij} = 0$.
But $\eta_j^{(r)} = \eta_j^{(1)} + \sum_{m=2}^{r} (\eta_j^{(m)} - \eta_j^{(m-1)})$, where $\eta_j^{(1)} \equiv \eta_j$. Since $\eta_j / r \to 0$ a.s., (as $r \to \infty$) and $(\eta_j^{(m)} - \eta_j^{(m-1)})$, $m \geq 2$, is an i.i.d. sequence, $\eta_j^{(r)} / r \to E_i(\eta_j^{(2)} - \eta_j^{(1)})$ a.s. (as $r \to \infty$). Since the limit (8.9) is positive, we have $E_j \eta_j = E_i(\eta_j^{(2)} - \eta_j^{(1)}) < \infty$. That is, j is positive recurrent. ∎

Exercise 8.2 Suppose a_r $(r = 1, 2, \ldots)$ is a sequence such that $c_r := (1/r) \sum_{m=1}^{r} a_m$ converges to a finite limit c, as $r \to \infty$. Show that $a_r / r \to 0$ as $r \to \infty$. [Hint: the sequence c_r is Cauchy and $\frac{a_r}{r} = (c_r - c_{r-1}) - (\frac{r-1}{r} - 1)c_{r-1}$.] ∎

Exercise 8.3 Find a Markov chain which is not irreducible but has a unique invariant probability.   ∎

The main result of this section is the following:

Theorem 8.1 *For an irreducible Markov chain the following statements are equivalent.*

(1) There exists a positive recurrent state.
(2) All states are positive recurrent.
(3) There exists an invariant distribution.
(4) There exists a unique invariant distribution.
(5) The quantities $\pi_j := 1/E_j\eta_j (j \in S)$ define a unique invariant distribution, and

$$\lim_{n \to \infty} \frac{1}{n} \sum_{m=1}^{n} p_{ij}^{(m)} = \pi_j \quad \forall i, j. \tag{8.10}$$

(6) $\pi_j := 1/E_j\eta_j (j \in S)$ define a unique invariant distribution, and

$$\lim_{n \to \infty} \sum_{j \in S} \left| \frac{1}{n} \sum_{m=1}^{n} p_{ij}^{(m)} - \pi_j \right| = 0 \quad \forall i. \tag{8.11}$$

Proof. (1) ⇔ (2) ⇒ (3). This follows from Proposition 8.1.

(2) ⇒ (5). To see this note that, if (2) holds then (8.1) holds $\forall i, j$, implying π is the unique invariant distribution (see Theorem 5.3).

(5) ⇒ (6). This follows from Scheffé's Theorem (see Lemma C 11.3 in Complements and Details) and (8.10).

Clearly, (6) ⇒ (5) ⇒ (4) ⇒ (3).

It remains to prove that (3) ⇒ (1). For this suppose π is an invariant distribution and j is a state such that $\pi_j > 0$. Then j is recurrent; for otherwise $p_{ij}^{(n)} \to 0 \ \forall i$ (by Theorem 7.1(i)), leading to the contradiction

$$0 < \pi_j = \sum_i \pi_i p_{ij}^{(n)} \to 0 \quad \text{as } n \to \infty.$$

By irreducibility, all states are recurrent. To prove that j is positive recurrent, consider the Markov chain $\{X_n : n = 0, 1, \ldots\}$ with some initial state i and define $N_n(j) = \sup\{r \geq 0 : \eta_j^{(r)} \leq n\} \ (r \geq 1), \ \eta_j^{(0)} = 0$. Note that $(\eta_j^{(r+1)} - \eta_j^{(r)}), \ r \geq 1$, are i.i.d. Suppose, if possible, j is not positive recurrent. Then

$$
\frac{\eta_j^{(N)}}{N} = \frac{\eta_j^{(1)} + \sum_{r=1}^{N-1}\left(\eta_j^{(r+1)} - \eta_j^{(r)}\right)}{N} \to E_i\left(\eta_j^{(2)} - \eta_j^{(1)}\right)
$$
$$
= E_j\eta_j = \infty \text{ a.s.}, \quad \text{as } N \to \infty,
$$

by the SLLN for the case of the infinite mean. In particular, with $N = N_n \equiv N_n(j)$,

$$
\frac{\eta_j^{(N_n)}}{N_n(j)} \to \infty \quad \text{and} \quad \frac{N_n(j)}{\eta_j^{(N_n)}} \to 0 \text{ a.s.}, \quad \text{as } n \to \infty.
$$

Since $\eta_j^{(N_n)} \leq n$, one then has

$$
\frac{N_n(j)}{n} \to 0 \text{ a.s.}, \quad \text{as } n \to \infty. \tag{8.12}
$$

But $N_n(j) = \sum_{m=1}^{n} 1_{[X_m=j]}$. Hence taking expectation in (8.12) we obtain

$$
\frac{\sum_{m=1}^{n} p_{ij}^{(m)}}{n} \to 0, \quad \text{as } n \to \infty. \tag{8.13}
$$

Since i is arbitrary, (8.13) holds for all i. Multiplying both sides of (8.13) by π_i and summing over i, we get

$$
\sum_i \pi_i \sum_{m=1}^{n} p_{ij}^{(m)}/n \to 0 \quad \text{as } n \to \infty.
$$

But the left-hand side above equals $\pi_j > 0 \ \forall n$, by invariance of π, leading to a contradiction. Hence j is positive recurrent. Thus $(3) \Rightarrow (1)$. ∎

Note that we have incidentally proved that π_j is the *ratio of the expected time spent in state j between two successive visits to state i, to the expected length of time between two successive visits to i*.

$$
\frac{E_i T_j^{(0)}}{E_i \eta_i} = \pi_j = \frac{1}{E_j \eta_j}. \tag{8.14}
$$

The arguments leading to (8.13) show that (8.13) holds for all null recurrent states i, j. As a corollary to Proposition 8.1 we then get

Corollary 8.1 *If an irreducible Markov chain has a null recurrent state then all states are null recurrent and (8.13) holds for all i, j.*

Example 8.1 (*Unrestricted birth-and-death chains on* $\mathbb{Z}$)   We take the state space $S = \{0, \pm 1, \pm 2, \ldots\} = \mathbb{Z}$. For a simple random walk on $\mathbb{Z}$, if the process is at state x at time n then at time $n + 1$ with probability p it is at the state $x + 1$ and with probability $q = 1 - p$ it is at the state $x - 1$. Thus the probabilities of moving one-step forward or one-step backward do not depend on the current state x. In general, however, one may construct a Markov chain with transition probabilities

$$p_{x,x+1} = \beta_x, \quad p_{x,x} = \alpha_x, \quad p_{x,x-1} = \delta_x$$
$$0 < \beta_x < 1, \quad 0 < \delta_x < 1, \quad \alpha_x \geq 0,$$
$$\alpha_x + \beta_x + \delta_x = 1 \quad (x \in \mathbb{Z}). \tag{8.15}$$

Such a Markov chain $\{X_n: n = 0, 1, \ldots\}$ is called a *birth-and-death chain* on $S = \mathbb{Z}$. Under the assumption $\beta_x > 0, \delta_x > 0$ for all x, this chain is clearly irreducible. Often one has $\alpha_x = 0$, $x \in \mathbb{Z}$, but we well consider the more general case $\alpha_x \geq 0$. In analogy with the simple random walk (see Section 2.6), we may calculate the probability $\varphi(x)$ that the birth-and-death chain will reach d before c, starting at $x, c < x < d$ $(c, d \in \mathbb{Z})$. By the Markov property,

$$\begin{aligned}
\varphi(x) &= P_x(\{X_n: n \geq 0\} \text{ reaches } d \text{ before } c, X_1 = x + 1) \\
&\quad + P_x(\{X_n: n \geq 0\} \text{ reaches } d \text{ before } c, X_1 = x) \\
&\quad + P_x(\{X_n: n \geq 0\} \text{ reaches } d \text{ before } c, X_1 = x - 1) \\
&= \beta_x \varphi(x + 1) + \alpha_x \varphi(x) + \delta_x \varphi(x - 1).
\end{aligned}$$

Writing $\alpha_x = 1 - \beta_x - \delta_x$, one obtains the *difference equation*

$$\beta_x(\varphi(x + 1) - \varphi(x)) = \delta_x(\varphi(x) - \varphi(x - 1)), \quad c < x < d,$$
$$\varphi(c) = 0, \quad \varphi(d) = 1. \tag{8.16}$$

The second line of (8.16) gives the *boundary conditions* for φ. We can rewrite the first line as

$$\varphi(x + 1) - \varphi(x) = \frac{\delta_x}{\beta_x}(\varphi(x) - \varphi(x - 1)), \quad c < x < d. \tag{8.17}$$

Iterating over $x = c + 1, \ldots, y - 1, y$, and using $\varphi(c) = 0$, one obtains

$$\varphi(y + 1) - \varphi(y) = \frac{\delta_y \delta_{y-1} \cdots \delta_{c+1}}{\beta_y \beta_{y-1} \cdots \beta_{c+1}} \varphi(c + 1), \quad c < y < d. \quad (8.18)$$

Summing this over $y = c + 1, c + 2, \ldots, d - 1$, and setting $\varphi(d) = 1$,

$$1 - \varphi(c + 1) = \left(\sum_{y=c+1}^{d-1} \frac{\delta_y \delta_{y-1} \cdots \delta_{c+1}}{\beta_y \beta_{y-1} \cdots \beta_{c+1}} \right) \varphi(c + 1),$$

$$\varphi(c + 1) = \frac{1}{1 + \sum_{y=c+1}^{d-1} \frac{\delta_y \delta_{y-1} \cdots \delta_{c+1}}{\beta_y \beta_{y-1} \cdots \beta_{c+1}}}.$$

Applying this to (8.17) and summing over $y = x - 1, x - 2, \ldots,$ $c + 1$,

$$\varphi(x) - \varphi(c + 1) = \sum_{y=c+1}^{x-1} \frac{\delta_y \delta_{y-1} \cdots \delta_{c+1}}{\beta_y \beta_{y-1} \cdots \beta_{c+1}} \varphi(c + 1),$$

$$\varphi(x) = \frac{1 + \sum_{y=c+1}^{x-1} \frac{\delta_y \delta_{y-1} \cdots \delta_{c+1}}{\beta_y \beta_{y-1} \cdots \beta_{c+1}}}{1 + \sum_{y=c+1}^{d-1} \frac{\delta_y \delta_{y-1} \cdots \delta_{c+1}}{\beta_y \beta_{y-1} \cdots \beta_{c+1}}} (c < x < d). \quad (8.19)$$

One may similarly obtain (Exercise) the probability $\psi(x)$ of reaching c before d, starting at x, $c \leq x \leq d$, as

$$\psi(x) = \frac{\sum_{y=x}^{d-1} \frac{\delta_y \delta_{y-1} \cdots \delta_{c+1}}{\beta_y \beta_{y-1} \cdots \beta_{c+1}}}{1 + \sum_{y=c+1}^{d-1} \frac{\delta_y \delta_{y-1} \cdots \delta_{c+1}}{\beta_y \beta_{y-1} \cdots \beta_{c+1}}}, \quad c < x < d,$$

$$\psi(c) = 1, \quad \psi(d) = 0. \quad (8.20)$$

Since $\varphi(x) + \psi(x) = 1$, starting from x, $c < x < d$, the chain reaches the *boundary* $\{c, d\}$ (in finite time) almost surely: $P_x(T_{\{c,d\}} < \infty) = 1$.

Another way of arriving at (8.20) is to show by a direct argument that $P_x(T_{\{c,d\}} < \infty) = 1$, $c < x < d$ (Exercise), which implies $\psi(x) = 1 - \varphi(x)$.

Exercise 8.4 In Example 8.1, let $c < x < d$. (a) Prove that $P(T_{\{c,d\}} \geq d - c - 1 \mid X_0 = x) \leq 1 - (\delta_{c+1} + \cdots + \delta_x + \beta_x + \beta_{x+1} + \cdots + \beta_{d-1}) \leq 1 - \lambda$, where $\lambda = \min \{\delta_{c+1} + \cdots + \delta_x + \beta_x + \cdots + \beta_{d-1} : c + 1 \leq$

$x \leq d - 1\} > 0$. (b) Use (a) to show that $P(T_{\{c,d\}} \geq n(d - c - 1) \mid X_0 = x) \leq (1 - \lambda)^n \; \forall n = 1, 2, \ldots$. ∎

Proposition 8.2 *The birth-and-death chain is recurrent if*

$$\sum_{y=-\infty}^{0} \frac{\beta_y \beta_{y+1} \cdots \beta_0}{\delta_y \delta_{y+1} \cdots \delta_0} = \infty \quad \text{and} \quad \sum_{y=0}^{\infty} \frac{\delta_0 \delta_1 \cdots \delta_y}{\beta_0 \beta_1 \cdots \beta_y} = \infty, \quad (8.21)$$

and it is transient if at least one of the two sums above converges.

Proof. By setting $d \uparrow \infty$ in (8.20) one gets $\rho_{xc} = 1$ for $c < x$ if and only if the limit of $\psi(x)$ in (8.20), as $d \uparrow \infty$, is 1. The latter holds if and only if

$$\sum_{y=c+1}^{\infty} \frac{\delta_{c+1} \delta_{c+2} \cdots \delta_{y-1} \delta_y}{\beta_{c+1} \beta_{c+2} \cdots \beta_{y-1} \beta_y} = \infty. \quad (8.22)$$

If $c + 1 = 0$, this is the second condition in (8.21). Suppose $c + 1 < 0$. Then first subtracting the finite sum from $y = c + 1$ through $y = -1$, and then dividing the remaining sum by $\delta_{c+1} \delta_{c+2} \cdots \delta_{-1}/(\beta_{c+1} \beta_{c+2} \cdots \beta_{-1})$, (8.22) is shown to be equivalent to the second condition in (8.21). Similarly, if $c + 1 > 0$, then by adding a finite sum and then multiplying by a fixed factor, (8.22) is shown to be equivalent to the second condition in (8.21). Thus the second condition in (8.21) is equivalent to $\rho_{xc} = 1$ for all x, c with $c < x$.

To show that the first condition in (8.21) is equivalent to $\rho_{yd} = 1$ for all y, d with $y < d$, one may take the limit of $\varphi(x)$ in (8.19) as $c \downarrow -\infty$. For this one may express $\varphi(x)$, on multiplying both the denominator and the numerator by $\beta_{c+1} \beta_{c+2} \cdots \beta_d/(\delta_{c+1} \delta_{c+2} \cdots \delta_d)$, in a more convenient form

$$\varphi(x) = \frac{\sum_{y=c}^{x-1} \frac{\beta_{y+1} \beta_{y+2} \cdots \beta_d}{\delta_{y+1} \delta_{y+2} \cdots \delta_d}}{\sum_{y=c}^{d-1} \frac{\beta_{y+1} \beta_{y+2} \cdots \beta_d}{\delta_{y+1} \delta_{y+2} \cdots \delta_d}}, \quad c < x < d. \quad (8.23)$$

Now letting $c \downarrow -\infty$ one obtains ρ_{yd}. An argument analogous to that given for ρ_{xc}, $c < x$, shows that $\rho_{yd} = 1$ for all $y < d$ if and only if the first condition in (8.21) holds. Since $\rho_{xx} = \delta_x \rho_{x,x-1} + \alpha_x \rho_{xx} + \beta_x \rho_{x,x+1}$, it follows that $\rho_{xx} \equiv (\delta_x \rho_{x,x-1} + \beta_x \rho_{x,x+1})/(\delta_x + \beta_x)$ equals 1 if and only if (8.21) holds. An alternative, and simpler, argument may be based on relabeling the states $x \to -x$, as indicated before the statement of the proposition (Exercise). ∎

In order to compute the expected time $E_x \eta_x$ of a birth-and-death chain to return to x, starting from x, or to decide when it is finite or infinite, consider again integers $c < x < d$. Let $m(x) = E_x T_{\{c,d\}}$. Then, again considering separately the cases $X_1 = +1, 0, -1$, we get

$$m(x) = 1 + \beta_x m(x + 1) + \alpha_x m(x) + \delta_x m(x - 1).$$

Writing $\alpha_x = 1 - \beta_x - \delta_x$, one obtains the *difference equation*

$$\beta_x(m(x + 1) - m(x)) + \delta_x(m(x) - m(x - 1)) = -1,$$
$$c < x < d, \quad m(c) = m(d) = 0. \tag{8.24}$$

The solution to (8.24) is left to Complements and Details. Letting $c \downarrow -\infty$ or $d \to \infty$ in $m(x)$, one computes $E_x \eta_d = E_x \eta_d(x < d)$ or $E_x T_c = E_x \eta_c$ ($c < x$). If both are finite for some x, c with $x > c$, and some x, d with $x < d$, then the chain is positive recurrent, implying, in particular, that

$$E_y \eta_y = 1 + \beta_y E_{y+1} \eta_y + \alpha_y E_y \eta_y + \delta_y E_{y-1} \eta_y < \infty$$
$$(\text{i.e., } E_y \eta_y = (1 - \alpha_y)^{-1}[1 + \beta_y E_{y+1} \eta_y + \delta_y E_{y-1} \eta_y] < \infty).$$

Else, the birth-and-death chain is null recurrent. We will now provide a different and simple derivation of the criteria for null and positive recurrence for this chain. For this, write

$$\theta(x) = \frac{\beta_0 \beta_1 \cdots \beta_{x-1}}{\delta_1 \delta_2 \cdots \delta_x}(x > 0), \quad \theta(x) = \frac{\delta_{x+1}\delta_{x+2} \cdots \delta_0}{\beta_x \beta_{x+1} cdots \beta_{-1}}(x < 0).$$
$$\tag{8.25}$$

Proposition 8.3

(a) A birth-and-death chain on $S = \mathbb{Z}$ is positive recurrent if and only if

$$\sum_{x=1}^{\infty} \theta(x) < \infty \quad \text{and} \quad \sum_{x=-\infty}^{-1} \theta(x) < \infty, \tag{8.26}$$

and, in this case it has the unique invariant distribution π given by

$$\pi_x \equiv \pi(\{x\}) = \frac{\theta(x)}{1 + \sum_{x=-\infty}^{-1} \theta(x) + \sum_{x=1}^{\infty} \theta(x)}, \quad (x \in \mathbb{Z}). \tag{8.27}$$

(b) A birth-and-death chain is null recurrent if (8.21) holds and at least one of the sums in (8.26) diverges.

Proof. To prove (a) we only need to find an invariant probability: π; namely, solve, for $x \in \mathbb{Z}$,

$$\pi_x p_{x,x} + \pi_{x+1} p_{x+1,x} + \pi_{x-1} p_{x-1,x} = \pi_x,$$

$$\text{or,} \ \pi_x \alpha_x + \pi_{x+1} \delta_{x+1} + \pi_{x-1} \beta_{x-1} = \pi_x,$$

$$\text{or,} \ \pi_{x+1} \delta_{x+1} + \pi_{x-1} \beta_{x-1} = \pi_x (\beta_x + \delta_x). \tag{8.28}$$

Writing $a(x) = \pi_x \delta_x$, $b(x) = \pi_x \beta_x$, the last relation may be expressed as $a(x+1) - a(x) = b(x) - b(x-1)$, which on summing over $x = y$, $y - 1, \ldots, z$ yields

$$a(y+1) - a(z) = b(y) - b(z-1) \quad (z < y).$$

Letting $z \downarrow -\infty$ (assuming $\pi_z \to 0$ as $|z| \to \infty$), one gets $a(y+1) = b(y)$ or $\pi_{y+1} \delta_{y+1} = \pi_y \beta_y$, or

$$\pi_{y+1} = \frac{\beta_y}{\delta_{y+1}} \pi_y \quad (y \in \mathbb{Z}), \tag{8.29}$$

yielding, for $x > 0$,

$$\pi_1 = \frac{\beta_0}{\delta_1} \pi_0, \quad \pi_2 = \frac{\beta_1}{\delta_2} \pi_1 = \frac{\beta_1 \beta_0}{\delta_2 \delta_1} \pi_0, \ldots,$$

$$\pi_x = \frac{\beta_{x-1} \cdots \beta_1 \beta_0}{\delta_x \cdots \delta_2 \delta_1} \pi_0 = \theta(x) \pi_0 \quad (x > 0). \tag{8.30}$$

Similarly, for $x < 0$, one gets, using (8.29) in the form $\pi_y = (\delta_{y+1}/\beta_y) \pi_{y+1}$, the following:

$$\pi_{-1} = \frac{\delta_0}{\beta_{-1}} \pi_0, \quad \pi_{-2} = \frac{\delta_{-1}}{\beta_{-2}} \pi_{-1} = \frac{\delta_{-1} \delta_0}{\beta_{-2} \beta_{-1}} \pi_0, \ldots,$$

$$\pi_x = \frac{\delta_{x+1} \cdots \delta_{-1} \delta_0}{\beta_x \cdots \beta_{-2} \beta_{-1}} \pi_0 = \theta(x) \pi_0 \quad (x < 0). \tag{8.31}$$

For π to be a probability, one must have

$$\left(\sum_{x>0} \theta(x) + \sum_{x<0} \theta(x) \right) \pi_0 + \pi_0 = 1,$$

so that

$$\pi_0 = \left(1 + \sum_{x \neq 0} \theta(x)\right)^{-1}, \quad \text{i.e.,} \quad \sum_{x>0} \theta(x) \text{ and } \sum_{x<0} \theta(x)$$

must both converge, which is the condition (8.26). Part (b) follows immediately. ∎

Example 8.2 (*Birth-and-death chain on* $\mathbb{Z}_+ = \{0, 1, 2, \ldots\}$ *with reflection at* 0). Here (8.15) holds for $x > 0$, and one has

$$p_{00} = \alpha_0, \quad p_{01} = \beta_0 \quad (\beta_0 > 0, \alpha_0 \geq 0, \alpha_0 + \beta_0 = 1). \quad (8.32)$$

Arguments entirely analogous to those for Example 8.1 lead to the following result.

Proposition 8.4

(a) *The birth-and-death chain on* $\mathbb{Z}_+$ *with the reflecting boundary condition* (8.32) *is recurrent if*

$$\sum_{y=1}^{\infty} \frac{\delta_1 \cdots \delta_y}{\beta_1 \cdots \beta_y} = \infty, \quad (8.33)$$

and it is transient if the above sum converges.

(b) *The chain is positive recurrent if and only if, with* $\theta(x)$ *as in* (8.25) *for* $x > 0$,

$$\sum_{x=1}^{\infty} \theta(x) < \infty. \quad (8.34)$$

In this case the unique invariant probability π *is given by*

$$\pi_0 = \left(1 + \sum_{x=1}^{\infty} \theta(x)\right)^{-1}, \quad \pi_x = \theta(x)\pi_0 \quad (x > 0). \quad (8.35)$$

(c) *The chain is null recurrent if and only if* (8.33) *holds and* $\sum_{x>0} \theta(x) = \infty$.

Exercise 8.5 For Example 8.2, show that (a) $\rho_{xd} = 1$ for $0 \leq x < d$ and (b) ρ_{xc} is the same as in Example 8.1, i.e., given by the limit of $\psi(x)$ in (8.20) as $d \uparrow \infty$, for $0 \leq c < x$. (Hint: (a) Let A_r be the event that the chain reaches d in rd units of time or less ($r \geq 1$). Then $P(A_r^c) = P(A_1^c)\ P(A_2^c \mid A_1^c) \cdots P(A_r^c \mid A_{r-1}^c) \leq (1-\vartheta)^r$, where $\vartheta := \beta_0 \beta_1 \cdots \beta_{d-1}$ is less than or equal to the probability that the chain reaches d in d steps or less, starting at a point $y < d$.)   ∎

Example 8.3 (*Ehrenfest model of heat exchange*) Physicists P. and T. Ehrenfest in 1907, and later Smoluchowski in 1916, used the following model of heat exchange between two bodies in order to resolve an apparent paradox that nearly wrecked *Boltzman's kinetic theory of matter* at the beginning of the twentieth century. In this model, the temperatures are assumed to change in steps of one unit and are represented by the numbers of balls in two boxes. The boxes are marked I and II, and there are $2d$ balls labeled $1, 2, \ldots, 2d$. We denote by X_n the *number of balls in box I* at time n (so that the number of balls in box II is $2d - X_n$). One considers $\{X_n : n \geq 0\}$ to be a Markov chain on the state space $S = \{0, 1, \ldots, 2d\}$ with transition probabilities

$$p_{i,i-1} = \frac{i}{2d}, \quad p_{i,i+1} = 1 - \frac{i}{2d} \quad (i = 1, 2, \ldots, 2d - 1);$$
$$p_{01} = 1, \quad p_{2d,2d-1} = 1, \quad p_{ij} = 0 \text{ otherwise.} \tag{8.36}$$

That is, at each time period, one of the boxes gives up a ball to the other. The probability that box j gives up a ball equals the fraction of the $2d$ balls it possesses ($j = $ I, II).

This is a *birth-and-death chain with two reflecting boundaries* 0 and $2d$. Note that

$$e_n := E_i(X_n - d) = E_i[X_{n-1} - d + (X_n - X_{n-1})]$$
$$= E_i(X_{n-1} - d) + E_i(X_n - X_{n-1})$$
$$= e_{n-1} + E_i \left(\frac{2d - X_{n-1}}{2d} - \frac{X_{n-1}}{2d} \right)$$
$$= e_{n-1} + E_i \left(\frac{d - X_{n-1}}{d} \right) = e_{n-1} - \frac{e_{n-1}}{d} = \left(1 - \frac{1}{d} \right) e_{n-1},$$

which yields on iteration, noting that $e_0 = i - d$,

$$e_n = \left(1 - \frac{1}{d} \right)^n (i - d) = (i - d)e^{-\nu t} \tag{8.37}$$

with $n = t\tau$ and $\nu = -\log[(1 - (1/d)]^\tau$, τ being the frequency of transitions per second in the physical model. The relation (8.37) is precisely *Newton's law of cooling*, so the model used is physically reasonable.

One can show as in Example 8.2, but more simply, that the chain has the invariant distribution π given by

$$\pi_j = \binom{2d}{j} 2^{-2d} \quad (j = 0, 1, \ldots, 2d), \tag{8.38}$$

a symmetric binomial distribution $Bin(2d, 1/2)$.

According to *thermodynamics*, heat exchange is an *irreversible* process and thermodynamic equilibrium is achieved when the temperatures of the two bodies equal, i.e., when boxes I and II both have (nearly) the same number of balls. In the random heat exchange model, however, the process of heat exchange is *not irreversible*, since the chain is (positive) recurrent and after reaching $j = d$, the process will go back to the states of extreme disequilibrium, namely, 0 and $2d$, in finite time. Historically, it was Poincaré who first pointed out that statistical mechanical systems have the recurrence property. Then the scientist Zermelo strongly argued that such a system is physically invalid since it contradicts the second law of thermodynamics, which asserts irreversibility. Finally, the Ehrenfests and Smoluchowski demonstrated what Boltzmann had argued earlier without providing compelling justification: *starting from a state far from equilibrium, the process approaches the equilibrium state $j = d$ fairly fast, while starting from the equilibrium state $j = d$ to reach 0 or $2d$, it takes an enormously large time even compared to cosmological times.* In other words, after reaching equilibrium the process will not go back to disequilibrium in physical time. Also, instead of considering the state $j = d$ as the thermodynamical equilibrium, kinetic theory implies that π is the state of thermodynamical equilibrium. Because of the enormity of the number of molecules in a physical system, the temperature of a body that one would feel or measure is given by the average with respect to π.

For an indication of the timescales involved, note that (See (8.1), (7.14), and (8.38))

$$E_d \eta_d = \frac{1}{\pi_d} = \frac{(d!)^2}{(2d)!} 2^{2d} = \sqrt{\pi} d^{1/2}(1 + 0(1)) \quad \text{as } d \to \infty,$$
$$E_0 \eta_0 = E_{2d} \eta_{2d} = 2^{2d}. \tag{8.39}$$

Also see Complements and Details. ∎

The final result of this section strengthens Theorem 8.1.

Theorem 8.2 *The n-step transition probability of an irreducible, positive recurrent, aperiodic Markov chain converges to its unique invariant distribution* π *in total variation distance, i.e.,*

$$\sum_{j\in S}\left|p_{ij}^{(n)}-\pi_j\right|\to 0 \quad as\ n\to\infty\ (\forall i\in S).\tag{8.40}$$

We will need a preliminary lemma to prove this theorem.

Lemma 8.1

   (a) Let B be a set of positive integers, which is closed under addition. If the greatest common divisor (g.c.d.) of B is 1, then there exists an integer n_0 such that B contains all integers larger than or equal to n_0.

   (b) Let j be an (essential) aperiodic state of a Markov chain with transition probability matrix **p**. *Then there exists n_0 such that $p_{jj}^{(n)}>0$ $\forall n\ge n_0$.*

Proof. (a) Let G be the set of all integers of the form $a-b$ with $a,b,\in B\cup\{0\}$. It is simple to check that G is the smallest subgroup of $\mathbb{Z}$, under addition, containing B (Exercise). Suppose the g.c.d. of the integers in B is 1. Then the smallest positive element of G must be 1 (i.e., $G=\mathbb{Z}$). For if this element is $d>1$, then G comprises 0, and all positive and negative multiples of d; but this would imply that all elements of B are multiples of d, a contradiction. Thus, either (i) $1=1-0\in B$ or (ii) there exist $a,b\in B$ such that $a-b=1$. In case (i), one may take $n_0=1$. In case (ii), one may take $n_0=(a+b)^2+1$. For if $n\ge n_0$, then it may be expressed as $n=q(a+b)+r$, where q and r are the integers satisfying $q\ge(a+b)$, $0\le r<a+b$. Thus $n=q(a+b)+r(a-b)=(q+r)a+(q-r)b\in B$. (b) Let $B=\{n\ge 1:p_{jj}^{(n)}>0\}$ and apply (a). ∎

Proof of Theorem 8.2. Consider two independent Markov chains $\{X_n:n\ge 0\}$ and $\{X_n':n\ge 0\}$ on S having the (same) transition probability matrix **p**. Then $\{(X_n,X_n'):n\ge 0\}$ is a Markov chain on $S\times S$, having the transition probability $(i,i')\to(j,j')$ given by $p_{ij}p_{i'j'}((i,i'),$ $(j,j')\in S\times S)$. Denote this probability matrix by $\mathbf{p}_2$. It is simple to check that the product probability $\pi\times\pi$ is *invariant* under this transition probability. We will now show that this transition probability $\mathbf{p}_2$ is *irreducible*. Fix (i,i') and (j,j') arbitrarily. Because **p** is irreducible, there exist positive integers n_1 and n_2 such that $p_{ij}^{(n_1)}>0$ and $p_{i'j'}^{(n_2)}>0$.

Because the states are aperiodic, there exist n_0, n_0' such that $p_{jj}^{(n)} > 0$
$\forall n \geq n_0$ and $p_{j'j'}^{(n)} > 0 \ \forall n \geq n_0'$ (by part (b) of the lemma). This implies $p_{ij}^{(n_1+n)} > 0 \ \forall n \geq n_0$ and $p_{i'j'}^{(n_2+n)} > 0 \ \forall n \geq n_0'$. If m is an integer
greater than both $n_1 + n_0$ and $n_2 + n_0'$, then it follows that $p_{ij}^{(m)} > 0$ and
$p_{i'j'}^{(m)} > 0$, so that the m-step transition probability

$$\left(p_2^{(m)}\right)_{(i,i'),(j,j')} \equiv p_{ij}^{(m)} p_{i'j'}^{(m)}$$
$$\left[= \text{Prob}((X_m, X_m') = (j, j') \mid (X_0, X_0') = (i, i'))\right] > 0,$$

establishing irreducibility of $\mathbf{p}_2$. Since $\mathbf{p}_2$ has an invariant probability
$\pi \times \pi$, it now follows that the Markov chain on $S \times S$ is *recurrent*.

Consider now the chain $\{(X_n, X_n'): n \geq 0\}$ as above, with $X_0 \equiv i$ and
X_0' having the distribution π. Define

$$\tau := \inf\{n \geq 0 : X_n = X_n'\}. \tag{8.41}$$

Then $\text{Prob}(\tau < \infty) = 1$, by recurrence and aperiodicity of the chain on
$S \times S$. Also, τ is a *stopping time* for $\{(X_n, X_n'): n \geq 0\}$, i.e., $\{\tau = m\}$ belongs to the sigmafield $\sigma(\{X_n, X_{n'}'\}: 0 \leq n \leq m\})$ $(m = 1, 2, \ldots)$. Now
modify $\{X_n': n \geq 0\}$ as follows:

$$Y_n := \begin{cases} X_n' & \text{if } n < \tau, \\ X_n & \text{if } n \geq \tau, \end{cases} \quad (n = 0, 1, \ldots). \tag{8.42}$$

By the *strong Markov property* of $\{(X_n, X_{n'}): n \geq 0\}$ (see Theorem 6.2),
$\{Y_n: n \geq 0\}$ has the same distribution as $\{X_n': n \geq 0\}$. For every $A \subset S$,
one has

$$\left| \sum_{j \in A} \left(p_{ij}^{(n)} - \pi_j \right) \right| = |P(X_n \in A) - P(Y_n \in A)|$$
$$= |P(X_n \in A, Y_n = X_n) + P(X_n \in A, Y_n \neq X_n)$$
$$- P(Y_n \in A, X_n = Y_n) - P(Y_n \in A, X_n \neq Y_n)|$$
$$= |P(X_n \in A, Y_n \neq X_n) - P(Y_n \in A, X_n \neq Y_n)|$$
$$\leq P(X_n \neq Y_n)$$
$$\leq P(\tau > n). \tag{8.43}$$

Letting A be in turn, $I^+ := \{j \in S: p_{ij}^{(n)} \geq \pi_j\}$, $I^- := \{j \in S: p_{ij}^{(n)} < \pi_j\}$, and adding up, one then gets (8.40). ∎

Exercise 8.6 Show that τ, defined by (8.41), is a stopping time for the chain $\{(X_n, X_n'): n \geqslant 0\}$. [Hint: $\tau = T_D$, where $D := \{(i, i): i \in S\}$ is the diagonal set in $S \times S$ and T_D is the first passage time to D.]   ∎

Example 8.4 (*A thermostat model of managerial behavior*)   Consider a manager supervising an activity, who decides at the beginning of each period $t = 0, 1, \ldots$ whether or not to allocate effort during the period to improving the operation of the activity. Let $U(t)$ denote a measure of performance of the activity in period t. The sequence $U(t)$ is assumed to be a stochastic process. During a period in which the manager makes an effort to improve the activity, performance can be expected to improve, at least on the average, whereas during a period in which the manager makes no such effort, performance can be expected to deteriorate. The situation is formally modeled as follows. Let $\{X(t)\}$ and $\{Y(t)\}$ each be a sequence of independent, identically distributed random variables, with the two sequences mutually independent. Let

$$E X(t) = \xi < 0$$
$$E Y(t) = \eta > 0. \tag{8.44}$$

If an improvement effort is made in period t, then the *increment* in performance is $Y(t)$, whereas if no such effort is made, the increment in performance is $X(t)$. Let $a(t)$ denote the manager's allocation of improvement (*action*) effort in period t, with $a(t) = 1$ denoting an effort and $a(t) = 0$ denoting no effort. Then

$$Z(t) \equiv U(t) - U(t - 1)$$
$$= a(t)Y(t) + [1 - a(t)]X(t). \tag{8.45}$$

A *behavior* is a sequence of functions A_t, which determine the manager's actions on the basis of past performance levels. Thus, A_t is a function of $U(0), \ldots, U(t - 1)$ and takes values 0 and 1 and

$$a(t) = A_t[U(0), \ldots, U(t - 1)], t = 1, 2, \ldots. \tag{8.46}$$

A simple behavior, which is similar to that of a thermostat with fixed upper and lower settings, say A and B, respectively, is the following. If the initial performance level $U(0)$ is less than A, then the manager makes an improvement effort until the first time that performance reaches or goes above A. At that time the manager ceases effort until the performance drops to or below B. The manager then renews his improvement effort until performance reaches or goes above A, etc. If the initial performance level $U(0)$ is at least A, then the manager starts with no effort and proceeds as above. This is called the *conservative* thermostat behavior, with settings A and $B(A > B)$. Formally, for $t \geq 1$,

$$\text{if } a(t-1) = 0, \quad \text{then } a(t) = \begin{Bmatrix} 0 \\ 1 \end{Bmatrix}$$

$$\text{as } U(t-1) = \begin{Bmatrix} > \\ \leq \end{Bmatrix} B$$

$$\text{if } a(t-1) = 1, \quad \text{then } a(t) = \begin{Bmatrix} 0 \\ 1 \end{Bmatrix}$$

$$\text{as } U(t-1) = \begin{Bmatrix} \geq \\ < \end{Bmatrix} A. \tag{8.47}$$

We will assume $A - B > 2$.

Given the manager's behavior, and the distributions of the random variables $X(t)$ and $Y(t)$, we are interested in the long-run properties of the sequence $\{U(t)\}$ of performance levels and the sequence $\{a(t)\}$ of actions, for example, the long-run average level of performance and the long-run frequency of periods in which an improvement effort is made. The following simplifying assumptions are made about the random variables $X(t)$ and $Y(t)$:

$X(t)$ and $Y(t)$ can take only the values -1, 0, and $+1$ with

$$\text{Prob}\{X(t) = +1\} = p_0, \quad \text{Prob}\{Y(t) = +1\} = p_1,$$
$$\text{Prob}\{X(t) = 0\} = r_0, \quad \text{Prob}\{Y(t) = 0\} = r_1,$$
$$\text{Prob}\{X(t) = -1\} = q_0, \quad \text{Prob}\{Y(t) = -1\} = q_1. \tag{8.48}$$

$$p_0 + q_0 + r_0 = p_1 + q_1 + r_1 = 1 \quad p_i > 0, \quad q_i > 0, \quad r_i > 0 \quad i = 0, 1.$$

Thus, performance can change by at most one unit in any time period, irrespective of improvement effort, and we have

$$\xi = EX(t) = p_0 - q_0 < 0,$$
$$\eta = EY(t) = p_1 - q_1 > 0. \tag{8.49}$$

The sequence of pairs $[a(t), U(t)]$ is a Markov chain. The state space of this chain in the set $S = \{(a, U): a = 0, 1, \text{ and } U \text{ an integer}, B \leq U \leq A\}$ $\cup \{(1, U): U \text{ an integer}, U < B\} \cup \{(0, U): U \text{ an integer}, U > A\}$.

Thus, the state space S is countably infinite and it is irreducible and aperiodic. Note that aperiodicity follows from $r_i > 0$ $(i = 0, 1)$.

We will show that $(0, B)$ is a *positive recurrent state*, which would then imply that all states are positive recurrent, in view of irreducibility (Theorem 8.1). Now, starting from $(0, B)$ one has $a(t) = 1$ $\forall t$ until the time $\eta_{(0, A)}$ of the first visit to $(0, A)$. During this period $U(t) = U(t-1) + Y(t)(1 \leq t \leq \eta_{(0, A)})$, behaving as the simple random walk of Example 6.3, with $E_{(0, B)}\eta_{(0, A)} = (A - B)/(p_1 - q_1)$. Similarly, starting from $(0, A)$, $U(t) = U(t-1) + X(t)$ until $U(t)$ reaches $(0, B)$ (i.e., for $1 \leq t \leq \eta_{(0, B)}$). Again, by the final computations in Example 6.3, $E_{(0, A)}\eta_{(0, B)} = (A - B)/(q_0 - p_0)$. Hence

$$E_{(0, B)}\eta_{(0, B)} \leq E_{(0, B)}\eta_{(0, A)} + E_{(0, A)}\eta_{(0, B)} < \infty,$$

proving the desired positive recurrence. By Theorem 8.2, there exists a unique invariant probability $\pi = (\pi_j : j \in S)$, and

$$\sum_{j \in S} \left| p_{ij}^{(n)} - \pi_j \right| \to 0 \quad \text{as } n \to \infty, \ \forall i \in S. \tag{8.50}$$

Note that if $r_0 = 0 = r_1$, the chain is still irreducible and the proof of positive recurrence remains unchanged. But the chain is no longer aperiodic. One then applies Theorem 8.1 to infer the existence of a unique invariant probability π and, in place of (8.50), the convergence

$$\lim_{n \to \infty} \sum_{j \in S} \left| \frac{1}{n} \sum_{m=1}^{n} p_{ij}^{(m)} - \pi_j \right| = 0 \quad \forall i \in S. \tag{8.51}$$

∎

Exercise 8.7 Write down the transition probabilities of the Markov chain in Example 8.4. ∎

2.9 Markov Processes on Measurable State Spaces: Existence of and Convergence to Unique Steady States

This section is devoted to the extension to general state spaces of two of the main results for Markov chains, namely, Theorem 5.1 (or Corollary 5.1) and Theorem 8.1.

Let the state space S be a nonempty set, and $\mathcal{S}$ a sigmafield on it, i.e., $(S, \mathcal{S})$ is a *measurable space*. For example, S may be $\mathbb{R}^k$ or $[0, \infty)^k$ $(k \geq 1)$, or a compact interval $[c, d]$, in which cases $\mathcal{S}$ is the Borel sigmafield of S; or S may be a countable set, and $\mathcal{S}$ the class of all subsets of S, as considered in the preceding sections. Let $p(x, A)$ $(x \in S, A \in \mathcal{S})$ be a transition probability on $(S, \mathcal{S})$. As before, we will denote by $\{X_n: n = 0, 1, 2, \ldots, \}$ a Markov process on $(S, \mathcal{S})$ with transition probability p, defined on some probability space $(\Omega, \mathcal{F}, P)$.

Definition 9.1 The *transition operator* T of a Markov process $\{X_n\}$ is the linear map on the *space* $\mathbf{B}(S)$ *of all real-valued bounded measurable functions on S* defined by $(Tf)(x) = E(f(X_{n+1}) \mid X_n = x)(n \geqslant 0)$. That is,

$$(Tf)(x) := \int_S f(y)p(x, dy), \quad (f \in \mathbf{B}(S)). \tag{9.1}$$

Its adjoint T^* is defined on the space $\mathcal{M}(S)$ of all finite signed measures on $(S, \mathcal{S})$ into itself by

$$T^*\mu(A) = \int_S p(x, A)\mu(dx). \tag{9.1$'$}$$

Note that T^* maps $\mathcal{P}(S)$, the set of all probability measures on $(S, \mathcal{S})$, into itself. For $m \geq 1$, T^{*m} is the mth *iterate* of T^*. Observe that X_m has the distribution $T^{*m}\mu$ if X_0 has distribution μ.

For the bilinear functional $(f, \mu) := \int f d\mu$,

$$(Tf, \mu) = (f, T^*\mu)(f \in \mathbf{B}(S), \quad \mu \in \mathcal{M}(S)) \tag{9.1$''$}$$

justifying the notation T^* as the *adjoint of* T.

Theorem 9.1 (*Doeblin minorization*) *Let $p(x, A)$ be a (one-step) transition probability on a measurable state space $(S, \mathcal{S})$ such that there exist $N \geq 1$ and some nonzero measure λ for which the N-step transition probability $p^{(N)}$ satisfies*

$$p^{(N)}(x, A) \geq \lambda(A) \quad \forall x \in S, \quad A \in \mathcal{S}. \tag{9.2}$$

Then there exists a unique invariant probability π for p, and one has

$$\sup_{x \in S, A \in \mathcal{S}} |p^{(n)}(x, A) - \pi(A)| \leq (1 - \bar{\chi})^{[\frac{n}{N}]}, \quad n \geq 1, \tag{9.3}$$

where $\bar{\chi} = \lambda(S) > 0$.

Proof. Note that $\lambda(S) \leq p^{(N)}(x, S) = 1$. If $\lambda(S) = 1$, then $p^{(N)}(x, A) = \lambda(A) \; \forall x \in S, A \in \mathcal{S}$. In other words, $\pi := \lambda$ is an invariant probability for $p^{(N)}$ (since $p^{(N)}(x, A)$ does not depend on x); also, $p^{(n)}(x, A) = \int_S p^{(N)}(y, A)p^{(n-N)}(x, dy) = \lambda(A) \, \forall n \geq N$. This establishes (9.3), with both sides equal to zero for $n \geq N$ and, interpreting $0^0 = 1$ takes care of the case $n < N$.

Consider now the important and nontrivial case $0 < \bar{\chi} < 1$. Define the *distance* d_{M1} on $\mathcal{P}(S)$ by

$$d_{M1}(\mu, v)$$
$$= \sup \left\{ \left| \int_S f d\mu - \int_S f dv \right| : f \text{ measurable, } \sup_x |f(x)| \leq 1 \right\}. \quad (9.4)$$

For the proof of the theorem, first note that

$$d_{M1}(T^{*m}\mu, T^{*m}v) \leq d_{M1}(\mu, v), \quad (m \geq 1, \mu, v \in \mathcal{P}(S)) \quad (9.5)$$

since, $x \to g(x) := \int f(y)p^{(m)}(x, dy)$ is bounded by 1, if f is bounded by one.

Next, write

$$p^{(N)}(x, A) = \lambda(A) + (p^N(x, A) - \lambda(A))$$
$$= \bar{\chi}\lambda_{\bar{\chi}}(A) + (1 - \bar{\chi})q_{\bar{\chi}}(x, A), \quad (9.6)$$

where $\lambda_{\bar{\chi}}$ is the probability measure $\lambda_{\bar{\chi}} = \lambda/\bar{\chi}$, and $q_{\bar{\chi}}(x, A) \equiv (p^{(N)}(x, A) - \lambda(A))/(1 - \bar{\chi})$ is a transition probability (i.e., $x \to q_{\bar{\chi}}(x, A)$ is a probability measure on $(S, \mathcal{S})$ for every $x \in S$, and $x \to q_{\bar{\chi}}(x, A)$ is measurable for every $A \in \mathcal{S}$). Then, for every measurable real-valued f on S, with $|f(x)| \leq 1 \, \forall x \in S$, one has

$$\int_S f(y)p^{(N)}(x, dy) = \bar{\chi} \int_S f(y)\lambda_{\bar{\chi}}(dy) + (1 - \bar{\chi}) \int_S f(y)q_{\bar{\chi}}(x, dy),$$
$$(9.7)$$

so that

$$\int_S f(x)(T^{*N}\mu)(dx) = \bar{\chi} \int_S \left\{ \int_S f(y)\lambda_{\bar{\chi}}(dy) \right\} \mu(dx)$$
$$+ (1 - \bar{\chi}) \int_S \left\{ \int_S f(y)q_{\bar{\chi}}(x, dy) \right\} \mu(dx),$$
$$(\mu \in \mathcal{P}(S)). \quad (9.8)$$

Therefore, writing $g(x) = \int_S f(y) q_{\tilde{x}}(x, dy)$, one has

$$\left| \int_S f d(T^{*N} \mu) - \int_S f d(T^{*N} v) \right|$$
$$= (1 - \bar{\chi}) \left| \int_S g(x) \mu(dx) - \int_S g(x) v(dx) \right|$$
$$\leq (1 - \bar{\chi}) d_{M1}(\mu, v), \quad (\mu, v \in \mathcal{P}(S)), \tag{9.9}$$

since $|g(x)| \leq 1$. Thus,

$$d_{M1}(T^{*N} \mu, T^{*N} v) \leq (1 - \bar{\chi}) d_{M1}(\mu, v), \quad (\mu, v \in \mathcal{P}(S). \tag{9.10}$$

Iterating (9.10) k times, one gets

$$d_{M1}(T^{*kN} \mu, T^{*kN} v) \leq (1 - \bar{\chi})^k d_{M1}(\mu, v), \quad (k = 1, 2, \ldots). \tag{9.11}$$

Combining this with (9.5) for $1 \leq m < n$, one arrives at

$$d_{M1}(T^{*n} \mu, T^{*n} v) \leq (1 - \bar{\chi})^{\left[\frac{n}{N}\right]} d_{M1}(\mu, v), \quad n \geq 1. \tag{9.12}$$

Now (9.10) says T^{*N} is a uniformly strict contraction so that, by the contraction mapping theorem (Theorem 2.2, Chapter 1), T^* has a unique fixed point π: $T^*\pi = \pi$. Therefore, by (9.12) with ($v = \pi$),

$$d_{M1}(T^{*n} \mu, \pi) \leq (1 - \bar{\chi})^{\left[\frac{n}{N}\right]} d_{M1}(\mu, \pi)(n \geq 1). \tag{9.13}$$

In particular, with $\mu = \delta_x$ in (9.13) (δ_x is the Dirac probability measure at x, $\delta_x(\{x\}) = 1$),

$$d_{M1}(p^{(n)}(x, dy), \pi) \leq (1 - \bar{\chi})^{\left[\frac{n}{N}\right]} d_{M1}(\delta_x, \pi). \tag{9.14}$$

Since (see Remark 9.2)

$$\sup_{A \in \mathcal{S}} |\mu(A) - v(A)| = \frac{1}{2} d_{M1}(\mu, v), \tag{9.15}$$

(9.3) follows from (9.14). ∎

Remark 9.1 An alternative proof based on a general theorem on "splitting" is given (for the case of S, a Borel subset of a Polish space) in Chapter 3 (Section 3.5.3).

Remark 9.2 To prove (9.15), let f be a simple function $\sum_{i=1}^{k} \alpha_i 1_{A_i}$ where $|\alpha_i| \leq 1 \ \forall i$, and $A_1, \ldots, A_k$ are disjoint. Then, writing I^+ for the set of i such that $\mu(A_i) \geq \nu(A_i)$ and I^- for those with $\mu(A_i) < \nu(A_i)$,

$$\left| \int_S f\,d\mu - \int_S f\,d\nu \right| = \left| \sum_{i \in I^+} \alpha_i(\mu(A_i) - \nu(A_i)) + \sum_{i \in I^-} \alpha_i(\mu(A_i) - \nu(A_i)) \right|$$

$$\leq \left| \sum_{i \in I^+} \alpha_i(\mu(A_i) - \nu(A_i)) \right| + \left| \sum_{i \in I^-} \alpha_i(\mu(A_i) - \nu(A_i)) \right|$$

$$\leq \sum_{i \in I^+} |\alpha_i|(\mu(A_i) - \nu(A_i)) + \sum_{i \in I^-} |\alpha_i|(\nu(A_i) - \mu(A_i))$$

$$\leq \sum_{i \in I^+} (\mu(A_i) - \nu(A_i)) + \sum_{i \in I^-} (\nu(A_i) - \mu(A_i))$$

$$= \mu(A^+) - \nu(A^+) + \nu(A^-) - \mu(A^-)$$

$$[A^+ := \cup_{i \in I^+} A_i, \; A^- := \cup_{i \in I^-} A_i]$$

$$\leq 2 \sup\{|\mu(A) - \nu(A)| : A \in \mathcal{S}\}. \tag{9.16}$$

Hence $d_{M1}(\mu, \nu) \leq 2 \sup\{|\mu(A) - \nu(A)| : A \in \mathcal{S}\}$. Conversely, for an arbitrary $A \in \mathcal{S}$, let $f = 1_A - 1_{A^c}$, i.e., $f = 1$ on A and -1 on A^c, so that

$$\left| \int f\,d\mu - \int f\,d\nu \right| = |\mu(A) - \mu(A^c) - \{\nu(A) - \nu(A^c)\}|$$

$$= |\mu(A) - \nu(A) - \{\mu(A^c) - \nu(A^c)\}|$$

$$= |\mu(A) - \nu(A) - \{1 - \mu(A) - (1 - \nu(A))\}|$$

$$= |2(\mu(A) - \nu(A))| = 2|\mu(A) - \nu(A)|. \tag{9.17}$$

Hence, taking supremum over $A \in \mathcal{S}$, we get

$$d_{M1}(\mu, \nu) \geq 2 \sup\{|\mu(A) - \nu(A)| : A \in \mathcal{S}\},$$

establishing (9.15). ∎

Remark 9.3 For the application of the contraction mapping theorem (Theorem 2.1, Chapter 1), one needs to show that the metric space $(\mathcal{P}(S), d_{M1})$ is complete. For this, let $\{\mu_n : n \geq 1\}$ be a Cauchy sequence for the metric d_{M1}, and define $\mu(A) := \lim_{n \to \infty} \mu_n(A)$, $A \in \mathcal{S}$ (this limit exists due to the completeness of the set $[0,1]$ of real numbers). From this the properties $0 \leq \mu(A) \leq 1$ ($A \in \mathcal{S}$), $\mu(S) = 1$, and finite additivity of

μ follow immediately. Since the convergence $\mu_n(A) \to \mu(A)$ is uniform for $A \in \mathcal{S}$, countable additivity of μ follows from that of the $\mu_n (n \geqslant 1)$. Thus $\mu \in \mathcal{P}(S)$ and $\mu_n \xrightarrow{d_{M_1}} \mu$.

Observe that Theorem 5.1 and Corollary 5.1 in Section 2.5 follow immediately from Theorem 9.1.

Another proof of Theorem 9.1 may be based on the following result, which is of independent interest. To state it we define

$$\Delta_{n,m} := \sup_{x,y \in S, A \varepsilon \mathcal{S}} \left| p^{(n)}(x, A) - p^{(m)}(y, A) \right|,$$

$$\Delta_n := \Delta_{n,n}. \tag{9.18}$$

Proposition 9.1 *A necessary and sufficient condition for the exponential convergence of $p^{(n)}(x, dy)$ to a probability measure π in the distance d_{M1}, uniformly for all $x \in S$, is that there exists some $N \geq 1$ such that $\Delta_N < 1$. In this case π is the unique invariant probability for $p(x, dy)$.*

Proof. To prove necessity, suppose $p^{(n)}(x, dy)$ converges to $\pi(dy)$ in the distance d_{M1} exponentially fast, uniformly for all $x \in S$. Then there exists $N \geq 1$ such that

$$d_{M1}(p^{(N)}(x, \cdot), \pi) < \frac{1}{2} \quad \forall x \in S. \tag{9.19}$$

By (9.15) and (9.19),

$$\left| p^{(N)}(x, A) - p^{(N)}(y, A) \right| \leq \left| p^{(N)}(x, A) - \pi(A) \right| + \left| p^{(N)}(y, A) - \pi(A) \right|$$

$$\leq \frac{1}{2} d_{M1}\left(p^{(N)}(x, \cdot), \pi \right) + \frac{1}{2} d_{M1}\left(p^{(N)}(y, \cdot), \pi \right) < \frac{1}{2} \forall x \in S \forall \in \mathcal{S}. \tag{9.20}$$

Hence $\Delta_N < \frac{1}{2}$.

To prove sufficiency, first write $\mu_n(A; x, y) = p^{(n)}(x, A) - p^{(n)}(y, A)$, and let $\mu_n^+(A; x, y) - \mu_n^-(A; x, y)$ be the Jordan–Hahn decomposition of the signed measure $\mu_n(\cdot; x, y)$ (see Loève 1963, pp. 86–87).

Recall that $\mu_n^+(S; x, y) = \sup\{\mu_n(A; x, y): A \in \mathcal{S}\} \le \Delta_n$. Also, since $\mu_n(S; x, y) = 0$, one has $\mu_n^+(S; x, y) = \mu_n^-(S; x, y)$. Therefore,

$$p^{(n+m)}(x, A) - p^{(n+m)}(y, A)$$

$$= \int_S p^{(m)}(z, A)\left[p^{(n)}(x, dz) - p^{(n)}(y, dz)\right]$$

$$= \int_S p^{(m)}(z, A)\left(\mu_n^+(dz; x, y) - \mu_n^-(dz; x, y)\right)$$

$$\le \sup\{p^{(m)}(z, A): z \in S\}\mu_n^+(S; x, y)$$

$$\quad - \inf\{p^{(m)}(z, A): z \in S\}\,\mu_n^-(S; x, y)$$

$$= \mu_n^+(S; x, y)\left[\sup\{p^{(m)}(z, A): z \in S\} - \inf\{p^{(m)}(z, A): z \in S\}\right]$$

$$= \mu_n^+(S; x, y)\left[\sup\{p^{(m)}(x, A) - p^{(m)}(y, A): x, y \in S\}\right]$$

$$\le \Delta_n\Delta_m. \tag{9.21}$$

Interchanging x and y, one can take the absolute value on the left. Taking supremum over x, y, A we arrive at the relation

$$\Delta_{n+m} \le \Delta_n\Delta_m, \quad \Delta_n \le \Delta_N^{[n/N]} \tag{9.22}$$

since $\Delta_r \le 1\ \forall r$. Hence, if $\Delta_N < 1$, $\Delta_n \to 0$ exponentially fast as $n \to \infty$. Next note that

$$p^{(n+m)}(x, A) - p^{(n)}(y, A) = \int (p^{(n)}(z, A) - p^{(n)}(y, A))p^{(m)}(x, dz)$$

$$\le \sup\{p^{(n)}(z, A) - p^{(n)}(y, A): z, y \in S\} \le \Delta_n.$$

Therefore,

$$\Delta_{n,n+m} \le \Delta_n. \tag{9.23}$$

Hence, for every x,

$$d_{M1}(p^{(n+m)}(x, \cdot), p^{(n)}(x, \cdot)) \le 2\Delta_{n,n+m} \le 2\Delta_n \to 0, \tag{9.24}$$

implying the existence of a limit π, say, of $p^{(n)}(x, \cdot)$ in total variation distance. Since $\Delta_n \to 0$, this limit is independent of x. On letting $m \to \infty$ in (9.24) one gets

$$d_{M1}(p^{(n)}(x, \cdot), \pi) \le 2\Delta_n \le 2\Delta_N^{[n/N]}. \tag{9.25}$$

Now take supremum of the left-hand side over $x \in S$ to obtain the desired result. ∎

Theorem 9.1 may be derived as an immediate corollary of Proposition 9.1. To see this, write

$$p^{(N)}(x, A) = \lambda(A) + q^{(N)}(x, A), \qquad (9.26)$$

where $q^{(N)}(x, A) = p^{(N)}(x, A) - \lambda(A) \geq 0$, $\quad q^{(N)}(x, S) = 1 - \lambda(S)$. Hence

$$
\begin{aligned}
\Delta_N &= \sup_{x \in S, A \in \mathcal{S}} \left| p^{(N)}(x, A) - p^{(N)}(y, A) \right| \\
&= \sup_{x \in S, A \in \mathcal{S}} \left| q^{(N)}(x, A) - q^{(N)}(y, A) \right| \\
&\leq \sup_{x \in S, A \in \mathcal{S}} q^{(N)}(x, A) = 1 - \lambda(S) < 1,
\end{aligned}
$$

verifying the hypothesis of Proposition 9.1.

Example 9.1 For a simple application of Theorem 9.1, consider a Markov process on a Borel subset S of $\mathbb{R}^k$ with a transition probability density $p(x, y)$ such that

$$g(y) := \inf_{x \in S} p(x, y) > 0 \text{ on a subset of } S \text{ of positive Lebesgue measure.}$$
$$(9.27)$$

Then one may take $\lambda(A) = \int_A g(y)dy$, $A \in \mathcal{S}$. The condition (9.27) holds, e.g., if S is a compact rectangle and $(x, y) \to p(x, y)$ is continuous and positive on $S \times S$. Then $g(y) \geq \delta := \min\{p(x, y): x, y \in S\} > 0$. ∎

The next result, which extends the convergence criterion of Theorem 8.1 to general state spaces, is a centerpiece of the general theory of irreducible Markov processes. Its proof is spelled out in Complements and Details.

Theorem 9.2 *Suppose there exists a set $A_0 \in \mathcal{S}$ with the following two properties:*

(1) A_0 is recurrent, i.e.,

$$P(\eta_{A_0} < \infty \mid X_0 = x) = 1 \quad \forall x \in S, \qquad (9.28)$$

and

(2) A_0 is locally minorized, i.e., there exist (i) a probability measure ν on $(S, \mathcal{S})$ with $\nu(A_0) = 1$, (ii) a number $\lambda > 0$, and (iii) a positive integer

N, such that

$$p^{(N)}(x, A) \geq \lambda v(A) \quad \forall x \in A_0, \quad A \in \mathcal{S}. \tag{9.29}$$

If, in addition,

(3) $\sup\{E(\eta_{A_0}|X_0 = x) : x \in A_0\} < \infty,$

then there exists a unique invariant probability π, and, for every $x \in S$,

$$\sup_{A \in \mathcal{S}} \left| \frac{1}{n} \sum_{m=1}^{n} p^{(m)}(x, A) - \pi(A) \right| \to 0, \quad as \; n \to \infty. \tag{9.30}$$

Definition 9.2 A Markov process $\{X_n : n = 0, 1, \ldots\}$ with transition probability p is said to be *irreducible* with respect to a nonzero sigma-finite measure v, or *v-irreducible*, if for each $x \in S$ and $B \in \mathcal{S}$ with $v(B) > 0$, one has

$$p^{(n)}(x, B) > 0 \quad \text{for some } n \geq 1. \tag{9.31}$$

If, in addition, it is *φ-recurrent*, i.e., for each $x \in S$ and each $A \in \mathcal{S}$ such that $\varphi(A) > 0$, one has $P(\eta_A < \infty \mid X_0 = x) = 1$ where $\eta_A = \eta_A^{(1)} = \inf\{n \geq 1 : X_n \in A\}$, then the Markov process is said to be *Harris recurrent*. If a Harris recurrent process satisfies $\sup\{E(\eta_{A_0} \mid X_0 = x) : x \in A_0\} < \infty$ for some φ-recurrent set A_0, then the process is said to be *positive Harris recurrent*.

Remark 9.4 It is known that if S is a Borel subset of a Polish space, with $\mathcal{S}$ its Borel sigmafield, then conditions (1) and (2) of Theorem 9.2 together are equivalent to the process being irreducible (with respect to v) and v-recurrent. Thus Theorem 9.2 says that, on such a state space, *a positive Harris recurrent process converges to a unique steady state π as time increases, in the sense (9.30).*

Remark 9.5 In the case $N = 1$, the Markov process in Theorem 9.2 is said to be *strongly aperiodic*. By an argument somewhat similar to that used for the countable state space case (Theorem 8.2), one can show that in this case one has

$$\sup_{A \in \mathcal{S}} |p^{(n)}(x, A) - \pi(A)| \to 0 \quad as \; n \to \infty, \tag{9.32}$$

strengthening (9.30) (see Complements and Details).

Example 9.2 (*Positive recurrent markov chain*)   Every irreducible positive recurrent Markov chain satisfies the hypothesis of Theorem 9.2, with $A_0 = \{x\}$, x an arbitrary state, N a sufficiently large multiple of the period of x, $\nu = \delta_x$ (i.e., $\nu(\{x\}) = 1$, $\nu(S\backslash\{x\}) = 0$, and $\lambda = p_{xx}^{(N)}$.   ∎

2.10 Strong Law of Large Numbers and Central Limit Theorem

In this section we derive the SLLN and the CLT for sums $\sum_{m=0}^{n} f(X_m)$ for a broad class of functions f on the state space S of a positive recurrent Markov chain $X_n (n \geq 0)$. The main tools for this are the classical SLLN and CLT for i.i.d. summands, together with the fact (see Theorem 7.3) that the random cycles $\mathbf{W}_r = (X_{\eta(r)}, X_{\eta(r)+1}, \ldots, X_{\eta(r+1)})$, $r \geq 1$, are i.i.d. Here $\eta^{(0)} \equiv 0$, $\eta^{(r)} := \eta_i^{(r)} = \inf\{n > \eta^{(r-1)}: X_n = i\}$ $(r \geq 1)$ with some (any) specified state i. For convenience we will consider the nonoverlapping cycles

$$\mathbf{W}_r' := (X_{\eta(r)}, X_{\eta(r)+1}, \ldots, X_{\eta(r+1)_{-1}})(r \geq 1). \qquad (10.1)$$

Since these are obtained simply by omitting the last term $X_{\eta(r+1)} \equiv i$ from the i.i.d. cycles $\mathbf{W}_r$, the cycles $\mathbf{W}_r'(r \geq 1)$ are also i.i.d.

Theorem 10.1 (*SLLN for Markov chains*)   *Let $X_n (n \geq 0)$ be an irreducible positive recurrent Markov chain on S with the unique invariant probability π. Then, if f is a real-valued function on S such that $\int |f| d\pi < \infty$, one has*

$$\lim_{n \to \infty} \frac{1}{n+1} \sum_{m=0}^{n} f(X_m) = \int_S f d\pi \ a.s., \qquad (10.2)$$

whatever the initial distribution (of X_0).

Proof.   Fix i and let $\eta^{(r)} = \eta_i^{(r)}$ as above. Recalling the notation $T_j^{(0)} = \#\{m: \eta^{(1)} \leq m \leq \eta^{(2)} - 1, X_m = j\}$ (see (8.1) and (8.7)), one has

$$E \sum_{m=\eta^{(1)}}^{\eta^{(2)}-1} |f(X_m)| = E \sum_{j \in S} T_j^{(0)} |f(j)| \qquad (10.3)$$

$$= \sum_{j \in S} \frac{\pi_j}{\pi_i} |f(j)| = (1/\pi_i) \int |f| d\pi < \infty.$$

Therefore, considering the i.i.d. sequence

$$Y_r := \sum_{\eta^{(r)} \le m \le \eta^{(r+1)}-1} f(X_m)(r \ge 1), \qquad (10.4)$$

one has, by the classical SLLN (see Appendix), with probability 1

$$\lim_{N \to \infty} \frac{1}{N} \sum_{m=\eta^{(1)}}^{\eta^{(N)}} f(X_m) = \lim_{N \to \infty} \frac{1}{N} \sum_{r=1}^{N} Y_r$$

$$= EY_1 = (1/\pi_i) \int f \, d\pi. \qquad (10.5)$$

Letting $N = N_n - 1$, where $N_n = \sup\{r \ge 0 : \eta^{(r)} \le n\}$, one has

$$\lim_{n \to \infty} \frac{1}{N_n - 1} \sum_{m=\eta^{(1)}}^{\eta^{(N_n)}} f(X_m) = (1/\pi_i) \int f \, d\pi \text{ a.s.} \qquad (10.6)$$

Write

$$\frac{1}{n+1} \sum_{m=0}^{n} f(X_m) = \frac{1}{n+1} \sum_{m=0}^{\eta^{(1)}-1} f(X_m)$$

$$+ \frac{N_n - 1}{n+1} \bullet \frac{1}{N_n - 1} \sum_{m=\eta^{(1)}}^{\eta^{(N_n)}-1} f(X_m)$$

$$+ \frac{1}{n+1} \sum_{m=\eta^{(N_n)}}^{n} f(X_m)$$

$$= I_1 + I_2 + I_3, \text{ say.} \qquad (10.7)$$

Now $I_1 \to 0$ a.s., as $n \to \infty$. I_3 is bounded by $\frac{1}{n+1} \sum_{m=\eta^{(N_n)}}^{\eta^{(N_n+1)}-1} |f(X_m)| \le$ $(1/N_n)Z_{N_n}$, where $Z_r := \sum_{\eta^{(r)} \le m \le \eta^{(r+1)}-1} |f(X_m)|$. By (10.3), $Z_r/r \to 0$ a.s., as $r \to \infty$ (see Exercise 8.2). Hence $Z_{N_n}/N_n \to 0$ a.s., as $n \to \infty$. Also, I_2 converges a.s. to $\int f \, d\pi$, as $n \to \infty$ by (10.6), and the fact $(n+1)/(N_n - 1) \to E_i \eta_i (= 1/\pi_i)$ a.s. (use (10.6) with $f \equiv 1$). Thus the left hand side of (10.7) converges a.s. to $\int f \, d\pi$. ∎

Remark 10.1 Note that (10.2) allows one to estimate π from observations, i.e.,

$$\pi_j \approx \frac{1}{n+1} \#\{m : 0 \le m \le n, X_m = j\} \forall j \in S, \text{ a.s.,} \qquad (10.8)$$

where $\approx$ indicates that the difference between the two sides goes to zero. A special case of the next theorem (Theorem 10.2) provides one with a fairly precise estimate of the error in the estimation (10.8).

Example 10.1 Consider the Ehrenfest model (Example 8.3). In view of (8.38), one has the limit (10.2) given by $\sum_{j=0}^{2d} f(j)\binom{2d}{j}2^{-2d}$. ∎

In order to prove the CLT we first need the following extension of the classical CLT (see Appendix) to the sum of a random number of i.i.d. summands.

Proposition 10.1 *Let $X_j(j \geq 1)$ be i.i.d. with $EX_j = 0$, $EX_j^2 = \sigma^2$ $(0 < \sigma^2 < \infty)$. Let v_n $(n \geq 1)$ be a sequence of nonnegative random variables satisfying*

$$\frac{v_n}{n} \to \alpha \text{ in probability}, \quad as \ n \to \infty, \tag{10.9}$$

for some constant $\alpha > 0$. Then $\sum_{j=1}^{v_n} X_j/\sqrt{v_n}$ converges in distribution to $N(0, \sigma^2)$ as $n \to \infty$.

A proof of this proposition is given under Complements and Details. We will write $\underset{\to}{\mathcal{L}}$ to denote *convergence in distribution*.

Theorem 10.2 (*CLT for Markov chains*) *In addition to the hypothesis of Theorem 10.1, assume $EY_1^2 < \infty$, where $Y_r(r \geq 1)$ is given by (10.4). Then, whatever be the distribution of X_0,*

$$\frac{1}{\sqrt{n+1}} \sum_{m=0}^{n} (f(X_m) - \int f \, d\pi) \underset{\to}{\mathcal{L}} N(0, \delta^2), \quad as \ n \to \infty, \tag{10.10}$$

where $\delta^2 = \pi_i \text{var}(\tilde{Y}_1)$, $\tilde{Y}_1 := \sum_{m=\eta^{(1)}}^{\eta^{(2)}-1}(f(X_m) - \int f \, d\pi)$.

Proof. Write $\tilde{f} = f - \int f \, d\pi$, $\tilde{Y}_r = \sum_{m=\eta^{(r)}}^{\eta^{(r+1)}-1} \tilde{f}(X_m)$, $S_N = \sum_{r=1}^{N} \tilde{Y}_r$. Let $N_n = \inf\{r \leq n : \eta^{(r)} \leq n\}$, where $\eta^{(r)} = \eta_i^{(r)}$ as above. Then $N_n/n \to \pi_i > 0$ a.s., as $n \to \infty$ (as in the proof of Theorem 10.1, or see (8.1) and

(8.4)). Therefore, by Proposition 10.1,

$$\frac{1}{\sqrt{N_n}} S_{N_n} \equiv \frac{1}{\sqrt{N_n}} \sum_{r=1}^{N_n} \tilde{Y}_r \xrightarrow{\mathcal{L}} N(0, \sigma^2), \quad \text{as } n \to \infty, \qquad (10.11)$$

where $\sigma^2 = E\tilde{Y}_1^2 \equiv \mathrm{var} Y_1$. Now write, as in (10.7),

$$\frac{1}{\sqrt{n+1}} \sum_{m=0}^{n} \tilde{f}(X_m) = \frac{1}{\sqrt{n+1}} \sum_{m=0}^{\eta^{(1)}-1} \tilde{f}(X_m) + \sqrt{\frac{N_n-1}{n+1}}$$

$$\cdot \frac{1}{\sqrt{N_n-1}} S_{N_n-1} + \frac{1}{\sqrt{n+1}} \sum_{m=\eta^{(N_n)}}^{n} \tilde{f}(X_m)$$

$$= J_1 + J_2 + J_3, \text{ say.} \qquad (10.12)$$

Now $J_1 \to 0$ a.s., as $n \to \infty$. J_2 converges in distribution to $N(0, \delta^2)$ by Proposition 10.1, noting that $\sqrt{(N_n-1)/(n+1)}$ converges a.s. to $\sqrt{\pi_i}$. Finally, $|J_3| \le (n+1)^{-1/2} [\sum_{m=0}^{\eta^{(1)}-1} |\tilde{f}(X_m)| + \max\{\sum_{m=\eta^{(N_n)}}^{\eta^{(N_n+1)}-1} |\tilde{f}(X_m)| : 1 \le r \le n-1\}]$. To prove that this converges to 0 in probability, it is enough to show that for an i.i.d. sequence U_r $(r \ge 1)$ with finite second moment, $n^{-1/2} \max\{|U_r| : 1 \le r \le n\} \to 0$ in probability. To prove this, write $M_n := \max\{|U_r| : 1 \le r \le n\}$. For any given $\varepsilon > 0$, writing *i.o.* for infinitely often or "infinitely many n," one has (Exercise 10.1),

$$P(n^{-1/2} M_n > \varepsilon \, i.o.) = P(M_n \ge n^{1/2} \varepsilon \, i.o.) = P(|U_n| \ge n^{1/2} \varepsilon \, i.o.)$$

$$\le \sum_{n=N}^{\infty} P(|U_n| \ge n^{1/2} \varepsilon) = \sum_{n=N}^{\infty} P(U_n^2/\varepsilon^2 \ge n) \to 0 \text{ as } N \to \infty.$$

$$\text{For,} \quad \sum_{n=1}^{\infty} P(U_1^2/\varepsilon^2 \ge n) \le E(U_1^2/\varepsilon^2) < \infty.$$

■

Exercise 10.1 (a) Show that $\{M_n \ge n^{1/2} \varepsilon \, i.o.\} = \{|U_n| \ge n^{1/2} \varepsilon \, i.o.\}$. (b) If V is nonnegative, then show that $EV \ge \sum_{n=1}^{\infty} P(V \ge n)$. [Hint:

$$EV \ge \sum_{n=1}^{\infty} (n-1) P(n-1 \le V < n)$$

$$= \sum_{n=1}^{\infty} (n-1) [P(V \ge n-1) - P(V \ge n)]$$

$$= \lim_{N \uparrow \infty} \sum_{n=2}^{N} (n-1)[P(V \geq n-1) - P(V \geq n)]$$

$$= \lim_{N \uparrow \infty} [\sum_{n=2}^{N} P(V \geq n)$$

$$+ \sum_{n=2}^{N} (n-1)P(V \geq n-1) - \sum_{n=2}^{N} n P(V \geq n)]$$

$$= \lim_{N \uparrow \infty} \left[\sum_{n=2}^{N} P(V \geq n) + P(V \geq 1) - NP(V \geq N) \right]$$

$$= \lim_{N \uparrow \infty} \left[\sum_{n=1}^{N} P(V \geq n) - NP(V \geq N) \right] = \sum_{n=1}^{\infty} P(V \geq n)$$

Note that $NP(V \geq N) \leq EV1_{\{V \geq N\}} \to 0$ as $N \to \infty$. ∎

Example 10.2 Consider the two-state model of Example 3.1. Let η be the first return time to 0. Then the distribution of η_0, starting at 0, is given by

$$P_0(\eta_0 = 1) = P(X_1 = 0 \mid X_0 = 0) = 1 - p,$$
$$P_0(\eta_0 = 2) = P(X_1 = 1, X_2 = 0 \mid X_0 = 0) = pq,$$
$$P_0(\eta_0 = k) = P(X_1 = 1, X_2 = 1, \ldots, X_{k-1} = 1, X_k = 0 \mid X_0 = 0)$$
$$= p(1-q)^{k-2}q = pq(1-q)^{k-2} \quad (k \geq 2). \tag{10.13}$$

Therefore (see Exercise 10.2),

$$E_0\eta_0 = 1(1-p) + \sum_{k=2}^{\infty} kpq(1-q)^{k-2}$$

$$= 1 - p + pq \sum_{k=2}^{\infty} \{1 + (k-1)\}(1-q)^{k-2}$$

$$= 1 - p + pq \left(\frac{1}{q}\right) + pq \left(\frac{1}{q^2}\right) = 1 + \frac{p}{q} = \frac{p+q}{q}, \tag{10.14}$$

which accords with the facts that $\pi_0 = 1/E_0\eta_0$ (see (8.14)) and $\pi_0 = q/(p+q)$ (see Example 3.1, or calculate directly from the invariance Equation (3.10)).

Now, let $f = \mathbf{1}_{\{0\}}$. Then, by Theorem 10.1,

$$\lim_{n\to\infty}\frac{1}{n+1}N_n = f(0)\pi_0 + f(1)\pi_1 = 1\pi_0 + 0\pi_1 = \pi_0$$

$$= \frac{q}{p+q} \quad \text{a.s.,} \tag{10.15}$$

where N_n is the number of visits to the state 0 in time n.

Next, we apply Theorem 10.2 to the same function $f = \mathbf{1}_{\{0\}}$. Let $\eta^{(r)}$ be the rth return time to 0. Then $E_\pi f = 1\pi_0 + 0\pi_1 = \pi_0 = q/(p+q)$, $\tilde{f} = \mathbf{1}_{\{0\}} - q/(p+q)$, $Y_1 \equiv 1$ (see (10.4)), and

$$\tilde{Y}_1 \equiv \sum_{m=\eta^{(1)}}^{\eta^{(2)}-1} \tilde{f}(X_m) = 1 - (q/(p+q))\left(\eta^{(2)} - \eta^{(1)}\right). \tag{10.16}$$

Since the distribution of $\eta^{(2)} - \eta^{(1)}$, starting from an arbitrary initial state, is the same as the distribution of η_0, starting from state 0,

$$\text{var}(\tilde{Y}_1) = \left(\frac{q}{p+q}\right)^2 \text{var}_0(\eta_0) = \left(\frac{q}{p+q}\right)^2 \left[E_0\eta_0^2 - (E_0\eta_0)^2\right]. \tag{10.17}$$

Using (10.13), we get (see Exercise 10.2)

$$E_0\eta_0^2 = 1^2(1-p) + \sum_{k=2}^{\infty} k^2 pq(1-q)^{k-2}$$

$$= 1 - p + pq[k(k-1) + (k-1) + 1](1-q)^{k-2}$$

$$= 1 - p + pq\left[\frac{2}{q^3} + \frac{1}{q^2} + \frac{1}{q}\right]$$

$$= 1 + \frac{2p+pq}{q^2} = \frac{q^2 + 2p + pq}{q^2}. \tag{10.18}$$

Therefore, by (10.14) and (10.17),

$$\text{var}_0(\eta_0) = \frac{q^2 + 2p + pq - (p+q)^2}{q^2} = \frac{p(2-p-q)}{q^2},$$

$$\text{var}(\tilde{Y}_1) = \frac{p(2-p-q)}{(p+q)^2}. \tag{10.19}$$

Since $\pi_0 = q/(p+q)$, one has, by Theorem 10.2,

$$\frac{1}{\sqrt{n+1}}\left[N_n - \left(\frac{n+1}{2}\right)q/(p+q)\right] \xrightarrow{\mathcal{L}} N\left(0, \frac{pq(2-p-q)}{(p+q)^3}\right), \tag{10.20}$$

as $n \to \infty$. Note that, in the special case $p = q = 1/2$, $f(X_m) \equiv \mathbf{1}_{\{X_m=0\}}$ ($m \geq 1$) are independent symmetric Bernoulli (since the transition probabilities are all equal to $1/2$, and they do not depend on the state). Therefore, $\sum_{m=1}^{n} f(X_m)$ is a symmetric binomial $B(n, 1/2)$. The limiting variance in this case is $1/2(1 - 1/2) = 1/4$, the same as given by (10.20) with $p = q = 1/2$. ∎

Exercise 10.2 Suppose $|a| < 1$. Then the geometric series equals $\sum_{r=0}^{\infty} a^r = (1 - a)^{-1}$. By term-by-term differentiation with respect to a show that (i)$\sum_{r=1}^{\infty} ra^{r-1} \equiv \sum_{k=2}^{\infty} (k - 1)a^{k-2} = (1 - a)^{-2}$, (ii) $\sum_{r=1}^{\infty} r(r - 1)a^{r-2} \equiv \sum_{r=2}^{\infty} r(r - 1)a^{r-2} = 2(1 - a)^{-3}$. ∎

2.11 Markov Processes on Metric Spaces: Existence of Steady States

We continue our study of Markov processes on general state spaces, but now with the relatively mild restriction that the spaces be metric. The focus here is to provide a criterion for the existence of an invariant probability, not necessarily unique, under simple topological assumptions.

Throughout this section, unless otherwise stated, (S, d) is a metric space and $\mathcal{S}$ the Borel sigmafield on S. We will denote by $C_b(S)$ the *class of all real-valued bounded continuous functions on S.*

Definition 11.1 A transition probability $p(x, dy)$ on $(S, \mathcal{S})$ is said to have the *Feller property*, or is *weakly continuous*, if for every continuous bounded real-valued function f on S the function

$$x \to Tf(x) := \int_S f(y)p(x, dy), \quad (f \in C_b(S)) \qquad (11.1)$$

is continuous.

Example 11.1 (*Markov chains*)   Let S be countable and $d(x, y) = 1$ if $y \neq x$, $d(x, x) = 0$ for all $x, y \in S$. Then the Borel sigmafield on S is the class $\mathcal{S}$ of all subsets of S, since every subset of S is open. Thus every bounded function on S is continuous. Hence every transition probability of a Markov chain has the Feller property. ∎

Example 11.2 (*Random dynamical systems*)   Let (S, d) be an arbitrary metric space and $\Gamma = \{g_1, g_2, \ldots, g_k\}$ a set of continuous functions on

S into S. Let $p_i > 0$ $(1 \le i \le k)$, $\sum_{i=1}^{k} p_i = 1$. For each $n \ge 1$, let α_n denote a random selection from Γ with $\text{Prob}(\alpha_n = g_i) = p_i (1 \le i \le k)$, and assume that $\{\alpha_n : n \ge 1\}$ is an i.i.d. sequence. For each initial X_0 independent of $\{\alpha_n : n \ge 1\}$ (e.g., $X_0 = x$ for some $s \in S$), define

$$X_1 = \alpha_1 X_0 \equiv \alpha_1(X_0), \quad X_2 = \alpha_2 \alpha_1 X_0 \equiv d_2(X_1), \ldots,$$
$$X_n = \alpha_n \alpha_{n-1} \cdots \alpha_1 X_0 \equiv \alpha_n(X_{n-1})(n \ge 1). \tag{11.2}$$

Then $\{X_n : n \ge 0\}$ is a Markov process on S, with the transition probability

$$p(x, B) = \text{Prob}(\alpha_1 x \in B) = \sum_{i=1}^{k} p_i \mathbf{1}_{g_i^{-1}(B)}(x). \tag{11.3}$$

If $f \in C_b(S)$, then

$$Tf(x) = Ef(\alpha_1 x) = \sum_{i=1}^{k} p_i f(g_i(x)) \equiv \sum_{i=1}^{k} p_i(f \circ g_i)(x), \tag{11.4}$$

which is clearly continuous on S. Hence $p(x, dy)$ has the Feller property. A more general version of this is considered in the next chapter. ■

Definition 11.2 A sequence of probability measures P_n $(n \ge 1)$ on a metric space (S, d) is said to *converge weakly* to a probability measure P on S if

$$\lim_{n \to \infty} \int_S f \, dP_n = \int_S f \, dP \quad \forall f \in C_b(S). \tag{11.5}$$

Example 11.3 (*Classical CLT*) The classical *CLT* says that the sequence of distributions P_n of $n^{-1/2}(X_1 + \cdots + X_n)$, where $X_n, n \ge 1$, are i.i.d. with mean vector $\mathbf{O}$ and a finite covariance matrix $\mathbf{V}$, converges weakly to the normal distribution $N(\mathbf{O}, \mathbf{V})$ with mean vector $\mathbf{O}$ and covariance matrix $\mathbf{V}$. ■

As a matter of notation, $p^{(0)}(x, dy) = \delta_x(dy)$, i.e., $p^{(0)}(x, \{x\}) = 1$. Correspondingly, T^0 is the identity map on $\mathbf{B}(S)$: $T^0 f = f$.

The basic result of this section may now be stated.

Theorem 11.1 *Let $p(x, dy)$ be a transition probability with the Feller property on a metric space (S, d).*

(a) Suppose there exist $x \in S$, a sequence of integers $1 \leq n_1 < n_2 < \cdots$, and a probability measure π_x on S such that

$$\frac{1}{n_k} \sum_{r=0}^{n_k-1} p^{(r)}(x, dy) \text{ converges weakly to } \pi_x \text{ as } k \to \infty. \quad (11.6)$$

Then π_x is invariant for p.

(b) If the limit (11.6) holds for all $x \in S$ with the same limit $\pi_x = \pi$ ($x \in S$), then π is the unique invariant probability for p.

Proof.

(a) By (11.6), recalling $(T^r f)(x) = \int f(y) p^{(r)}(x, dy)$, one has

$$\frac{1}{n_k} \sum_{r=0}^{n_k-1} (T^r f)(x) \underset{k \to \infty}{\to} \int_S f(y) \pi_x(dy) \quad \forall f \in C_b(S). \quad (11.7)$$

Applying this to Tf, which belongs to $C_b(S)$ by virtue of the Feller property of $p(x, dy)$, one gets

$$\frac{1}{n_k} \sum_{r=1}^{n_k} (T^r f)(x) \underset{k \to \infty}{\to} \int_S (Tf)(y) \pi_x(dy). \quad (11.8)$$

But the left-hand sides of (11.7) and (11.8) have the same limits. Hence

$$\int_S (Tf)(y) \pi_x(dy) = \int_S f(y) \pi_x(dy) \quad \forall f \in C_b(S),$$

or (see (9.1')),

$$\int_S f(y) (T^* \pi_x)(dy) = \int_S f(y) \pi_x(dy) \quad \forall f \in C_b(S),$$

proving the invariance of π_x: $T^* \pi_x = \pi_x$.

(b) By (a), π is invariant, and (11.7) holds with π in place of π_x. If ν is another invariant probability, then integrating both sides of (11.7) w.r.t. ν, one obtains $\int f \, d\nu = \int f \, d\pi \ \forall f \in C_b(S)$, proving $\nu = \pi$. ∎

Remark 11.1 One can define a metric on the space $\mathcal{P}(S)$ of all probability measures on (a metric space) S such that the convergence $P_n \to P$ in that metric is equivalent to (11.5) (see Complements and Details). When

S is a compact metric space, one such metric is given by

$$\tilde{d}(P, Q) = \sum_{k=1}^{\infty} 2^{-k}(1 + \|f_k\|)^{-1} \left| \int_S f_k dP - \int f_k dQ \right|, \quad P, Q \in \mathcal{P}(S),$$

(11.9)

where $\{f_k: k \geq 1\}$ is dense in the *sup-norm* topology of $C(S) \equiv C_b(S)$. [Here $\|f\| := \max\{|f(x)|: x \in S\}$]. Such a dense sequence exists because $C(S)$ is a (complete) separable metric space in the sup-norm distance $\|f - g\|$ ($f, g \in C(S)$). By a Cantor diagonal argument one can then derive the following result, (whose proof is given in Complements and Details):

Lemma 11.1 *If S is a compact metric space then the space $\mathcal{P}(S)$ of all probability measures on $(S, \mathcal{S})$ is a compact metric space under the metric (11.9), which metrizes the topology of weak convergence on $\mathcal{P}(S)$.*

As a consequence of this lemma we can prove the following.

Theorem 11.2 *Suppose on a compact metric state space S, $p(x, dy)$ is a transition probability having the Feller property. Then there exists at least one invariant probability for p.*

Proof. Fix $x \in S$. In view of the Lemma above, there exists a subsequence of the sequence $\frac{1}{n} \sum_{r=0}^{n-1} p^{(r)}(x, dy)$, $n \geq 1$, say, $(1/n_k) \sum_{r=0}^{n_k-1} p^{(r)}(x, dy)$, $k \geq 1$, which converges to some probability measure π_x on $(S, \mathcal{S})$, as $k \to \infty$. By Theorem 11.1 (a), π_x is invariant for p. ∎

Corollary 11.1 *The transition probability of a Markov chain on a finite state space S has at least one invariant probability.*

Proof. S is compact metric under the metric $d(x, y) = 1$ if $x \neq y$, and $d(x, x) = 0$. In this case every transition probability has the Feller property (see Example 11.1). ∎

That the Feller property is generally indispensable in Theorems 11.1 and 11.2 is illustrated by the following example due to Liggett.

Example 11.4 Let $S = \{m/(m+1): m = 0, 1, \ldots\} \cup \{1\}$. Define $p(m/(m+1), \{(m+1)/(m+2)\}) = 1 \; \forall m = 0, 1, 2, \ldots$, and $p(1, \{0\}) = 1$. Then there does not exist any invariant probability, although S is compact (Exercise). This is, of course, the case of an extremely degenerate transition probability given by a deterministic dynamical system: $f(m/m+1) = (m+1)/(m+2) (m = 0, 1, 2, \ldots)$, $f(1) = 0$. One may, however, easily perturb this example to construct one in which the transition probability is not degenerate. For example, let $p(m/(m+1), \{(m+1)/(m+2)\}) = \theta$, $p(m/(m+1); \{(m+2)/(m+3)\}) = 1 - \theta$ $(m = 0, 1, 2, \ldots)$, $p(1, \{0\}) = \theta$, $p(1, \{1/2\}) = 1 - \theta$, for some $\theta \in (0, 1)$. One may show that there does not exist an invariant probability in this case as well (Exercise). ∎

It may be noted that the results on the existence and uniqueness of an invariant probability in Section 2.9 did not require topological assumptions. However, the assumptions were strong enough to imply the convergence of the sequence $(1/n) \sum_{r=0}^{n-1} p^{(r)}(x, B)$ to a probability $\pi(B)$ uniformly for all $B \in \mathcal{S}$. The existence of an invariant probability may be demonstrated without the assumption of uniform convergence, as shown below.

Proposition 11.1 *Let $(S, \mathcal{S})$ be a measurable state space. If for some $x \in S$, there exist a probability measure π_x and a sequence $1 \leq n_1 < n_2 < \cdots$ such that*

$$\frac{1}{n_k} \sum_{r=0}^{n_k-1} p^{(r)}(x, B) \xrightarrow[k\to\infty]{} \pi_x(B) \quad \forall B \in \mathcal{S}, \tag{11.10}$$

then π_x is an invariant probability for p. If, in addition, π_x does not depend on x, then it is the unique invariant probability.

Proof. By standard measure theoretic arguments for the approximation of bounded measurable functions by the so-called simple functions of the form $\sum_{i=1}^{m} a_i \mathbf{1}_{A_i}$ ($A_1, \ldots, A_m$ disjoint sets in $\mathcal{S}$, and $a_1, a_2, \ldots, a_m$ nonzero reals), (11.10) may be shown to be equivalent to

$$\frac{1}{n_k} \sum_{r=0}^{n_k-1} (T^r f)(x) \xrightarrow[k\to\infty]{} \int f(y)\pi_x(dy) \quad \forall f \in \mathbf{B}(S). \tag{11.11}$$

Now follow the same line of argument as used in the proof of Theorem 11.1, but for all $f \in \mathbf{B}(S)$. ∎

Remark 11.2 The property (a) in Theorem 11.1 is implied by the *tightness* of the sequence of probability measures $\{\frac{1}{n}\sum_{r=1}^{m}p^{(r)}(x,dy):$ $n \geq 1\}$ (see Complements and Details). Here *a sequence* $\{P_n : n \geq 1\}$ $\subset \mathcal{P}(S)$ *is tight* if $\forall \varepsilon > 0$ there exists a compact set $\mathcal{K}_\varepsilon$ such that $P_n(\mathcal{K}_\varepsilon) \geq 1 - \varepsilon \; \forall n$. By combining this fact with Theorem C9.3 in Complements and Details, we get the following result:

Proposition 11.2 *Suppose a Markov process on a metric space* (S, d) *has the Feller property, and it is* ν*-irreducible (with respect to some nonzero sigmafinite measure* ν*). If, in addition,* $\{\frac{1}{n}\sum_{r=1}^{n}p^{(r)}(x,dy):$ $n \geq 1\}$ *is tight for some* x*, then the process is positive Harris recurrent and* $\frac{1}{n}\sum_{r=1}^{n}p^{(r)}(z,dy)$ *converges in variation distance to a unique invariant probability* π*, for all* $z \in S$.

2.12 Asymptotic Stationarity

We have focused much attention in the preceding sections on the *stability in distribution* of Markov processes $\{X_n : n \geq 0\}$, i.e., convergence of the distribution X_n to a steady state, or invariant probability π, whatever be the initial distribution as $n \to \infty$. One may suspect that this means that the *process* $X_n^+ := (X_n, X_{n+1}, \ldots)$ converges in distribution, as $n \to \infty$, to the stationary Markov process $(X_0^*, X_1^*, \ldots)$ whose initial distribution is π and which has the same transition probability as $\{X_n : n \geq 0\}$. In this section we demonstrate that this is indeed true under broad conditions.

Definition 12.1 By the *distribution of a process* $\{Y_k : k = 0, 1, 2, \ldots\}$ on a measurable state space $(S, \mathcal{S})$ we mean a probability measure Q on the Kolmogorov (or product) sigmafield $(S^\infty, \mathcal{S}^{\otimes\infty})$, such that $\mathrm{Prob}(\{Y_k : k \geq 0\} \in B) = Q(B) \; \forall B \in \mathcal{S}^{\otimes\infty}$. Here S^∞ is the set of all sequences $(x_0, x_1, \ldots)$ in S (i.e., $x_j \in S \; \forall j \geq 0$), and $\mathcal{S}^{\otimes\infty}$ is the sigmafield on S^∞ generated by all finite dimensional measurable sets of the form $\{\mathbf{x} \in S^\infty : (x_0, x_1, \ldots, x_k) \in A\} \; \forall A \in \mathcal{S}^{\otimes\{0,1,\ldots,k\}} \equiv \mathcal{S} \otimes \mathcal{S} \otimes \cdots \otimes \mathcal{S}(k+1$ terms).

Definition 12.2 If (S, d) is a separable metric space and $\mathcal{S}$ its Borel sigmafield, a process $\{X_n : n \geq 0\}$ on $(S, \mathcal{S})$ is said to be *asymptotically stationary* if $X_n^+ := (X_n, X_{n+1}, X_{n+2}, \ldots)$ converges in distribution, as $n \to \infty$, to $(X_0^*, X_1^*, X_2^*, \ldots)$, where the latter is a stationary process. Here distributions of X_n^+ and $(X_0^*, X_1^*, \ldots)$ are probability measures Q_n,

Q, say, on the space $(S^\infty, S^{\otimes\infty})$, where S^∞ has the product topology and $S^{\otimes\infty}$ is its Borel sigmafield. One can show that S^∞ is a separable metric space with the metric (see Appendix)

$$d^\infty(\mathbf{x}, \mathbf{y}) = \sum_{n=0}^{\infty} \frac{1}{2^n}(d(x_n, y_n) \wedge 1) \quad \forall \mathbf{x}, \mathbf{y} \in S^\infty. \tag{12.1}$$

On the product space $S^m \equiv S^{\{0,1,\dots,m-1\}}$ we will use the metric

$$d^m(\mathbf{x}, \mathbf{y}) = \sum_{n=0}^{m-1} \frac{1}{2^n}(d(x_n, y_n) \wedge 1) \quad \forall \mathbf{x} = (x_0, \dots, x_{m-1}),$$

$$\mathbf{y} = (y_0, \dots, y_{m-1}) \in S^m. \tag{12.2}$$

Lemma 12.1 *Let (S, d) be a separable metric space, and S^∞ the space of all sequence $\mathbf{x} = (x_0, x_1, x_2, \dots)$ of elements of S, endowed with the metric d^∞ in (12.1). Let $C_F(S^\infty)$ be the set of all real-valued bounded continuous functions on S^∞ depending only on finitely many coordinates. If $P_n(n \geq 1)$, P are probability measures on the Borel sigmafield $\mathcal{B}(S^\infty)$ such that*

$$\int h\, dP_n \to h\, dP \quad \text{for all } h \in C_F(S^\infty), \quad \text{as } n \to \infty, \tag{12.3}$$

then P_n converges weakly to P.

Proof. Let g be an arbitrary real-valued bounded uniformly continuous function on S^∞. Given $\varepsilon > 0$, there exists $h \in C_F(S^\infty)$ such that

$$\sup_{\mathbf{x}} |g(\mathbf{x}) - h(\mathbf{x})| \equiv \|g - h\|_\infty < \varepsilon/2. \tag{12.4}$$

To see this note that there exists $\delta \equiv \delta(\varepsilon) > 0$ such that $|g(\mathbf{x}) - g(\mathbf{y})| < \varepsilon/2$ if $d^\infty(\mathbf{x}, \mathbf{y}) < \delta$. Find $k \equiv k(\varepsilon)$ such that $\sum_{n=k+1}^{\infty} \frac{1}{2^n} < \delta$. Then $d^\infty(\mathbf{x}, \mathbf{y}) < \delta$ if $x_n = y_n$ for all $n \leq k$. Now fix $z \in S$ and define

$$h(\mathbf{x}) := g(x_0, x_1, \dots, x_k, z, z, z, \dots). \tag{12.5}$$

Then (12.4) holds. In view of this, one has

$$\left| \int g\, dP_n - \int h\, dP_n \right| < \varepsilon/2 \quad \text{and} \quad \left| \int g\, dP - \int h\, dP \right| < \varepsilon/2,$$

so that

$$\left| \int g \, dP_n - \int g \, dP \right| < \left| \int g \, dP_n - \int h \, dP_n \right|$$

$$+ \left| \int h \, dP_n - \int h \, dP \right| + \left| \int h \, dP - \int g \, dP \right|$$

$$< \varepsilon + \left| \int h \, dP_n - \int h \, dP \right|.$$

By (12.3) this implies

$$\lim_{n \to \infty} \left| \int g \, dP_n - \int g \, dP \right| \leq \varepsilon.$$

Since ε is arbitrary, it follows that

$$\int g \, dP_n \to \int g \, dP.$$

This implies that P_n converges weakly to P. (See Theorem C 11.1 (ii) in Complements and Details.) ∎

We now state and prove one of the main results on the asymptotic stationarity of Markov processes $\{X_n \colon n \geq 1\}$.

Theorem 12.1 *Assume that X_n has a Feller continuous transition probability and converges in distribution to a probability measure π as $n \to \infty$. Then (a) π is an invariant probability and (b) $X_n^+ = (X_n, X_{n+1}, \ldots)$ converges in distribution to Q_π as $n \to \infty$, where Q_π is the distribution of the stationary Markov process $(X_0^*, X_1^*, \ldots)$ having the initial distribution π and the same transition probability $p(x, dy)$ as that of $X_n (n \geq 0)$.*

Proof. (a) Let ν_n denote the distribution of X_n. For every bounded continuous $f \colon S \to \mathbb{R}$, one has

$$\int f(y) \pi(dy) = \lim_{n \to \infty} E f(X_{n+1}) = \lim_{n \to \infty} E E[f(X_{n+1}) \mid X_n]$$

$$= \lim_{n \to \infty} \int_S \left\{ \int_S f(y) p(x, dy) \right\} \nu_n(dx)$$

$$= \int_S \left\{ \int_S f(y) p(x, dy) \right\} \pi(dx),$$

since, by Feller continuity, $x \to \int f(y)p(x, dy)$ is continuous (and bounded) on S. (b) We will show that, for all $m \geq 0$, and for all *bounded continuous* $f : S^{\{0,1,\ldots,m\}} \to \mathbb{R}$, one has

$$\lim_{n \to \infty} E f(X_n, X_{n+1}, \ldots, X_{n+m})$$

$$= \int \int \cdots \int f(x_0, x_1, \ldots, x_m)p(x_{m-1}, dx_m) \cdots p(x_0, dx_1)\pi(dx_0).$$
(12.6)

This is clearly enough, by Lemma 12.1. Fix $m \geq 0$ and f as above. Now using the Markov property,

$$E f(X_n, X_{n+1}, \ldots, X_{n+m})$$

$$= \int \int \cdots \int f(x_0, x_1, \ldots, x_m)p(x_{m-1}, dx_m) \cdots p(x_0, dx_1)v_n(dx_0)$$

$$= \int g(x_0)v_n(dx_0),$$
(12.7)

where

$$g(x_0) := \int \cdots \int f(x_0, x_1, \ldots, x_m)p(x_{m-1}, dx_m) \cdots p(x_0, dx_1).$$
(12.8)

To prove continuity of g, we first show that the function

$$h_m(x_0, x_1, \ldots, x_{m-1}) := \int f(x_0, x_1, \ldots, x_{m-1}, x_m)p(x_{m-1}, dx_m)$$
(12.9)

is continuous on $S^{\{0,1,\ldots,m-1\}}$ and bounded. Boundedness follows from $\|h_m\|_\infty \leq \|f\|_\infty$. Now fix $(x_0, x_1, \ldots, x_{m-1})$, and let $(y_0^{(n)}, \ldots, y_{m-1}^{(n)})$ converge to $(x_0, x_1, \ldots, x_{m-1})$ as $n \to \infty$. Then, by Lebesgue's Dominated Convergence Theorem,

$$\limsup_{n \to \infty} \left| h_m \left(y_0^{(n)}, \ldots, y_{m-1}^{(n)} \right) - h_m (x_0, \ldots, x_{m-1}) \right|$$

$$\leq \limsup \left| \int \left[f \left(y_0^{(n)}, \ldots, y_{m-1}^{(n)}, x_m \right) - f(x_0, \ldots, x_{m-1}, x_m) \right] p(x_{m-1}, dx_m) \right|$$

$$+ \limsup \left| \int f \left(y_0^{(n)}, \ldots, y_{m-1}^{(n)}, x_m \right) p \left(y_{m-1}^{(n)}, dx_m \right) \right.$$

$$\left. - \int f \left(y_0^{(n)}, \ldots, y_{m-1}^{(n)}, x_m \right) p(x_{m-1}, dx_m) \right|$$

$$= \limsup \left| \int f \left(y_0^{(n)}, \ldots, y_{m-1}^{(n)}, x_m \right) p \left(y_{m-1}^{(n)}, dx_m \right) \right.$$

$$\left. - \int f \left(y_0^{(n)}, \ldots, y_{m-1}^{(n)}, x_m \right) p(x_{m-1}, dx_m) \right|.$$
(12.10)

The set of functions $\mathcal{F} := \{x_m \to f(y_0^{(n)}, \ldots, y_{m-1}^{(n)}, x_m): n \geq 1\}$ is a *uniformity class* (see Proposition C11.1 in Complements and Details). Therefore,

$$\sup_{h \in \mathcal{F}} \left| \int h(x_m) p\left(y_{m-1}^{(n)}, dx_m\right) - \int h(x_m) p(x_{m-1}, dx_m) \right| \to 0$$

$$\text{as } n \to \infty. \quad (12.11)$$

The same proof applies to show that

$$h_{m-1}(x_0, x_1, \ldots, x_{m-2})$$
$$:= \int h_m(x_0, x_1, \ldots, x_{m-2}, x_{m-1}) p(x_{m-2}, dx_{m-1}) \quad (12.12)$$

is continuous (and bounded). Proceeding in this manner, we finally show that

$$g(x_0) = h_1(x_0) \equiv \int h_2(x_0, x_1) p(x_0, dx_1) \quad (12.13)$$

is continuous.   ∎

Remark 12.1 As an application of Theorem 12.1, let $X_n (n \geq 0)$ be a Markov process on $S = \mathbb{R}^1$ with an arbitrary initial distribution, and $L_n :=$ length of run of values $> c$, beginning with time n. Under the hypothesis of Theorem 12.1, L_n converges in distribution to the length of runs under Q_π, beginning with time 0, provided $\pi(\{c\}) = 0$. To see this, write

$$P(L_n = k) = P(X_{n+j} > c \text{ for } 0 \leq j \leq k-1, X_{n+k} \leq c)$$
$$\to Q_\pi(\{(x_0, x_1, \ldots): x_j > c \text{ for } 0 \leq j \leq k-1, x_k \leq c\})(k > 0),$$
$$P(L_n = 0) = P(X_n \leq c) \to \pi(\{x: x \leq c\}). \quad (12.14)$$

Note that the topological boundary of the set $A = \{(x_0, x_1, \ldots): x_j > c$ for $0 \leq j \leq k-1, x_k \leq c\}$ is contained in the union of the $k+1$ sets $B(j) = \{(x_0, x_1, \ldots): x_j = c\}, 0 \leq j \leq k$, and $Q_\pi(B(j)) = \pi(\{c\}) = 0$, by assumption.

A stronger notion of asymptotic stationarity applies when the distribution of X_n converges in variation norm, as in the case of Harris positive recurrent cases considered in Section 2.9.

Definition 12.3 Let $X_n(n \geq 0)$ be a Markov process on a measurable state space $(S, \mathcal{S})$. If the distribution of $X_n^+ = (X_n, X_{n+1}, \ldots)$ converges in total variation distance to that of a stationary process $(X_0^*, X_1^*, \ldots)$ with the same transition probability, then we will say $X_n(n \geq 0)$ is *asymptotically stationary in the strong sense.*

Theorem 12.2 *Let $p(x, dy)$ be a transition probability on $(S, \mathcal{S})$. Suppose the n-step transition probability $p^{(n)}(x, dy)$ converges to $\pi(dy)$ in total variation distance for every $n \in S$. Then a Markov process $X_n(n \geq 0)$ having the transition probability $p(x, dy)$ is asymptotically stationary in the strong sense, no matter what its initial distribution is.*

Proof. Let $B \in \mathcal{S}^{\otimes\infty}$ be arbitrary. Then,

$$P_\mu(X_n^+ \in B) = \int \text{Prob}(X_n^+ \in B \mid X_0 = x)\mu(dx),$$

$$\text{Prob}(X_n^+ \in B \mid X_0 = x) = \int_S P_y(B)p^{(n)}(x, dy),$$

where P_μ, P_y denote the distributions of the Markov process under the initial distribution μ and the initial state y, respectively. Hence

$$\sup_{B \in \mathcal{S}^{\otimes\infty}} \left| P_\mu\left(X_n^+ \in B\right) - P_\pi(B) \right|$$

$$= \sup_{B \in \mathcal{S}^{\otimes\infty}} \left| \int_S \left\{ \int_S P_y(B)\left[p^{(n)}(x, dy) - \pi(dy)\right] \right\} \mu(dx) \right|$$

$$\leq \int \sup \left\{ \left| \int f(y)\left[p^{(n)}(x, dy) - \pi(dy)\right] \right| : \right.$$

$$\left. |f| \leq 1, f \text{ measurable} \right\} \mu(dx) \to 0 \quad \text{as } n \to \infty. \qquad \blacksquare$$

2.13 Complements and Details

Section 2.2. A complete measure theoretic proof of *Kolmogorov's existence theorem* may be found in Billingsley (1986, pp. 506–517). For Kolmogorov's original proof, see Kolmogorov (1950). We sketch here an elegant functional analytic proof due to Nelson (1959). First, let S be a *compact metric space* and take $\mathbb{Z}_+$ as the parameter set. The set C_F of continuous functions on the compact metric space $S^{\mathbb{Z}_+}$ depending only on finitely many coordinates is dense (with respect to the supremum norm) in the set $C(S^{\mathbb{Z}+})$ of all continuous functions on $S^{\mathbb{Z}+}$ by the *Stone–Weirstrass theorem* (see Royden 1968, p. 174). On C_F a continuous linear functional ℓ is defined, by integration of $f \in C_F$ with respect to the finite-dimensional distribution on the coordinates on which f depends. This functional is *well defined* by the *consistency condition*. The unique extension of this to $C(S^{\mathbb{Z}+})$, also denoted ℓ, then has the representation $\ell(f) = \int f \, d\mu$, for a uniquely defined probability measure μ on the Borel sigmafield of $S^{\mathbb{Z}+}$, by the *Riesz representation theorem* (see Royden 1968, p. 310). In case S is noncompact, but Polish, S may be mapped homeomorphically onto a Borel subset of $[0, 1]^\infty$, and thereby we may regard S as a Borel subset of $[0, 1]^\infty$ (see Royden 1968, p. 326). Then the closure $\bar{S}$ of S in $[0, 1]^\infty$ is compact. This leads to a unique probability measure μ on $\bar{S}^{\mathbb{Z}+}$, whose finite-dimensional distributions are as those specified. It is simple to check that $\mu(S^{\mathbb{Z}+}) = 1$, completing the construction.

An important extension of Kolmogorov's Existence Theorem to *arbitrary measurable state spaces* $(S, \mathcal{S})$ holds for Markov processes in general and, independent sequences in particular. This extension is due to Tulcea (see, e.g., Neveu 1965, pp. 161–167). Here the probability is constructed on $(S^{\mathbb{Z}+}, \mathcal{S}^{\otimes \mathbb{Z}+})$, where $\mathcal{S}^{\otimes \mathbb{Z}+}$ is the usual product sigmafield. We sometimes write $(S^\infty, \mathcal{S}^{\otimes \infty})$, instead, for this infinite product space.

Section 2.4. Suppose a state i of a Markov chain has period d, and $i \leftrightarrow j$. Then there exist positive integers a and b such that $p_{ij}^{(a)} > 0$ and $p_{ji}^{(b)} > 0$. If m is a positive integer such that $p_{jj}^{(m)} > 0$, then one has $p_{jj}^{(2m)} \geq p_{jj}^{(m)} p_{jj}^{(m)} > 0$, so that $p_{ii}^{(a+m+b)} \geq p_{ij}^{(a)} p_{jj}^{(m)} p_{ji}^{(b)} > 0$ and $p_{ii}^{(a+2m+b)} \geq p_{ij}^{(a)} p_{jj}^{(2m)} p_{ji}^{(b)} > 0$. Hence $a + 2m + b$ and $a + m + b$ are both multiples of d, as is their difference m. This implies that the period of j is greater than or equal to the period of i, namely d. By symmetry of $\leftrightarrow$, the period of i is greater than or equal to that of j. Thus all essential states belonging to the same equivalence class of communicating states have the same period.

Section 2.5. The following result was stated without proof in Section 2.5.

Proposition C5.1 *For an irreducible aperiodic Markov chain on a space S there exists for each pair (i, j) a positive integer $v = v_{ij}$ such that $p_{ij}^{(v)} > 0$. If S is finite, one may choose $v = max\{v_{ij} : i, j \in S\}$ independent of (i, j).*

Proof. Fix a state j, and let $B_j = \{r \geq 1: p_{jj}^{(r)} > 0\}$. Observe that B_j is *closed under addition*. By hypothesis, g.c.d. of B_j is 1. The set $G_j = \{m - n: m, n \in B_j\}$ is an *additive subgroup of* $\mathbb{Z}$ and is, therefore, of the from $d\mathbb{Z} \equiv \{dn: n \in \mathbb{Z}\}$ for some positive integer d. Since $G_j \supset B_j$, $d = 1$. For all elements of B_j are divisible by d, so that g.c.d. of B_j is no less than d. Hence $1 \in G_j$. Therefore, there exists an integer $r \geq 1$, such that both r and $r + 1$ belong to B_j (so that $(r + 1) - r = 1 \in G_j$). If $n > (2r + 1)^2 + 1$, write $n = t(2r + 1) + q$, where $q = 0, 1, \ldots, 2r$ (and $t \geq 2r + 1$). Then n may be expressed as $n = (t + q)(r + 1) + (t - q)r$, with both $(t + q)(r + 1)$ (a multiple of $r + 1$) and $(t - q)r$ (a multiple of r) belong to B_j. Hence $n \in B_j$.

Finally, for each $i \in S$, there exists n_i such that $p_{ij}^{(n_i)} > 0$. It follows that $p_{ij}^{(n+n_i)} > 0 \; \forall n > (2r + 1)^2 + 1$. Let $v_{ij} = (2r + 1)^2 + 2 + \max\{n_i: i \in S\backslash\{j\}\}$. ∎

Section 2.6. The class $\mathcal{G}$, say, of sets F in the Kolmogorov sigmafield $\mathcal{F} = S^{\otimes\infty}$ for which (6.2) holds is easily seen to satisfy the following: (i) $\Omega \in \mathcal{G}$, (ii) $A, B \in \mathcal{G}, A \subset B \Rightarrow B\backslash A \in \mathcal{G}$, and (iii) $\mathcal{G}$ is closed under monotone increasing and decreasing limits. Since the class $\mathcal{A}$ of finite-dimensional sets in $S^{\otimes\infty}$ is a field, and $\mathcal{A} \subset \mathcal{G}$, it follows by a standard monotone class theorem that $\sigma(\mathcal{A}) \subset \mathcal{G}$. But $\sigma(\mathcal{A}) = S^{\otimes\infty}$, by definition, so that $\mathcal{G} = S^{\otimes\infty}$. (Note that this argument holds for Markov processes on more general state spaces $(S, \mathcal{S})$, in which case $\mathcal{A}$ is the class of all *measurable* finite-dimensional sets.) See Billingsley (1986, pp. 36–39) for the above monotone class theorem known as *Dynkin's π–λ theorem.* Also see Dynkin (1961).

Section 2.7. The statement that the random cycles, or blocks, $\mathbf{W}_r = (X_{\eta^{(r)}}, X_{\eta^{(r)}+1}, \ldots, X_{\eta^{(r+1)}})$, $r \geq 1$, are i.i.d. is proved in detail as follows. Given any $m \geq 1$ and states $i_1, i_2, \ldots, i_{m-1} \in S\backslash\{x\}$, the event $A_r := \{\mathbf{W}_r = (x, i_1, i_2, \ldots, i_{m-1}, x)\}$ is the same as the event $\{\eta^{(r+1)} = \eta^{(r)} + m, (X_{\eta^{(r)}}^+)_n = i_n$ for $n = 1, 2, \ldots, m - 1\}$. By the strong Markov property, the conditional probability of the last event, given the pre-$\eta^{(r)}$ sigmafield $\mathcal{F}_{\eta^{(r)}}$, is $P_x(X_1 = i_1, X_2 = i_2, \ldots, X_{m-1} = i_{m-1}, X_m = x)$. Since the latter (conditional) probability is a constant, i.e., nonrandom, A_r is independent of $\mathcal{F}_{\eta^{(r)}}$ (and, therefore, of A_j; $0 \leq j \leq r - 1$), and since this probability does not depend on r, we have established that $\mathbf{W}_r$'s are independent and have the same distribution ($r \geq 1$).

Theorem 7.2 was proved by Polya (1921). For a proof of Stirling's formula, see Feller (1968, pp. 52–54).

Section 2.8. We will solve Equation (8.24) for the expected time $m(x)$ to reach the boundary $\{c, d\}$, starting from x, $c \le x \le d$. For this introduce $u_x = 1 + \sum_{y=c+1}^{x-1} \delta_{c+1} \cdots \delta_y/(\beta_{c+1} \cdots \beta_y)$, $c + 2 \le x \le d$, $u(c) = 0$, $u(c + 1) = 1$. Then (8.24) may be expressed as

$$\frac{m(x + 1) - m(x)}{u_{x+1} - u_x} - \frac{m(x) - m(x - 1)}{u_x - u_{x-1}}$$

$$= -\frac{\beta_{c+1} \cdots \beta_x}{\delta_{c+1} \cdots \delta_x} \cdot \frac{1}{\beta_x}, \quad c + 1 \le x \le d - 1.$$

Summing over $x = c + 1, \ldots, y$, yields

$$\frac{m(y + 1) - m(y)}{u_{y+1} - u_y} - m(c + 1)$$

$$= -\sum_{x=c+1}^{y} \frac{\beta_{c+1} \cdots \beta_x}{\delta_{c+1} \cdots \delta_x} \cdot \frac{1}{\beta_x}, \quad c + 1 \le y \le d - 1. \quad (C8.1)$$

Multiplying both sides by $u_{y+1} - u_y$ and then summing over $y = d - 1, \ldots, c + 1$, one computes

$$m(c + 1) = \frac{\sum_{y=c+1}^{d-1} \sum_{x=c+1}^{y} \frac{\delta_x \cdots \delta_y}{\beta_x \cdots \beta_y} \cdot \frac{1}{\delta_x}}{1 + \sum_{y=c+1}^{d-1} \frac{\delta_{c+1} \cdots \delta_y}{\beta_{c+1} \cdots \beta_y}}. \quad (C8.2)$$

Multiplying (C8.1) by u_{y+1} and summing (C8.1) over $y = d - 1, \ldots, x$, one finally arrives at

$$m(x) = \sum_{y=x}^{d-1} \sum_{z=c+1}^{y} \frac{\delta_z \cdots \delta_y}{\beta_z \cdots \beta_y} \cdot \frac{1}{\delta_z} - m(c + 1)$$

$$\cdot \sum_{y=x}^{d-1} \frac{\delta_{c+1} \cdots \delta_y}{\beta_{c+1} \cdots \beta_y}, \quad c + 1 \le x \le d - 1. \quad (C8.3)$$

Turning to the *Ehrenfest heat exchange problem* in Example 8.3, let $\bar{m}(x) := E_x T_d$, when $T_d = \inf\{n \ge 0 : X_n = d\}$. Then, as in the preceding case, one has

$$(\bar{m}(x + 1) - \bar{m}(x))\beta_x - (\bar{m}(x) - \bar{m}(x - 1))\delta_x = -1(1 \le x \le 2d - 1),$$

but with boundary conditions $\bar{m}(d) = 0$, $\bar{m}(2d) - \bar{m}(2d - 1) = 1$. Solving it, in a manner analogous to that described above, one obtains

$$m(x) = \sum_{y=1}^{x} \frac{\beta_y \beta_{y+1} \cdots \beta_{2d-1}}{\delta_y \delta_{y+1} \cdots \delta_{2d-1}} + \sum_{y=1}^{x}$$

$$\sum_{z=y}^{2d-1} \frac{\beta_z \cdots \beta_y}{\delta_z \cdots \delta_y} \cdot \frac{1}{\beta_z}, \quad 1 \le x \le 2d-1,$$

$$m(0) = 0, \quad m(2d) = 1 + m(2d-1). \qquad (C8.4)$$

Here $\beta_y = 1 - y/2d$, $\delta_y = y/2d(1 \le y \le 2d - 1)$, $\beta_{2d} = 0, \delta_{2d} = 1$.

In particular, the expected time to reach disequilibrium (namely, the state 0 or $2d$), starting from the equilibrium (namely, the state d) is easily seen to be

$$m(d) = \frac{1}{2d} 2^{2d} \left(1 + 0 \left(\frac{1}{d} \right) \right), \quad \text{as } d \to \infty. \qquad (C8.5)$$

On the other hand, the expected time $\bar{m}(0)$ of reaching equilibrium (state d) starting from the extreme disequilibrium state 0 (or $2d$) may be shown, in the same manner as above, to satisfy

$$\bar{m}(0) \le d \log d + d + 0(1) \quad \text{as } d \to \infty. \qquad (C8.6)$$

For $d = 10,000$ balls, and rate of transition 1 ball per second,

$$m(d) \text{ is nearly } 10^{6000} \text{ years}, \quad \bar{m}(0) < 29 \text{ hours.} \qquad (C8.7)$$

A proof of Theorem 8.2 may be found in Kolmogorov (1936). Our proof is adapted from Durrett (1996, pp. 82–83). The idea of meeting together, or *coupling*, of two independent positive recurrent aperiodic Markov chains $\{X_n\}_{n\ge0}$, $\{X'_n\}_{n\ge0}$, having the same transition probability, but one with initial state x and the other with the invariant initial distributions, is due to Doeblin (1937). For the history behind the Ehrenfest model, see Kac (1947). Also see Waymire (1982).

For comprehensive and standard treatments of Markov chains, we refer to Doob (1953), Feller (1968), Chung (1967), and Karlin and Taylor (1975, 1981). Our nomenclature of Markov chains as Markov processes with countable state spaces follows these authors.

Finally, specializing (C8.2) and (C8.3) to the case of the simple random walk of Example 6.4, we get

$$
m(x) = \begin{cases} \dfrac{(d-x)(x-c)}{p+q} & \text{for } c \leq x \leq d, \text{ if } p = q, \\[2ex] \dfrac{d-c}{p-q}\left\{\dfrac{1-\left(\frac{q}{p}\right)^{x-c}}{1-\left(\frac{q}{p}\right)^{d-c}}\right\} - \dfrac{x-c}{p-q}, & c \leq x \leq d, \text{ if } p \neq q. \end{cases} \tag{C8.8}
$$

Section 2.9. Consider a sequence $\{Y_n: n \geq 1\}$ of i.i.d. positive integer-valued random variables and Y_0 a nonnegative integer-valued random variable independent of $\{Y_n: n \geq 1\}$. The partial sum process $S_0 = Y_0$, $S_k = Y_0 + \cdots + Y_k (k \geq 1)$ defines the so-called *delayed renewal process* $\{N_n: n \geq 1\}$:

$$
N_n := \inf\{k \geq 0: S_k \geq n\}(n = 0, 1, \ldots), \tag{C9.1}
$$

with Y_0 as the *delay* and its distribution as the *delay distribution*. In the case $Y_0 \equiv 0$, $\{N_n: n \geq 0\}$ is simply referred to as a *renewal process*. This nomenclature is motivated by classical renewal theory in which components subject to failure (e.g., light bulbs) are instantly replaced upon failure, and $Y_1, Y_2, \ldots$, represent the random durations or *lifetimes* of the successive replacements. The delay random variable Y_0 represents the length of time remaining in the life of the initial component with respect to some specified time origin. For a more general context consider, as in Example 1, a Markov chain $\{X_n: n \geq 0\}$ on a (countable) state space S, and fix a state y such that the first passage time $T_y \equiv \eta_y^{(0)}$ to state y is finite a.s., as are the successive return times $\eta_y^{(k)}$ to $y(k \geq 1)$. By the strong Markov property, $Y_0 := \eta_y^{(0)}$, $Y_k := \eta_y^{(k)} - \eta_y^{(k-1)}(k \geq 1)$, are independent, with $\{Y_k : k \geq 1\}$ i.i.d. The renewal process may now be defined with $S_k = \eta_y^{(k)}(k \geq 1)$, the kth *renewal* occurring at time $\eta_y^{(k)}$ if $\eta_y^{(0)} = 0$, i.e., y is the initial state. Let $N_n = \inf\{k \geq 0: \eta^{(k)} \geq n\}$.

It may be helpful to think of the partial sum process $\{S_k: k \geq 0\}$ to be a *point process* $0 \leq S_0 < S_1 < S_2 < \cdots$ realized as a randomly located increasing sequence of *renewal times* or *renewal epochs* on $\mathbb{Z}_+ = \{0, 1, 2, \ldots\}$. The processes $\{S_k: k \geq 0\}$ and $\{N_n: n \geq 0\}$ may be thought of as (approximate) *inverse functions* of each other on $\mathbb{Z}_+$. Another related process of interest is the *residual lifetime process* defined by

$$
R_n := S_{N_n} - n = \inf\{S_k - n: k \text{ such that } S_k - n \geq 0\}. \tag{C9.2}
$$

The indices n for which $R_n = 0$ are precisely the renewal times.

Definition C9.1 A random variable τ taking values in $\mathbb{Z}_+ \cup \{\infty\}$ is said to be a $\{\mathcal{F}_k\}_{k \geq 0}$-*stopping time* with respect to an increasing sequence of sigmafields $\{\mathcal{F}_k: k \geq 0\}$ if $\{\tau \leq k\} \in \mathcal{F}_k \ \forall k \geq 0$. The *pre-$\tau$ sigmafield* $\mathcal{F}_\tau$ is defined as the class of all events $\{A: A \cap [\tau = k]\} \in \mathcal{F}_k \ \forall k \geq 0\}$.

Proposition C9.1 *Let f be the probability mass function (p.m.f.) of Y_1. Then the following hold:*

(a) $\{R_n: n \geq 0\}$ is a Markov chain on $S = \mathbb{Z}_+$ with the transition probability matrix $\mathbf{p} = (p_{i,j})$ given by

$$p_{0,j} = f(j+1) \quad \text{for } j \geq 0, \quad p_{j,j-1} = 1 \quad \text{for } j \geq 1. \quad (C9.3)$$

(b) If $\mu \equiv EY_1 < \infty$, then there exists a unique invariant probability $\pi = \{\pi_j: j \geq 0\}$ for $\{R_n: n \geq 0\}$ given by

$$\pi_j = \sum_{i=j+1}^{\infty} f(i)/\mu \quad (j = 0, 1, 2, \ldots). \quad (C9.4)$$

(c) If $EY_1 = \infty$, then $\bar{\pi}_j = \sum_{i=j+1}^{\infty} f(i) \, (j \geq 0)$ provides an invariant measure $\bar{\pi} = \{\bar{\pi}_j: j \geq 0\}$ for $\mathbf{p}$, which is unique up to a multiplicative constant.

Proof.

(a) Observe that $\{N_n: n \geq 0\}$ are $\{\mathcal{F}_k: k \geq 0\}$ – stopping times, where $\mathcal{F}_k = \sigma\{Y_j: 0 \leq j \leq k\}$. We will first show that $V_n := S_{N_n} (n \geq 0)$ has the (inhomogeneous) Markov property. For this note that if $S_{N_n} > n$, then $N_{n+1} = N_n$ and $S_{N_{n+1}} = S_{N_n}$, and if $S_{N_n} = n$, then $N_{n+1} = N_n + 1$ and $S_{N_{n+1}} = S_{N_n} + Y_{N_n+1}$. Hence

$$\begin{aligned}
S_{N_{n+1}} &= S_{N_n} \mathbf{1}_{\{S_{N_n} > n\}} + (S_{N_n} + Y_{N_n+1}) \mathbf{1}_{\{S_{N_n} = n\}} \\
&= S_{N_n} \mathbf{1}_{\{S_{N_n} > n\}} + (S_{N_n}^+)_1 \mathbf{1}_{\{S_{N_n} = n\}}, \quad (C9.5)
\end{aligned}$$

where $S_{N_n}^+$ is the after-N_n process $\{(S_{N_n}^+)_k := S_{N_n+k} : k = 0, 1, 2, \ldots\}$. It follows from (C9.5) and the strong Markov property that the conditional distribution of $V_{n+1} \equiv S_{N_{n+1}}$, given the pre-$N_n$ sigmafield $\mathcal{G}_n \equiv \mathcal{F}_{N_n}$, depends only on $V_n = S_{N_n}$. Since V_n is $\mathcal{G}_n$-measurable, it follows that $\{V_n: n \geq 0\}$ has the Markov property, and that its time-dependent transition probabilities are

$$q_n(n, n + j) \equiv P(V_{n+1} = n + j \mid V_n = n) = P(Y_1 = j)$$
$$= f(j), \quad j \geq 1,$$
$$q_n(m, m) = 1 \quad m > n. \tag{C9.6}$$

Since $R_n = V_n - n(n \geq 0)$, $\{R_n : n \geq 0\}$ has the Markov property, and its transition probabilities are $P(R_{n+1} = j \mid R_n = 0) \equiv P(V_{n+1} = n + 1 + j \mid V_n = n) = f(j + 1)(j \geq 0)$, $P(R_{n+1} = j - 1 \mid R_n = j) \equiv P(V_{n+1} = n + j \mid V_n = n + j) = 1(j \geq 1)$. Thus $\{R_n : n \geq 0\}$ is a time-homogeneous Markov process on $S = \mathbb{Z}_+$ with transition probabilities given by (C9.3).

(b) Assume $\mu \equiv EY_1 < \infty$. If $\pi = \{\pi_j : j \geq 0\}$ is an invariant probability for $\mathbf{p}$ then one must have

$$\pi_0 = \sum_{j=0}^{\infty} \pi_j p_{j,0} = \pi_0 p_{0,0} + \pi_1 p_{1,0} = \pi_0 f(1) + \pi_1,$$

$$\pi_i = \sum_{j=0}^{\infty} \pi_j p_{j,i} = \pi_0 p_{0,i} + \pi_{i+1} p_{i+1,i}$$
$$= \pi_0 f(i + 1) + \pi_{i+1}(i \geq 1). \tag{C9.7}$$

Thus

$$\pi_i - \pi_{i+1} = \pi_0 f(i + 1)(i \geq 0). \tag{C9.8}$$

Summing (C9.8) over $i = 0, 1, \ldots, j - 1$, one gets $\pi_0 - \pi_j = \pi_0 \sum_{i=1}^{j} f(i)$, or $\pi_j = \pi_0 \sum_{i=j+1}^{\infty} f(i)$ $(j \geq 0)$. Summing over j, one finally obtains $\pi_0 = 1/\mu$, since $\sum_{j=0}^{\infty}(1 - F(j)) = \mu$, with $F(j) = \sum_{i=0}^{j} f(i)$.

(c) If $\bar{\pi} = \{\bar{\pi}_j : j \geq 0\}$ is an invariant measure for $\mathbf{p}$ then it satisfies (C9.7) and (C9.8) with $\bar{\pi}_i$ replacing $\pi_i(i \geq 0)$. Hence, by the computation above, $\bar{\pi}_j = \bar{\pi}_0(1 - F(j))$ $(j \geq 0)$. One may choose $\bar{\pi}_0 > 0$ arbitrarily, for example, $\bar{\pi}_0 = 1$.   ∎

Thus the residual lifetime process $\{R_n : n \geq 0\}$ is a stationary Markov process if and only if (i) $\mu \equiv EY_1 < \infty$ and (ii) the delay distribution is given by π in (C9.4). This proposition plays a crucial role in providing a *successful coupling* for a proof of *Feller's renewal theorem* below. Define the *lattice span* d of the p.m.f. f on the set $\mathbb{Z}_{++}$ of positive integers as the g.c.d. of $\{j \geq 1 : f(j) > 0\}$.

Theorem C9.1 (*Feller's renewal theorem*) *Let the common p.m.f.* f *of* Y_j *on* $\mathbb{Z}_{++}(j \geq 1)$ *have span 1 and* $\mu \equiv EY_j < \infty$. *Then whatever the (delay) distribution of* Y_0 *one has for every positive integer* m,

$$\lim_{n \to \infty} E(N_{n+m} - N_n) = \frac{m}{\mu}. \tag{C9.9}$$

Proof. On a common probability space $(\Omega, \mathcal{F}, P)$, construct two independent sequences of random variables $\{Y_0, Y_1, Y_2, \ldots\}$ and $\{\tilde{Y}_0, \tilde{Y}_1, \tilde{Y}_2, \ldots\}$ such that (i) Y_k $(k \geq 1)$ are i.i.d. with common p.m.f. f and the same holds for $\tilde{Y}_k$ $(k \geq 1)$, (ii) Y_0 is independent of $\{Y_k : k \geq 1\}$ and has an arbitrary delay distribution, while $\tilde{Y}_0$ is independent of $\{\tilde{Y}_k : k \geq 1\}$ and has the equilibrium delay distribution π of (C9.4). Let $\{S_k : k \geq 0\}$, $\{\tilde{S}_k : k \geq 0\}$ be the partial sum processes, and $\{N_n : n \geq 0\}$, $\{\tilde{N}_n : n \geq 0\}$ the renewal processes, corresponding to $\{Y_k : k \geq 0\}$ and $\{\tilde{Y}_k : k \geq 0\}$, respectively. The residual lifetime processes $\{R_n := S_{N_n} - n : n \geq 0\}$ and $\{\tilde{R}_n := \tilde{S}_{\tilde{N}_n} - n : n \geq 0\}$ are independent Markov chains on $\mathbb{Z}_+ = \{0, 1, 2, \ldots\}$ each with transition probabilities given by (C9.3). Since the span of f is 1, it is simple to check that these are aperiodic and irreducible. Hence, by Theorem 8.1, the Markov chain $\{(R_n, \tilde{R}_n) : n \geq 0\}$ on $\mathbb{Z}_+ \times \mathbb{Z}_+$ is positive recurrent, so that

$$T := \inf\{n \geq 0 : (R_n, \tilde{R}_n) = (0, 0)\} < \infty \text{ a.s.} \tag{C9.10}$$

Define

$$R'_n := \begin{cases} R_n & \text{if } T > n, \\ \tilde{R}_n & \text{if } T \leq n. \end{cases} \tag{C9.11}$$

Then $\{R'_n : n \geq 0\}$ has the same distribution as $\{R_n : n \geq 0\}$. Note also that

$$N_{n+m} - N_n = \sum_{j=n+1}^{n+m} \mathbf{1}_{\{R_j=0\}}, \quad \tilde{N}_{n+m} - \tilde{N}_n = \sum_{j=n+1}^{n+m} \mathbf{1}_{\{\tilde{R}_j=0\}}, \tag{C9.12}$$

$$E(\tilde{N}_{n+m} - \tilde{N}_n) = \sum_{j=n+1}^{n+m} P(\tilde{R}_j = 0) = m\pi_0 = m/\mu. \tag{C9.13}$$

Now

$$\begin{aligned} E(N_{n+m} - N_n) &= E((N_{n+m} - N_n)\mathbf{1}_{\{T>n\}}) + E((\tilde{N}_{n+m} - \tilde{N}_n)\mathbf{1}_{\{T\leq n\}}) \\ &= E((N_{n+m} - N_n)\mathbf{1}_{\{T>n\}}) - E((\tilde{N}_{n+m} - \tilde{N}_n)\mathbf{1}_{\{T>n\}}) \\ &\quad + E(\tilde{N}_{n+m} - \tilde{N}_n). \end{aligned} \tag{C9.14}$$

Since the first two terms on the right-hand side of the last equality are each bounded by $mP(T > n) \to 0$ as $n \to \infty$, the proof is complete.   ∎

Corollary C9.1 *If f has a lattice span $d > 1$ and $\mu \equiv EY_1 < \infty$ then, whatever the delay distribution on the lattice $\{kd: k = 1, 2, \ldots\}$,*

$$\lim_{n \to \infty} E(N_{nd+md} - N_{nd}) = \frac{md}{\mu}(m = 1, 2, \ldots). \qquad (C9.15)$$

Proof. Consider the renewal process for the sequence of lifetimes $\{Y_k/d: k = 1, 2, \ldots\}$, and apply Theorem 13.1, noting that $EY_1/d = \mu/d$.   ∎

Definition C9.2 Suppose that $\{X_n\}_{n=0}^{\infty}$ is a Markov process with values in a measurable state space $(S, \mathcal{S})$ having transition probabilities $p(x, dy)$. If there is a set $A_0 \in \mathcal{S}$, a probability ν on $(S, \mathcal{S})$ with $\nu(A_0) = 1$, a positive integer N, and a positive number λ such that

(i) $P(X_n \in A_0 \text{ for some } n \geq 1 \mid X_0 = x) = 1$ for all $x \in S$,
(ii) (local minorization) $p^{(N)}(x, A) \geq \lambda \cdot \nu(A)$ for all $A \in \mathcal{S}, x \in A_0$,

then we say $\{X_n\}_{n=0}^{\infty}$ is A_0-*recurrent and locally minorized.* Such a set A_0 is called $\nu(N)$-*small* or, simply, a *small set.*
We first consider the case $N = 1$. Let

$$\eta^{(0)} \equiv \eta_{A_0}^{(0)} = 0, \quad \eta^{(j)} \equiv \eta_{A_0}^{(j)}$$
$$= \inf\left\{n > \eta_{A_0}^{(j-1)} : X_n \in A_0\right\}, \quad j = 1, 2, \ldots, \qquad (C9.16)$$

denote the successive times at which the process $\{X_n\}_{n=0}^{\infty}$ visits A_0.
We will now show that, in view of the local minorization on A_0, cycles between visits to A_0 may be defined to occur at times $\rho^{(1)}, \rho^{(2)}, \ldots$, called *regeneration times* so as to make the cycles between these times i.i.d. As a result of this *regenerative structure,* one may then obtain convergence to the invariant probability by independent coupling.

Proposition C9.2 (*Regeneration lemma*) *Let $\{X_n\}_{n=0}^{\infty}$ be a Markov process on a Polish state space $(S, \mathcal{S})$, which is A_0-recurrent and locally minorized on A_0 with $N = 1$. Then (a)$\{X_n\}_{n=0}^{\infty}$ has a representation by i.i.d. random cycles between regeneration times $\rho^{(1)}, \rho^{(2)}, \ldots,$*

namely,

$$U_j := (X_{\rho^{(j)}+1}, \ldots, X_{\rho^{(j+1)}}), \quad j = 0, 1, 2, \ldots, \tag{C9.17}$$

which are independent for $j \geq 0$ and identically distributed for $j \geq 1$.
(b) If, in addition, $c := \sup\{E(\eta^{(1)} \mid X_0 = x) : x \in A_0\} < \infty$, then
$E(\rho^{(j)} - \rho^{(j-1)}) \leq c/\lambda$.

Proof. Assume without loss of generality that $0 < \lambda < 1$. One may express $p(x, dy)$ as

$$p(x, B) = [\lambda v(B) + (1 - \lambda)q(x, B)]1_{A_0}(x) + p(x, B)1_{S \setminus A_0}(x), \tag{C9.18}$$

where, for $x \in A_0$, $q(x, dy)$ is the probability measure on $(S, \mathcal{S})$ given by

$$q(x, B) := (1 - \lambda)^{-1}[p(x, B) - \lambda v(B)] \quad (x \in A_0, B \in \mathcal{S}). \tag{C9.19}$$

This suggests that if $X_n = x$ is in A_0 then X_{n+1} may be constructed as $\theta Z + (1 - \theta)Y$, where θ, Z, Y are independent, with (i) θ *Bernoulli* $B_e(\lambda)$: $P(\theta = 1) = \lambda$, $P(\theta = 0) = 1 - \lambda$, (ii) Z has distribution v on $(S, \mathcal{S})$ and Y has distribution $q(x, dy)$. If $X_n = x \notin A_0$, then X_{n+1} is taken to have distribution $p(x, dy)$. We now make this precise.

Let $\{(X_0, \theta_0, \theta_n, Z_n, \alpha_n): n = 1, 2, \ldots\}$ be an independent family on $(\Omega, \mathcal{F}, P)$: (1) X_0 has an arbitrary (initial) distribution on $(S, \mathcal{S})$, (2) $\{\theta_n: n \geq 0\}$ is an i.i.d. Bernoulli $B_e(\lambda)$ sequence, (3) $\{Z_n: n \geq 1\}$ is an i.i.d. sequence with common distribution v on $(S, \mathcal{S})$, and (4) $\{\alpha_n: n \geq 1\}$ is an i.i.d. sequence of random maps on $(S, \mathcal{S})$ such that

$$P(\alpha_n x \in B) = q(x, B)1_{A_0}(x) + p(x, B)1_{S \setminus A_0}(x), \quad (x \in S, B \in \mathcal{S}). \tag{C9.20}$$

Note that the right-hand side is a transition probability on $(S, \mathcal{S})$, so that an i.i.d. sequence $\{\alpha_n: n \geq 1\}$ satisfying (C9.21) exists (see Chapter 3, Proposition C1.1 in Complements and Details. Now define recursively

$$X_{n+1} = f(X_n, Z_{n+1}; \theta_n, \alpha_{n+1})(n \geq 0), \tag{C9.21}$$

where

$$f(x, z; \theta, \gamma) = \begin{cases} z & \text{if } x \in A_0, \theta = 1, \\ \gamma(x) & \text{if } x \in A_0, \theta = 0, \\ \gamma(x) & \text{if } x \notin A_0. \end{cases} \tag{C9.22}$$

Observe that $\{\theta_n, Z_{n+1}, \alpha_{n+1}\}$ is an independent triple, which is independent of $\{X_0, X_1, \ldots, X_n\}$ since, by (C9.21), the latter are functions of $X_0; \theta_0, \theta_1, \ldots, \theta_{n-1}; Z_1, Z_2, \ldots, Z_n; \alpha_1, \alpha_2, \ldots, \alpha_n$. Hence the conditional distribution of X_{n+1}, given $\{X_0, \ldots, X_n\}$, is the distribution of

$$[\theta_n Z_{n+1} + (1 - \theta_n)\alpha_{n+1}(x)]\mathbf{1}_{A_0}(x) + a_{n+1}(x)\mathbf{1}_{S \setminus A_0}(x) \qquad \text{(C9.23)}$$

on the set $\{X_n = x\}$. The distribution of (C9.23) is $p(x, dy)$ in view of (C9.18). Thus $\{X_n : n \geq 0\}$ is Markov with transition probability $p(x, dy)$.

The above argument also shows that $\{W_n := (X_n, \theta_n) : n \geq 0\}$ is also a Markov process, as θ_n is independent of $(X_0, \theta_0), \ldots, (X_{n-1}, \theta_{n-1})$ and X_n. Consider the stopping times (for $\{W_n : n \geq 0\}$) $\eta^{(j)}$ defined by (C9.16), and $\rho^{(j)}$ defined by

$$\rho^{(0)} := \inf\{n \geq 0 : X_n \in A_0, \theta_n = 1\},$$
$$\rho^{(j)} := \inf\{n > \rho^{(j-1)} : X_n \in A_0, \theta_n = 1\}(j \geq 1). \qquad \text{(C9.24)}$$

To show that $\rho^{(j)} < \infty$ a.s., we will first establish that $\{\theta_{\eta^{(j)}} : j \geq 0\}$ is an i.i.d. $B_e(\lambda)$ sequence. For this define $\mathcal{F}_j = \sigma\{(X_0, \theta_0), \ldots, (X_j, \theta_j)\}$ $(j \geq 0)$ and the pre-$\eta^{(j)}$ sigmafield $\mathcal{F}_{\eta^{(j)}} = [A \in \mathcal{F} : A \cap \{\eta^{(j)} \leq n\} \in \mathcal{F}_n$ $\forall n]$. By the strong Markov property, the conditional distribution of $\{(X_{\eta^{(j)}+m}, \theta_{\eta^{(j)}+m}) : m \geq 0\}$, given $\mathcal{F}_{\eta^{(j)}}$, is the distribution of $\{(X_m, \theta_m) : m \geq 0\}$ under the initial state $(x, \theta) = (X_{\eta^{(j)}}, \theta_{\eta^{(j)}})$. In particular, the conditional distribution of $\theta_{\eta^{(j+1)}}$, given $\mathcal{F}_{\eta^{(j)}}$, is the same as the distribution of $\theta_{\eta^{(1)}}$ for the process $\{(X_m, \theta_m) : m \geq 0\}$ evaluated at the initial state (x, θ) set at $(X_{\eta^{(j)}}, \theta_{\eta^{(j)}})$. But whatever the initial state (x, θ), $\text{Prob}(\theta_{\eta^{(1)}} = 1) = \sum_{m=1}^{\infty} \text{Prob}(\eta^{(1)} = m, \theta_m = 1) = \sum_{m=1}^{\infty} \text{Prob}(\eta^{(1)} = m) \cdot \text{Prob}(\theta_m = 1) = \sum_{m=1}^{\infty} \text{Prob}(\eta^{(1)} = m)\lambda = \lambda$, since θ_m is independent of $\mathcal{G}_m \equiv \sigma\{X_0, X_1, \ldots, X_m\}$ and $\{\eta^{(1)} = m\} \in \mathcal{G}_m$.

This shows that $\theta_{\eta^{(j+1)}}$ is independent of $\mathcal{F}_{\eta^{(j)}}$ and it is $B_e(\lambda)$. It follows that $\{\theta_{\eta^{(j)}} : j \geq 0\}$ is an i.i.d. $B_e(\lambda)$ sequence. The above argument actually shows that $\theta_{\eta^{(j+1)}}$ is independent of $\mathcal{F}_{\eta^{(j)}}$ and $(X_{\eta^{(j+1)}}, \eta^{(j+1)})$, since $\{X_m \in B, \eta^{(j+1)} = m\} \in \mathcal{G}_m$ and θ_m is independent of $\mathcal{G}_m$. Now the event $\{\rho^{(1)} < \infty\}$ equals the event $\cup_{j=1}^{\infty}\{\eta^{(j)} < \infty, \theta_{\eta^{(j)}} = 1\}$ whose probability is the same as that of $\cup_{j=1}^{\infty}\{\theta_{\eta^{(j)}} = 1\}$ since $\eta^{(j)} < \infty$ a.s. for all j, by hypothesis. But the probability of $\cup_{j=1}^{\infty}\{\theta_{\eta^{(j)}} = 1\}$ is the same as that of the event $\{\theta_n = 1 \text{ for some } n = 1, 2, \ldots\}$, where θ_n are i.i.d.

$B_e(\lambda)$, namely, 1. Thus we have proved $\rho^{(1)} < \infty$ a.s., no matter what the initial state (x, θ) of the process $\{W_n : n \geq 0\}$ is.

To see that the blocks (C9.17) are i.i.d., note that if $X_j \in A_0$ and $\theta_j = 1$, then $X_{j+1} = Z_{j+1}$, so that $W_{j+1} = (Z_{j+1}, \theta_{j+1})$, which is independent of $\{W_0, W_1, \ldots, W_j\}$. Hence, by the strong Markov property of $\{W_n : n \geq 0\}$ applied with the stopping time $\rho^{(j)}$, the conditional distribution of $\{W_{\rho^{(j)}+n} : n \geq 1\}$, given the pre-$\rho^{(j)}$ sigmafield $\mathcal{F}_{\rho^{(j)}}$, is the distribution of a $\{W_n : n \geq 0\}$ process with an initial random vector $W_0 = (Z, \theta)$: Z and θ independent having distributions ν and $B_e(\lambda)$, respectively. In particular, this conditional distribution does not depend on $W_{\rho^{(j)}}$. Hence $\{W_{\rho^{(j)}+n} : n \geq 1\}$ is independent of $\mathcal{F}_{\rho^{(j)}}$. It follows that the blocks U_j $(j \geq 1)$ are independent and each has the same distribution as that of $\{X_n : 1 \leq n \leq \rho^{(1)}\}$ when $W_0 = (Z, \theta)$ as above. This completes the proof of part (a).

To prove part (b), we need to show that $E\rho^{(1)} < \infty$ when $X_0 \in A_0$ and $\theta_0 = 1$ a.s. Write

$$\rho^{(1)} = \sum_{j=1}^{\infty} \eta^{(j)} \mathbf{1}_{\{\theta_{\eta^{(i)}}=0 \text{ for } 1 \leq i < j, \theta_{\eta^{(j)}}=1\}}, \tag{C9.25}$$

the first term on the right-hand side being $\eta^{(1)} \mathbf{1}_{\{\theta_{\rho^{(1)}}=1\}}$. Taking expectations and noting that (i) $\theta_{\eta^{(j)}}$ is independent of $\mathcal{F}_{\eta^{(j-1)}}$ and of $\eta^{(j)}$, (ii) $E(\eta^{(j)} - \eta^{(j-1)} \mid \mathcal{F}_{\eta^{(j-1)}}) \leq c := \sup\{E(\eta^{(1)} \mid X_0 = x) : x \in A_0\}$, one has

$$E\rho^{(1)} = \sum_{j=1}^{\infty} E\left[\mathbf{1}_{\{\theta_{\eta^{(i)}}=0, 1 \leq i \leq j-1\}} (\eta^{(j)} P(\theta_{\eta^{(j)}} = 1))\right]$$

$$= \lambda \sum_{j=1}^{\infty} E\left[\mathbf{1}_{\{\theta_{\eta^{(i)}}=0, 1 \leq i \leq j-1\}} \left\{\eta^{(j-1)} + E\left(\eta^{(j)} - \eta^{(j-1)} \mid \mathcal{F}_{\eta^{(j-1)}}\right)\right\}\right]$$

$$\leq \lambda \sum_{j=1}^{\infty} E\left[\mathbf{1}_{\{\theta_{\eta^{(i)}}=0, 1 \leq i \leq j-1\}} \left\{\eta^{(j-1)} + c\right\}\right]$$

$$= \lambda \sum_{j=1}^{\infty} c(1-\lambda)^{j-1} + \lambda \sum_{j=1}^{\infty} E\left[\mathbf{1}_{\{\theta_{\eta^{(i)}}=0, 1 \leq i \leq j-1\}} \eta^{(j-1)}\right]$$

$$= c + \lambda \sum_{j=2}^{\infty} (1-\lambda) E\left[\mathbf{1}_{\{\theta_{\eta^{(i)}}=0, 1 \leq i \leq j-2\}}\right.$$

$$\left. \left\{\eta^{(j-2)} + E(\eta^{(j-1)} - \eta^{(j-2)} \mid \mathcal{F}_{\eta^{(j-2)}})\right\}\right]$$

$$\leq c + c\lambda \sum_{j=2}^{\infty}(1-\lambda)^{j-1} + \lambda(1-\lambda)\sum_{j=3}^{\infty} E\left[\mathbf{1}_{\{\theta_{\eta^{(i)}}=0,\,1\leq i\leq j-2\}}\eta^{(j-2)}\right]$$

$$\leq c + c(1-\lambda) + c(1-\lambda)^2 + \cdots = c/\lambda. \qquad\blacksquare$$

Theorem C9.2 (*Ergodicity of Harris recurrent processes*)   *Let $p(x, dy)$ be a transition probability on $(S, \mathcal{S})$, a Borel subset of a Polish space. Assume that a Markov process $\{X_n: n \geq 0\}$ with this transition probability is A_0-recurrent and locally minorized, according to Definition C9.2, and that the process is strongly aperiodic, i.e., $N = 1$. If, in addition, $\sup\{E_x\eta^{(1)}: x \in A_0\} < \infty$, there exists a unique invariant probability π and, whatever the initial distribution of $\{X_n: n \geq 0\}$,*

$$\sup_{A\in\mathcal{S}^{\otimes\infty}} |Q^{(n)}(A) - Q_\pi(A)| \to 0 \quad \text{as } n \to \infty, \qquad \text{(C9.26)}$$

where $Q^{(n)}$ is the distribution of the after-n process $X_n^+ := \{X_{n+m} : m \geq 0\}$, and Q_π is the distribution of the process with initial distribution π. In particular, $\sup\{|p^{(n)}(x, B) - (\pi(B)|: B \in \mathcal{S}\} \to 0$ for every $x \in S$.

Proof.   Define the finite measure π_1 on $(S, \mathcal{S})$ by

$$\pi_1(B) := E_v \sum_{n=\rho^{(0)}+1}^{\rho^{(1)}} \mathbf{1}_{\{X_n\in B\}} \quad (B \in \mathcal{S}). \qquad \text{(C9.27)}$$

We will first show that

$$\pi(B) := \pi_1(B)/\pi_1(S) \quad (B \in \mathcal{S}) \qquad \text{(C9.28)}$$

is the unique invariant probability for $p(x, dy)$. For this note that, for every bounded measurable f,

$$V_j(f) := \sum_{m=\rho^{(j-1)}+1}^{\rho(j)} f(X_m), \quad (j \geq 1) \qquad \text{(C9.29)}$$

is, by Proposition C9.2, an i.i.d. sequence with finite expectation $\int f\, d\pi_1$. By the SLLN,

$$N^{-1}\sum_{j=1}^{N} V_j(f) \to \int_S f\pi_1 \text{ a.s.} \quad \text{as } N \to \infty, \qquad \text{(C9.30)}$$

whatever be the initial distribution. In particular, taking $f \equiv 1$ one has, with probability 1,

$$\frac{\rho^{(N)}}{N} = \frac{\rho^{(0)}}{N} + \frac{\rho^{(N)} - \rho^{(0)}}{N} \to E\left(\rho^{(1)} - \rho^{(0)}\right) \equiv \pi_1(S), \quad \text{as} N \to \infty.$$

(C9.31)

Now define

$$M_n := \sup\{j \geq 1 \colon \rho^{(j)} \leq n\} \quad (n = 1, 2, \ldots).$$

Then, by (C9.31),

$$\rho^{(M_n)}/M_n \to \pi_1(S).$$

(C9.32)

Using the fact that $0 \leq n - \rho^{(M_n)} \leq \rho^{(M_n+1)} - \rho^{(M_n)}$ and that $(\rho^{(M_n+1)} - \rho^{(M_n)})/M_n \to 0$ a.s. as $n \to \infty$ (see Exercise 8.2),

$$\frac{n}{M_n} \to \pi_1(S) \text{ a.s.} \quad \text{as } n \to \infty.$$

(C9.33)

In the same manner, for all bounded measurable f, and for all $n > \rho^{(1)}$,

$$\frac{1}{n}\sum_{m=1}^{n} f(X_m) = \frac{1}{n}\sum_{m=1}^{\rho^{(1)}} f(X_m) + \frac{1}{n}\sum_{j=1}^{M_n} V_j(f) + \frac{1}{n}\sum_{\rho^{(M_n)} < m \leq n} f(X_m)$$

$$\approx \frac{M_n}{n} \cdot \frac{1}{M_n}\sum_{j=1}^{M_n} V_j(f) \to \frac{1}{\pi_1(S)}\int f\, d\pi_1, \quad \text{(C9.34)}$$

since $(1/n)(n - \rho^{(M_n)}) = (M_n/n)(n - \rho^{(M_n)})/M_n \to 0$, as shown above. Thus, with probability 1,

$$\frac{1}{n}\sum_{m=1}^{n} f(X_m) \to \int f\, d\pi \; \forall \text{ bounded measurable } f,$$

(C9.35)

which implies

$$\frac{1}{n}\sum_{m=1}^{n} f(X_{m+1}) \to \int f\, d\pi \text{ a.s.} \quad \text{as } n \to \infty.$$

(C9.36)

Taking expectations term by term after conditioning the mth term with respect to X_m, one gets

$$E\left[\frac{1}{n}\sum_{m=1}^{n}(Tf)(X_m)\right] \to \int f\, d\pi.$$

(C9.37)

But the random sequence within brackets on the left-hand side converges a.s. to $\int Tf\,d\pi$, by (C9.35). Hence $\int Tf\,d\pi = \int f\,d\pi$ for all bounded measurable f, proving the invariance of π. Uniqueness of the invariant probability follows on integrating $g_n(x) := (1/n)\sum_{m=1}^{n}\int f(y)p^{(m)}(x,dy)$ with respect to an invariant probability $\tilde{\pi}$, say, to get $\int f\,d\tilde{\pi}\ \forall n$, and noting that $g_n(x) \to \int f\,d\pi$ as $n \to \infty$.

We will complete the proof by a coupling argument. One may construct a (common) probability space $(\Omega', \mathcal{F}', P')$, say, on which are defined two independent families $\{X_0, \theta_0, (\theta_n, Z_n, \alpha_n)_{n\geq 1}\}$ and $\{\tilde{X}_0, \tilde{\theta}_0, (\tilde{\theta}_n, \tilde{Z}_n, \tilde{\alpha}_n)_{n\geq 1}\}$ as above, with $\{\theta_0, (\theta_n, Z_n, \alpha_n)_{n\geq 1}\}$ and $\{\tilde{\theta}_0, (\tilde{\theta}_n, \tilde{Z}_n, \tilde{\alpha}_n)_{n\geq 1}\}$ independent of X_0 and $\tilde{X}_0$. Let X_0 have (an arbitrary initial) distribution μ and let $\tilde{X}_0$ have the invariant distribution. Let $\{\rho^{(j)}: j \geq 0\}$ and $\{\tilde{\rho}^{(j)}: j \geq 0\}$ denote the corresponding independent sequences of regeneration times, and let $\{Y_0 = \rho^{(0)}, Y_k = \rho^{(k)} - \rho^{(k-1)}(k \geq 1)\}$, $\{\tilde{Y}_0 = \tilde{\rho}^{(0)}, \tilde{Y}_k = \tilde{\rho}^{(k)} - \tilde{\rho}^{(k-1)}(k \geq 1)\}$ be the corresponding sequences of lifetimes of two independent renewal processes. Under the hypothesis, Y_k and $\tilde{Y}_k$ $(k \geq 1)$ each has a lattice span 1, and $EY_k = E\tilde{Y}_k = E(\rho^{(2)} - \rho^{(1)}) < \infty$. Hence, as in the proof of Feller's renewal theorem C9.1 (see (C9.10)), $\rho := \inf\{n \geq 0: R_n = \tilde{R}_n = 0\} < \infty$ a.s., where $\{R_n: n \geq 0\}$ and $\{\tilde{R}_n: n \geq 0\}$ are the two residual lifetime processes corresponding to $\{Y_k: k \geq 0\}$ and $\{\tilde{Y}_k: k \geq 0\}$, respectively. Note that ρ is the first common renewal epoch, i.e., there are $m, \tilde{m}$ such that $\rho_m = \tilde{\rho}_{\tilde{m}} = \rho$. Now define

$$X_n' := \begin{cases} X_n & \text{if } \rho > n, \\ \tilde{X}_n & \text{if } \rho \leq n. \end{cases} \tag{C9.38}$$

Since ρ is a stopping time for the Markov process $\{(W_n, \tilde{W}_n): n \geq 1\}$, with $W_n := (X_n, \theta_n)$, $\tilde{W}_n = (\tilde{X}_n, \tilde{\theta}_n)$, one may use the strong Markov property to see that $\{X_n: n \geq 0\}$ and $\{X_n': n \geq 0\}$ have the same distribution. Hence, for all $A \in \mathcal{S}^{\otimes\infty}$,

$$\left| P\left(X_n^+ \in A\right) - P\left(\tilde{X}_n^+ \in A\right) \right|$$
$$\leq \left| P\left(X_n'\right)^+ \in A, \rho \leq n) - P\left(\tilde{X}_n^+ \in A, \rho \leq n\right) \right|$$
$$+ \left| P\left((X_n')^+ \in A, \rho > n\right) - P\left(\tilde{X}_n^+ \in A, \rho > n\right) \right| \leq P(\rho > n). \tag{C9.40}$$

Here we have used the fact $(X_n')^+ = X_n^+$ on $\{\rho \leq n\}$. ∎

If one lets $N > 1$ in the local minorization condition in Theorem C9.2 (or Theorem 9.2), one must allow the possibility of cycles. For example,

every (irreducible) positive recurrent Markov chain satisfies the hypothesis of Theorem 9.2. with A_0 a singleton $\{x_0\}$ and $N = $ period of the chain.
In particular, for positive recurrent birth–death chains with $p_{x,x+1} = \beta_x$,
$p_{x,x-1} \equiv \delta_x = 1 - \beta_x$ $(0 < \beta_x < 1)$, which are periodic of period 2,
clearly, the conclusion of Theorem C9.2 does not hold; in particular,
$p^{(n)}(x, y)$ does not converge as $n \to \infty$. It is known, however, that, under
the hypothesis of Theorem 9.2, there exists a unique invariant probability
π, and the state space has a decomposition $S = (\cup_{i=1}^d D_i) \cup M$, where
$D_1, D_2, \ldots, D_d, M$ are disjoint sets in S such that (i) $p(x, D_{i+1}) = 1$
$\forall x \in D_i, 1 \le i \le d$ (with $D_{d+1} := D_1$) and (ii) M is π-null: $\pi(M) = 0$.
The hypothesis of Theorem C9.2 holds on each D_i as the state space,
with the transition probability $q(x, dy) = p^{(d)}(x, dy)$ where N is a
multiple of d. In view of the recurrence property, the convergence of
$\frac{1}{n} \sum_{m=1}^n p^{(m)}(x, dy)$ to $\pi(dy)$ (in variation distance) follows for every
$x \in S$, using Theorem C9.2.

Theorems 9.2, C9.2 were obtained independently by Athreya and Ney
(1978) and Nummelin (1978a). For a general analysis of the structure of
ϕ-irreducible and ϕ-recurrent processes, we refer to the original work by
Jain and Jamison (1967) and Orey (1971). For a comprehensive account
of stability in distribution for ϕ-irreducible and ϕ-recurrent Markov processes, we refer to Meyn and Tweedie (1993). For the sake of completeness, we state the following useful result (see Meyn and Tweedie, loc.
cit. Chapter 10).

Theorem C9.3 *If a Markov process is ϕ-irreducible (with respect to a
nonzero sigmafinite measure ϕ) and admits an invariant probability π,
then (i) the process is positive Harris (π-) recurrent, (ii) π is the unique
invariant probability, and (iii) the convergence (9.30) holds.*

The original work of Harris appears in Harris (1956). Our treatment
follows that of Bhattacharya and Waymire (2006b, Chapter 4, Section 16).

Section 2.10. In order to prove Proposition 10.1, we will make use of
the following inequality, which is of independent interest. Write $\mathcal{F}_k$ for
the sigmafield generated by $\{X_j: 0 \le j \le k\}$.

Lemma C10.1 (*Kolmogorov's maximal inequality*) *Let $X_0, X_1, \ldots, X_n$
be independent random variables, $E X_j = 0$ for $1 \le j \le n$, $E X_j^2 < \infty$
for $0 \le j \le n$. Write $S_k := X_0 + \cdots + X_k$, $M_n := \max\{|S_k|: 0 \le k \le n\}$.*

Then for every $\lambda > 0$, one has

$$P(M_n \geq \lambda) \leq \frac{1}{\lambda^2} \int_{\{M_n \geq \lambda\}} S_n^2 \, dP \leq \frac{1}{\lambda^2} E S_n^2.$$

Proof. Write $\{M_n \geq \lambda\} = \cup_{k=0}^n A_k$, where $A_0 = \{|S_0| \geq \lambda\}$, $A_k = \{|S_j| < \lambda$ for $0 \leq j \leq k - 1, |S_k| \geq \lambda\}$, $1 \leq k \leq n$. Then

$$P(M_n \geq \lambda) = \sum_{k=0}^n P(A_k) = \sum_{k=0}^n E\left(1_{A_k}\right) \leq \sum_{k=0}^n E\left(1_{A_k}\left(S_k^2/\lambda^2\right)\right)$$

$$= \frac{1}{\lambda^2} \sum_{k=0}^n E\left(1_{A_k} E\left(S_k^2 \mid \mathcal{F}_k\right)\right) \leq \frac{1}{\lambda^2} \sum_{k=0}^n E\left(1_{A_k} E\left(S_n^2 \mid \mathcal{F}_k\right)\right),$$

(C10.1)

since $E(S_n^2 \mid \mathcal{F}_k) = E[S_k^2 + 2S_k(S_n - S_k) + (S_n - S_k)^2 \mid \mathcal{F}_k] = E(S_k^2 \mid \mathcal{F}_k) + 2S_k E(S_n - S_k \mid \mathcal{F}_k) + E\{(S_n - S_k)^2 \mid \mathcal{F}_k)\} = E(S_k^2 \mid \mathcal{F}_k) + 2S_k \cdot 0 + E(S_n - S_k)^2 \geq E(S_k^2 \mid \mathcal{F}_k)$. Now the last expression in (C10.1) equals

$$\frac{1}{\lambda^2} \sum_{k=0}^n E\left(1_{A_k} S_n^2\right) = \frac{1}{\lambda^2} E\left(1_{\{M_n \geq \lambda\}} S_n^2\right) \leq \frac{1}{\lambda^2} E S_n^2. \qquad \blacksquare$$

Proof of Proposition 10.1. Without loss of generality, let $\sigma = 1$. Write $S_n = X_1 + \cdots + X_n$. For any given $\varepsilon > 0$, one has

$$P\left(\left|S_{\nu_n} - S_{[n\alpha]}\right| \geq \varepsilon[n\alpha]^{1/2}\right) \leq P(|\nu_n - [n\alpha]| \geq \varepsilon^3[n\alpha])$$

$$+ P\left(\max_{\{m:|m-[n\alpha]| < \varepsilon^3[n\alpha]\}} |S_m - S_{[n\alpha]}| \geq \varepsilon([n\alpha])^{1/2}\right). \qquad \text{(C10.2)}$$

Thus $(S_{\nu_n} - S_{[n\alpha]})/[n\alpha]^{1/2} \to 0$ in probability, as $n \to \infty$. Since $S_{[n\alpha]}/[n\alpha]^{1/2}$ converges to $N(0, 1)$ in distribution, and $\nu_n/[nd] \to 1$ in probability, $S_{\nu_n}/\sqrt{\nu_n} = (S_{\nu_n}/[n_\alpha]^{1/2})([n_\alpha]/\nu_n)^{1/2}$ converges to $N(0, 1)$ in distribution.

The first term on the right-hand side goes to zero, by assumption (10.9). By Lemma C10.1, the second term on the right-hand side is no more than $2(\varepsilon([n\alpha]^{1/2})^{-2}(\varepsilon^3[n\alpha]) = 2\varepsilon$. The factor 2 here arises from the maximum of two maxima, the first over $\{m: [n\alpha] - \varepsilon^3[n\alpha] < m \leq [n\alpha]\}$ (in which case Lemma C10.1 is applied to summands $X'_j := X_{[n\alpha]-j}, 1 \leq j < \varepsilon^3[n\alpha]$), while the second maximum is taken over $\{m : [n\alpha] < m <$

$[n\alpha] + \varepsilon^3[n\alpha]\}$ (in which case Lemma C10.1 is applied to summands $X_j'' := X_{[n\alpha]+j}, 1 \leq j < \varepsilon^3[n\alpha])$. ∎

Section 2.11. Among references to the weak convergence theory of probability measures, we cite Billingsley (1968), Parthasarathy (1967), Bhattacharya and Ranga Rao (1976), Dudley (1989), and Pollard (1984). Our treatment is also influenced by that of Bhattacharya and Waymire (2006(a)).

A basic result on weak convergence of probability measures on a metric space (S, d) is Theorem C11.1. For its statement, recall that the *boundary* $\partial A := \bar{A} \backslash A^o$, where $\bar{A}$ is the *closure* of A and A^o, is the *interior* of A. Finally, $\mathcal{S}$ is the Borel sigmfield of S. A set $A \in \mathcal{S}$ is said to be a *P-continuity set* (for some given probability measure P on $(S, \mathcal{S})$) if $P(\partial A) = 0$. Unless stated otherwise, all probability measures on S are defined on $\mathcal{S}$.

Theorem C11.1 (*Alexandrov's theorem*) *Let P, $P_n (n \geq 1)$ be probability measures on S. The following statements are equivalent:*

(i) P_n converges weakly to P.

(ii) $\lim_n \int_S f \, dP_n = \int_S f \, dP$ for all bounded, uniformly continuous real f.

(iii) $\lim \sup_n P_n(F) \leq P(F)$ for all closed F.

(iv) $\lim \inf_n P_n(G) \geq P(G)$ for all open G.

(v) $\lim_n P_n(A) = P(A)$ for all P-continuity sets A.

Proof. The plan is to first prove: (i) implies (ii), (ii) implies (iii), (iii) implies (i), and hence that (i), (ii), and (iii) are equivalent. We then directly prove that (iii) and (iv) are equivalent and that (iii) and (v) are equivalent. *(i) implies (ii):* This follows directly from the definition. *(ii) implies (iii):* Let F be a closed set and $\delta > 0$. For a sufficiently small but fixed value of ε, $G_\varepsilon = \{x : d(x, F) < \varepsilon\}$ satisfies $P(G_\varepsilon) < P(F) + \delta$, by continuity of the probability measure P from above, since the sets G_ε decrease to $F = \cap_{\varepsilon \downarrow 0} G_\varepsilon$. Construct a uniformly continuous function h on S such that $h(x) = 1$ on F, $h(x) = 0$ on the complement G_ε^c of G_ε, and $0 \leq h(x) \leq 1$ for all x. For example, define θ_ε on $[0, \infty)$ by $\theta_\varepsilon(u) = 1 - u/\varepsilon, 0 \leq u \leq \varepsilon, \theta_\varepsilon(u) = 0$ for $u > \varepsilon$, and let $h(x) = \theta_\varepsilon(d(x, F))$. In view of (ii) one has $\lim_n \int_S h \, dP_n = \int_S h \, dP$.

In addition,

$$P_n(F) = \int_F h \, dP_n \le \int_S h \, dP_n$$

and

$$\int_S h \, dP = \int_{G_\varepsilon} h \, dP \le P(G_\varepsilon) < P(F) + \delta. \tag{C11.1}$$

Thus

$$\limsup_n P_n(F) \le \lim_n \int_S h \, dP_n = \int_S h \, dP < P(F) + \delta.$$

Since δ is arbitrary, this proves (iii).

(iii) implies (i): Let $f \in C_b(S)$. It suffices to prove

$$\limsup_n \int_S f \, dP_n \le \int_S f \, dP. \tag{C11.2}$$

For then one also gets $\inf_n \int_S f \, dP_n \ge \int_S f \, dP$, and hence (i), by replacing f by $-f$. But in fact, for (C11.2) it suffices to consider $f \in C_b(S)$ such that $0 < f(x) < 1$, $x \in S$, since the more general $f \in C_b(S)$ can be reduced to this by translating and rescaling f. Fix an integer k and let F_i be the closed set $F_i = \{x : f(x) \ge i/k\}$, $i = 0, 1, \dots, k$. Then taking advantage of $0 < f < 1$, one has

$$\sum_{i=1}^k \frac{i-1}{k} P\left(\left\{x : \frac{i-1}{k} \le f(x) < \frac{i}{k}\right\}\right)$$

$$\le \int_S f \, dP \le \sum_{i=1}^k \frac{i}{k} P\left(\left\{x : \frac{i-1}{k} \le f(x) < \frac{i}{k}\right\}\right)$$

$$= \sum_{i=1}^k \frac{i}{k}[P(F_{i-1}) - P(F_i)] = \frac{1}{k} + \frac{1}{k}\sum_{i=1}^k P(F_i). \tag{C11.3}$$

The sum on the extreme left may be reduced similarly to obtain

$$\frac{1}{k}\sum_{i=1}^k P(F_i) \le \int_S f \, dP < \frac{1}{k} + \frac{1}{k}\sum_{i=1}^k P(F_i). \tag{C11.4}$$

In view of (iii) $\limsup_n P_n(F_i) \le P(F_i)$ for each i. So, using the upper bound in (C11.4) with P_n in place of P and the lower bound with P, it follows that

$$\limsup_n \int_S f \, dP_n \le \frac{1}{k} + \int_S f \, dP.$$

Now let $k \to \infty$ to obtain the asserted inequality (C11.2) to complete the proof of (i) from (iii).

(iii) iff (iv): This is simply due to the fact that open and closed sets are complementary.

(iii) implies (v): Let A be a P-continuity set. Since (iii) implies (iv), one has

$$P(\bar{A}) \ge \limsup_n P_n(\bar{A}) \ge \limsup_n P_n(A)$$

$$\ge \liminf_n P_n(A) \ge \liminf_n P_n(A^o) \ge P(A^o). \quad (C11.5)$$

Since $P(\partial A) = 0$, $P(\bar{A}) = P(A^o)$ so that the inequalities squeeze down to $P(A)$ and $\lim_n P_n(A) = P(A)$ follows.

(v) implies (iii): Let F be a closed set. Since $\partial\{x: d(x, F) \le \delta\} \subseteq \{x: d(x, F) = \delta\}$, these boundaries are disjoint for distinct δ. Thus at most countably many of them can have positive P-measure (Exercise), all others being P-continuity sets. In particular there is a sequence of positive numbers $\delta_k \to 0$ such that the sets $F_k = \{x: d(x, F) \le \delta_k\}$ are P-continuity sets. From (v) one has $\limsup_n P_n(F) \le \lim_n P_n(F_k) = P(F_k)$ for each k. Since F is closed, one also has $F_k \downarrow F$, so that (iii) follows from continuity of the probability P from above. ∎

In the special finite dimensional case $S = \mathbb{R}^k$, the following theorem provides some alternative useful conditions for weak convergence.

Theorem C11.2 *(Finite dimensional weak convergence)* Let $P_1, P_2, \ldots$, P be probability measures on $\mathbb{R}^k$. The following are equivalent statements.

(a) $\{P_n: n = 1, 2, \ldots\}$ *converge weakly to* P.

(b) $\int_{\mathbb{R}^k} f \, dP_n \to \int_{\mathbb{R}^k} f \, dP$ *for all (bounded) continuous* f *vanishing outside a bounded set.*

(c) $\int_{\mathbb{R}^k} f \, dP_n \to \int_{\mathbb{R}^k} f \, dP$ *for all infinitely differentiable functions* f *vanishing outside a bounded set.*

(d) Let $F_n(x) := P_n((-\infty, x_1] \times \cdots \times (-\infty, x_k])$, *and* $F(x) := P((-\infty, x_1] \times \cdots \times (-\infty, x_k])$, $x \in R^k$, $n = 1, 2, \ldots$ *be such that* $F_n(x) \to F(x)$ *as* $n \to \infty$, *for every point of continuity* x *of* F.

Proof. We give the proof for the one-dimensional case $k = 1$. First let us check that (b) is *sufficient*. It is obviously *necessary* by definition of weak convergence. Assume (b) and let f be an arbitrary bounded continuous function, $|f(x)| \leq c$, for all x. For notational convenience, write $\{x \in \mathbb{R}^1 : |x| \geq N\} = \{|x| \geq N\}$, etc. Given $\varepsilon > 0$, there exists N such that $P(\{|x| \geq N\}) < \varepsilon/2c$. Define θ_N by $\theta_N(x) = 1$ for $|x| \leq N, \theta_N(x) = 0$ for $|x| \geq N + 1$, and linearly interpolate for $N \leq |x| \leq N + 1$. Then,

$$\underline{\lim}_{n \to \infty} P_n(\{|x| \leq N + 1\}) \geq \underline{\lim}_{n \to \infty} \int \theta_N(x) \, dP_n(x)$$

$$= \int \theta_N(x) \, dP(x)$$

$$\geq P(\{|x| \leq N\}) > 1 - \frac{\varepsilon}{2c},$$

so that

$$\overline{\lim}_{n \to \infty} P_n(\{|x| > N + 1\}) \equiv 1 - \underline{\lim}_{n \to \infty} P_n(\{|x| \leq N + 1\}) < \frac{\varepsilon}{2c}.$$
$$(\text{C11.6})$$

Now let $\theta_N' := \theta_{N+1}$ and define $f_N := f\theta_N'$. Noting that $f = f_N$ on $\{|x| \leq N + 1\}$ and that on $\{|x| > N + 1\}$, one has $|f(x)| \leq c$, we have

$$\overline{\lim}_{n \to \infty} \left| \int_{\mathbb{R}^1} f \, dP_n - \int_{\mathbb{R}^1} f \, dP \right| \leq \overline{\lim}_{n \to \infty} \left| \int_{\mathbb{R}^1} f_N \, dP_n - \int_{\mathbb{R}^1} f_N \, dP \right|$$

$$+ \overline{\lim}_{n \to \infty} (c P_n(\{|x| > N + 1\}) + c P(\{|x| > N + 1\}))$$

$$< c\frac{\varepsilon}{2c} + c\frac{\varepsilon}{2c} = \varepsilon.$$

Since $\varepsilon > 0$ is arbitrary, $\int_{\mathbb{R}^1} f \, dP_n \to \int_{\mathbb{R}^1} f \, dP$. So (a) and (b) are equivalent. Let us now show that (b) and (c) are equivalent. It is enough to prove (c) implies (b). For each $\varepsilon > 0$ define the function

$$\rho_\varepsilon(x) = d(\varepsilon) \exp\left\{ -\frac{1}{1 - x^2/\varepsilon^2} \right\} \quad \text{for } |x| < \varepsilon,$$

$$= 0 \quad \text{for } |x| \geq \varepsilon, \qquad\qquad (\text{C11.7})$$

where $d(\varepsilon)$ is so chosen as to make $\int \rho_\varepsilon(x) \, dx = 1$. One may check that $\rho_\varepsilon(x)$ is infinitely differentiable in x. Now let f be a continuous function that vanishes outside a finite interval. Then f is uniformly continuous and, therefore, $\delta(\varepsilon) := \sup\{|f(x) - f(y)| : |x - y| \le \varepsilon\} \to 0$ as $\varepsilon \downarrow 0$. Define

$$f^\varepsilon(x) = f * \rho^\varepsilon(x) := \int_{-\varepsilon}^{\varepsilon} f(x - y)\rho^\varepsilon(y)\,dy, \qquad (\text{C}11.8)$$

and note that, since $f^\varepsilon(x)$ is an average over values of f within the interval $(x - \varepsilon, x + \varepsilon)$, $|f^\varepsilon(x) - f(x)| \le \delta(\varepsilon)$ for all x. Hence,

$$\left| \int_{\mathbb{R}^1} f\,dP_n - \int_{\mathbb{R}^1} f^\varepsilon\,dP_n \right| \le \delta(\varepsilon) \quad \text{for all } n,$$

$$\left| \int_{\mathbb{R}^1} f\,dP - \int_{\mathbb{R}^1} f^\varepsilon\,dP \right| \le \delta(\varepsilon),$$

$$\left| \int_{\mathbb{R}^1} f\,dP_n - \int_{\mathbb{R}^1} f\,dP \right| \le \left| \int_{\mathbb{R}^1} f\,dP_n - \int_{\mathbb{R}^1} f^\varepsilon\,dP_n \right|$$

$$+ \left| \int_{\mathbb{R}^1} f^\varepsilon\,dP_n - \int_{\mathbb{R}^1} f^\varepsilon\,dP \right| + \left| \int_{\mathbb{R}^1} f\,dP - \int_{\mathbb{R}^1} f\,dP \right|$$

$$\le 2\delta(\varepsilon) + \left| \int_{\mathbb{R}^1} f^\varepsilon\,dP_n - \int_{\mathbb{R}^1} f^\varepsilon\,dP \right| \to 2\delta(\varepsilon) \quad \text{as } n \to \infty.$$

Since $\varepsilon > 0$ is arbitrary and $\delta(\varepsilon) \to 0$ as $\varepsilon \to 0$, it follows that $\int_{\mathbb{R}^1} f\,dP_n \to \int_{\mathbb{R}^1} f\,dP$, as claimed. Next let F_n, F be the distribution functions of P_n, P, respectively ($n = 1, 2, \ldots$), and suppose (a) holds and observe that $(-\infty, x]$ is a P−continuity set if and only if $0 = P(\partial(-\infty, x]) \equiv P(\{x\})$. That is, x must be a continuity point of F. Therefore, that (a) implies (c), follows from Alexandrov's theorem. To show that the converse is also true, suppose $F_n(x) \to F(x)$ at all points of continuity of a distribution function (d.f.) F. Consider a continuous function f that vanishes outside $[a, b]$ where a, b are points of continuity of F. Partition $[a, b]$ into a finite number of small subintervals whose end points are all points of continuity of F, and approximate f by a step function constant over each subinterval. Then the integral of this step function with respect to P_n converges to that with respect to P and, therefore, one gets (a) by a triangle inequality. ∎

Remark C11.1 It is straightforward to extend the proof of the equivalence of (a), (b), and (c) to $\mathbb{R}^k$, $k \ge 1$. The equivalence of (c) and (d) for the case $\mathbb{R}^k$, $k > 1$, is similar, but one needs to use P-continuities for approximating finite and infinite rectangles. See Billingsley (1968, pp. 17–18).

The weak convergence of probability measures defined in Section 2.11 (Definition 11.2) and discussed above is in fact a metric notion as we now show. First we define the weak topology on the space of probability measures.

Definition C11.1 The *weak topology* on the set $\mathcal{P}(S)$ of all probability measures on the Borel sigmafield $\mathcal{S}$ of a metric space (S, d) is defined by the following system of neighborhoods:

$$N(\mu_0; f_1, f_2, \ldots, f_m; \varepsilon)$$

$$:= \left\{ \mu \in \mathcal{P}(S) : \left| \int_S f_i \, d\mu - \int_S f_i \, d\mu_0 \right| < \varepsilon \, \forall i = 1, \ldots, m \right\},$$

$(\mu_0 \in \mathcal{P}(S); m \geq 1; f_i$ are bounded and continuous on S,

$1 \leq i \leq m; \varepsilon > 0).$ (C11.9)

In the case (S, d) is a compact metric space, $C(S) \equiv C_b(S)$ is a *complete separable metric space* under the "*sup*" norm $\|f\|_\infty := \max\{|f(x)| : x \in S\}$, i.e., under the distance $d_\infty(f, g) := \|f - g\|_\infty \equiv \max\{|f(x) - g(x)| : x \in S\}$ (see Dieudonné 1960, p. 134). In this case the weak topology is metrizable, i.e., $\mathcal{P}(S)$ is a metric space with the metric

$$d_W(\mu, v) := \sum_{n=1}^{\infty} 2^{-n} \left| \int_S f_n \, d\mu - \int_S f_n \, dv \right|,$$ (C11.10)

where $\{f_n: n \geq 1\}$ is a dense subset of $\{f \in C(S): \|f\|_\infty \leq 1\}$.

Exercise C11.1 Check that d_W metrizes the weak topology. [Hint: Given a ball $B(\mu_0, r)$ in the d_W metric with center μ_0 and radius $r > 0$, there exists a neighborhood $N(\mu_0: f_1, f_2, \ldots, f_k, \varepsilon)$ contained in it, and vice versa.]   ∎

We now give a proof of Lemma 11.1 (see Section 2.11), namely, $(\mathcal{P}(S), d_W)$ is compact if (S, d) is compact metric.

Proof of Lemma 11.1. Let $\{P_n: n \geq 1\}$ be a sequence of probability measures on a compact metric space (S, d). Note that $\mathcal{P}(S)$ is metrized by the metric d_W (C11.10). Let $n_{1j}(j \geq 1)$ be an increasing sequence of integers such that $\int f_1 \, dP_{n_{1j}}$ converges to some number, say, $\lambda(f_1)$ as $j \to \infty$. From the sequence $\{n_{1j}: j \geq 1\}$ select a subsequence $n_{2j}(j \geq 1)$

such that $\int f_2 \, dP_{n_{2j}}$ converges to same number $\lambda(f_2)$ as $j \to \infty$. In this manner one has a nested family of subsequences $\{P_{n_{ij}} : j \geq 1\}$, $i \geq 1$, each contained in the preceding, and such that $\int f_k \, dP_{n_{ij}} \to \lambda(f_k)$, as $j \to \infty, \forall k = 1, 2, \ldots, i$. For the *diagonal sequence* $\{P_{n_{ii}} : i = 1, 2, \ldots\}$ one has

$$\int f_k \, dP_{n_{ii}} \to \lambda(f_k) \quad \text{as } i \to \infty (k = 1, 2, \ldots).$$

Since $\{f_k : k \geq 1\}$ is dense in $C(S)$ in the sup-norm distance, λ extends to $C(S)$ as a continuous nonnegative linear functional such that $\lambda(\mathbf{1}) = 1$, where $\mathbf{1}$ is the constant function whose value is 1 on all of S. By the Riesz representation theorem (see Royden 1968, p. 310), there exists a unique probability measure P on S such that $\lambda(f) = \int f \, dP \; \forall f \in C(S)$, and $\lim_{i \to \infty} \int f \, dP_{n_{ii}} = \int f \, dP \; \forall f \in C(S)$. ∎

For our next result we need the following lemmas. Let $H = [0, 1]^N$ be the space of all sequences in $[0, 1]$ with the product topology, referred to as the *Hilbert cube*.

Lemma C11.1 *Let (S, d) be a separable metric space. There exists a map h on S into the Hilbert cube $H \equiv [0, 1]^N$ with the product topology, such that h is a homeomorphism of S onto $h(S)$, in the relative topology of $h(S)$.*

Proof. Without loss of generality, assume $d(x, y) \leq 1 \; \forall \, x, y \in S$. Let $\{z_k : k = 1, 2, \ldots\}$ be a dense subset of S. Define the map (on S into H):

$$h(x) = (d(x, z_1), d(x, z_2), \ldots, d(x, z_k), \ldots) \quad (x \in S). \quad \text{(C11.11)}$$

If $x_n \to x$ in S, then $d(x_n, z_k) \to d(x, z_k) \; \forall \, k$, so that $h(x_n) \to h(x)$ in the (metrizable) product topology (of pointwise convergence) on $h(S)$. Also, h is one-to-one. For if $x \neq y$, one may find z_k such that $d(x, z_k) < \frac{1}{3} d(x, y)$, which implies $d(y, z_k) \geq d(y, x) - d(z_k, x) > \frac{2}{3} d(x, y)$, so that $d(x, z_k) \neq d(y, z_k)$. Finally, let $\tilde{a}_n \equiv (a_{n1}, a_{n2}, \ldots) \to \tilde{a} = (a_1, a_2, \ldots)$ in $h(S)$, and let $x_n = h^{-1}(\tilde{a}_n)$, $x = h^{-1}(\tilde{a})$. One then has $(d(x_n, z_1), d(x_n, z_2), \ldots) \to (d(x, z_1), d(x, z_2), \ldots)$. Hence $d(x_n, z_k) \to d(x, z_k) \forall k$, implying $x_n \to x$, since $\{z_k : k \geq 1\}$ is dense in S. ∎

Lemma C11.2 *The weak topology is defined by the system of neighborhoods of the form (C11.9) with $f_1, f_2, \ldots, f_m$ bounded and uniformly continuous.*

Proof. Fix $P_0 \in \mathcal{P}(S)$, $f \in C_b(S)$, $\varepsilon > 0$. We need to show that the set $\{Q \in \mathcal{P}(S): |\int f \, dQ - \int f \, dP_0| < \varepsilon\}$ contains a set of the from (C11.9) (with $\mu_0 = P_0$, $\mu = Q$), but with f_i's *uniformly continuous* and bounded. Without essential loss of generality, assume $0 < f < 1$. As in the proof of Theorem C11.2 (iii) $\Rightarrow$ (i) (see the relations (C11.3)), one may choose and fix a large k such that $1/k < \varepsilon/4$ and consider the sets F_i in that proof. Next, as in the proof of (ii) $\Rightarrow$ (iii) (of Theorem C11.3), there exist uniformly continuous functions g_i, $0 \leq g_i \leq 1$, such that $|\int g_i \, dP_0 - P_0(F_i)| < \varepsilon/4k$, $1 \leq i \leq k$. Then on the set

$$\left\{ Q: \left| \int g_i \, dQ - \int g_i \, dP_0 \right| < \varepsilon/4k, \, 1 \leq i \leq k \right\},$$

one has (see (C11.3))

$$
\begin{aligned}
\int f \, dQ &\leq \sum_{i=1}^{k} Q(F_i) + \frac{1}{k} \leq \sum_{i=1}^{k} \int g_i \, dQ + \frac{1}{k} \\
&< \sum_{i=1}^{k} \int g_i \, dP_0 + \frac{\varepsilon}{4} + \frac{1}{k} \\
&\leq \sum_{i=1}^{k} P_0(F_i) + \frac{\varepsilon}{4} + \frac{\varepsilon}{4} + \frac{1}{k} \\
&< \int f \, dP_0 + \frac{1}{k} + \frac{2\varepsilon}{4} + \frac{1}{k} \\
&\leq \int f \, dP_0 + \varepsilon.
\end{aligned}
$$

Similarly, replacing f by $1 - f$ in the above argument, one may find uniformly continuous h_i, $0 \leq h_i \leq 1$, such that, on the set

$$\left\{ Q : \left| \int h_i \, dQ - \int h_i \, dP_0 \right| < \varepsilon/4k, \, 1 \leq i \leq k \right\},$$

one has

$$\int (1 - f) dQ < \int (1 - f) dP_0 + \varepsilon.$$

Therefore,

$$\left\{ Q: \left| \int f \, dQ - \int f \, dP_0 \right| < \varepsilon \right\} \supset$$

$$\left\{ Q: \left| \int g_i \, dQ - \int g_i \, dP_0 \right| < \varepsilon/4k, \left| \int h_i \, dQ - \int h_i \, dP_0 \right| < \varepsilon/4k, \right.$$

$$\left. \text{for } 1 \le i \le k \right\}.$$

By taking intersections over m such sets, it follows that a neighborhood $N(P_0)$ of P_0 of the form (C11.9) (with $\mu_0 = P_0$ and $f_i \in C_b(S)$, $1 \le i \le m$) contains a neighborhood of P_0 defined with respect to bounded uniformly continuous functions. In particular, $N(P_0)$ is an open set defined by the latter neighborhood system. Since the latter neighborhood system is a subset of the system (C11.9), the proof is complete. ∎

Our main result on metrization is the following.

Theorem C11.3 *Let (S, d) be a separable metric space, then $\mathcal{P}(S)$ is a separable metric (i.e., metrizable) space under the weak topology.*

Proof. By Lemma C11.1, S may be replaced by its homeomorphic image $S_h \equiv h(S)$ in $[0, 1]^{\mathbb{N}}$. The product topology on $[0, 1]^{\mathbb{N}}$ makes it compact by Tychonoff's theorem and is metrizable with the metric $\tilde{d}(\tilde{a}, \tilde{b}) := \sum_{n=1}^{\infty} 2^{-n} |a_n - b_n| (\tilde{a} = (a_1, a_2, \ldots), \tilde{b} = (b_1, b_2, \ldots))$. Every uniformly continuous (bounded) f on S_h (uniform continuity being in the $\tilde{d}$-metric) has a unique extension $\bar{f}$ to $\bar{S}_h (\equiv$ closure of S_h in $[0, 1]^{\mathbb{N}})$: $\bar{f}(\tilde{a}) := \lim_{k \to \infty} f(\tilde{a}^k)$, where $\tilde{a}^k \in S_h$, $\tilde{a}^k \to \tilde{a}$. Conversely, the restriction of every $g \in C(\bar{S}_h)$ is a uniformly continuous bounded function of S_h. In other words, the set of all uniformly continuous real-valued bounded functions on S_h, namely, $UC_b(S_h)$ may be identified with $C(\bar{S}_h)$ as sets and as metric spaces under the supremum distance d_∞ between functions. Since $\bar{S}_h$ is compact, $C_b(\bar{S}_h) \equiv C(\bar{S}_h)$ is a separable metric space under the supremum distance d_∞, and, therefore, so is $UC_b(S_h)$. Letting $\{f_n : n = 1, 2, \ldots\}$ be a dense subset of $UC_b(S_h)$, one now defines a metric d_W on $\mathcal{P}(S_h)$ as in (C11.10). This proves metrizability of $\mathcal{P}(S_h)$.

To prove separability of $\mathcal{P}(S_h)$, for each $k = 1, 2, \ldots$, let $D_k := \{x_{ki} : i = 1, \ldots, n_k\}$ be a finite $(1/k)$-net of S_h (i.e., every point of S_h is within a distance $1/k$ from some point in this net). Let $D = \{x_{ki} :$

$i = 1, \ldots, n_k, k \geq 1\} = \cup_{k=1}^{\infty} D_k$. Consider the set $\mathcal{E}$ of all probabilities with finite support contained in D, having rational mass at each point of support. Then $\mathcal{E}$ is countable and is dense in $\mathcal{P}(S_h)$. To prove this last assertion, fix $P_0 \in \mathcal{P}(S_h)$. Consider the partition generated by the set of open balls $\{x \in S_h : d(x, x_{ki}) < \frac{1}{k}\}$, $1 \leq i \leq n_k$. Let P_k be the probability measure defined by letting the mass of P_0 on each nonempty set of the partition be assigned to a singleton $\{x_{ki}\}$ in D_k, which is at a distance of at most $1/k$ from the set. Now construct $Q_k \in \mathcal{E}$, where Q_k has the same support as P_k but the point masses of Q_k are rational and are such that the sum of the absolute differences between these masses of P_k and corresponding ones of Q_k is less than $1/k$. Then it is simple to check that $d_W(P_0, Q_k) \to 0$ as $k \to \infty$, that is, $\int_{S_h} g \, dQ_k \to \int_{S_h} g \, dP_0$ for every uniformly continuous and bounded g on S_h.   ∎

Our next task is to derive a number of auxiliary results concerning convergence of probability measures, which have been made use of in the text.

In the following lemma we deal with (signed) measures on a sigmafield $\mathcal{E}$ of subsets of an abstract space Ω. For a finite signed measure μ on this space the *variation norm* $\|\mu\|$ is defined by $\|\mu\| := \sup\{|\mu(A)| : A \in \mathcal{E}\}$.

Lemma C11.3 *(Scheffé's Theorem)   Let $(\Omega, \mathcal{E}, \lambda)$ be a measure space. Let $Q_n (n = 1, 2, \ldots)$, Q be probability measures on $(\Omega, \mathcal{E})$ that are absolutely continuous with respect to λ and have densities (i.e., Radon–Nikodym derivatives) $q_n (n = 1, 2, \ldots)$, q, respectively, with respect to λ. If $\{q_n\}$ converges to q almost everywhere (λ), then*

$$\lim_n \|Q_n - Q\| = 0.$$

Proof.   Let $h_n = q - q_n$. Clearly,

$$\int h_n \, d\lambda = 0 \quad (n = 1, 2, \ldots),$$

so that

$$2\|Q_n - Q\| = \int_{\{h_n > 0\}} h_n \, d\lambda - \int_{\{h_n \leq 0\}} h_n \, d\lambda$$
$$= 2\int_{\{h_n > 0\}} h_n \, d\lambda = 2\int h_n \cdot \mathbf{1}_{\{h_n > 0\}} d\lambda. \quad \text{(C11.12)}$$

The last integrand in (C11.12) is nonnegative and bounded above by q. Since it converges to zero almost everywhere, its integral converges to zero.   ∎

Lemma C11.4 *Let S be a separable metric space, and let Q be a probability measure on S. For every positive ε there exists a countable family* $\{A_k : k = 1, 2, \ldots\}$ *of pairwise disjoint Borel subsets of S such that* (i) $\cup\{A_k : k = 1, 2, \ldots\} = S$, (ii) *the diameter of* A_k *is less than ε for every k, and* (iii) *every* A_k *is a Q-continuity set.*

Proof. For each x in S there are uncountably many balls $\{B(x{:}\delta): 0 < \delta < \varepsilon/2\}$ (perhaps not all distinct). Since

$$\partial B(x{:}\delta) \subset \{y{:}\rho(x, y) = \delta\},$$

the collection $\{\partial B(x{:}\delta): 0 < \delta < \varepsilon/2\}$ is pairwise disjoint. But given any pairwise disjoint collection of Borel sets, those with positive Q-probability form a countable subcollection. Hence there exists a ball $B(x)$ in $\{B(x{:}\delta): 0 < \delta < \varepsilon/2\}$ whose boundary has zero Q-probability. The collection $\{B(x): x \in S\}$ is an open cover of S and, by separability, admits a countable subcover $\{B(x_k): k = 1, 2, \ldots\}$, say. Now define

$$A_1 = B(x_1),\ A_2 = B(x_2)\backslash B(x_1),\ \ldots,\ A_k = B(x_k)\backslash \left(\cup_{i=1}^{k-1} B(x_i)\right),\ \ldots\ .$$
(C11.13)

Clearly each A_k has diameter less than ε. Since

$$\partial A = \partial(S\backslash A),\ \partial(A \cap B) \subset (\partial A) \cup (\partial B),$$
(C11.14)

for arbitrary subsets A, B of S, it follows that each A_k is a Q-continuity set. ∎

To state the next lemma, we define, the *oscillation* $\omega_f(A)$ *of* a complex-valued function f *on a set* A by

$$\omega_f(A) = \sup\{|f(x) - f(y)|: x, y \in A\} \quad (A \subset S).$$
(C11.15)

In particular, let

$$\omega_f(x : \varepsilon) := \omega_f(B(x : \varepsilon)) \quad (x \in S,\ \varepsilon > 0).$$
(C11.16)

Definition C11.2 Let Q be a probability measure on a metric space S. A class $\mathcal{G}$ of real-valued functions on S is said to be a *Q-uniformity class* if

$$\limsup_{\substack{n \\ f \in \mathcal{G}}} \left| \int f\, dQ_n - \int f\, dQ \right| = 0$$
(C11.17)

for every sequence $Q_n (n \geq 1)$ of probability measures converging weakly to Q.

Theorem C11.4 *Let S be separable metric space, and Q a probability measure on it. A family $\mathfrak{F}$ of real-valued, bounded, Borel measurable functions on S is a Q-uniformity class if and only if*

$$\text{(i)}\quad \sup_{f \in \mathfrak{F}} \omega_f(S) < \infty, \tag{C11.18}$$

$$\text{(ii)}\quad \lim_{\varepsilon \downarrow 0} \sup_{f \in \mathfrak{F}} \{ Q(\{ \omega_f(x \,:\, \varepsilon) > \delta \}) \} = 0 \text{ for every positive } \delta.$$

Proof. We will only prove the *sufficiency* part, as the *necessity* part is not needed for the development in Chapter 2.

Sufficiency: Assume that (C11.18) holds. Let c, ε be two positive numbers. Let $\{ A_k : k = 1, 2, \ldots \}$ be a partition of S by Borel-measurable Q-continuity sets each of diameter less than ε. Lemma C11.4 makes such a choice possible. Define the class of functions $\mathfrak{F}_{c,\varepsilon}$ by

$$\mathfrak{F}_{c,\varepsilon} = \left\{ \sum_k c_k I_{A_k} : |c_k| \leqslant c \text{ for all } k \right\}. \tag{C11.19}$$

Then $\mathfrak{F}_{c,\varepsilon}$ is a Q-uniformity class. To prove this, suppose that $\{ Q_n \}$ is a sequence of probability measures converging weakly to Q. Define functions $q_n (n = 1, 2, \ldots)$, q on the set of all positive integers by

$$q_n(k) = Q_n(A_k), \quad q(k) = Q(A_k), \quad (k = 1, 2, \ldots). \tag{C11.20}$$

The functions q_n, q are densities (i.e., Radon–Nikodym derivatives) of probability measures on the set of all positive integers (endowed with the sigmafield of all subsets), with respect to the counting measure that assigns mass 1 to each singleton. Since each A_k is a Q-continuity set, $q_n(k)$ converges to $q(k)$ for every k. Hence, by Scheffé's theorem (Lemma C11.3),

$$\lim_n \sum_k |q_n(k) - q(k)| = 0. \tag{C11.21}$$

Therefore

$$
\sup\left\{\left|\int f\,dQ_n - \int f\,dQ\right| : f \in \mathfrak{F}_{c,\varepsilon}\right\}
$$
$$
= \sup\left\{\left|\sum_k c_k q_n(k) - \sum_k c_k q(k)\right| \,|c_k| \leqslant c \quad \text{for all } k\right\}
$$
$$
\leqslant c \sum_k |q_n(k) - q(k)| \to 0 \quad \text{as } n \to \infty. \tag{C11.22}
$$

Thus we have shown that $\mathfrak{F}_{c,\varepsilon}$ is a Q-uniformity class. We now assume, without loss of generality, that every function f in $\mathfrak{F}$ is centered; that is,

$$
\inf_{x\in S} f(x) = -\sup_{x\in S} f(x), \tag{C11.23}
$$

by subtracting from each f the midpoint c_f of its range. This is permissible because

$$
\sup_{f\in\mathfrak{F}}\left|\int f\,d(Q_n - Q)\right| = \sup_{f\in\mathfrak{F}}\left|\int (f - c_f)\,d(Q_n - Q)\right| \tag{C11.24}
$$

whatever the probability measure Q_n. For each $f \in \mathfrak{F}$ define

$$
g_f(x) = \inf_{x\in A_k} f(x) \quad \text{for } x \in A_k \ (k = 1, 2, \ldots),
$$
$$
h_f(x) = \sup_{x\in A_k} f(x) \quad \text{for } x \in A_k \ (k = 1, 2, \ldots).
$$
$$\tag{C11.25}$$

Note that $g_f, h_f \in \mathfrak{F}_{c,\varepsilon}$, where $c = \sup\{\omega_f(s) : f \in \mathfrak{F}\}$, and

$$
g_f \leqslant f \leqslant h_f, \quad h_f(x) - g_f(x) = \omega_f(A_k), \quad \text{for } x \in A_k \ (k = 1, 2, \ldots).
$$
$$\tag{C11.26}$$

It follows that

$$
\int g_f\,d(Q_n - Q) - \int (h_f - g_f)\,dQ = \int g_f\,dQ_n - \int h_f\,dQ
$$
$$
\leqslant \int f\,dQ_n - \int f\,dQ \leqslant \int h_f\,dQ_n - \int g_f\,dQ
$$
$$
= \int h_f\,d(Q_n - Q) + \int (h_f - g_f)\,dQ, \tag{C11.27}
$$

so that

$$\sup_{f\in\mathfrak{F}}\left|\int f\,d(Q_n-Q)\right|\leqslant \sup_{f'\in\mathfrak{F}_{c,\varepsilon}}\left|\int f'\,d(Q_n-Q)\right|+\sup_{f\in\mathfrak{F}}\int (h_f-g_f)\,dQ.$$

$$\text{(C11.28)}$$

Since $\mathfrak{F}_{c,\varepsilon}$ is a Q-uniformity class,

$$\begin{aligned}
\overline{\lim_{n}}\sup_{f\in\mathfrak{F}}\left|\int f\,d(Q_n-Q)\right| &\leqslant \sup_{f\in\mathfrak{F}}\int (h_f-g_f)\,dQ \\
&= \sup_{f\in\mathfrak{F}}\sum_k \omega_f(A_k)Q(A_k) \\
&\leqslant \sup_{f\in\mathfrak{F}}\left[c\sum_{w_f(A_k)>\delta}Q(A_k)+\delta\right] \\
&\leqslant c\sup_{f\in\mathfrak{F}}Q(\{x\colon \omega_f(x:\varepsilon)>\delta\})+\delta,
\end{aligned}$$

$$\text{(C11.29)}$$

since the diameter of each A_k is less than ε. First let $\varepsilon\downarrow 0$ and then let $\delta\downarrow 0$. This proves sufficiency. ∎

Proposition C11.1 *Let S be a separable metric space. A class $\mathfrak{F}$ of bounded functions is a Q-uniformity class for every probability measure Q on S if and only if* (i) $\sup\{\omega_f(S)\colon f\in\mathfrak{F}\}<\infty$, *and* (ii) $\mathfrak{F}$ *is equicontinuous at every point of S, that is,*

$$\lim_{\varepsilon\downarrow 0}\sup_{f\in\mathfrak{F}}\omega_f(x:\varepsilon)=0 \quad \text{for all } x\in S. \qquad \text{(C11.30)}$$

Proof. We assume, without loss of generality, that the functions in $\mathfrak{F}$ are real-valued. Suppose that (i) and (ii) hold. Let Q be a probability measure on S. Whatever the positive numbers δ and ε are,

$$\sup_{f\in\mathfrak{F}}Q(\{x\colon \omega_f(x:\varepsilon)>\delta\})\leqslant Q\left(\left\{x\colon \sup_{f\in\mathfrak{F}}\omega_f(x:\varepsilon)>\delta\right\}\right). \text{(C11.31)}$$

For every positive δ the right-hand side goes to zero as $\varepsilon\downarrow 0$. Therefore, by Theorem C11.4, $\mathfrak{F}$ is a Q-uniformity class. ∎

An interesting metrization of $\mathcal{P}(S)$ is provided by the next corollary. For every pair of positive numbers a,b, define a class $L(a,b)$ of

Lipschitzian functions on S by

$$L(a, b) = \{f: \omega_f(S) \leqslant a, |f(x) - f(y)| \leqslant bd(x, y) \text{ for all } x, y \in S\}.$$
(C11.32)

Now define the *bounded Lipschitzian distance* d_{BL} by

$$d_{\text{BL}}(Q_1, Q_2) = \sup_{f \in L(1,1)} \left| \int f \, dQ_1 - \int f \, dQ_2 \right| \quad (Q_1, Q_2 \in \mathcal{P}(S)).$$
(C11.33)

Corollary C11.1 *If S is a separable metric space, then d_{BL} metrizes the weak topology on $\mathcal{P}(S)$.*

Proof. By Proposition C11.1, $L(1, 1)$ is a Q-uniformity class for every probability measure Q on S. Hence if $\{Q_n\}$ is a sequence of probability measures converging weakly to Q, then

$$\lim_n d_{\text{BL}}(Q_n, Q) = 0.$$
(C11.34)

Conversely, suppose (C11.34) holds. We shall show that $\{Q_n\}$ converges weakly to Q. It follows from (C11.33) that

$$\lim_n \left| \int f \, dQ_n - \int f \, dQ \right| = 0 \text{ for every bounded Lipschitzian function } f.$$
(C11.35)

For, if $|f(x) - f(y)| \leqslant bd(x, y)$ for all $x, y \in S$,

$$\int f \, dQ_n - \int f \, dQ = c \left(\int f' \, dQ_n - \int f' \, dQ \right),$$

where $c = \max\{\omega_f(S), b\}$ and $f' = f/c \in L(1, 1)$. Let F be any nonempty closed subset of S. We now prove

$$\overline{\lim_n} Q_n(F) \leqslant Q(F).$$
(C11.36)

For $\varepsilon > 0$ define the real-valued function f_ε on S by

$$f_\varepsilon(x) = \psi(\varepsilon^{-1} d(x, F)) \quad (x \in S),$$
(C11.37)

where ψ is defined on $[0, \infty)$ by

$$\psi(t) = \begin{cases} 1 - t & \text{if } 0 \leqslant t \leqslant 1, \\ 0 & \text{if } t > 1. \end{cases}$$
(C11.38)

Note that f_ε is, for every positive ε, a bounded Lipschitzian function satisfying

$$\omega_{f_\varepsilon}(S) \leqslant 1, \quad |f_\varepsilon(x) - f_\varepsilon(y)| \leqslant \left| \frac{1}{\varepsilon} d(x, F) - \frac{1}{\varepsilon} d(y, F) \right|$$

$$\leqslant \frac{1}{\varepsilon} d(x, y) \quad (x, y \in S),$$

so that, by (C11.35)

$$\lim_n \int f_\varepsilon \, dQ_n = \int f_\varepsilon \, dQ \quad (\varepsilon > 0). \tag{C11.39}$$

Since $1_F \leqslant f_\varepsilon$ for every positive ε,

$$\overline{\lim_n} Q_n(F) \leqslant \overline{\lim_n} \int f_\varepsilon \, dQ_n = \int f_\varepsilon \, dQ \quad (\varepsilon > 0). \tag{C11.40}$$

Also, $\lim_{\varepsilon \downarrow 0} f_\varepsilon(x) = I_F(x)$ for all x in S. Hence

$$\lim_{\varepsilon \downarrow 0} \int f_\varepsilon \, dQ = Q(F). \tag{C11.41}$$

By Theorem C11.1, $\{Q_n\}$ converges weakly to Q. Finally, it is easy to check that d_{BL} is a distance function on $\mathcal{P}(S)$. ∎

Remark C11.2 Note that in the second part of the proof of the above corollary, we have shown that a sequence Q_n $(n \geq 1)$ of probability measures on a metric space (S, d) converges weakly to a probability measure Q if $\int f \, dQ_n \to \int f \, dQ \, \forall \, f \in L(1, 1)$. (That is, uniform convergence over the class $L(1, 1)$ was not used for this fact.)

The next result is of considerable importance in probability. To state it we need a notion called "tightness."

Definition C11.3 A subset Λ of $\mathcal{P}(S)$ is said to be *tight* if, for every $\varepsilon > 0$, there exists a compact subset K_ε of S such that

$$P(K_\varepsilon) \geq 1 - \varepsilon \quad \forall P \in \Lambda. \tag{C11.42}$$

Theorem C11.5 *(Prokhorov's theorem)*
(a) *Let (S, d) be a separable metric space. If $\Lambda \subset \mathcal{P}(S)$ is tight, then its weak closure $\bar{\Lambda}$ is compact (metric) in the weak topology.*

(b) *If (S, d) is Polish, then the converse is true: for a set Λ to be conditionally compact (i.e., $\bar{\Lambda}$ compact) in the weak topology, it is necessary that Λ is tight.*

Proof.

(a) Let $\tilde{S} = \cup_{j=1}^{\infty} K_{1/j}$, where $K_{1/j}$ is a compact set determined from (C11.42) with $\varepsilon = 1/j$. Then $P(\tilde{S}) = 1 \ \forall \ P \in \Lambda$. Also $\tilde{S}$ is σ-compact and so is its image $\tilde{S}_h = \cup_{j=1}^{\infty} h(K_{1/j})$ under the map h (appearing in the proofs of Lemma C11.1 and Theorem C11.3), since the image of a compact set under a continuous map is compact. In particular, $\tilde{S}_h$ is a Borel subset of $[0, 1]^{\mathbb{N}}$ and, therefore, of $\bar{\tilde{S}}_h$. Let Λ be tight and let Λ_h be the image of Λ in $\tilde{S}_h$ under h, i.e., $\Lambda_h = \{P_0 h^{-1} \colon P \in \Lambda\} \subset \mathcal{P}(\tilde{S}_h)$. In view of the homeomorphism $h \colon \tilde{S} \to \tilde{S}_h$, it is enough to prove that Λ_h is conditionally compact as a subset of $\mathcal{P}(\tilde{S}_h)$.

Since $\tilde{S}_h$ is a Borel subset of $\bar{\tilde{S}}_h$, one may take $\mathcal{P}(\tilde{S}_h)$ as a subset of $\mathcal{P}(\bar{\tilde{S}}_h)$, extending P in $\mathcal{P}(\tilde{S}_h)$ by setting $P(\bar{\tilde{S}}_h \backslash \tilde{S}_h) = 0$. Thus $\Lambda_h \subseteq \mathcal{P}(\tilde{S}_h) \subseteq \mathcal{P}(\bar{\tilde{S}}_h)$. By Lemma C11.1, $\mathcal{P}(\bar{\tilde{S}}_h)$ is compact metric (in the weak topology). Hence every sequence $\{P_n \colon n = 1, 2, \ldots\}$ in Λ_h has a subsequence $\{P_{n_k} \colon k = 1, 2, \ldots\}$ converging weakly to some $Q \in \mathcal{P}(\bar{\tilde{S}}_h)$. We need to show $Q \in \mathcal{P}(\tilde{S}_h)$, that is $Q(\tilde{S}_h) = 1$. By Theorem C11.1, $Q(h(K_{1/j})) \geq \lim \sup_{k \to \infty} P_{n_k}(h(K_{1/j})) \geq 1 - 1/j$ (by hypothesis, $P(h(K_{1/j})) \geq 1 - 1/j \ \forall \ P \in \Lambda_h$). Letting $j \to \infty$, one gets $Q(\tilde{S}_h) = 1$.

(b) Assume (S, d) is separable and complete, and Λ is relatively compact in the weak topology. We first show that *given any sequence of open sets $G_n \uparrow S$, there exists, for every $\varepsilon > 0$, $n = n(\varepsilon)$ such that $P(G_{n(\varepsilon)}) \geq 1 - \varepsilon \ \forall P \in \Lambda$.* If this is not true, then there exists $\varepsilon > 0$ and a sequence $P_n \in \Lambda (n = 1, 2, \ldots)$ such that $P_n(G_n) < 1 - \varepsilon \ \forall n$. Then, by assumption, there exists a subsequence $1 \leq n(1) < n(2) < \cdots$ such that $P_{n(k)}$ converges weakly to some $Q \in \mathcal{P}(S)$ as $k \to \infty$. But this would imply, by Theorem C11.3, $Q(G_n) \leq \lim \inf_{k \to \infty} P_{n(k)}(G_n) \leq \lim \inf_{k \to \infty} P_{n(k)}(G_{n(k)}) < 1 - \varepsilon, \ \forall n = 1, 2, \ldots$ (Note that, for sufficiently large k, $n \leq n(k)$ so that $G_n \subset G_{n(k)}$). This leads to the contradiction $Q(S) = \lim_{n \to \infty} Q(G_n) \leq 1 - \varepsilon$.

To prove Λ is tight, fix $\varepsilon > 0$. Due to separability of S, there exists for each $k = 1, 2, \ldots$ a sequence of open balls $B_{n,k}$ of radius less than $1/k$ ($n = 1, 2, \ldots$) such that $\cup_{n=1}^{\infty} B_{n,k} = S$. Denote $G_{n,k} = \cup_{m=1}^{n} B_{m,k}$. Then $G_{n,k} \uparrow S$ as $n \uparrow \infty$, and, by the italicized statement in the preceding paragraph, one may find $n = n(k)$ such that $P(G_{n(k),k}) \geq 1 - \varepsilon/2^k \ \forall P \in \Lambda$. Let $A = \cap_{k=1}^{\infty} G_{n(k),k}$. Then $\bar{A}$ is *totally bounded* since, for each k, there exists a finite cover of $\bar{A}$ by $n(k)$ balls $\bar{B}_{n,k}$ ($1 \leq n \leq n(k)$) each of diameter less than $1/k$. Hence, by completeness of S, $\bar{A}$ is compact (see Diendonné 1960, p. 56). On the other hand, $P(\bar{A}) \geq P(A) \geq 1 - \sum_{k=1}^{\infty} \varepsilon/2^k = 1 - \varepsilon \ \forall P \in \Lambda$. ∎

The following corollary (left as an exercise) is an easy consequence of Theorem C11.5(b).

Corollary C11.2 *Let (S, d) be a Polish space. Then* (i) *every finite set $\Lambda \subset \mathcal{P}(S)$ is tight and* (ii) *if Λ comprises a Cauchy sequence in $(\mathcal{P}(S), d_{BL})$ then it is tight.*

Perhaps the most popular metric on $(\mathcal{P}(S)$ is the *Prokhorov metric* d_π defined by

$$d_\pi(P, Q) := \inf\{\varepsilon > 0 : P(B) \leq Q(B^\varepsilon) + \varepsilon, \ Q(B) \leq P(B^\varepsilon) + \varepsilon \ \forall B \in \mathcal{S}\}. \tag{C11.43}$$

Proposition C11.2 *Let (S, d) be a metric space. Then the following will hold:* (a) *d_π is a metric on $\mathcal{P}(S)$,* (b) *if $d_\pi(P_n, P) \to 0$, then P_n converges weakly to P, and* (c) *if (S, d) is a separable metric space, then if P_n converges weakly to P, $d_\pi(P_n, P) \to 0$, i.e., d_π metrizes the weak topology on $\mathcal{P}(S)$.*

Proof.

(a) To check that (i) $d_\pi(P, Q) = 0 \Rightarrow P = Q$, consider the inequalities (within curly brackets in (C11.43)) for $B = F$ closed. Letting $\varepsilon \downarrow 0$, one gets $P(F) \leq Q(F)$ and $Q(F) \leq P(F)$ ∀ closed F. (ii) Clearly, $d_\pi(P, Q) = d_\pi(Q, P)$. (iii) To check the *triangle inequality* $d_\pi(P_1, P_2) + d_\pi(P_2, P_3) \geq d_\pi(P_1, P_3)$, let $d_\pi(P_1, P_2) < \varepsilon_1$ and $d_\pi(P_2, P_3) < \varepsilon_2$. Then $P_1(B) \leq P_2(B^{\varepsilon_1}) + \varepsilon_1 \leq P_3((B^{\varepsilon_1})^{\varepsilon_2}) + \varepsilon_2 + \varepsilon_1 \leq P_3(B^{\varepsilon_1+\varepsilon_2}) + \varepsilon_1 + \varepsilon_2 \ \forall B \in \mathcal{S}$ and, similarly, $P_3(B) \leq P_1(B^{\varepsilon_1+\varepsilon_2}) + \varepsilon_1 + \varepsilon_2 \ \forall B \in \mathcal{S}$. Hence $d_\pi(P_1, P_3) \leq \varepsilon_1 + \varepsilon_2$. This provides $d_\pi(P_1, P_3) \leq d_\pi(P_1, P_2) + d_\pi(P_2, P_3)$.

(b) Suppose $d_\pi(P_n, P) \to 0$. Let $\varepsilon_n \to 0$ be such that $d_\pi(P_n, P) < \varepsilon_n$. Then, by the definition (C11.43), $P_n(F) \leq P(F^{\varepsilon_n}) + \varepsilon_n \ \forall$ closed F. Hence $\limsup P_n(F) \leq P(F) \ \forall$ closed F, proving that P_n converges weakly to P.

(c) We will show that on a metric space (S, d),

$$d_\pi(P, Q) \le (d_{BL}(P, Q))^{1/2} \quad \forall P, Q \in \mathcal{P}(S). \tag{C11.44}$$

Fix $\varepsilon > 0$. For a Borel set B, the function $f(x) := \max\{0, 1 - \varepsilon^{-1} d(x, B)\}$ is Lipschitz: $|f(x) - f(y)| \le \varepsilon^{-1} d(x, y)$, $\|f\|_\infty \le 1$. Hence $\varepsilon f \in L(1, 1)$. For $P, Q \in \mathcal{P}(S)$, $Q(B) \le \int f \, dQ$ (since $f \ge 0$, and $f = 1$ on B) $\le \int f \, dP + \varepsilon^{-1} d_{BL}(P, Q)$ (since $d_{BL}(P, Q) \ge |\int \varepsilon f \, d(P - Q)|) \le P(B^\varepsilon) + \varepsilon^{-1} d_{BL}(P, Q)$ (since $f = 0$ outside B^ε). One may interchange P, Q in the above relations. Thus, if $d_{BL}(P, Q) \le \varepsilon^2$, $d_\pi(P, Q) \le \varepsilon$. Since $\varepsilon > 0$ is arbitrary, (C11.44) follows.

Now if P_n converges weakly to P, one has $d_{BL}(P_n, P) \to 0$ (by Corollary C11.1) and, in view of (C11.44) $d_\pi(P_n, P) \to 0$. ∎

Theorem C11.6 *Let (S, d) be a complete separable metric space. Then $(\mathcal{P}(S), d_{BL})$ is a complete separable metric space.*

Proof. Suppose $\{P_n: n \ge 1\}$ is a d_{BL}-Cauchy sequence. Then, by Corollary C11.2, it is tight. Now Theorem C11.5 (Prokhorov's theorem) implies there exists a subsequence $1 \le n(1) < n(2) < \dots$ such that $P_{n(k)}$ converges weakly to some $P \in \mathcal{P}(S)$. Therefore, $d_{BL}(P_{n(k)}, P) \to 0$ as $k \to \infty$, which in turn implies $d_{BL}(P_n, P) \to 0$. ∎

Remark C11.3 Theorem C11.3 holds with d_π in place of d_{BL}. For this one needs to show that a d_π-Cauchy sequence is tight (see Dudley 1989, p. 317).

The final result in this section concerns the Kolmogorov distance.

Definition C11.4 Let P, Q be probability measures on (the Borel sigmafield of) $\mathbb{R}^k$, with distribution functions F, G, respectively. The *Kolmogorov distance* d_K between P and Q is defined by

$$d_K(P, Q) = \sup_{\mathbf{x} \in \mathbb{R}^k} |F(\mathbf{x}) - G(\mathbf{x})|. \tag{C11.45}$$

Proposition C11.3

(a) *Convergence in the Kolmogorov distance implies weak convergence on $\mathcal{P}(\mathbb{R}^k)$.*

(b) $(\mathcal{P}(\mathbb{R}^k), d_K)$ *is a complete metric space.*

Proof. Part (a) follows from Theorem C11.2. To prove part (b), let $P_n(n \geq 1)$ be a Cauchy sequence in the metric d_K, with corresponding distribution functions $F_n(n \geq 1)$. Then F_n converges uniformly to a function G, say, on $\mathbb{R}^k$, which is right-continuous. We need to show that G is the distribution function of a probability measure. Now, $\{P_n: n \geq 1\}$ is tight. For given $\varepsilon > 0$, one can find n_ε such that $|F_n(\mathbf{x}) - F_{n_\varepsilon}(\mathbf{x})| < \varepsilon/4k \ \forall \mathbf{x}$, if $n \geq n_\varepsilon$. One may find $b_1, b_2, \ldots, b_k$ large and $a_1, a_2, \ldots, a_k$ small such that $F_{n_\varepsilon}(b_1, b_2, \ldots, b_k) > 1 - \varepsilon/4k$, and $F_{n_\varepsilon}(b_1, b_2, \ldots, b_{i-1}, a_i, b_{i+1}, \ldots, b_k) < \varepsilon/4k \ \forall i = 1, 2, \ldots, k$. Then, writing $\mathbf{a} = (a_1, \ldots, a_k)$, $\mathbf{b} = (b_1, \ldots, b_k)$, $[\mathbf{a}, \mathbf{b}] = (a_1, b_1) \times \cdots \times (a_k, b_k)$ one has $\forall n \geq n_\varepsilon$,

$$P_n([\mathbf{a}, \mathbf{b}]) \geq F_n(\mathbf{b}) - \sum_{i=1}^{k} F_n(b_1, \ldots, b_{i-1}, a_i, b_{i+1}, \ldots, b_k)$$

$$\geq F_{n_\varepsilon}(\mathbf{b}) - \frac{\varepsilon}{4k} - \sum_{i=1}^{k} \left(F_{n_\varepsilon}(b_1, \ldots, b_{i-1}, a_i, b_{i+1}, \ldots, b_k) + \frac{\varepsilon}{4k} \right)$$

$$\geq 1 - \frac{\varepsilon}{4k} - \frac{\varepsilon}{4k} - k \left(\frac{\varepsilon}{4k} + \frac{\varepsilon}{4k} \right) = 1 - \frac{\varepsilon}{2k} - \frac{\varepsilon}{2} \geq 1 - \varepsilon.$$

Hence $\{P_n: n \geq 1\}$ is tight. By Prokhorov's theorem (Theorem C11.5), there exists a sequence $n_k(k \geq 1)$ such that P_{n_k} converges weakly to some probability measure P with distribution function F, say. Then $F_{n_k}(\mathbf{x}) \to F(\mathbf{x})$ at all points of continuity $\mathbf{x}$ of F. Then $F(\mathbf{x}) = G(\mathbf{x})$ for all such $\mathbf{x}$. Since both F and G are right continuous, and since for every $\mathbf{x} \in R^k$ there exists a sequence $\mathbf{x}_j$ of continuity points of F such that $\mathbf{x}_j \downarrow \mathbf{x}$, it follows that $F(\mathbf{x}) = G(\mathbf{x}) \ \forall \mathbf{x}$. Thus G is the distribution function of P, and $d_K(P_n, P) \to 0$. ∎

Bibliographical Notes. On precise "rate of convergence to" the invariant distribution see Diaconis (1988) and the articles by Diaconis and Strook (1991), and Fill (1991). An extension to continuous time Markov process is in Bhattacharya (1999). For a useful survey of some results on invariant distributions of Markov processes and their applications to economics, see Futia (1982). To go beyond Example 8.1, see Radner and Rothschild (1975), Radner (1986), Rothschild (1974), and Simon (1959, 1986). For results on characterization and stability of invariant distributions of diffusion process, see Khas'minskii (1960), Bhattacharya (1978), and Bhattacharya and Majumdar (1980).

2.14 Supplementary Exercises

(1) Consider the following transition probability matrix:

$$\mathbf{p} = \begin{pmatrix} 0.25 & 0.25 & 0.5 \\ 0.5 & 0.25 & 0.25 \\ 0.2 & 0.3 & 0.5 \end{pmatrix}.$$

(a) Show that the invariant distribution of the matrix $\mathbf{p}$ is given by

$$\pi = \begin{pmatrix} 0.2963 \\ 0.2716 \\ 0.4321 \end{pmatrix}.$$

(b) Note that

$$\mathbf{p}^{100} = \begin{pmatrix} 0.2963 & 0.2716 & 0.4321 \\ 0.2963 & 0.2716 & 0.4321 \\ 0.2963 & 0.2716 & 0.4321 \end{pmatrix}.$$

(2) Consider the following transition probability matrix:

$$\mathbf{p} = \begin{pmatrix} 0.4 & 0.6 & 0 \\ 0.2 & 0.4 & 0.4 \\ 0 & 0.7 & 0.3 \end{pmatrix}.$$

(a) Show that the invariant distribution of the matrix $\mathbf{p}$ is given by

$$\pi = \begin{pmatrix} 0.175 \\ 0.525 \\ 0.3 \end{pmatrix}.$$

(b) Verify that

$$\mathbf{p}^2 = \begin{pmatrix} 0.28 & 0.48 & 0.24 \\ 0.16 & 0.56 & 0.28 \\ 0.14 & 0.49 & 0.37 \end{pmatrix}.$$

(c) Note that

$$\mathbf{p}^{100} = \begin{pmatrix} 0.175 & 0.525 & 0.3 \\ 0.175 & 0.525 & 0.3 \\ 0.175 & 0.525 & 0.3 \end{pmatrix}.$$

(3) Consider the following transition probability matrix:

$$\mathbf{p} = \begin{pmatrix} 0 & 0 & 0.6 & 0.4 \\ 0 & 0 & 0.2 & 0.8 \\ 0.25 & 0.75 & 0 & 0 \\ 0.5 & 0.5 & 0 & 0 \end{pmatrix}.$$

(a) Show that the invariant distribution of the matrix $\mathbf{p}$ is given by

$$\pi = \begin{pmatrix} 0.2045 \\ 0.2955 \\ 0.1818 \\ 0.3182 \end{pmatrix}.$$

(b) Verify that

$$\mathbf{p}^2 = \begin{pmatrix} 0.35 & 0.65 & 0 & 0 \\ 0.45 & 0.55 & 0 & 0 \\ 0 & 0 & 0.3 & 0.7 \\ 0 & 0 & 0.4 & 0.6 \end{pmatrix}.$$

(c) The invariant distribution of $\mathbf{p}^2$ is given by

$$\widehat{\pi} = \begin{pmatrix} 0.4091 \\ 0.5909 \\ 0 \\ 0 \end{pmatrix}.$$

(d) Note that

$$\mathbf{p}^{99} = \begin{pmatrix} 0 & 0 & 0.3636 & 0.6364 \\ 0 & 0 & 0.3636 & 0.6364 \\ 0.4091 & 0.5909 & 0 & 0 \\ 0.4091 & 0.5909 & 0 & 0 \end{pmatrix},$$

$$\mathbf{p}^{100} = \begin{pmatrix} 0.4091 & 0.5909 & 0 & 0 \\ 0.4091 & 0.5909 & 0 & 0 \\ 0 & 0 & 0.3636 & 0.6364 \\ 0 & 0 & 0.3636 & 0.6364 \end{pmatrix}.$$

(4) Let $Q = [q_{ij}]$ be a matrix with rows and columns indexed by the elements of a finite or countable set U. Suppose Q is *substochastic* in

the sense that

$$q_{ij} \geq 0 \quad \text{and} \quad \text{for all } i, \quad \sum_j q_{ij} \leq 1.$$

Let $Q^n = [q_{ij}^{(n)}]$ be the nth power so that

$$q_{ij}^{(n+1)} = \sum_v q_{iv}q_{vj}^{(n)}$$

$$q_{ij}^{(0)} = \delta_{ij} = \begin{cases} 1 & \text{if } i = j, \\ 0 & \text{if } i \neq j. \end{cases}$$

(a) Show that the row sums

$$\sigma_i^{(n)} = \sum_j q_{ij}^{(n)}$$

satisfy $\sigma_i^{(n+1)} \leq \sigma_i^{(n)}$. Hence,

$$\sigma_i = \lim_{n \to \infty} \sigma_i^{(n)} = \lim_{n \to \infty} \sum_j q_{ij}^{(n)} \tag{14.1}$$

exist, and, show that

$$\sigma_i = \sum_j q_{ij}\sigma_j. \tag{14.2}$$

[Hint: (i) $\sigma_i^{(n+1)} = \sum_j q_{ij}\sigma_j^{(n)}$. (ii) If $\lim_{n \to \infty} x_{nk} = x_k$ and $|x_{nk}| \leq M_k$ where $\sum_k M_k < \infty$, then $\lim_{n \to \infty} \sum_k x_{nk} = \sum_k x_k$.]

(b) Thus, σ_i solve the system

$$x_i = \sum_{j \in U} q_{ij}x_j, \quad i \in U,$$

$$0 \leq x_i \leq 1, \quad i \in U. \tag{14.3}$$

For an arbitrary solution (x_i) of (14.3), show that $x_i \leq \sigma_i$. Thus, the σ_i give the *maximal solution* to (14.3).

[Hint: $x_i = \sum_j q_{ij}x_j \leq \sum_j q_{ij} = \sigma_i^{(1)}$, and $x_i \leq \sigma_i^{(n)}$ for all i implies that $x_i \leq \sum_j q_{ij}\sigma_j^{(n)} = \sigma_i^{(n+1)}$. Thus, $x_i \leq \sigma_i^{(n)}$ for all n by induction.]

To summarize:

[R.1] For a substochastic matrix Q, the limits (14.1) are the maximal solution of (14.3).

(c) Now suppose that U is a subset of the state space S. Then $[p_{ij}]$ for i and j in U give a substochastic matrix Q. Then, $\sigma_i^{(n)} = P_i[X_t \in U, t \leq n]$.
Letting $n \to \infty$,

$$\sigma_i = P_i[X_t \in U, t = 1, 2, \ldots], \quad i \in U \qquad (14.4)$$

Then, [R.1] can be restated as follows:
[R.2] For $U \subset S$, the probabilities (σ_i) in (14.4) are the maximal solution of the system

$$x_i = \sum_{j \in U} p_{ij} x_j, \quad i \in U,$$
$$0 \leq x_i \leq 1, \quad i \in U. \qquad (14.5)$$

Now consider the system (14.5) with $U = S - \{i_0\}$ for a single state i_0:

$$x_i = \sum_{j \neq i_0} p_{ij} x_j, \quad i \neq i_0,$$
$$0 \leq x_i \leq 1, \quad i \neq i_0. \qquad (14.6)$$

There is always a trivial solution to (14.6), namely $x_i = 0$.
Now prove the following:
[R.3] An irreducible chain is transient if and only if (14.6) has a nontrivial solution.
[Hint: The probabilities

$$1 - \rho_{ii_0} = P_i[X_n \neq i_0, n \geq 1] \quad i \neq i_0 \qquad (14.7)$$

are, by [R.2], the maximal solution of (14.6) and hence vanish identically if and only if (14.6) has no nontrivial solution.
Verify that

$$\rho_{i_0 i_0} = p_{i_0 i_0} + \sum_{i \neq i_0} p_{i_0 i_0} \rho_{i i_0}$$

If the chain is transient, $\rho_{i_0 i_0} < 1$, hence $\rho_{i i_0} < 1$ for some $i \neq i_0$]
Since the Equations (14.6) are homogeneous, the central question is *whether they have a nonzero solution that is nonnegative and bounded*. If they do, $0 \leq x_i \leq 1$ is ensured by normalization.
(d) Review Example 6.2. For a random walk with state space $\mathbb{Z}$, consider the set $U = \mathbb{Z}_+ = \{0, 1, 2, \ldots\}$. Then, consider the system (14.5)

restated as

$$x_i = px_{i+1} + qx_{i-1} \quad \text{for } i \geq 1,$$
$$x_0 = px_1.$$

The solution (recall from Chapter 1) is given by

$$x_n = A + An \qquad\qquad \text{if } p = q$$
$$x_n = A - A(q/p)^{n+1} \quad \text{if } p \neq q$$

Verify that if $q \geq p$, the only bounded solution is $x_n = 0$. If $q < p$, $A = 1$ gives the maximal solution $x_n = 1 - (q/p)^{n+1}$.

(e) Consider a Markov chain with state space $\mathbb{Z}_+ = \{0, 1, 2, \ldots\}$ and transition matrix of the form

$$\mathbf{p} = \begin{pmatrix} q & p & r & 0 & 0 & \cdots \\ q & p & r & 0 & 0 & \cdots \\ 0 & q & p & r & 0 & \cdots \\ 0 & 0 & q & p & r & \cdots \\ \cdot\cdot & \cdot\cdot & \cdot\cdot & \cdot\cdot & \cdot\cdot & \cdots \end{pmatrix}.$$

We can interpret this as a queuing model: if there are i customers in the queue and $i \geq 1$, the customer at the front is served and leaves, and then 0, 1, 2 new customers arrive with probabilities q, p, r. This creates a queue of length $i - 1, i$, or $i + 1$. If $i = 0$, no one is served, and the new arrivals bring the queue length to 0, 1, 2. Assume that $q > 0, r > 0$, so that the chain is irreducible.

For $i_0 = 0$, the system (14.6) is

$$x_1 = px_1 + rx_2,$$
$$x_k = qx_{k-1} + px_k + rx_{k+1}, \quad k \geq 2. \tag{14.7}$$

Show that the second line of (14.7) has the form

$$x_k = \alpha x_{k+1} + \beta x_{k-1}, \quad k \geq 2. \tag{14.8}$$

Verify that the solution to (14.8) is

$$x_k = \begin{cases} B((\beta/\alpha)^k - 1) & \text{if } q \neq r \\ Bk & \text{if } q = r, \end{cases} \tag{14.9}$$

where $\alpha = r/(q + r)$, $\beta = q/(q + r)$.

Hence, show that if $q < r$, the chain is transient and the queue size goes to infinity.

Now consider the problem of computing an invariant distribution $\pi = (\pi_k)$ satisfying $\pi' p = \pi$. Using the structure of p we get $\pi_k = \alpha \pi_{k-1} + \beta \pi_{k+1} (k \geq 1)$.

$$\pi_k = \begin{cases} A + B(\alpha/\beta)^k & \text{if } r \neq q \\ A + Bk & \text{if } r = q \end{cases}$$

for $k \geq 2$. Show that (a) if $q < r$, there is no stationary distribution, (b) if $q = r$, there is again no stationary distribution, and (c) if $q > r$, compute $\pi = (\pi_k)$. Show that there is a "small" probability of a large queue length.

(f) Suppose that the states are $0, 1, 2, \ldots$ and the transition matrix is

$$p = \begin{pmatrix} q_0 & p_0 & 0 & 0 & \cdots \\ q_1 & 0 & p_1 & 0 & \cdots \\ q_2 & 0 & 0 & p_2 & \cdots \\ .. & .. & .. & .. & \cdots \end{pmatrix},$$

where p_i and q_i are positive. The state i represents the length of a success run, the conditional chance of a further success being p_i. A solution of the system (14.6) for checking transience must have the form $x_k = x_1/p_1 \cdots p_{k-1}$, and so there is a bounded, nontrivial solution if and only if $\Pi_{k=1}^n (1 - q_k)$ is bounded away from 0, which holds if and only if $\sum_k q_k < \infty$. Thus the chain is recurrent if and only if $\sum_k q_k = \infty$.

Any solution of the steady state equations $\pi' p = \pi$ has the form $\pi_k = \pi_0 p_0 \cdots p_{k-1}$, and so there is a stationary distribution if and only if $\sum_k p_0 \cdots p_k$ converges.

(5) Analyze the Markov chain with the following transition matrix ($p_i > 0$ for $i = 0, 1, 2, \ldots$).

$$p = \begin{bmatrix} p_0 & p_1 & p_2 & \cdots \\ 1 & 0 & 0 & \cdots \\ 0 & 1 & 0 & \cdots \\ 0 & 0 & 1 & \cdots \\ \cdots & \cdots & \cdots & \cdots \end{bmatrix}.$$

[Hint: Concentrate on the case when the process starts from state 0.]

3

Random Dynamical Systems

One way which I believe is particularly fruitful and promising is to study what would become of the solution of a deterministic dynamic system if it were exposed to a stream of erratic shocks that constantly upsets the evolution.

Ragnar Frisch

3.1 Introduction

A random dynamical system is described by a triplet (S, Γ, Q) where S is the state space, Γ an appropriate family of maps from S into itself (interpreted as the set of all admissible laws of motion), and Q is a probability distribution on (some sigmafield of) Γ. The evolution of the system is depicted informally as follows: initially, the system is in some state x in S; an element α_1 of Γ is chosen by Tyche according to the distribution Q, and the system moves to the state $X_1 = \alpha_1(x)$ in period 1. Again, independently of α_1, Tyche chooses α_2 from Γ according to the same Q, and the state of the system in period 2 is obtained as $X_2 = \alpha_2(X_1)$, and the story is repeated. The initial state x can also be a random variable X_0 chosen independently of the maps α_n. The sequence X_n so generated is a Markov process. It is an interesting and significant mathematical result that every Markov process on a *standard* state space may be represented as a random dynamical system. Apart from this formal proposition (see Complements and Details, Proposition C1.1), many important Markov models in applications arise, and are effectively analyzed, as random dynamical systems. It is the purpose of this chapter to study the long-run behavior of such systems from a unified point of view. We identify conditions under which there is a unique invariant distribution, and, furthermore, study the stability of the process. The formal definitions are collected in Sections 3.2 and 3.3. We move on to discuss the role of uncertainty in influencing the evolution of the process through simple examples in Section 3.4. Two general themes – splitting and average contraction – are developed in later sections. Applications and examples are treated in some detail.

This chapter is largely self-contained, that is, independent of Chapter 2.

3.2 Random Dynamical Systems

Let S be a metric space and $\mathcal{S}$ be the Borel sigmafield of S. Endow Γ, a family of maps from S into S, with a sigmafield Σ such that the map $(\gamma, x) \to (\gamma(x))$ on $(\Gamma \times S, \Sigma \otimes \mathcal{S})$ into $(S, \mathcal{S})$ is measurable. Let Q be a probability measure on (Γ, Σ).

On some probability space $(\Omega, \mathcal{F}, P)$, let $(\alpha_n)_{n=1}^{\infty}$ be a sequence of random functions from Γ with a common distribution Q. For a given random variable X_0 (with values in S), independent of the sequence $(\alpha_n)_{n=1}^{\infty}$, define

$$X_1 \equiv \alpha_1(X_0) \equiv \alpha_1 X_0, \qquad (2.1)$$

$$X_{n+1} = \alpha_{n+1}(X_n) \equiv \alpha_{n+1}\alpha_n \cdots \alpha_1 X_0 \ (n \geq 0). \qquad (2.2)$$

We write $X_n(z)$ for the case $X_0 = z$. Then X_n is a Markov process with a stationary transition probability $p(x, dy)$ given as follows: for $x \in S$, $C \in \mathcal{S}$,

$$p(x, C) = Q(\{\gamma \in \Gamma \colon \gamma(x) \in C\}). \qquad (2.3)$$

Recall that (see Definition 11.1 of Chapter 2) the transition probability $p(x, dy)$ is said to be *weakly continuous* or to have the *Feller property* if for any sequence x_n converging to x, the sequence of probability measures $p(x_n, \cdot)$ converges weakly to $p(x, \cdot)$. One can show that if Γ consists of a family of continuous maps, $p(x, dy)$ has the Feller property (recall Example 11.2 of Chapter 2 dealing with a finite Γ): if $x_n \to x$, then $Tf(x_n) \equiv Ef(\alpha_1 x_n) \to Ef(\alpha_1, x) \equiv Tf(x)$, for every bounded continuous real-valued function f on S. To see this, note that $\alpha_1(\omega)$ is a continuous function on S (into S) for every $\omega \in \Omega$, so that $\alpha_1(\omega)x_n \to \alpha_1(\omega)x$. This implies $f(\alpha_1(\omega)x_n) \to f(\alpha_1(\omega)x)$. Taking expectation, the result follows.

3.3 Evolution

To study the evolution of the process (2.2), it is convenient to recall the definitions of a transition operator and its adjoint. Let $\mathbf{B}(S)$ be the linear space of all bounded real-valued measurable functions on S. The transition operator T on $\mathbf{B}(S)$ is given by (see Chapter 2, Definition 9.1)

$$(Tf)(x) = \int_S f(y)p(x, dy), \quad f \in \mathbf{B}(S), \tag{3.1}$$

and T^* is defined on the space $\mathcal{M}(S)$ of all finite signed measures on $(S, \mathcal{S})$ by

$$T^*\mu(C) = \int_S p(x, C)\mu(dx) = \int_\Gamma \mu(\gamma^{-1}C)Q(d\gamma), \quad \mu \in \mathcal{M}(S). \tag{3.2}$$

As before, let $\mathcal{P}(S)$ be the set of all probability measures on $(S, \mathcal{S})$. An element π of $\mathcal{P}(S)$ is *invariant* for $p(x, dy)$ (or for the Markov process X_n) if it is a fixed point of T^*, i.e.,

$$\pi \text{ is invariant} \quad \text{iff} \quad T^*\pi = \pi. \tag{3.3}$$

Now write $p^{(n)}(x, dy)$ for the n-step transition probability with $p^{(1)} \equiv p(x, dy)$. Then $p^{(n)}(x, dy)$ is the distribution of $\alpha_n \cdots \alpha_1 x$. Recall that T^{*n} is the nth iterate of T^*:

$$T^{*n}\mu = T^{*(n-1)}(T^*\mu)(n \geq 2), \quad T^{*1} = T^*, \quad T^{*0} = \text{identity.} \tag{3.4}$$

Then for any $C \in \mathcal{S}$,

$$(T^{*n}\mu)(C) = \int_S p^{(n)}(x, C)\mu(dx), \tag{3.5}$$

so that $T^{*n}\mu$ is the distribution of X_n when X_0 has distribution μ. To express T^{*n} in terms of the common distribution Q of the i.i.d. maps, let Γ^n denote the usual Cartesian product $\Gamma \times \Gamma \times \cdots \times \Gamma$ (n terms), and let Q^n be the product probability $Q \times Q \times \cdots \times Q$ on $(\Gamma^n, \sum^{\otimes n})$, where $\sum^{\otimes n}$ is the product sigmafield on Γ^n. Thus Q^n is the (joint) distribution of $\alpha = (\alpha_1, \alpha_2, \ldots, \alpha_n)$. For $\gamma = (\gamma_1, \gamma_2, \ldots, \gamma_n) \in \Gamma^n$, let $\tilde{\gamma}$ denote the composition

$$\tilde{\gamma} := \gamma_n \gamma_{n-1} \cdots \gamma_1. \tag{3.6}$$

Then, since $T^{*n}\mu$ is the distribution of $X_n = \alpha_n \cdots \alpha_1 X_0$, one has, $(T^{*n}\mu)(A) = \text{Prob}(X_n \in A) = \text{Prob}(X_0 \in \tilde{\alpha}^{-1}A)$, where $\tilde{\alpha} = \alpha_n$

$\alpha_{n-1} \cdots \alpha_1$, and by the independence of $\tilde{\alpha}$ and X_0,

$$(T^{*n}\mu)(A) = \int_{\Gamma^n} \mu(\tilde{\gamma}^{-1}A)Q^n(d\gamma) \quad (A \in \mathcal{S}, \ \mu \in \mathcal{P}(\mathcal{S})). \quad (3.7)$$

Finally, we come to the definition of *stability*.

Definition 3.1 A Markov process X_n is *stable in distribution* if there is a unique invariant probability measure π such that $X_n(x)$ converges in distribution to π irrespective of the initial state x, i.e., if $p^{(n)}(x, dy)$ converges weakly to the same probability measure π for all x. In the case one has $(1/n) \sum_{m=1}^{n} p^{(m)}(x, dy)$ converging weakly to the same invariant π for all x, we may define the Markov process to be *stable in distribution on the average*.

3.4 The Role of Uncertainty: Two Examples

Our first example contrasts the steady state of a random dynamical system (S, Γ, Q) with the steady states of deterministic laws of Γ.

Example 4.1 Let $S = [0, 1]$ and consider the maps $\bar{f}$ and $\bar{\bar{f}}$ on S into S defined by

$$\bar{f}(x) = x/2$$
$$\bar{\bar{f}}(x) = x/2 + \tfrac{1}{2}.$$

Now, if we consider the deterministic dynamical systems $(S, \bar{f})$ and $(S, \bar{\bar{f}})$, then for each system all the trajectories converge to a unique fixed point (0 and 1, respectively) independently of initial condition.

Think of the random dynamical system (S, Γ, Q) where $\Gamma = \{\bar{f}, \bar{\bar{f}}\}$ and $Q(\{\bar{f}\}) = p > 0$, $Q(\{\bar{\bar{f}}\}) = 1 - p > 0$. It follows from Theorem 5.1 that, irrespective of the initial x, the distribution of $X_n(x)$ converges in the Kolmogorov metric (see (5.3)) to a unique invariant distribution π, which is nonatomic (i.e., the distribution function of π is continuous). ∎

Exercise 4.1 If $p = \tfrac{1}{2}$, then the uniform distribution over $[0, 1]$ is the unique invariant distribution.   ∎

The study of existence of a unique invariant distribution and its stability is relatively simple for those cases in which the transition probability $p(x, dy)$ has a density $p(x, y)$, say, with respect to some reference measure $\mu(dy)$ on the state space S. We illustrate a dramatic difference between the cases when such a density exists and when it does not.

Example 4.2 Let $S = [-2, 2]$ and consider the Markov process

$$X_{n+1} = f(X_n) + \varepsilon_{n+1} \quad n \geq 0,$$

where X_0 is independent of $\{\varepsilon_n\}$, and $\{\varepsilon_n\}$ is an i.i.d. sequence with values in $[-1, 1]$ on a probability space $(\Omega, \mathcal{F}, \mathcal{P})$, and

$$f(x) = \begin{cases} x + 1 & \text{if } -2 \leq x \leq 0, \\ x - 1 & \text{if } 0 < x \leq 2. \end{cases}$$

First, let ε_n be Bernoulli, i.e.,

$$P(\varepsilon_n = 1) = \tfrac{1}{2} = P(\varepsilon_n = -1).$$

By direct computation (see Bhattacharya and Waymire 1990, p. 181 for details), one can verify that if $x \in (0, 2]$, $\{X_n(x), n \geq 1\}$ is i.i.d. with a common (two-point) distribution π_x. In particular, π_x is an invariant distribution. It assigns mass $\tfrac{1}{2}$ each to $\{x - 2\}$ and $\{x\}$. On the other hand, if $x \in [-2, 0]$, then $\{X_n(x), n \geq 1\}$ is i.i.d. with a common distribution π_{x+2}, assigning mass $\tfrac{1}{2}$ to $\{x + 2\}$ and $\{x\}$. Thus, there is an uncountable family of mutually singular invariant distributions $\{\pi_x : 0 < x < 1\} \cup \{\pi_{x+2} : -1 \leq x \leq 0\}$.

On the other hand, suppose ε_n is uniform over $[-1, 1]$, i.e., has the density $\tfrac{1}{2}$ on $[-1, 1]$ and 0 outside. One can check that $\{X_{2n}(x), n \geq 1\}$ is an i.i.d. sequence whose common distribution does not depend on x, and has a density

$$\pi(y) = \frac{2 - |y|}{4}, \quad -2 \leq y \leq 2.$$

The same is true of the sequence $\{X_{2n+1}(x) : n \geq 1\}$. Thus, $\pi(y)dy$ is the unique invariant probability and stability holds. ∎

Remark 4.1 Example 4.2 illustrates the dramatic difference between the cases when ε_n assumes a discrete set of values, and when it has a density (with respect to the Lebesgue measure), in the case of a nonlinear transfer function f on an interval S. One may modify the above function f near zero, so as to make it *continuous*, and still show that the above phenomena persist. See Bhattacharya and Waymire (1990, Exercise II.14.7, p. 212).

3.5 Splitting

3.5.1 Splitting and Monotone Maps

Let S be a nondegenerate interval (finite or infinite, closed, semiclosed, or open) and Γ a set of *monotone* maps from S into S; i.e., each element of Γ is either a nondecreasing function on S or a nonincreasing function.

We will assume the following *splitting condition*:

    (H)   *There exist $z_0 \in S$, $\tilde{\chi}_i > 0$ ($i = 1, 2$) and a positive N such that*

(1) $P(\alpha_N \alpha_{N-1} \cdots \alpha_1 x \leqslant z_0 \; \forall x \in S) \geqslant \tilde{\chi}_1$,
(2) $P(\alpha_N \alpha_{N-1} \cdots \alpha_1 x \geqslant z_0 \; \forall x \in S) \geqslant \tilde{\chi}_2$.

Note that conditions (1) and (2) in **(H)** may be expressed, respectively, as

$$Q^N(\{\gamma \in \Gamma^N : \tilde{\gamma}^{-1}[x \in S : x \leq z_0] = S\}) \geqslant \tilde{\chi}_1, \qquad (5.1)$$

and

$$Q^N(\{\gamma \in \Gamma^N : \tilde{\gamma}^{-1}[x \in S : x \geq z_0] = S\}) \geqslant \tilde{\chi}_2. \qquad (5.2)$$

Here $\tilde{\gamma} = \gamma_N \gamma_{N-1} \cdots \gamma_1$.

The splitting condition **(H)** remains meaningful when applied to a Borel subset S of $\mathbb{R}^\ell$ with the *partial order* $\mathbf{x} \leq \mathbf{y}$ if $x_i \leq y_i$ for all $i = 1 \ldots, \ell$ ($\mathbf{x} = (x_1, \ldots, x_\ell)$, $\mathbf{y} = (y_1 \ldots, y_\ell) \in S$). One writes $\mathbf{x} \geq \mathbf{y}$ if $\mathbf{y} \leq \mathbf{x}$. Thus a *monotone function* $f : S \to S$ is one such that either (i) $f(\mathbf{x}) \leq f(\mathbf{y})$ if $\mathbf{x} \leq \mathbf{y}$ (monotone nondecreasing), or (ii) $f(\mathbf{x}) \geq f(\mathbf{y})$ if $\mathbf{x} \leq \mathbf{y}$ (monotone nonincreasing).

The following remarks clarify the role of the splitting condition:

Remark 5.1 Let $S = [a, b]$ and $\alpha_n (n \geqslant 1)$ a sequence of i.i.d. continuous nondecreasing maps on S into S. Suppose that π is the unique invariant distribution of the Markov process (see Theorem 11.1 in Chapter 2). *If π is not degenerate, then the splitting condition holds.* We verify this claim. Since $p^{(n)}(x, dy)$ is the distribution of $X_n(x) = \alpha_n \cdots \alpha_1 x$, it is therefore also the distribution of $Y_n(x) = \alpha_1 \cdots \alpha_n x$. Note that $Y_{n+1}(a) = \alpha_1 \alpha_2 \cdots \alpha_n \alpha_{n+1} a \geqslant \alpha_1 \alpha_2 \cdots \alpha_n a = Y_n(a)$ (since $\alpha_{n+1} a \geqslant a$). Let $\underline{Y}$ be the limit of the nondecreasing sequence $Y_n(a)(n \geqslant 1)$. Then $p^{(n)}(a, dy)$ converges weakly to the distribution $\underline{\pi}$, say, of $\underline{Y}$. Similarly, $Y_{n+1}(b) \leqslant Y_n(b) \; \forall n$, and the nonincreasing sequence $Y_n(b)$ converges to a limit $\bar{Y}$, so that $p^{(n)}(b, dy)$ converges weakly to the distribution $\bar{\pi}$ of $\bar{Y}$. By uniqueness of the invariant π, $\underline{\pi} = \bar{\pi} = \pi$. Since $\underline{Y} \leqslant \bar{Y}$, it follows that $\underline{Y} = \bar{Y}$ (a.s.) $= Y$, say. Assume π is not degenerate. Then there exist $a', b' \in (a, b)$, $a' < b'$ such that $P(Y < a') > 0$ and $P(Y > b') > 0$. Since $Y_n(a) \uparrow Y$ and $Y_n(b) \downarrow Y$, there exists a positive integer N such that (i) $\tilde{\chi}_1 := P(Y_N(b) < a') > 0$, and (ii) $\tilde{\chi}_2 := P(Y_N(a) > b') > 0$. But then $P(X_N(x) < a' \ \forall x \in [a, b]) = P(Y_N(x) < a' \ \forall x \in [a, b]) = P(Y_N(b) < a') = \tilde{\chi}_1$, and $P(X_N(x) > b' \ \forall x \in [a, b]) = P(Y_N(x) > b'$, $\forall x \in [a, b]) = P(Y_N(a) > b') = \tilde{\chi}_2$. Choose any z_0 in (a', b'). The splitting condition is now verified. The assumption of continuity of α_n may be dispensed with here since, as shown in Complements and Details (see Lemma C5.2), if α_n are monotone nondecreasing and $p^{(n)}(x, dy)$ converges weakly to $\pi(dy)$ for every $x \in S$, then π is the unique invariant probability on $[c, d]$.

Remark 5.2 We say that a Borel subset A of S is *closed* under p if $p(x, A) = 1$ for all x in A. If the state space S has two disjoint closed subintervals that are closed under p or under $p^{(n)}$ for some n, then the splitting condition does not hold. This is the case, for example, when these intervals are invariant under α almost surely.

Denote by $d_K(\mu, \nu)$ the *Kolmogorov distance* on $\mathcal{P}(S)$. That is, if F_μ, F_ν denote the distribution functions (d.f.) of μ and ν, respectively, then

$$d_K(\mu, \nu) := \sup_{x \in \mathbb{R}} |\mu((-\infty, x] \cap S) - \nu(-\infty, x] \cap S)|$$

$$\equiv \sup_{x \in \mathbb{R}} |F_\mu(x) - F_\nu(x)|, \ \mu, \nu \in \mathcal{P}((S)). \tag{5.3}$$

Remark 5.3 It should be noted that convergence in the distance d_K on $\mathcal{P}(S)$ implies weak convergence in $\mathcal{P}(S)$. (See Theorem C11.2(d) of Chapter 2.)

Theorem 5.1 *Assume that the splitting condition* (**H**) *holds on a nondegenerate interval S.*
Then

(a) the distribution $T^{*n}\mu$ of $X_n := \alpha_n \cdots \alpha_1 X_0$ converges to a probability measure π on S exponentially fast in the Kolmogorov distance d_K irrespective of X_0. Indeed,

$$d_K(T^{*n}\mu, \pi) \leqslant (1 - \tilde{\chi})^{[n/N]} \quad \forall \mu \in \mathcal{P}(S), \tag{5.4}$$

where $\tilde{\chi} := \min\{\tilde{\chi}_1, \tilde{\chi}_2\}$ and $[y]$ denotes the integer part of y.

(b) π *in (a) is the unique invariant probability of the Markov process* X_n.

Proof. Suppose **(H)** holds, i.e., (5.1), (5.2) hold. Fix γ monotone, and $\mu, \nu \in \mathcal{P}(S)$ arbitrarily. We will first show that, for all x,

$$|\mu(\gamma^{-1}(-\infty, x]) - \nu(\gamma^{-1}(-\infty, x])| \leq d_K(\mu, \nu). \qquad (5.5)$$

For this, note that the only possibilities are the following:

(i) $\gamma^{-1}(-\infty, x] = \phi$ (i.e., $\gamma(z) > x \forall z \in S$),

(ii) $\gamma^{-1}(-\infty, x] = (-\infty, y] \cap S$,

(iii) $\gamma^{-1}(-\infty, x] = (-\infty, y) \cap S$,

(iv) $\gamma^{-1}(-\infty, x] = [y, \infty) \cap S$,

(v) $\gamma^{-1}(-\infty, x] = (y, \infty) \cap S$.

If $\gamma^{-1}(-\infty, x] \neq \phi$, then cases (ii) and (iii) may arise when γ is nondecreasing, and cases (iv) and (v) may arise when γ is nonincreasing. In case (i), the left-hand side of (5.5) vanishes. In case (ii), (5.5) holds by definition. In case (v), (5.5) holds by complementation:

$$|\mu((y, \infty) \cap S) - \nu((y, \infty) \cap S)|$$
$$= |1 - \mu((-\infty, y] \cap S) - (1 - \nu((-\infty, y] \cap S))|$$
$$= |\nu((-\infty, y] \cap S) - \mu((-\infty, y] \cap S)| \leq d_K(\mu, \nu).$$

To derive (5.5) in case (iii), let $y_n \uparrow y$ ($y_n < y$) $\forall n$. Then

$$|\mu((-\infty, y) \cap S) - \nu((-\infty, y) \cap S)|$$
$$= \lim_{n \to \infty} |\mu((-\infty, y_n] \cap S) - \nu((-\infty, y_n] \cap S)| \leq d_K(\mu, \nu).$$

For the remaining case (iv), use the result for the case (iii) and complementation. Thus (5.5) holds in all cases. Taking supremum over x in (5.5), we get (see (3.7))

$$d_K(T^*\mu, T^*\nu) = \sup_{x \in R} \left| \int_\Gamma (\mu(\gamma^{-1}(-\infty, x]) - \nu(\gamma^{-1}(-\infty, x]))Q(d\gamma) \right|$$
$$\leq d_K(\mu, \nu). \qquad (5.6)$$

Let $x \geqslant z_0$, and denote by Γ_1 the set of all $\gamma \in \Gamma^N$ appearing within the curly brackets in (5.1). Then, for all $\gamma \in \Gamma_1$, $\tilde{\gamma}^{-1}(-\infty, x] = S$, so that for all μ and ν in $\mathcal{P}(S)$, one has $\mu(\tilde{\gamma}^{-1}(-\infty, x]) - \nu(\tilde{\gamma}^{-1}(-\infty, x]) = 1 - 1 = 0$. Therefore,

$$
\begin{aligned}
&\left| (T^{*N}\mu)(-\infty, x] - (T^{*N}\nu)(-\infty, x] \right| \\
&= \left| \int_{\Gamma^N} [\mu(\tilde{\gamma}^{-1}(-\infty, x]) - \nu(\tilde{\gamma}^{-1}(-\infty, x])] Q^N(d\gamma) \right| \\
&= \left| \int_{\Gamma_1} + \int_{\Gamma^N \setminus \Gamma_1} \right| \\
&= \left| \int_{\Gamma^N \setminus \Gamma_1} [(\mu \circ \tilde{\gamma}^{-1})(-\infty, x] - (\nu \circ \tilde{\gamma}^{-1})(-\infty, x]] Q^N(d\gamma) \right| \\
&\leqslant (1 - \chi_1) d_K(\mu, \nu) \quad (\mu, \nu \in \mathcal{P}(S)).
\end{aligned}
\tag{5.7}
$$

Note that, for the last inequality, we have used the facts

(i) $Q^N(\Gamma^N \setminus \Gamma_1) \leqslant 1 - \tilde{\chi}_1$ and
(ii) $|\mu(\tilde{\gamma}^{-1}(-\infty, x]) - \nu(\tilde{\gamma}^{-1}(-\infty, x])| \leqslant d_K(\mu, \nu)$ (use (5.5) with $\tilde{\gamma}$ in place of γ).

Similarly, if $x < z_0$ then, letting Γ_2 denote the set in curly brackets in (5.2), we arrive at the same relations as in (5.7), expecting that the extreme right-hand side is now $(1 - \tilde{\chi}_2) d_K(\mu, \nu)$. For this, note that $(\mu \circ \tilde{\gamma}^{-1})(-\infty, x] = \mu(\phi) = 0 = (\nu \circ \tilde{\gamma}^{-1})(-\infty, x] \, \forall \gamma \in \Gamma_2$. Combining these two inequalities, we get

$$
d_K(T^{*N}\mu, T^{*N}\nu) \leqslant (1 - \tilde{\chi}) d_K(\mu, \nu) \quad (\mu, \nu \in \mathcal{P}(S)).
\tag{5.8}
$$

That is, T^{*N} is a uniformly strict contraction and T^* is a contraction. As a consequence, $\forall n > N$, one has

$$
\begin{aligned}
d_K(T^{*n}\mu, T^{*n}\nu) &= d_K(T^{*N}(T^{*(n-N)}\mu), T^{*N}(T^{*(n-N)}\nu)) \\
&\leqslant (1 - \tilde{\chi}) d_K(T^{*(n-N)}\mu, T^{*(n-N)}\nu) \leq \cdots \\
&\leqslant (1 - \tilde{\chi})^{[n/N]} d_K(T^{*(n-[n/N]N)}\mu, T^{*(n-[n/N]N)}\nu) \\
&\leqslant (1 - \tilde{\chi})^{[n/N]} d_K(\mu, \nu).
\end{aligned}
\tag{5.9}
$$

Now suppose that we are able to show that $\mathcal{P}(S)$ is a *complete* metric space under d_K. Then by using (5.8) and Theorem 2.2 of Chapter 1, T^*

has a unique fixed point π in $\mathcal{P}(S)$. Now take $\nu = \pi$ in (5.9) to get the desired relation (5.4).

It remains to prove the completeness of $(\mathcal{P}(S), d_K)$. Let μ_n be a Cauchy sequence in $\mathcal{P}(S)$, and let F_{μ_n} be the d.f. of μ_n. Then $\sup_{x \in \mathbb{R}} |F_{\mu_n}(x) - F_{\mu_m}(x)| \to 0$ as $n, m \to \infty$. By the completeness of $\mathbb{R}$, there exists a function $H(x)$ such that $\sup_{x \in \mathbb{R}} |F_{\mu_n}(x) - H(x)| \to 0$. It is simple to check that H is the d.f. of a probability measure ν, say, on $\mathbb{R}$ (Exercise). We need to check that $\nu(S) = 1$. For this, given $\varepsilon > 0$ find $n(\varepsilon)$ such that $\sup_x |F_{\mu_n}(x) - H(x)| < \frac{\varepsilon}{2} \; \forall n \geqslant n(\varepsilon)$. Now find $y_\varepsilon \in S$ such that $F_{\mu_{n(\varepsilon)}}(y_\varepsilon) > 1 - \frac{\varepsilon}{2}$. It follows that $H(y_\varepsilon) > 1 - \varepsilon$. In the same manner, find $x_\varepsilon \in S$ such that $F_{\mu_{n(\varepsilon)}}(x_\varepsilon -) < \varepsilon/2$, so that $H(x_\varepsilon -) < \varepsilon$. Then $\nu(S) \geqslant H(y_\varepsilon) - H(x_\varepsilon -) > 1 - 2\varepsilon \; \forall \varepsilon > 0$, so that $\nu(S) = 1$. ∎

Remark 5.4 Suppose that α_n are strictly monotone a.s. Then if the initial distribution μ is nonatomic (i.e., $\mu(\{x\}) = 0 \; \forall x$ or, equivalently, the d.f. of μ is continuous), $\mu \circ \gamma^{-1}$ is nonatomic $\forall \gamma \in \Gamma$ (outside a set of zero P-probability). It follows that if X_0 has a continuous d.f., then so has X_1 and in turn X_2 has a continuous d.f., and so on. Since, by Theorem 5.1, this sequence of continuous d.f.s (of $X_n (n \geqslant 1)$) converges uniformly to the d.f. of π, the latter is continuous. Thus π is *nonatomic* if α_n are strictly monotone a.s.

Example 5.1 Let $S = [0, 1]$ and Γ be a family of monotone nondecreasing functions from S into S. As before, for any $z \in S$, let

$$X_n(z) = \alpha_n \cdots \alpha_1 z.$$

One can verify the following two results, following an argument in Remark 5.1:

[R.1] $P[X_n(0) \leq x]$ *is nonincreasing in n and converges for each $x \in S$.*

[R.2] $P[X_n(1) \leq x]$ *is nondecreasing in n and converges for each $x \in S$.*

Write

$$F_0(x) \equiv \lim_{n \to \infty} P(X_n(0) \leq x)$$

and

$$F_1(x) = \lim_{n \to \infty} P(X_n(1) \leq x).$$

Note that $F_1(x) \leq F_0(x)$ for all x. Consider the case when $\Gamma \equiv \{f\}$, where

$$f(x) = \begin{cases} \frac{1}{4} + \frac{x}{4} & \text{if } 0 \leq x < \frac{1}{3} \\ \frac{1}{3} + \frac{x}{3} & \text{if } \frac{1}{3} \leq x \leq \frac{2}{3} \\ \frac{1}{3} + \frac{x}{2} & \text{if } \frac{2}{3} < x \leq 1 \end{cases}.$$

Exercise 5.1 Verify that f is a monotone increasing map from S into S, but f is *not* continuous. Calculate that

$$F_0(x) = \begin{cases} 0 & \text{if } 0 \leq x < \frac{1}{3} \\ 1 & \text{if } \frac{1}{3} \leq x \leq 1 \end{cases} \qquad F_1(x) = \begin{cases} 0 & \text{if } 0 \leq x < \frac{2}{3} \\ 1 & \text{if } \frac{2}{3} \leq x \leq 1. \end{cases}$$

Neither F_0 nor F_1 is a stationary distribution function. ∎

Example 5.2 Let $S = [0, 1]$ and $\Gamma = \{f_1, f_2\}$. In each period f_i is chosen with probability $\frac{1}{2}$. f_1 is the function f defined in Example 5.1, and $f_2(x) = \frac{1}{3} + \frac{x}{3}$, for $x \in S$.
  Then

$$F_0(x) = F_1(x) = \begin{cases} 0 & \text{if } 0 \leq x < \frac{1}{2} \\ 1 & \text{if } \frac{1}{2} \leq x \leq 1 \end{cases}$$

and $F_0(x)$ is the unique stationary distribution. Note that $f_1\left(\frac{1}{2}\right) = f_2\left(\frac{1}{2}\right) = \frac{1}{2}$, i.e., f_1 and f_2 have a common fixed point. Examples 5.1 and 5.2 are taken from Yahav (1975). ∎

Exercise 5.2 Show that splitting **(H)** does not occur in Example 5.2

3.5.2 Splitting: A Generalization

Let S be a Borel subset of a complete separable metric space (or a Polish space) and $\mathcal{S}$ its Borel sigmafield. Let $\mathcal{A} \subset \mathcal{S}$, and define

$$d_A(\mu, v) := \sup_{A \in \mathcal{A}} |\mu(A) - v(A)| \quad (\mu, v \in \mathcal{P}(S)). \qquad (5.10)$$

Let Γ be a set of measurable maps on $\mathcal{S}$, with a sigmafield $\sum$ such that the measurability assumption stated at the beginning of Section 3.2 is satisfied. Consider a sequence of i.i.d. maps $\{\alpha_n : n \geq 1\}$ with distribution Q on $(\Gamma, \sum)$.

Consider the following hypothesis (**H**$_1$):

(1) $(\mathcal{P}(S), d_A)$ is a complete metric space; (5.11)

(2) there exists a positive integer N such that for all $\gamma \in \Gamma^N$, one has

$$d_A(\mu\,\tilde{\gamma}^{-1}, v\tilde{\gamma}^{-1}) \le d_A(\mu, v)\,(\mu, v \in \mathcal{P}(S)), \qquad (5.12)$$

where $\tilde{\gamma}$ is defined in (3.6) with N in place of n;

(3) there exists $\hat{\chi} > 0$ such that $\forall A \in \mathcal{A}$, and with N as in (2), one has

$$P((\alpha_N \cdots \alpha_1)^{-1} A = S \text{ or } \varphi) \ge \hat{\chi}. \qquad (5.13)$$

We shall sometimes refer to $\mathcal{A}$ as the *splitting class*.

Theorem 5.2 *Assume the hypothesis* (**H**$_1$). *Then there exists a unique invariant probability π for the Markov process $X_n := \alpha_n \cdots \alpha_1 X_0$, where X_0 is independent of $\{\alpha_n := n \ge 1\}$. Also, one has*

$$d_A(T^{*n}\mu, \pi) \le (1 - \hat{\chi})^{[n/N]} \quad (\mu \in \mathcal{P}(S)), \qquad (5.14)$$

where $T^{*n}\mu$ is the distribution of X_n when X_0 has distribution μ, and $[n/N]$ is the integer part of n/N.

Proof. Let $A \in \mathcal{A}$. Then (5.13) holds, which one may express as

$$Q^N(\{\gamma \in \Gamma^N: \tilde{\gamma}^{-1} A = S \text{ or } \varphi\}) \ge \hat{\chi}. \qquad (5.15)$$

Then, $\forall \mu, v \in \mathcal{P}(S)$,

$$\left| (T^{*N}\mu)(A) - (T^{*N}v)(A) \right|$$
$$= \left| \int_{\Gamma^N} (\mu(\tilde{\gamma}^{-1} A) - v(\tilde{\gamma}^{-1} A)) Q^N(d\gamma) \right|. \qquad (5.16)$$

Denoting the set in curly brackets in (5.15) by Γ_1, one then has

$$|(T^{*N}\mu)(A) - (T^{*N}v)(A)|$$
$$= \left| \begin{array}{l} \int_{\Gamma_1} (\mu(\tilde{\gamma}^{-1} A) - v(\tilde{\gamma}^{-1} A)) Q^N(d\gamma) \\ + \int_{\Gamma^N \backslash \Gamma_1} (\mu(\tilde{\gamma}^{-1} A) - v(\tilde{\gamma}^{-1} A)) Q^N(d\gamma) \end{array} \right|$$
$$= \left| \int_{\Gamma^N \backslash \Gamma_1} (\mu(\tilde{\gamma}^{-1} A) - v(\tilde{\gamma}^{-1} A)) Q^N(d\gamma) \right| \qquad (5.17)$$

since on Γ_1 the set $\tilde{\gamma}^{-1}A$ is S or φ, so that $\mu(\tilde{\gamma}^{-1}A) = 1 = \nu(\tilde{\gamma}^{-1}A)$, or $\mu(\tilde{\gamma}^{-1}A) = 0 = \nu(\tilde{\gamma}^{-1}A)$. Hence, using (5.12) and (5.13),

$$|(T^{*N}\mu)(A) - (T^{*N}\nu)(A)| \le (1 - \hat{\chi})d_A(\mu, \nu). \qquad (5.18)$$

Thus

$$d_A(T^{*N}\mu, T^{*N}\nu) \le (1 - \hat{\chi})d_A(\mu, \nu). \qquad (5.19)$$

Since $(\mathcal{P}(S), d_A)$ is a complete metric space by assumption $(\mathbf{H}_1)(1)$, and T^{*N} is a uniformly strict contraction on $\mathcal{P}(S)$ by (5.19), there exists by Theorem 2.2 of Chapter 1, a unique *fixed point* π of T^*, i.e., $T^*\pi = \pi$, and

$$\begin{aligned} d_A(T^{*kN}\mu, \pi) &= d_A(T^{*N}(T^{*(k-1)N}\mu), T^{*N}\pi) \\ &\le (1 - \hat{\chi})d_A(T^{*(k-1)N}\mu, \pi) \le \cdots \\ &\le (1 - \hat{\chi})^k d_A(\mu, \pi). \end{aligned} \qquad (5.20)$$

Finally, since $d_A(\mu, \nu) \le 1$ one has, with $n = [n/N]N + r$,

$$d_A(T^{*n}\mu, \pi) = d_A(T^{*[n/N]N}T^{*r}\mu, \pi) \le (1 - \hat{\chi})^{[n/N]}. \qquad (5.21)$$

This completes the proof. ∎

We now state two corollaries of Theorem 5.2 applied to i.i.d. monotone maps.

Corollary 5.1 *Let S be a closed subset of $\mathbb{R}$. Suppose $\alpha_n(n \ge 1)$ is a sequence of i.i.d. monotone maps on S satisfying the splitting condition* $(\mathbf{H}')$:

$(\mathbf{H}')$ *There exist $z_0 \in S$, a positive integer N, and a constant $\chi'' > 0$ such that*

$$P(\alpha_N\alpha_{N-1}\cdots\alpha_1 x \le z_0 \,\forall x \in S) \ge \chi'',$$
$$P(\alpha_N\alpha_{N-1}\cdots\alpha_1 x \ge z_0 \,\forall x \in S) \ge \chi''.$$

Then

(a) the distribution $T^{*n}\mu$ of $X_n := \alpha_n \cdots \alpha_1 X_0$ converges to a probability measure π on S exponentially fast in the Kolmogorov distance d_K

irrespective of X_0. Indeed,

$$d_K(T^{*n}\mu, \pi) \leq (1 - \chi'')^{[n/N]} \quad \forall \mu \in \mathcal{P}(S), \qquad (5.22)$$

where $[y]$ denotes the integer part of y.

(b) π in (a) is the unique invariant probability of the Markov process X_n.

Proof. To apply Theorem 5.2, let $\mathcal{A}$ be the class of all sets $A = (-\infty, y] \cap S$, $y \in \mathbb{R}$. Completeness of $(\mathcal{P}(S), d_A)$ is established directly (see Remark 5.5).

To check condition (2) of $(\mathbf{H}_1)$, note that if γ is monotone nondecreasing and $A = (-\infty, y] \cap S$, then $\gamma^{-1}((-\infty, y] \cap S) = (-\infty, x] \cap S$ or $(-\infty, x) \cap S$, where $x = \sup\{z: \gamma(z) \leq y\}$. Thus,

$$|\mu(\gamma^{-1}A) - \nu(\gamma^{-1}A)| = |\mu((-\infty, x] \cap S) - \nu((-\infty, x] \cap S)$$

or $|\mu((-\infty, x) \cap S) - \nu((-\infty, x) \cap S)|$.

In either case, $|\mu(\gamma^{-1}A) - \nu(\gamma^{-1}A)| \leq d_K(\mu, \nu)$, since $\mu((-\infty, x - 1/n] \cap S) \uparrow \mu((-\infty, x) \cap S)$ (and the same holds for ν). If γ is monotone nonincreasing, then $\gamma^{-1}A$ is of the form $[x, \infty) \cap S$ or $(x, \infty) \cap S$, where $x := \inf\{z: \gamma(z) \leq y\}$. Again it is easily shown, $|\mu(\gamma^{-1}A) - \nu(\gamma^{-1}A)| \leq d_K(\mu, \nu)$. Finally, (5.13) holds for all $A = (-\infty, y] \cap S$, by $(\mathbf{H}')$.

This completes the proof. ∎

Remark 5.5 Let $\{\mu_n: n \geq 1\}$ be a Cauchy sequence in $\mathcal{P}(S)$ with respect to the metric d_A. This means, for $\mu_n (n \geq 1)$ considered as probability measures on $\mathbb{R}$, the sequence of d.f.s $\{F_{\mu_n}: n \geq 1\}$ is Cauchy with respect to the supremum distance (or the Kolmogorov distance) on $\mathbb{R}$. Their uniform limit H is a d.f. of a probability measure μ on $\mathbb{R}$. This implies μ_n converges weakly to μ on $\mathbb{R}$. By Alexandrov's theorem (see Theorem C11.1, Chapter 2), $\mu(S) \geq \limsup_{n \to \infty} \mu_n(S) = 1$. That is, $\mu(S) = 1$. In particular, $\mu_n(S \cap (-\infty, x]) = F_{\mu_n}(x) \to H(x) = \mu(S \cap (-\infty, x])$ uniformly in $x \in \mathbb{R}$. Hence $d_A(\mu_n, \mu) \to 0$. A more general proof of completeness of $(\mathcal{P}(S), d_A)$ may be found in Complements and Details.

To state the next corollary, let S be a closed subset of $\mathbb{R}^\ell$. Define $\mathcal{A}$ to be the class of all sets of the form

$$A = \{\mathbf{y} \in S: \phi(\mathbf{y}) \leq \mathbf{x}\},$$

$$\phi \text{ continuous and monotone on } S \text{ into } \mathbb{R}^\ell, \ \mathbf{x} \in \mathbb{R}^\ell. \quad (5.23)$$

Again by " $\leq$ " we mean the *partial order*: $\mathbf{x} = (x_1, \ldots, x_\ell) \leq \mathbf{y} = (y_1, \ldots, y_l)$ iff $x_i \leq y_i \ \forall i = 1, 2, \ldots, \ell$. In the following corollary we will interpret (**H**) to hold with this partial order $\leq$, and with "$\mathbf{x} \geq \mathbf{y}$" meaning $\mathbf{y} \leq \mathbf{x}$.

Corollary 5.2 *Let S be a closed subset of $\mathbb{R}^\ell$ and $\mathcal{A}$ be the class of sets (5.23). If $\alpha_n(n \geq 1)$ is a sequence of i.i.d. monotone maps which are continuous Q a.s., and the splitting condition (**H**) holds, then there exists a unique invariant probability π and*

$$d_A(T^{*n}\mu, \pi) \leq (1 - \tilde{\chi})^{[n/N]} \quad \forall \mu \in \mathcal{P}(S), \ n \geq 1, \quad (5.24)$$

where $d_A(\mu, \nu) := \sup\{|\mu(A) - \nu(A)|: A \in \mathcal{A}\}$, $\tilde{\chi} = \min\{\tilde{\chi}_1, \tilde{\chi}_2\}$, and $[n/N]$ is the integer part of n/N.

Proof. The completeness of $(\mathcal{P}(S), d_A)$, i.e., (5.11) is proved in Complements and Details (see Lemma C5.1).

Condition (2) in (**H$_1$**), or (5.12), is immediate. For if $A = \{\mathbf{y}: \phi(\mathbf{y}) \leq \mathbf{x}\}$, γ is continuous and monotone, then $\tilde{\gamma}^{-1}A = \{\mathbf{y}: (\phi \circ \tilde{\gamma})\mathbf{y} \leq \mathbf{x}\} \in \mathcal{A}$ since $\phi \circ \tilde{\gamma}$ is monotone and continuous.

It remains to verify (**H$_1$**) (3), i.e., (5.13). We write $\mathbf{x}_0$ in place of $\mathbf{z}_0$ in (**H**). Let A in (5.23) be such that ϕ is monotone nondecreasing. If $\phi(\mathbf{x}_0) \leq \mathbf{x}$, then, by the splitting condition (**H**),

$$\tilde{\chi} \leq P(\alpha_N \cdots \alpha_1 \mathbf{z} \leq \mathbf{x}_0 \forall \mathbf{z} \in S) \leq P(\phi\alpha_N \cdots \alpha_1 \mathbf{z} \leq \phi(\mathbf{x}_0)\forall \mathbf{z} \in S)$$
$$\leq P(\phi\alpha_N \cdots \alpha_1 \mathbf{z} \leq \mathbf{x}\forall \mathbf{z} \in S) = P(a_N \cdots \alpha_1 \mathbf{z} \in A\forall \mathbf{z} \in S)$$
$$= P((\alpha_N \cdots \alpha_1)^{-1}A = S). \quad (5.25)$$

If $\mathbf{x}$ in the definition of A in (5.23) is such that $\phi(\mathbf{x}_0) \not\leq \mathbf{x}$ (i.e., at least one coordinate of $\phi(\mathbf{x}_0)$ is larger than the corresponding coordinate of $\mathbf{x}$), then

$$\tilde{\chi} \leq P(\alpha_N \cdots \alpha_1 \mathbf{z} \geq \mathbf{x}_0 \ \forall \mathbf{z} \in S) \leq P(\phi\alpha_N \cdots \alpha_1 \mathbf{z} \geq \phi(\mathbf{x}_0)\forall \mathbf{z} \in S)$$
$$\leq P(\phi\alpha_N \cdots \alpha_1 \mathbf{z} \not\leq \mathbf{x}\forall \mathbf{z} \in S) \leq P(\alpha_N \cdots \alpha_1 \mathbf{z} \in A^c\forall \mathbf{z} \in S)$$
$$= P((\alpha_N \cdots \alpha_1)^{-1}A = \varphi). \quad (5.26)$$

Now let ϕ in the definition of A in (5.23) be monotone decreasing. If $\phi(\mathbf{x}_0) \leq \mathbf{x}$, then

$$\tilde{\chi} \leq P(\alpha_N \cdots \alpha_1 \mathbf{z} \geq \mathbf{x}_0 \forall \mathbf{z} \in S)$$
$$\leq P(\phi \alpha_N \cdots \alpha_1 \mathbf{z} \leq \phi(\mathbf{x}_0) \forall \mathbf{z} \in S) \leq P(\phi \alpha_N \cdots \alpha_1 \mathbf{z} \leq \mathbf{x} \forall \mathbf{z} \in S)$$
$$= P(\alpha_N \cdots \alpha_1 \mathbf{z} \in A \forall \mathbf{z} \in S). \tag{5.27}$$

If $\phi(\mathbf{x}_0) \not\leq \mathbf{x}$, then

$$\tilde{\chi} \leq P(\alpha_N \cdots \alpha_1 \mathbf{z} \leq \mathbf{x}_0 \forall \mathbf{z} \in S) \leq P(\phi \alpha_N \cdots \alpha_1 \mathbf{z} \geq \phi(\mathbf{x}_0) \forall \mathbf{z} \in S)$$
$$\leq P(\phi \alpha_N \cdots \alpha_1 \mathbf{z} \not\leq \mathbf{x} \forall \mathbf{z} \in S) = P(\alpha_N \cdots \alpha_1 \mathbf{z} \in A^c \forall \mathbf{z} \in S).$$
$$\tag{5.28}$$

Thus $(\mathbf{H}_1)$ (3) is verified for all $A \in \mathcal{A}$. This completes the proof. ∎

Remark 5.6 Throughout Subsections 3.5.1, 3.5.2, we have tacitly assumed that the sets within parenthesis in (5.15) (and (5.1) and (5.2)) are measurable, i.e., they belong to $\sum^{\otimes N}$. If this does not hold, we assume that these sets have measurable subsets for which the inequality (5.15) holds.

Remark 5.7 In Remark 5.1, it was shown that for monotone nondecreasing maps α_n on a compact interval $S = [a, b]$, splitting is *necessary* for the existence of a unique nontrivial (i.e., nondegenerate) invariant probability. A similar result holds in multidimension (see Bhattacharya and Lee 1988).

3.5.3 The Doeblin Minorization Theorem Once Again

As an application of Theorem 5.2, we provide a proof of the important Doeblin minorization theorem (Chapter 2, Theorem 9.1) when the state space S is a Borel subset of the complete separable metric space.

Corollary 5.3 *Let $p(x, A)$ be a transition probability on a Borel subset S of a complete, separable metric space with $\mathcal{S}$ as the Borel sigmafield of S. Suppose there exists a nonzero measure λ on S and a positive integer m such that*

$$p^{(m)}(x, A) \geq \lambda(A) \quad \forall x \in S, \quad A \in \mathcal{S}. \tag{5.29}$$

Then there exists a unique invariant probability π for $p(\cdot, \cdot)$ such that

$$\sup_{x, A} |p^{(n)}(x, A) - \pi(A)| \leq (1 - \bar{\chi}')^{[n/m]}, \quad \bar{\chi}' := \lambda(S). \qquad (5.30)$$

Proof. We will show that (5.30) is an almost immediate consequence of Theorem 5.2. For this, express $p^{(m)}(x, A)$ as

$$p^{(m)}(x, A) = \lambda(A) + (p^{(m)}(x, A) - \lambda(A))$$
$$= \bar{\chi}' \lambda_\delta(A) + (1 - \bar{\chi}') q_\delta(x, A), \qquad (5.31)$$

where, assuming $0 < \lambda(S) \equiv \bar{x}' < 1$ (see Exercise 5.3),

$$\lambda_\delta(A) := \frac{\lambda(A)}{\bar{\chi}'}, \quad q_\delta(x, A) = \frac{p^{(m)}(x, A) - \lambda(A)}{1 - \bar{\chi}'}. \qquad (5.32)$$

Let $\beta_n (n \geq 1)$ be an i.i.d. sequence of maps on S constructed as follows. For each n, with probability $\bar{\chi}'$ let $\beta_n \equiv Z_n$, where Z_n is a random variable with values in S and distribution λ; and with probability $1 - \bar{\chi}'$ let $\beta_n = \alpha_n$, where α_n is a random map on S such that $P(\alpha_n x \in A) = q_\delta(x, A)$ (see Proposition C1.1, Complements and Details). Then Theorem 5.2 applies to the transition probability $p^{(m)}(x, A)$ (for $p(x, A)$) with $\mathcal{A} = S$, $N = 1$. Note that $P(\beta_1^{-1} A = S \text{ or } \phi) \geq P(\beta_1(\cdot) \equiv Z_1) = \bar{\chi}'$. Hence (5.13) holds. Since $\mathcal{A} = S$ in this example, completeness of $(\mathcal{P}(S), d_A)$ and the condition (5.12) obviously hold. ∎

Exercise 5.3 Suppose $\lambda(S) = 1$ in (5.29). Show that λ is the unique invariant probability for p, and that (5.30) holds, with the convention $0° = 1$. [Hint: $p^{(m)}(x, A) = \lambda(A) \forall A \in S, \forall x \in S$. For every $k \geq 0$, $p^{m+k}(x, A) = \lambda(A) \forall A \in S, \forall x \in S$, since the conditional distribution of X_{m+k}, given X_k, is λ (no matter what X_k is).]

As the corollaries indicate, the significance of Theorem 5.2 stems from the fact that it provides geometric rates of convergence in appropriate metrics for different classes of irreducible as well as nonirreducible Markov processes. The metric d_A depends on the structure of the process.

3.6 Applications

3.6.1 First-Order Nonlinear Autoregressive Processes (NLAR(1))

There is a substantial and growing literature on nonlinear time series models (see Tong 1990; Granger and Terasvirta 1993; and Diks 1999). A stability property of the general model

$$X_{n+1} = f(X_n, \varepsilon_{n+1})$$

(where ε_n is an i.i.d. sequence) can be derived from contraction assumptions (see Lasota and Mackey 1989). As an application of Corollary 5.3, we consider a simple *first-order nonlinear autoregressive model* (NLAR(1)).

Let $S = \mathbb{R}$, and $f : \mathbb{R} \to [a, b]$ a (bounded) measurable function. Consider the Markov process

$$X_{n+1} = f(X_n) + \varepsilon_{n+1}, \tag{6.1}$$

where

[A.1] $\{\varepsilon_n\}$ is an i.i.d. sequence of random variables whose common distribution has a density $\hat{\phi}$ (with respect to the Lebesgue measure), which is bounded away from zero on interval $[c, d]$ with $d - c > b - a$.

Let X_0 be a random variable (independent of $\{\varepsilon_n\}$). The transition probability of the Markov process X_n has the density

$$p(x, y) = \hat{\phi}(y - f(x)).$$

We define $\psi(y)$ as

$$\psi(y) \equiv \begin{cases} \chi' \equiv \inf\{\hat{\phi}(z): c \leq z \leq d\} & \text{for } y \in [c + b,\, d + a], \\ 0 & \text{for } y \in [c + d,\, d + a]^c. \end{cases} \tag{6.2}$$

It follows that for all $x \in \mathbb{R}$

$$p(x, y) \equiv \hat{\phi}(y - f(x)) \geq \psi(y). \tag{6.3}$$

Proposition 6.1 *If [A.1] holds then there exists a unique invariant probability π and (5.30) holds with $m = 1$ and $\bar{\chi}' = \int \psi(y)dy > 0$.*

Proof. In view of (6.3) and the fact that $\psi(y) \geq \chi' > 0$ on $[c + b, d + a]$, one has $\bar{\chi}' := \int \psi(y)dy \geq (d - c - (b - a))\chi' > 0$. Since (5.29) holds with λ having density ψ, the Doeblin stability result Corollary 5.3 applies directly, with $m = 1$. ∎

For monotone f we can relax the assumption on the distribution of the noise term ε_{n+1} in (6.1). Let supp(Q) denote the *support* of the distribution Q of ε_n.

Proposition 6.2 *Suppose f is monotone on $\mathbb{R}$ into $[a, b]$. Let the infimum and supremum of* supp(Q) *be c and d, respectively. If $d - c > b - a$, then $X_n(n \geq 0)$ has a unique invariant probability π, and the distribution of X_n converges to π exponentially fast in the Kolmogorov distance as $n \to \infty$, uniformly with respect to all initial distributions.*

Proof. Choose $z_0 \in (c + b, d + a)$. Then $X_1(x) \equiv f(x) + \varepsilon_1$ satisfies

$$P(X_1(x) \leq z_0 \; \forall x \in \mathbb{R}) = P(\varepsilon_1 \leq z_0 - f(x) \; \forall x \in \mathbb{R})$$
$$\geq P(\varepsilon_1 \leq z_0 - b) = \tilde{\chi}_1, \quad \text{say,} \quad \tilde{\chi}_1 > 0,$$

since $z_0 - b > c$. Similarly,

$$P(X_1(x) \geq z_0 \; \forall x \in \mathbb{R}) \geq P(\varepsilon_1 \geq z_0 - a) = \tilde{\chi}_2, \quad \text{say,} \quad \tilde{\chi}_2 > 0,$$

since $z_0 - a < d$. Hence Theorem 5.1 applies with $N = 1$. ∎

Generalizations of Proposition 6.1 and 6.2 to higher order NLAR(k), $k > 1$, are given in Chapter 4.

3.6.2 Stability of Invariant Distributions in Models of Economic Growth

Models of descriptive as well as optimal growth under uncertainty have led to random dynamical systems that are stable in distribution. We look at a "canonical" example and show how Theorem 5.1 can be applied. A complete list of references to the earlier literature, which owes much to the pioneering efforts of Brock, Mirman, and Zilcha (Brock and Mirman (1972, 1973); Mirman and Zilcha (1975)), is in Majumdar, Mitra, and Nyarko (1989).

Recall that, for any function h on S into S, we write h^n for the nth iterate of h.

Example 6.1 Consider the case where $S = \mathbb{R}_+$, and $\Gamma = \{F_1, F_2, \ldots, F_i, \ldots, F_N\}$, where the distinct laws of motion F_i satisfy

[F.1] F_i *is strictly increasing, continuous, and there is some* $r_i > 0$ *such that* $F_i(x) > x$ *on* $(0, r_i)$ *and* $F_i(x) < x$ *for* $x > r_i$. *Note that* $F_i(r_i) = r_i$ *for all* $i = 1, \ldots, N$. *Next, assume*
[F.2] $r_i \neq r_j$ *for* $i \neq j$.

In other words, the unique positive fixed points r_i of distinct laws of motion are all distinct. We choose the indices $i = 1, 2, \ldots, N$ so that $r_1 < r_2 < \cdots < r_N$. Let $\mathrm{Prob}(\alpha_n = F_i) = p_i > 0$ $(1 \leq i \leq N)$.

Consider the Markov process $\{X_n(x)\}$ with the state space $(0, \infty)$. If $y \geq r_1$, then $F_i(y) \geq F_i(r_1) > r_1$ for $i = 2, \ldots N$, and $F_1(r_1) = r_1$, so that $X_n(x) \geq r_1$ for all $n \geq 0$ if $x \geq r_1$. Similarly, if $y \leq r_N$, then $F_i(y) \leq F_i(r_N) < r_N$ for $i = 1, \ldots, N - 1$ and $F_N(r_N) = r_N$, so that $X_n(x) \leq r_N$ for all $n \geq 0$ if $x \leq r_N$. Hence, if the initial state x is in $[r_1, r_N]$, then the process $\{X_n(x) \colon n \geq 0\}$ remains in $[r_1, r_N]$ forever. We shall presently see that for a long-run analysis, we can consider $[r_1, r_N]$ as the effective state space.

We shall first indicate that on the state space $[r_1, r_N]$ the splitting condition **(H)** is satisfied. If $x \geq r_1$, $F_1(x) \leq x$, $F_1^2(x) \equiv F_1(F_1(x)) \leq F_1(x)$, etc. The limit of this decreasing sequence $F_1^n(x)$ must be a fixed point of F_1, and therefore must be r_1. Similarly, if $x \leq r_N$, then $F_N^n(x)$ increases to r_N. In particular,

$$\lim_{n \to \infty} F_1^n(r_N) = r_1, \qquad \lim_{n \to \infty} F_N^n(r_1) = r_N.$$

Thus, there must exist a positive integer n_0 such that

$$F_1^{n_0}(r_N) < F_N^{n_0}(r_1).$$

This means that if $z_0 \in [F_1^{n_0}(r_N), F_N^{n_0}(r_1)]$, then

$$\mathrm{Prob}(X_{n_0}(x) \leq z_0 \;\; \forall x \in [r_1, r_N])$$
$$\geq \mathrm{Prob}\,(\alpha_n = F_1 \text{ for } 1 \leq n \leq n_0) = p_1^{n_0} > 0,$$
$$\mathrm{Prob}(X_{n_0}(x) \geq z_0 \;\; \forall x \in [r_1, r_N])$$
$$\geq \mathrm{Prob}(\alpha_n = F_N \text{ for } 1 \leq n \leq n_0) = p_N^{n_0} > 0.$$

Hence, considering $[r_1, r_N]$ as the state space, and using Theorem 5.1, there is a unique invariant probability π, with the stability property holding for all initial $x \in [r_1, r_N]$.

Now fix the initial state $x \in (0, r_1)$, and define $m(x) = \min_{i=1,\ldots,N} F_i(x)$.

One can verify that (i) m is continuous, (ii) m is strictly increasing, and (iii) $m(r_1) = r_1$ and $m(x) > x$ for $x \in (0, r_1)$, and $m(x) < x$ for $x > r_1$. Let $x \in (0, r_1)$. Clearly $m^n(x)$ increases with n, and $m^n(x) \le r_1$. The limit of the sequence $m^n(x)$ must be a fixed point of m and is, therefore, r_1. Since $F_i(r_1) > r_1$ for $i = 2, \ldots, N$, there exists some $\varepsilon > 0$ such that $F_i(y) > r_1 (2 \le i \le N)$ for all $y \in [r_1 - \varepsilon, r_1]$. Clearly there is some n_ε such that $m^{n_\varepsilon}(x) \ge r_1 - \varepsilon$. If $\tau_1 = \inf\{n \ge 1: X_n(x) > r_1\}$ then it follows that for all $k \ge 1$

$$\text{Prob}(\tau_1 > n_\varepsilon + k) \le p_1^k.$$

Since p_1^k goes to zero as $k \to \infty$, it follows that τ_1 is finite almost surely. Also, $X_{\tau_1}(x) \le r_N$, since for $y \le r_1$, (i) $F_i(y) < F_i(r_N)$ for all i and (ii) $F_i(r_N) < r_N$ for $i = 1, 2, \ldots, N - 1$ and $F_N(r_N) = r_N$. (In a single period, it is not possible to go from a state less than or equal to r_1 to one larger than r_N.) By the strong Markov property, and our earlier result, $X_{\tau+m}(x)$ converges in distribution to π as $m \to \infty$ for all $x \in (0, r_1)$. Similarly, one can check that as $n \to \infty$, $X_n(x)$ converges in distribution to π for all $x > r_N$. ∎

Exercise 6.1 Consider the case where $S = \mathbb{R}_+$ and $\Gamma = \{F, G\}$ where

(i) F is strictly increasing, continuous, $F(0) > 0$ and there is $\bar{x}$ such that $F(x) > x$ on $(0, \bar{x})$ and $F(x) < x$ for $x > \bar{x}$.

(ii) G is strictly increasing, continuous, $G(x) < x$ for all $x > 0$.

Let $\text{Prob}(\alpha_n = F) = p$, $\text{Prob}(\alpha_n = G) = 1 - p$. Analyze the long-run behavior of the process (2.2). In this context, see the paper by Foley and Hellwig (1975). In their model uncertainty arises out of the possibility of unemployment. ∎

The assumption that Γ is finite can be dispensed with if one has additional structures in the model.

Example 6.2 (*Multiplicative shocks*)   Let $F : \mathbb{R}_+ \to \mathbb{R}_+$ satisfy

[F.1] F is strictly increasing and continuous.

We shall keep F fixed.

Consider $\Theta = [\theta_1, \theta_2]$, where $0 < \theta_1 < \theta_2$, and assume the following concavity and "end point" conditions:

[F.2] $F(x)/x$ is strictly decreasing in $x > 0$, $\frac{\theta_2 F(x'')}{x''} < 1$ for some $x'' > 0$, $\frac{\theta_1 F(x')}{x'} > 1$ for some $x' > 0$.

Since $\frac{\theta F(x)}{x}$ is also strictly decreasing in x, [F.1] and [F.2] imply that for each $\theta \in \Theta$, there is a unique $x_\theta > 0$ such that $\frac{\theta F(x_\theta)}{x_\theta} = 1$, i.e., $\theta F(x_\theta) = x_\theta$. Observe that $\frac{\theta F(x)}{x} > 1$ for $0 < x < x_\theta$, $\frac{\theta F(x)}{x} < 1$ for $x > x_\theta$. Now, $\theta' > \theta''$ implies $x_{\theta'} > x_{\theta''}$: $\frac{\theta' F(x_{\theta''})}{x_{\theta''}} > \frac{\theta'' F(x_{\theta''})}{x_{\theta''}} = 1 = \frac{\theta' F(x_{\theta'})}{x_{\theta'}}$, implying "$x_{\theta'} > x_{\theta''}$."

Write $\Gamma = \{f : f = \theta F, \ \theta \in \Theta\}$, and $f_1 \equiv \theta_1 F$, $f_2 \equiv \theta_2 F$.

Assume that $\theta^{(n)}$ is chosen i.i.d. according to a density function $g(\theta)$ on Θ, which is positive and continuous on Θ.

In our notation, $f_1(x_{\theta_1}) = x_{\theta_1}$; $f_2(x_{\theta_2}) = x_{\theta_2}$. If $x \geq x_{\theta_1}$, and $\tilde{\theta}$ has a density g, then $f(x) \equiv \tilde{\theta} F(x) \geq f(x_{\theta_1}) \geq f_1(x_{\theta_1}) = x_{\theta_1}$. Hence $X_n(x) \geq x_{\theta_1}$ for all $n \geq 0$ if $x \geq x_{\theta_1}$. If $x \leq x_{\theta_2}$

$$f(x) \leq f(x_{\theta_2}) \leq f_2(x_{\theta_2}) = x_{\theta_2}.$$

Hence, if $x \in [x_{\theta_1}, x_{\theta_2}]$ then the process $X_n(x)$ remains in $[x_{\theta_1}, x_{\theta_2}]$ forever. Now, $\lim_{n\to\infty} f_1^n(x_{\theta_2}) = x_{\theta_1}$ and $\lim_{n\to\infty} f_2^n(x_{\theta_1}) = x_{\theta_2}$, and there must exist a positive integer n_0 such that $f_1^{n_0}(x_{\theta_2}) < f_2^{n_0}(x_{\theta_1})$, by the arguments given for F_1, F_N in the preceding example. Choose some $z_0 \in (f_1^{n_0}(x_{\theta_2}), f_2^{n_0}(x_{\theta_1}))$. There exist intervals $[\theta_1, \theta_1 + \delta]$, $[\theta_2 - \delta, \theta_2]$ such that for all $\theta \in [\theta_1, \theta_1 + \delta]$ and $\hat{\theta} \in [\theta_2 - \delta, \theta_2]$,

$$(\theta F)^{n_0}(x_{\theta_2}) < z_0 < (\hat{\theta} F)^{n_0}(x_{\theta_1}).$$

Then the splitting condition holds on the state space $[x_{\theta_1}, x_{\theta_2}]$. Now fix x such that $0 < x < x_{\theta_1}$, then

$$\tilde{\theta} F(x) \geq \theta_1 F(x) > x.$$

Let m be any given positive integer. Since $(\theta_1 F)^n(x) \to x_{\theta_1}$ as $n \to \infty$, there exists $n' \equiv n'(x)$ such that $(\theta_1 F)^n(x) > x_{\theta_1} - \frac{1}{m}$ for all $n \geq n'$. This implies that $X_n(x) > x_{\theta_1} - \frac{1}{m}$ for all $n \geq n'$. Therefore, lim inf$_{n\to\infty} X_n(x) \geq x_{\theta_1}$. We now argue that with probability 1, lim inf$_{n\to\infty} X_n(x) > x_{\theta_1}$. For this, note that if we choose $\eta = \frac{\theta_2-\theta_1}{2}$ and $\varepsilon > 0$ such that $x_{\theta_1} - \varepsilon > 0$, then min$\{\theta F(y) - \theta_1 F(y): \theta_2 \geq \theta \geq \theta_1 + \eta, y \geq x_{\theta_1} - \varepsilon\} = \eta F(x_{\theta_1} - \varepsilon) > 0$. Write $\eta' \equiv \eta F(x_{\theta_1} - \varepsilon) > 0$. Since with probability 1, the i.i.d. sequence $\{\theta^{(n)}: n = 1, 2, \ldots\}$ takes values in $[\theta_1 + \eta, \theta_2]$ infinitely often, one has lim inf$_{n\to\infty} X_n(x) > x_{\theta_1} - \frac{1}{m} + \eta'$. Choose m so that $\frac{1}{m} < \eta'$. Then with probability 1 the sequence $X_n(x)$ exceeds x_{θ_1}. Since $x_{\theta_2} = f_2^{(n)}(x_{\theta_2}) \geq X_n(x_{\theta_2}) \geq X_n(x)$ for all n, it follows that with probability 1 $X_n(x)$ reaches the interval $[x_{\theta_1}, x_{\theta_2}]$ and remains in it thereafter. Similarly, one can prove that if $x > x_{\theta_2}$ then with probability 1, the Markov process $X_n(x)$ will reach $[x_{\theta_1}, x_{\theta_2}]$ in finite time and stay in the interval thereafter. ∎

Remark 6.1 Recall from Section 1.5 that in growth models, the condition [F.1] is often derived from appropriate "end point" or *Uzawa–Inada condition*. It should perhaps be stressed that convexity assumptions have not appeared in the discussion of this section so far.

3.6.3 Interaction of Growth and Cycles

We go back to the end of Section 1.9.5 where we indicated that some variations of the Solow model may lead to complex behavior.

One can abstract from the specific "sources" of complexity and from variations of the basic ingredients of the model, and try to study the long-run behavior of an economy in which qualitatively distinct laws of motion arise with positive probability. This is the motivation behind the formal analysis of a rather special case.

Consider a random dynamical system with $S = \mathbb{R}_+$ and with two possible laws of motion, denoted by F and G (i.e., $\Gamma = \{F, G\}$) occurring with probabilities p and $1 - p$ ($0 < p < 1$). The law of motion F has the strong monotonicity and stability property with an attracting positive fixed point (see [G.1]); however, the other law G triggers cyclical forces and has a pair of locally attracting periodic points of period 2 (and a repelling fixed point: see [P.1]–[P.4]). One may interpret F as the dominant long-run growth law (the probability p is "large"), while G represents short-run cyclical interruptions. Our main result identifies conditions under which Theorems 5.1 and 5.2 are applicable.

The law of motion that generates the growth process is represented by a continuous increasing function $F : [0, 1] \rightarrow [0, 1]$. We assume that

[G.1] *F has a fixed point $r > 1/2$ such that*
$$F(x) > x \text{ for } 0 < x < r,$$
$$F(x) < x \text{ for } x > r.$$

Whether or not $F(0) = 0$ is not relevant for our subsequent analysis. Note that the trajectory from any initial $x_0 > 0$ converges to r; indeed, if $0 < x_0 < r$, the sequence $F^n(x_0)$ increases to r; whereas if $x_0 > r$, the sequence $F^n(x_0)$ decreases to r.

The law of motion that triggers cyclical forces is denoted by a continuous map $G : [0, 1] \rightarrow [0, 1]$. We assume

[P.1] *G is increasing on $[0, 1/2]$ and decreasing on $[1/2, 1]$.*
[P.2] *$G(x) > x$ on $[0, 1/2]$.*

Suppose that the law of motion F appears with probability $p > 0$ and G appears with probability $1 - p > 0$. We can prove the following boundedness property of the trajectories:

Lemma 6.1 *Let $x \in (0, 1)$ and $x_0 = x$. Write $M = G(1/2)$ and*
$$\tilde{x} = \max[x, M, F(1)],$$
$$\underset{\sim}{x} = \min[x, 1/2, G(\tilde{x})].$$

Then for all $n \geqslant 0$

$$\underset{\sim}{x} \leqslant X_n \leqslant \tilde{x}. \tag{6.4}$$

Proof. Note that $\underset{\sim}{x} \leq \frac{1}{2}$, $\tilde{x} > \frac{1}{2}$. The range of F on $[\underset{\sim}{x}, \tilde{x}]$ is $[F(\underset{\sim}{x}), F(\tilde{x})] \subset [\underset{\sim}{x}, \tilde{x}]$, since $F(x) \geq x$ (recall that $F(x) \geqslant x$ on $[0, \frac{1}{2}]$) and $F(\tilde{x}) \leq F(1) \leqslant \tilde{x}$. The range of G on $[\underset{\sim}{x}, \tilde{x}] \equiv [\underset{\sim}{x}, \frac{1}{2}] \cup [\frac{1}{2}, \tilde{x}]$ is $[G(\underset{\sim}{x}), M] \cup [G(\tilde{x}), M]$. Now $G(\underset{\sim}{x}) \geq \underset{\sim}{x}$ (since $G(x) > x$ on $[0, \frac{1}{2}]$ and $G(\tilde{x}) \geq \underset{\sim}{x}$ (by definition of $\underset{\sim}{x}$). Therefore, the range of G on $[\underset{\sim}{x}, \tilde{x}]$ is contained in $[\underset{\sim}{x}, \tilde{x}]$. Note that $M \leq \tilde{x}$. Thus if (6.4) holds for some n, then $\underset{\sim}{x} \leq F(X_n) \leq \tilde{x}$, $\underset{\sim}{x} \leq G(X_n) \leq \tilde{x}$, i.e., $\underset{\sim}{x} \leq X_{n+1} \leq \tilde{x}$. Since (6.4) holds for $n = 0$, it holds for all n. ∎

The simplest way of capturing "cyclical forces" is to make the following assumption:

[P.3] *G has two periodic points of period 2 denoted by $\{\beta_1, \beta_2\}$, and a repelling fixed point x^*, and no other fixed point or periodic point. Moreover, $\{\beta_1, \beta_2\}$ are locally stable fixed points of $G^2 \equiv G \circ G$.*

We shall also assume

[P.4] *G has an invariant interval $[a, b](1/2 \le a < b < 1)$; $a < \beta_1 < x^* < \beta_2 < b$. Also $r \in (a, b)$, $r \notin \{\beta_1, x^*, \beta_2\}$.*

Note that G^2 is increasing on $[a, b]$. By invariance of $[a, b]$, $G^2(a) \ge a$, $G^2(b) \le b$.

By using the fact that G^2 is increasing on $[a, b]$, we get that the sequence $G^{2n}(a)$ is nondecreasing and converges to some limit, say, z_1. By continuity of G^2, z_1 is a fixed point of G^2. Since $a \le \beta_1$, it follows that $z_1 \le \beta_1$ and by assumption [P.3], z_1 must be β_1. Hence, $G^{2n}(a)$ increases to the fixed point β_1. Similarly, $G^{2n}(b)$ decreases to the fixed point β_2. Since neither a nor b is a fixed point of G^2, it is in fact the case that $G^2(a) - a > 0$, $G^2(b) - b < 0$. Hence, $G^2(x) - x > 0$ for all $x \in [a, \beta_1)$ and, $G^2(x) - x < 0$ for all $x \in (\beta_2, b]$. (If, for some z in $[a, \beta_1)$, $G^2(z) - z \le 0$, by continuity of G^2, there is a fixed point of G^2 between a and z, a contradiction.)

It follows that for all x in $[a, \beta_1)$, $G^{2n}(x)$ increases to β_1 and for all x in $(\beta_2, b]$, $G^{2n}(x)$ decreases to β_2 as $n \to \infty$.

Observe, next, that $G^2x - x$ cannot change sign on either (β_1, x^*) or (β_2, x^*). For otherwise, G^2 has a fixed point other than β_1, β_2, x^*.

Suppose, if possible, that $G^2(x) - x > 0$ on (β_1, x^*); then from any $y \in (\beta_1, x^*)$, $G^{2n}(y)$ will increase to some fixed point; this fixed point must be x^*; but x^* is repelling, a contradiction; hence, $G^2(x) - x < 0$ on (β_1, x^*). A similar argument ensures that $G^2(x) - x > 0$ on (x^*, β_2). It follows that $G^{2n}(x)$ decreases to β_1 for all x in (β_1, x^*) and $G^{2n}(x)$ increases to β_2 for all x in (x^*, β_2) (as $n \to \infty$). To summarize

For all x in $[a, \beta_1)$, $G^2(x) - x > 0$ and $G^{2n}(x)$ increases to β_1, as $n \uparrow \infty$. For all x in $(\beta_2, b]$, $G^2(x) - x < 0$ and $G^{2n}(x)$ decreases to β_2, as $n \uparrow \infty$. Also, $G^2(x) - x < 0$ on (β_1, x^*) and $G^2(x) - x > 0$ on (x^*, β_2). Hence, $G^{2n}(x)$ decreases to β_1 on (β_1, x^*) and increases to β_2 on (x^*, β_2) as $n \uparrow \infty$.

Further, from [G.1] and [P.4], $[a, b]$ is invariant under F. Consider $X_{n+1} = \alpha_{n+1}(X_n)$, where $\alpha_n = F$ with probability p and $\alpha_n = G$ with probability $1 - p$. In what follows, the initial $X_0 = x \in (0, 1)$. Using Lemma 6.1 and the structure of F, it is clear that the process $X_n(x)$ lands in $[a, b]$ with probability 1 after a finite number of steps. For the process lands in $[a, b]$ from any initial $x \in (0, 1)$, F occurs in a finite (but sufficiently large) number of consecutive periods; i.e., there is a long "run" of F. It is useful to distinguish four cases depending on the location of the attractive fixed point r of F.

Case (i). $a < r < \beta_1 < x^* < \beta_2 < b$. Here, choose δ small enough so that $r + \delta < \beta_1$; there is some N_δ such that $F^n[a, b] \subset (r - \delta, r + \delta)$ for all $n \geqslant N_\delta$. Choose $\varepsilon > 0$ so that $\beta_1 - \varepsilon > r + \delta$. Recall that $G^{2n}(a)$ increases to β_1; let N_1 be such that $G^{2n}(a) > \beta_1 - \varepsilon$ for all $n \geqslant N_1$. Then $G^{2n}(x) > \beta_1 - \varepsilon$ for all $x \in [a, b]$ and for all $n \geqslant N_1$.

Now choose some z_0 satisfying $r + \delta < z_0 < \beta_1 - \varepsilon$. Let N be some *even* integer greater than $\max(2N_1, N_\delta)$. Then

$$\text{Prob}(\alpha_N \cdots \alpha_1 x \geqslant z_0 \quad \forall x \in [a, b]) \geqslant (1 - p)^N.$$

This follows from the possibility of G occurring in N consecutive periods. Also,

$$\text{Prob}(\alpha_N \cdots \alpha_1 x \leqslant z_0 \quad \forall x \in [a, b]) \geqslant p^N,$$

which follows from the possibility of F occurring in N consecutive periods. This verifies the splitting condition (**H**), in Subsection 3.5.1.

Case (ii). $a < \beta_1 < r < x^* < \beta_2 < b$. Here choose ε, δ positive so that $a < \beta_1 + \varepsilon < r - \delta < r + \delta < x^* < \beta_2 < b$. Again, there is some N_δ such that $F^n[a, b] \subset (r - \delta, r + \delta)$ $\forall n \geqslant N_\delta$. There is N_1 such that $G^{2n}(r + \delta) \leqslant \beta_1 + \varepsilon$ for all $n \geqslant N_1$, so that $G^{2n}[a, r + \delta] \subset [a, \beta_1 + \varepsilon]$ $\forall n \geqslant N_1$.

Consider the composition $G^{2N_1} \circ F^{N_\delta}$. It is clear that for all $x \in [a, b]$, $G^{2N_1} \circ F^{N_\delta}(x) \leqslant \beta_1 + \varepsilon$. Let N be $2N_1 + N_\delta$ and choose some z_0 satisfying $\beta_1 + \varepsilon < z_0 < r - \delta$. Then $F^N[a, b] \subset [z_0, b]$, $G^{2N_1} \circ F^{N_\delta}[a, b] \subset [a, z_0]$, so that

$$\text{Prob}(\alpha_N \cdots \alpha_1 x \geqslant z_0 \quad \forall x \in [a, b]) \geqslant p^N,$$
$$\text{Prob}(\alpha_N \cdots \alpha_1 x \leqslant z_0 \quad \forall x \in [a, b]) \geqslant p^{N_\delta}(1 - p)^{2N_1}.$$

Thus the splitting condition is verified in this case also.

Case (iii). $a < \beta_1 < x^* < r < \beta_2 < b$. Choose $\delta > 0$, $\varepsilon > 0$ such that $x^* < r - \delta < r + \delta < \beta_2 - \varepsilon$. There exists N_2 such that $F^n[a, b] \subset (r - \delta, r + \delta)$, $\forall n \geq N_2$ and $G^{2N_2}(r - \delta) > \beta_2 - \varepsilon$ so that $G^{2N_2}[r - \delta, b] \subset (\beta_2 - \varepsilon, b]$. Hence, splitting occurs with $z_0 \in (r + \delta, \beta_2 - \varepsilon)$ and $N = 3N_2$: $P(\alpha_N \cdots \alpha_1 x \leq z_0 \, \forall x \in [a, b]) \geq p^N$, $P(\alpha_N \cdots \alpha_1 x \geq z_0 \, \forall x \in [a, b]) \geq p^{N_2}(1 - p)^{2N_2}$.

Case (iv). $a < \beta_1 < x^* < \beta_2 < r < b$. Choose $\delta > 0$, $\varepsilon > 0$ such that $\beta_2 + \varepsilon < r - \delta$. There exists N_3 such that $F^{2N_3}[a, b] \subset (r - \delta, r + \delta)$, and $G^{2N_3}(b) < \beta_2 + \varepsilon$ so that $G^{2N_3}[a, b] \subset [a, \beta_2 + \varepsilon)$. Hence splitting occurs with $z_0 \in (\beta_2 + \varepsilon, r - \delta)$ and $N = 2N_3$:

$$P(\alpha_N \cdots \alpha_1 x \leq z_0 \quad \forall x \in [a, b]) \geq (1 - p)^N,$$
$$P(\alpha_N \cdots \alpha_1 x \geq z_0 \quad \forall x \in [a, b]) \geq p^N.$$

We now prove the following result:

Theorem 6.1 *Under the hypotheses* [G.1], [P.1]–[P.4], *the distribution of* $X_n(x)$ *converges exponentially fast in the Kolmogorov distance to a unique invariant probability, for every initial* $x \in (0, 1)$.

Proof. In view of the preceding arguments, Theorem 5.1 guarantees exponential convergence of the distribution function $H_n(z) := \text{Prob}(X_n(x) \leq z)$ to the distribution function $F_\pi(z)$ of the invariant probability π with an error no more than $(1 - \delta)^{[n/N]}$, provided the initial state $x \in [a, b]$. If $x \in (0, 1) \cap [a, b]^c$ then, by Lemma 6.1, there exists $0 < x < \tilde{x} < 1$ such that $X_n(x) \in [x, \tilde{x}] \forall n \geq 0$. Since $F^n(y) \uparrow r$ as $n \uparrow \infty$ for $0 < y < r$, and $F^n(y) \downarrow r$ as $n \uparrow \infty$ for $r < y < 1$, it follows that there exists an integer n_0 such that $F^n(y) \in [a, b] \forall n \geq n_0$ and $\forall y \in [x, \tilde{x}]$. Therefore, $\text{Prob}(X_m(x) \notin [a, b]) \leq (1 - p^{n_0})^{[m/n_0]} \forall m \geq n_0$. For the probability of not having an n_0-long run of F in any of the consecutive $[m/n_0]$ periods of length, n_0 is $(1 - p^{n_0})^{[m/n_0]}$. Now let $n \geq 2n_0$, $m_1 = [n/2]$, $m_2 = n - m_1$. Then $\forall z \in \mathbb{R}$ one has

$$H_n(z) = \text{Prob}(X_n(x) \leq z, X_{m_1}(x) \in [a, b])$$
$$+ \text{Prob}(X_n(x) \leq z, X_{m_1}(x) \notin [a, b])$$
$$= \int_{[a,b]} \text{Prob}(X_{n-m_1}(y) \leq z) p^{(m_1)}(x, dy)$$
$$+ \text{Prob}(X_n(x) \leq z, X_{m_1}(x) \notin [a, b]).$$

Since splitting (**H**) holds, there exist $\tilde{\chi} > 0$ and N such that

(1) the integrand above lies between $F_\pi(z) \pm (1 - \tilde{\chi})^{[m_2/N]}$,
(2) $1 \geqslant \int_{[a,b]} p^{(m_1)}(x, dy) \geqslant 1 - (1 - p^{n_0})^{[m_1/n_0]}$, and
(3) $\text{Prob}(X_n(x) \leqslant z, X_{m_1}(x) \notin [a, b]) \leqslant (1 - p^{n_0})^{[m_1/n_0]}$,

it follows that

$$|H_n(z) - F_\pi(z)| \leqslant (1 - \tilde{\chi})^{[m_2/N]} + 2(1 - p^{n_0})^{[m_1/n_0]}.$$ ∎

Example 6.3 We now consider an interesting example of $\{F, G\}$ satisfying the hypotheses [G.1], [P.1]–[P.4], so that Theorem 6.1 holds.

Consider the family of maps $F_\theta : [0, 1] \to [0, 1]$ defined as

$$F_\theta(x) = \theta x(1 - x).$$

For $1 < \theta \leq 4$, the fixed points of F_θ are 0, $x_\theta^* = 1 - \frac{1}{\theta}$. For $\theta > 2$, $x_\theta^* > 1/2$. Let us try to find an interval $[a, b]$ on which F_θ is monotone and which is left invariant under F_θ, i.e., $F_\theta([a, b]) \subset [a, b]$. One must have $1/2 \leq a < b \leq 1$, and one can check that $[1/2, \theta/4]$ is such an interval, provided $F_\theta(\theta/4) \equiv F_\theta^2(1/2) \geq 1/2$. This holds if $2 \leq \theta \leq 1 + \sqrt{5}$. For any such θ, F_θ is strictly decreasing and F_θ^2 is strictly increasing on $[1/2, \theta/4]$.

Consider $3 < \theta \leq 1 + \sqrt{5}$. In this case, $|F_\theta'(x_\theta^*)| = \theta - 2 > 1$ so that x_θ^* is a repelling fixed point of F_θ. One can show that there are two locally attracting periodic points of period 2, say, β_1 and β_2, and $\beta_1 < x_\theta^* < \beta_2$ (see Section 1.7). Since F_θ^2 is a fourth-degree polynomial, $\{0, \beta_1, x_\theta^*, \beta_2\}$ are the only fixed points of F_θ^2. Also, $F_\theta^2(x) - x$ does not change sign on (β_1, x_θ^*) or on (x_θ^*, β_2). Take $\theta = 3.1$. Then the invariant interval is $[1/2, 0.775]$ and $x_\theta^* = 0.677$. One can compute (see, e.g., Sandefur 1990, pp. 172–181), that when $\theta \in (3, 1 + \sqrt{6})$, β_1 and β_2 are given by $\sqrt{\theta + 1}[(\sqrt{\theta + 1} \pm \sqrt{\theta - 3}/2\theta]$. For $\theta = 3.1$, one has $\beta_1 = 0.558014125$, $\beta_2 = 0.76456652$, $F_\theta'(\beta_1) \cong -0.3597$, $F_\theta'(\beta_2) = -1.6403$, $|F_\theta'(\beta_1)F_\theta'(\beta_2)| \cong 0.5904 < 1$.

As an example of functions of the kind considered in this section, let $F(x) = (3/4)x^{1/2}$ and $G(x) = F_\theta(x)$ as above, with $\theta = 3.1$. Let $\text{Prob}(\alpha_n = F) = p$, $\text{Prob}(\alpha_n = G) = 1 - p(0 < p < 1)$. Then with $a = 1/2$, $b = \theta/4 = (3.10)/4$, $[a, b]$ is an invariant interval for F and G. The fixed point of F in $[a, b]$ is $(3/4)^2 = 9/16 = r$. The repelling fixed point of G is $x_\theta^* = 0.677\ldots$ and the period-two orbit is $\{\beta_1 = 0.558\ldots, \beta_2 = 0.764\ldots\}$. Thus the hypotheses imposed on

F, G, namely, [G.1] and [P.1]–[P.4] are satisfied here and, by Theorem 6.1, the distribution of $X_n(x)$ converges exponentially fast in the Kolmogorov distance to a unique invariant probability, for every initial $x \in (0, 1)$. ∎

3.6.4 Comparative Dynamics

Let $\hat{F}$ be a continuous map on $[c, d]$ (into $[c, d]$). For each $\varepsilon \in (0, 1)$ consider the Markov process $X_0, X_n^\varepsilon = \alpha_n^\varepsilon \alpha_{n-1}^\varepsilon \cdots \alpha_1^\varepsilon X_0 (n \geq 1)$, where $\{\alpha_n^\varepsilon : n \geq 1\}$ is an i.i.d. sequence of maps on $[c, d]$ independent of X_0 such that

$$\text{Prob}(\alpha_n^\varepsilon = \hat{F}) = 1 - \varepsilon. \tag{6.5}$$

We say a probability measure μ is $\hat{F}$-*invariant* if

$$\int h(x)\mu(dx) = \int h(\hat{F}(x))\mu(dx) \quad \forall \text{ continuous } h \text{ on } [c, d]. \tag{6.6}$$

Note that (6.6) implies $\int h(x)\mu(dx) = \int h(\hat{F}^n(x))\mu(dx)$ for all $n \geq 1$ and all continuous h on $[c, d]$. In particular, if z_0 is a unique fixed point of $\hat{F}$ and it is attractive, in the sense that $\hat{F}^n(x) \to z_0 \ \forall x \in [c, d]$, then iterating (6.6), one gets $\int h(x)\mu(dx) = \int h(\hat{F}^n(x))\mu(dx) \to h(z_0)$ as $n \to \infty$. This implies $\mu = \delta_{z_0}$. Conversely, for μ to be a unique $\hat{F}$-invariant probability, there must exist a unique fixed point z_0, and no other fixed or periodic points; and in this case $\mu = \delta_{z_0}$.

Theorem 6.2 *Suppose that for each $\varepsilon \in (0, 1)$, π_ε is an invariant probability of X_n^ε. Then every weak limit point π of π_ε as $\varepsilon \downarrow 0$ is $\hat{F}$-invariant. If $\mu = \delta_{z_0}$ is the unique $\hat{F}$-invariant probability, then π_ε converges weakly to μ as $\varepsilon \downarrow 0$.*

Proof. Since the set of probability measures on $[c, d]$ is a compact metric space under the weak topology (see Lemma 11.1 in Chapter 2), consider a limit point π of π_ε as $\varepsilon \downarrow 0$. Then there exists a sequence $\varepsilon_j \downarrow 0$ as $j \uparrow \infty$ such that $\pi_{\varepsilon_j} \to \pi$ weakly. By invariance of π_{ε_j}, for every real-valued continuous function h on $[c, d]$, one has

$$\int_{[c,d]} h(x)\pi_{\varepsilon_j}(dx) = \int_{[c,d]} \left(\int_{[c,d]} h(y) p_{\varepsilon_j}(x, dy) \right) \pi_{\varepsilon_j}(dx), \tag{6.7}$$

where $p_{\varepsilon_j}(x, dy)$ is the transition probability of the Markov process $X_n^\varepsilon (n \geq 0)$ above with $\varepsilon = \varepsilon_j$. Note that $p_{\varepsilon_j}(x, \{\hat{F}(x)\}) \geq 1 - \varepsilon_j$. Hence the right-hand side in (6.7) equals

$$\int_{[c,d]} [(1 - \varepsilon_j)h(\hat{F}(x)) + \delta_j(x)]\pi_{\varepsilon_j}(dx), \qquad (6.8)$$

where $\delta_j(x) = \varepsilon_j E[h(\alpha_1 x) \mid \alpha_1 \neq \hat{F}]$, $|\delta_j(x)| \leq \varepsilon_j(\max_y |h(y)|)$.
 Therefore,

$$\int_{[c,d]} h(x)\pi_{\varepsilon_j}(dx) = (1 - \varepsilon_j) \int_{[c,d]} h(\hat{F}(x))\pi_{\varepsilon_j}(dx)$$

$$+ \int_{[c,d]} \delta_j(x)\pi_{\varepsilon_j}(dx)$$

$$\rightarrow \int_{[c,d]} h(\hat{F}(x))\pi(dx) \quad \text{as } \varepsilon_j \rightarrow 0, \quad (6.9)$$

since $\pi_{\varepsilon_j} \rightarrow \pi$ weakly. On the other hand, again by this weak convergence, the left-hand side of (6.9) converges to $\int_{[c,d]} h(x)\pi(dx)$. Therefore,

$$\int_{[c,d]} h(x)\pi(dx) = \int_{[c,d]} h(\hat{F}(x))\pi(dx). \qquad (6.10)$$

In other words, π is $\hat{F}$-invariant. If there is a unique $\hat{F}$-invariant probability μ, one has $\pi = \mu$. Since this limit is the same whatever be the sequence $\varepsilon_j \downarrow 0$, the proof is complete. ∎

Remark 6.2 Suppose $\hat{F}$ and $\hat{G}$ are monotone and continuous, $\text{Prob}(\alpha_1 = \hat{F}) = 1 - \varepsilon$ and $\text{Prob}(\alpha_1 = \hat{G}) = \varepsilon$, and that the process X_n in (6.5) satisfies the splitting condition (**H**). Then the invariant probability π_ε of X_n is unique. Suppose ε is small. Then, whatever be the initial state X_0, the distribution of X_n is approximated well (in the Kolmogorov distance) by π_ε, if n is large. On the other hand, π_ε is close to δ_{z_0} (in the weak topology) if z_0 is the unique attractive fixed point of $\hat{F}$ in the sense of Theorem 6.2. Thus X_n would be close to z_0 in probability if ε is small and n is large.

Remark 6.3 Theorem 6.2 holds on arbitrary compact metric spaces in place of $[c, d]$. There is a broad class of dynamical systems F on compact metric spaces S, each with a distinguished F-invariant probability

π whose support is large and to which the invariant probability π_ε of the perturbed (random) dynamical system converges as the perturbation, measured by ε, goes to 0. Such measures π are sometimes called *Kolmogorov measures*, or *Sinai–Ruelle–Bowen* (SRB) measures. See, for example, Kifer (1988) for precise statements and details. For similar results for the case of quadratic maps $F_\theta(x) = \theta x(1 - x)(x \in [0, 1])$, see Katok and Kifer (1986).

3.7 Contractions

3.7.1 Iteration of Random Lipschitz Maps

In this section (S, d) is a complete separable metric space. A map f from S into S is *Lipschitz* if for some finite $L > 0$,

$$d(f(x), f(y)) \leq Ld(x, y)$$

for all $x, y \in S$. Let Γ be a family of Lipschitz maps from S into S.

Let $\{\alpha_n : n \geq 1\}$ be an i.i.d. sequence of random Lipschitz maps from Γ. Denote by $L_j^k(k \geq j)$, the *random Lipschitz coefficient* of $\alpha_k \cdots \alpha_j$, i.e.,

$$L_j^k(\cdot) := \sup \{d(\alpha_k \cdots \alpha_j x, \alpha_k \cdots \alpha_j y)/d(x, y): x \neq y\}. \quad (7.1)$$

As before (see Remark 4.1), $X_n(x) = \alpha_n \cdots \alpha_1 x$, and let $Y_n(x) = \alpha_1 \cdots \alpha_n x$ denote the *backward iteration* ($n \geq 1$; $x \in S$). Note that, for every $n \geq 1$, $X_n(x)$ and $Y_n(x)$ have the same distribution for each x, and we denote this by

$$X_n(x) \overset{\mathcal{L}}{=} Y_n(x).$$

Before deriving the general result of this section, namely, Theorem 7.2, we first state and prove the following simpler special case, which suffices for a number of applications.

Theorem 7.1 *Let (S, d) be compact metric, and assume*

$$-\infty \leq E \log L_1^r < 0, \quad \text{for some } r \geq 1. \quad (7.2)$$

Then the Markov process X_n has a unique invariant probability and is stable in distribution.

Proof.   First assume (7.2) with $r = 1$. A convenient distance metrizing the weak topology of $\mathcal{P}(S)$ is the *bounded Lipschitzian distance* d_{BL} (see Complements and Details, Chapter 2, (C11.33)),

$$d_{\text{BL}}(\mu, \nu) = \sup\left\{\left|\int f \, d\mu - \int f \, d\nu\right| : f \in \mathbf{L}\right\},$$

$$\mathbf{L} := \{f : S \to \mathbb{R}, |f(x) - f(y)| \le 1 \forall x, y,$$

$$|f(x) - f(y)| \le d(x, y) \, \forall x, y\}. \tag{7.3}$$

By Prokhorov's theorem (see Complements and Details, Chapter 2, Theorem C11.5), the space $(\mathcal{P}(S), d_{BL})$ is a compact metric space. Since T^* is a continuous map on this latter space, due to Feller continuity of $p(x, dy)$, it has a fixed point π – an invariant probability. To prove uniqueness of π and stability, write $L_j = L_j^j$ (see (7.1)) and note that for all $x, y \in S$, one has

$$d(X_n(x), X_n(y)) \equiv d(\alpha_n \cdots \alpha_1 x, \alpha_n \cdots \alpha_1 y)$$

$$\le L_n d(\alpha_{n-1} \cdots \alpha_1 x, \alpha_{n-1} \cdots \alpha_1 y)$$

$$\cdots$$

$$\le L_n L_{n-1} \cdots L_1 d(x, y) \le L_n L_{n-1} \cdots L_1 M, \quad (7.4)$$

where $M = \sup\{d(x, y) : x, y \in S\}$. By the strong law of large numbers, the logarithm of the last term in (7.4) goes to $-\infty$ a.s. since, in view of (7.4) (with $r = 1$), $\frac{1}{n} \sum_{j=1}^{n} \log L_j \to E \log L_1 < 0$. Hence

$$\sup_{x, y \in S} d(X_n(x), X_n(y)) \to 0 \text{ a.s.} \quad \text{as } n \to \infty. \tag{7.5}$$

Let X_0 be independent of $(\alpha_n)_{n \ge 1}$ and have distribution π. Then (7.5) yields

$$\sup_{x \in S} d(X_n(x), X_n(X_0)) \to 0 \text{ a.s.} \quad \text{as } n \to \infty. \tag{7.6}$$

Since $X_n(x)$ has distribution $p^{(n)}(x, dy)$ and $X_n(X_0)$ has distribution π, one then has (see (7.3))

$$d_{\text{BL}}(p^{(n)}(x, \cdot), \pi) = \sup\{|Ef(X_n(x)) - Ef(X_n(X_0))| : f \in L\}$$

$$\le Ed(X_n(x), X_n(X_0))\} \to 0 \quad \text{as } n \to \infty. \tag{7.7}$$

This completes the proof for the case $r = 1$.

Let now $r > 1$ (in (7.2)). The above argument applied to $p^{(r)}(x, dy)$, regarded as a one-step transition probability, shows that for every $x \in S$, as $k \to \infty$, $p^{(kr)}(x, \cdot)$ converges in d_{BL}-distance (and, therefore, weakly) to a probability π that is invariant for $p^{(r)}(\cdot, \cdot)$. This implies that $T^{*kr}\mu \to \pi$ (weakly) for every $\mu \in \mathcal{P}(S)$. In particular, for each $j = 1, 2, \ldots, r - 1$,

$$p^{(kr+j)}(x, \cdot) \equiv T^{*kr}p^{(j)}(x, \cdot) \to \pi \quad \text{as} \quad k \to \infty. \quad \text{That is, for every}$$
$x \in S$,

$$p^{(n)}(x, \cdot) \text{ converges weakly to } \pi \text{ as } n \to \infty. \quad (7.8)$$

In turn, (7.8) implies $T^{*n}\mu \to \pi$ weakly for every $\mu \in \mathcal{P}(S)$. Therefore, π is the unique invariant probability for $p(\cdot, \cdot)$ and stability holds. ∎

Remark 7.1 The last argument above shows, in general, that if for some $r > 1$, the *skeleton Markov process* $\{X_{kr}: k = 0, 1, \ldots\}$ has a unique invariant probability π and is stable, then the entire Markov process $\{X_n: n = 0, 1, \ldots\}$ has the unique invariant probability π and is stable.

Exercise 7.1 A direct application of Theorem 7.1 leads to a strengthening of the stability results in Green and Majumdar (Theorems 5.1–5.2, 1975) who introduced (i.i.d.) random shocks in the Walras-Samuelson tatonnement.

We now turn to the main result of this section, which extends Theorem 7.1 and which will find an important application in Chapter 4.

Theorem 7.2 *Let S be a complete separable metric space with metric d. Assume that (7.2) holds for some $r \geq 1$ and there exists a point $x_0 \in S$ such that*

$$E \log^+ d(\alpha_r \cdots \alpha_1 x_0, x_0) < \infty. \quad (7.9)$$

Then the Markov process $X_n(x) := \alpha_n \cdots \alpha_1 x$ ($n \geq 1$) has a unique invariant probability and is stable in distribution.

Proof. First consider the case $r = 1$ for simplicity. For this case, we will show that the desired conclusion follows from the following two assertions:

(a) $\sup\{d(Y_n(x),\ Y_n(y)): d(x, y) \leq M\} \to 0$ in probability as $n \to \infty$, for every $M > 0$.

(b) The sequence of distributions of $d(X_n(x_0), x_0)$, $n \geq 1$, is relatively weakly compact.

Assuming (a), (b), note that for every real-valued Lipschitz $f \in L$ (see (7.3)), one has for all $M > 0$,

$$\sup_{\rho(x,y)\leq M} |Ef(X_n(x)) - Ef(X_n(y))|$$
$$\leq \sup_{\rho(x,y)\leq M} E[\rho(X_n(x),\ X_n(y)) \wedge 1] \to 0 \qquad (7.10)$$

as $n \to \infty$, by (a) and Lebesgue's dominated convergence theorem. Also,

$$|Ef(X_{n+m}(x_0)) - Ef(X_n(x_0))|$$
$$= |Ef(Y_{n+m}(x_0)) - Ef(Y_n(x_0))|$$
$$\equiv |Ef(\alpha_1 \cdots \alpha_n \alpha_{n+1} \cdots \alpha_{n+m} x_0) - Ef(\alpha_1 \cdots \alpha_n x_0)|$$
$$\leq E[d(\alpha_1 \cdots \alpha_n \alpha_{n+1} \cdots \alpha_{n+m} x_0,\ \alpha_1 \cdots \alpha_n x_0) \wedge 1]. \quad (7.11)$$

For a given n and arbitrary $M > 0$, divide the sample space into the disjoint sets:

(i) $B_1 = \{d(\alpha_{n+1} \cdots \alpha_{n+m} x_0,\ x_0) \geq M\}$

and

(ii) $B_2 = \{d(\alpha_{n+1} \cdots \alpha_{n+m} x_0,\ x_0) < M\}$.

The last expectation in (7.11) is no more than $P(B_1)$ on B_1. On B_2, denoting for the moment $X = \alpha_{n+1} \cdots \alpha_{n+m} x_0$, one has $d(\alpha_1 \cdots \alpha_n \alpha_{n+1} \cdots \alpha_{n+m} x_0,\ \alpha_1 \cdots \alpha_n x_0) \equiv d(\alpha_1 \cdots \alpha_n X,\ \alpha_1 \cdots \alpha_n x_0)$ $\equiv d(Y_n(X),\ Y_n(x_0)) \leq \sup\{d(Y_n(x),\ Y_n(y)): d(x, y) \leq M\} = Z$, say. The last inequality holds since, on B_2, $d(X, x_0) \leq M$. Now, for every $\delta > 0$, $E(Z \wedge 1) = E(Z \wedge 1 \cdot 1_{\{Z>\delta\}}) + E(Z \wedge 1 \cdot 1_{\{Z\leq\delta\}}) \leq P(Z > \delta) + \delta$. Hence,

$$|Ef(X_{n+m}(x_0)) - Ef(X_n(x_0))| \leq P(d(\alpha_{n+1} \cdots \alpha_{n+m} x_0,\ x_0) \geq M)$$
$$+ P(\sup\{d(Y_n(x),\ Y_n(y)): d(x, y) \leq M\} > \delta) + \delta. \qquad (7.12)$$

Fix $\varepsilon > 0$, and let $\delta = \varepsilon/3$. Choose $M = M_\varepsilon$ such that

$$P(d(\alpha_1 \cdots \alpha_m x_0,\ x_0) \geq M_\varepsilon) < \varepsilon/3 \quad \text{for all } m = 1, 2, \ldots. \qquad (7.13)$$

This is possible by (b). Now (a) implies that the middle term on the right in (7.12) goes to zero as $n \to \infty$. Hence (7.12), (7.13), and (a) imply that there exists n_ε such that for all $m = 1, 2, \ldots$, and $f \in L$,

$$|Ef(X_{n+m}(x_0)) - Ef(X_n(x_0))| < \varepsilon \quad \text{for all } n \geq n_\varepsilon. \quad (7.14)$$

In other words, recalling definition (7.3) of the bounded Lipschitzian distance $d_{BL}(\mu, \nu)$,

$$\sup_{m \geq 1} d_{BL}(p^{(n+m)}(x_0, dy), p^{(n)}(x_0, dy)) \to 0 \quad \text{as } n \to \infty. \quad (7.15)$$

This implies that $p^{(n)}(x_0, dy)$ is Cauchy in $(\mathcal{P}(S), d_{BL})$ and, therefore, converges weakly to a probability measure $\pi(dy)$ as $n \to \infty$, by the *completeness* of $(\mathcal{P}(S), d_{BL})$ (Theorem C11.6 in Complements and Details, Chapter 2). Since, $p(x, dy)$ has the *Feller property*, it follows that π is an invariant probability. In view of (7.10) and (a), $p^{(n)}(x, dy)$ converges weakly to π, for every $x \in S$. This implies that π is the unique invariant probability.

It remains to prove (a) and (b).

To prove (a), recall $L_n = \sup\{d(\alpha_n x, \alpha_n y)/d(x, y): x \neq y\}$, and as in (7.4),

$$\begin{aligned}
d(Y_n(x), Y_n(y)) &= d(\alpha_1 \alpha_2 \cdots \alpha_n x, \alpha_1 \alpha_2 \cdots \alpha_n y) \\
&\leq L_1 d(\alpha_2 \cdots \alpha_n x_1, \alpha_2 \cdots \alpha_n y) \\
&\leq \cdots \leq L_1 L_2 \cdots L_n \rho(x, y) \\
&\leq L_1 L_2 \cdots L_n M \text{ for all } (x, y) \text{ such that } d(x, y) \leq M.
\end{aligned}$$

Take logarithms to get

$$\sup_{\{(x,y):d(x,y)\leq M\}} \log d(Y_n(x), Y_n(y)) \leq \log L_1 + \cdots$$
$$+ \log L_n + \log M \to -\infty \text{ a.s.} \quad \text{as } n \to \infty, \quad (7.16)$$

by the strong law of large numbers and (7.2). This proves (a). To prove (b), it is enough to prove that $d(\alpha_1 \ldots \alpha_n x_0, x_0)$ converges almost surely and, therefore, in distribution, to some a.s. finite limit (use the fact that $d(Y_n(x_0), x_0) = d(\alpha_1 \cdots \alpha_n x_0, x_0)$ and $d(X_n(x_0), x_0) = d(\alpha_n \cdots \alpha_1 x_0, x_0)$ have the same distribution). For this use the triangle inequality to write

$$
\begin{aligned}
&|d(\alpha_1 \cdots \alpha_{n+m} x_0, x_0) - d(\alpha_1 \cdots \alpha_n x_0, x_0)| \\
&\leq d(\alpha_1 \cdots \alpha_{n+m} x_0, \alpha_1 \cdots \alpha_n x_0) \\
&\leq \sum_{j=1}^{m} d(\alpha_1 \cdots \alpha_{n+j} x_0, \alpha_1 \cdots \alpha_{n+j-1} x_0) \\
&\leq \sum_{j=1}^{m} L_1 \cdots L_{n+j-1} \, d(\alpha_{n+j} x_0, x_0).
\end{aligned}
\tag{7.17}
$$

Let $c := -E \log L_1 (> 0)$. First assume c is finite and fix ε, $0 < \varepsilon < c$. It follows from (7.2) and the strong law of large numbers that there exists $n_1(\omega)$ such that, outside a P-null set N_1,

$$
\log (L_1 \cdots L_{n'})^{\frac{1}{n'}} = \frac{1}{n'} \sum_{k=1}^{n'} \log L_k < -(c - \varepsilon/2),
$$
$$
(L_1 \cdots L_{n'})^{\frac{1}{n'}} < e^{-(c-\varepsilon/2)}, \quad L_1 \cdots L_{n'} < e^{-n'(c-\varepsilon/2)}
$$
$$
\forall n' \geq n_1(\cdot).
\tag{7.18}
$$

In view of (7.9), with $r = 1$,

$$
\begin{aligned}
\sum_{k=1}^{\infty} P\left(\frac{\log^+ d(\alpha_k x_0, x_0)}{\varepsilon/2} > k \right) &= \sum_{k=1}^{\infty} P\left(\frac{\log^+ d(\alpha_1 x_0, x_0)}{\varepsilon/2} > k \right) \\
&\leq \frac{E \log^+ d(\alpha_1 x_0, x_0)}{\varepsilon/2} < \infty,
\end{aligned}
\tag{7.19}
$$

since $\sum_{k=1}^{\infty} P(V > k) \leq EV$ for every nonnegative random variable V (see Exercise 10.1, Chapter 2).

Therefore, by the Borel–Cantelli lemma, there exists $n_2(\omega)$ such that, outside a P-null set N_2,

$$
d(\alpha_k x_0, x_0) \leq e^{k\varepsilon/2} \quad \text{for all } k \geq n_2(\cdot).
\tag{7.20}
$$

Using (7.18) and (7.20) in (7.17), one gets

$$\sup_{m \geq 1} |d(\alpha_1 \cdots \alpha_n \alpha_{n+1} \cdots \alpha_{n+m} x_0, x_0) - d(\alpha_1 \cdots \alpha_n x_0, x_0)|$$

$$\leq \sum_{j=1}^{\infty} e^{-(n+j-1)(c-\varepsilon/2)} e^{(n+j)\varepsilon/2}$$

$$= e^{c-\varepsilon/2} e^{-n(c-\varepsilon)} \sum_{j=1}^{\infty} e^{-j(c-\varepsilon)}$$

$$\text{for all } n \geq n_0(\cdot) := \max\{n_1(\cdot), n_2(\cdot)\}, \quad (7.21)$$

outside the P-null set $N = N_1 \cup N_2$. Thus the left-hand side of (7.21) goes to 0 almost surely as $n \to \infty$, showing that $d(\alpha_1 \cdots \alpha_n x_0, x_0)$ $(n \geq 1)$ is Cauchy and, therefore, converges almost surely as $n \to \infty$. The proof of (b) is now complete if $-E \log L_1 < \infty$. If $-E \log L_1 = \infty$, then (7.18) and (7.21) hold for any $c > 0$. This concludes the proof of the theorem for $r = 1$ (in (7.2)). Now, the extension to the case $r > 1$ follows exactly as in the proof of Theorem 7.1 (see Remark 7.1). ∎

Remark 7.2 The theorem holds if $P(L_1 = 0) > 0$. The proof (for the case $E \log L_1 = -\infty$) includes this case. But the proof may be made simpler in this case by noting that for all sufficiently large n, say $n \geq n(\omega)$, $X_n(x) = X_n(y) \ \forall x, y$. (Exercise: Use Borel–Cantelli lemma to show this.)

Remark 7.3 The moment condition (7.9) under which Theorem 7.2 is derived here is nearly optimal, as seen from its specialization to the affine linear case in the next section. Under the additional moment assumption $EL_1^1 < \infty$, the stability result is proved in Diaconis and Freedman (1999) (for the case $r = 1$), and a speed of convergence is derived. The authors also provide a number of important applications as well as an extensive list of references to the earlier literature.

3.7.2 A Variant Due to Dubins and Freedman

Stepping back for a moment, we see that the general criterion for the existence of a unique invariant probability for the Markov process X_n described by (2.1) and for its stability in distribution is given by the following two conditions, which only require α_n to be continuous and not necessarily Lipschitz. We have proved

Proposition 7.1 *On a complete separable metric space, let $\alpha_n (n \geq 1)$ be i.i.d. continuous. The Markov process $X_n(x) \equiv \alpha_n \cdots \alpha_1 x$ is stable in distribution if*

(a) $\sup\{d(\alpha_n \alpha_{n-1} \cdots \alpha_1 x, \ \alpha_n \alpha_{n-1} \cdots \alpha_1 y) : d(x, y) \leq M\} \to 0$ *in probability, as $n \to \infty$, for every $M > 0$,*
and
(b) for some $x_0 \in S$, the sequence of distributions of $d(X_n(x_0), x_0) \equiv d(\alpha_n \cdots \alpha_1 x_0, x_0)$ is relatively weakly compact.

When S is *compact*, then (b) is automatic, since the sequence $d(X_n(x_0), x_0)$ is bounded by the diameter of S, whatever be x_0 and n. Obviously here one may simplify the statement of (a) by taking just one M, namely, the diameter of S:

(a)': $\text{diam}(\alpha_n \alpha_{n-1} \cdots \alpha_1 S) \to 0$ in probability as $n \to \infty$.

The criterion (a)' is now used to prove the following basic result of Dubins and Freedman (1966).

Theorem 7.3 *Let (S, d) be compact metric and Γ the set of all contractions on S. Let Q be a probability measure on the Borel sigmafield of Γ (with respect to the "supremum" distance) such that the support of Q contains a strict contraction. Then the Markov process X_n defined by (2.1) and (2.2) has a unique invariant probability and is stable in distribution.*

Proof. Let γ be a *strict contraction* in the support of Q, i.e., $d(\gamma x, \gamma y) < d(x, y) \forall x \neq y$.

As before, the jth iterate of γ is denoted by γ^j. Since $\gamma^{j+1} S = \gamma^j (\gamma S) \subset \gamma^j S$, it follows that $\gamma^j S$ decreases as j increases. Indeed

$$\gamma^j S \text{ decreases to } \cap_{j=1}^{\infty} \gamma^j S = \text{ a singleton } \{x_0\}. \qquad (7.22)$$

To see this, first note that the limit is nonempty, by the finite intersection property of (S, d), and the fact that $\gamma^j S$ is the continuous image of a compact set S and therefore, closed (compact). Assume, if possible, that there are points x_0, y_0 in the limit set, $x_0 \neq y_0$. Let

$\delta := d(x_0, y_0)$. The continuous function $(x, y) \to d(\gamma x, \gamma y)/d(x, y)$ on the compact set $\mathcal{K}_\delta = \{(x, y): d(x, y) \geq \delta\}(\subset S \times S)$ attains a maximum $c < 1$. Let x_1, y_1 be two preimages of x_0, y_0, respectively, under γ, i.e., $\gamma x_1 = x_0$, $\gamma y_1 = y_0$. Since γ, is a contraction, $d(x_1, y_1) \geq d(\gamma x_1, \gamma y_1) = d(x_0, y_0) = \delta$. Therefore, $(x_1, y_1) \in \mathcal{K}_\delta$ and $d(x_0, y_0) \leq cd(x_1, y_1)$, or $d(x_1, y_1) \geq \delta/c$. In general, let x_j, y_j be two preimages under γ of x_{j-1}, y_{j-1}, respectively. Then, by induction, $d(x_j, y_j) \geq d(x_{j-1}, y_{j-1})/c \geq \cdots \geq d(x_0, y_0)/c^j \to \infty$ as $j \to \infty$, which contradicts the fact that S is bounded.

By the same kind of reasoning, one shows that

$$\text{diam}\,(\gamma^j S) \downarrow 0 \quad \text{as } j \uparrow \infty. \tag{7.23}$$

For, if (7.23) is not true, there exists $\delta > 0$ such that $\text{diam}\,(\gamma^j S) > \delta > 0$ for all j. Thus there exist $x_j, y_j \in \gamma^j S$ such that $d(x_j, y_j) \geq \delta$, which implies, using preimages, that there exist $x_0, y_0 \in S$ satisfying $d(x_0, y_0) \geq \delta/c^j \to \infty$ as $j \to \infty$, a contradiction.

Now fix j and let $\gamma_1, \gamma_2, \ldots, \gamma_n$, $n > j$, be such that j consecutive members, say $\gamma_{i+1}, \ldots, \gamma_{i+j}$, are within a distance ε from γ, i.e., $\|\gamma_{i+k} - \gamma\|_\infty < \varepsilon$, $k = 1, \ldots, j$, where $1 \leq i+1 < i+j \leq n$. We shall show

$$\text{diam}\,(\gamma_n \gamma_{n-1} \cdots \gamma_1 S) \leq \text{diam}\,(\gamma^j S) + 2j\varepsilon. \tag{7.24}$$

For this note that the contraction $\gamma_n \gamma_{n-1} \cdots \gamma_{i+j+1}$ does not increase the diameter of the set $\gamma_{i+j} \cdots \gamma_{i+1} \gamma_i \cdots \gamma_1 S$. Also, $\gamma_{i+j} \cdots \gamma_{i+1}(\gamma_i \cdots \gamma_1 S) \subset \gamma_{i+j} \cdots \gamma_{i+1} S$.
Hence the left-hand side of (7.24) is no more than $\text{diam}\,(\gamma_{i+j} \cdots \gamma_{i+1} S)$.

Now, whatever be x, y, one has, by the triangle inequality,

$$d(\gamma_{i+j} \cdots \gamma_{i+2} \gamma_{i+1} x, \gamma_{i+j} \cdots \gamma_{i+2} \gamma_{i+1} y)$$
$$\leq 2\varepsilon + d(\gamma_{i+j} \cdots \gamma_{i+2} \gamma x, \gamma_{i+j} \cdots \gamma_{i+2} \gamma y).$$

Applying the same argument with γx, γy in place of x, y, and replacing γ_{i+2} by γ, and so on, one arrives at (7.24).

We are now ready to verify (a)′. Choose $\delta > 0$ arbitrarily. Let j be such that $\text{diam}\,(\gamma^j S) < \delta/2$. Let $\varepsilon = \delta/4j$, so that the right-hand side of (7.24) is less than δ. Define the events

$$A_m = \{\|\alpha_{j(m-1)+k} - \gamma\|_\infty < \varepsilon \; \forall k = 1, 2, \ldots, j\}, \quad (m = 1, 2, \ldots). \tag{7.25}$$

Note that $P(A_m) = b^j$, where $b := P(\|\alpha_1 - \gamma\|_\infty < \varepsilon)$. Since γ is in the support of Q, $b > 0$. The events $A_m (m = 1, 2, \ldots)$ form an independent sequence, and $\sum_{m=1}^\infty P(A_m) = \infty$. Hence, by the second Borel–Cantelli lemma, $P(A_m$ occurs for some $m) \geq P(A_m$ occurs for infinitely many $m) = 1$. But on A_m, diam $(\alpha_{mj}\alpha_{mj-1} \cdots \alpha_{mj-j+1}S) < \delta$ so that, with probability 1, diam $(\alpha_n \cdots \alpha_1 S) < \delta$ for all sufficiently large n. Since $\delta > 0$ is arbitrary, (a)' holds.   ∎

Remark 7.4 The proof given above is merely an amplification of the derivation given in the original article by Dubins and Freedman (1966). It may be noted that this result does not follow from Theorem 7.1, since one may easily construct strict contractions with the Lipschitz constant 1. On the other hand, Theorems 7.1 and 7.2 apply to sets Γ, which contain (noncontracting) maps with one Lipschitz constant larger than 1 in the support of Q, but still satisfy the criteria (a) and (b) for stability in distribution.

3.8 Complements and Details

Sections 3.1, 3.2.
Proposition C1.1. *Let S be a Borel subset of a Polish space and $\mathcal{S}$ its Borel sigmafield. Let $p(x, B)$ $(x \in S, B \in \mathcal{S})$ be a transition probability on $(S, \mathcal{S})$. Then there exists a set Γ of functions on S into S and a probability measure Q on a sigmafield $\sum$ on Γ such that (i) the map $(\gamma, x) \to \gamma(x)$ on $\Gamma \times S$ into S is measurable with respect to the sigmafields $\sum \otimes \mathcal{S}$ (on $\Gamma \times S$) and $\mathcal{S}$ (on S), and (ii) (2.3) holds: $p(x, C) = Q(\{\gamma \in \Gamma: \gamma(x) \in C\}) \forall x \in S, C \in \mathcal{S}$.*

Proof. By a well-known theorem of Kuratowski (see Royden 1968, p. 326), there exists a one-to-one Borel measurable map h on S onto a Borel subset D of the unit interval $[0, 1]$ such that h^{-1} is Borel measurable on D onto S. Hence one may, without loss of generality, consider D in place of S and $\mathcal{D} = D \cap \mathcal{B}([0, 1])$ in place of $\mathcal{S}$. Since one may define the equivalent (or corresponding) transition probability $\tilde{p}(u, B) := p(h^{-1}(u), h^{-1}(B)) \forall u \in D, B \in \mathcal{D}$, it suffices to prove the desired representation for $\tilde{p}(u, B)$ on the state space $(D, \mathcal{D})$. For this purpose, let the distribution function of $\tilde{p}(u, dy)$ be $F_u(y') := \tilde{p}(u, D \cap [0, y']), 0 \leq y' \leq 1, u \in D$. Define its *inverse* $F_u^{-1}(v) := \inf\{y' \in [0, 1]; F_u(y') > v\}$. Let **m** denote the Lebesgue measure on

[0, 1]. Then we check that

$$\mathbf{m}(\{v \in [0, 1]: F_u^{-1}(v) \le y\}) \ge \mathbf{m}(\{v \in [0, 1]: F_u(y) > v\})$$
$$= F_u(y), \quad (u \in D, \ y \in [0, 1]) \tag{C1.1}$$

and

$$\mathbf{m}(\{v \in [0, 1]: F_u^{-1}(v) \le y\}) \le \mathbf{m}(\{v \in [0, 1]: F_u(y) \ge v\})$$
$$= F_u(y), \quad (u \in D, \ y \in [0, 1]). \tag{C1.2}$$

To check the inequality in (C1.1) note that if $F_u(y) > v$ then (by the definition of the *inverse*), $F_u^{-1}(v) \le y$ (since y *belongs to the set* over which the infimum is taken to define $F_u^{-1}(v)$). The equality in (C1.1) simply gives the Lebesgue measure of the interval $[0, F_u(y))$. Conversely, to derive (C1.2), note first that if $v \ne 1$, then there exists in [0, 1] a sequence $y_n \downarrow F_u^{-1}(v)$, $y_n > F_u^{-1}(v)$ $\forall n$. Then $F_u(y_n) > v$ $\forall n$ and by the right-continuity of F_u, $F_u(F_u^{-1}(v)) \ge v$. Hence $y \ge F_u^{-1}(v)$ implies $F_u(y) \ge v$, proving the inequality in (C1.2), since $\mathbf{m}(\{1\}) = 0$.

Together (C1.1) and (C1.2) imply

$$\mathbf{m}(\{v \in [0, 1]: F_u^{-1}(v) \le y\}) = F_u(y) \ (u \in D, y \in [0, 1]). \tag{C1.3}$$

Define $\Gamma = \{\gamma_v : v \in [0, 1]\}$, where $\gamma_v(u) = F_u^{-1}(v), u \in D$. One may then identify Γ with the label set [0, 1], and let $\sum$ be the Borel sigmafield of [0, 1], $Q = \mathbf{m}$. With this identification, one obtains $\tilde{p}(u, [0, y] \cap D) = Q(\{\gamma : \gamma(u) \in [0, y] \cap D\}), y \in [0, 1], u \in D$.

For complete details on measurability (i.e., property(i)) see Blumenthal and Corson (1972). ∎

An interesting and ingenious application of i.i.d. iteration of random maps in image encoding was obtained by Diaconis and Shashahani (1986), Barnsley and Elton (1988), and Barnsley (1993). An image is represented by a probability measure v, which assigns mass $1/b$ to each of the b black points. This measure v is then approximated by the invariant probability of a random dynamical system generated by two-dimensional affine maps chosen at random from a finite set using an appropriate matching algorithm.

Section 3.5. (Nondecreasing Maps on $\mathbb{R}_+^\ell$) For generalizations of Theorem 5.1 to multidimension, we will use a different route, which is useful in other contexts, for example, in proving a central limit theorem. In the rest of this section, we present a brief exposition of the method given in Bhattacharya and Lee (1988), along with some amendments and extensions, for an appropriate generalization of Theorem 5.1 to $\mathbb{R}^\ell (\ell > 1)$.

Theorem C5.1 is not a complete generalization of Corollary 5.2. For the latter applies to i.i.d (continuous) monotone maps, while Theorem C5.1 assumes the maps α_n to be i.i.d. (measurable) monotone *nondecreasing*. The proof given below easily adapts to i.i.d. (measurable) monotone *nonincreasing* maps as well, but not quite to the case in which α_n may be nondecreasing with positive probability and, at the same time, nonincreasing with positive probability. See Bhattacharya and Lee (1988), also Bhattacharya and Lee 1997 (correction)) and Bhattacharya and Majumdar (1999b, 2004).

Let S be a Borel subset of $\mathbb{R}^\ell$ and α_n an i.i.d. sequence of measurable *nondecreasing* random maps on S into itself. Recall that for two vectors $\mathbf{x} = (x_1, \ldots, x_\ell)$ and $\mathbf{y} = (y_1, \ldots, y_\ell)$ in $\mathbb{R}^\ell$, we write $\mathbf{x} \geq \mathbf{y}$ of $x_i \geq y_i$ for all i. A map $\gamma = (\gamma_i)$ is *nondecreasing* if each coordinate map γ_i is nondecreasing, i.e., $\mathbf{x} \geq \mathbf{y}$ implies $\gamma_i(\mathbf{x}) \geq \gamma_i(\mathbf{y})$. The significant distinction and the main source of difficulty in analysis here is that, unlike the one-dimensional case, the relation $\geq$ is only a partial order. It is somewhat surprising then that one still has an appropriate and broad generalization of Theorem 5.1. To this effect let $\mathcal{A}$ be the class of all sets A of the form

$$A = \{\mathbf{y} \in S \colon \phi(\mathbf{y}) \leq \mathbf{x}\}$$

ϕ continuous and nondecreasing on S into $\mathbb{R}^\ell$, $\quad \mathbf{x} \in \mathbb{R}^\ell$.

(C5.1)

A generalization on $\mathcal{P}(S)$ of the Kolmogorov distance (5.3) is

$$d_A(\mu, \nu) := \sup_{A \in \mathcal{A}} |\mu(A) - \nu(A)|, \quad (\mu, \nu \in \mathcal{P}(S)). \qquad (C5.2)$$

Consider the following splitting condition:

(H₂) There exist $\chi_i^0 > 0$ $(i = 1, 2)$, F_i measurable subsets of $\Gamma^N (i = 1, 2)$, $\mathbf{x}_0 \in S$, and a positive integer N such that

(1) $F_1 \subset \{\gamma \in \Gamma^N \colon \tilde{\gamma}\mathbf{x} \leq \mathbf{x}_0 \ \forall \mathbf{x} \in S\}, \quad Q^N(F_1) = \chi_1^0$

(2) $F_2\{\gamma \in \Gamma^N \colon \tilde{\gamma}\mathbf{x} \geq \mathbf{x}_0 \ \forall \mathbf{x} \in S\}, \quad Q^N(F_2) = \chi_2^0. \qquad (C5.3)$

As before for $\gamma = (\gamma_1, \gamma_2, \ldots, \gamma_N) \in \Gamma^N, \tilde{\gamma} = \gamma_N \gamma_{N-1} \cdots \gamma_1$.

We will show that, under $(\mathbf{H_2})$, Theorem 5.1 extends to a large class of Borel sets $S \subset \mathbb{R}^\ell$. The problems that one faces here are

(a) T^{*^N} may not be a strict contraction on $\mathcal{P}(S)$ for the metric d_A,
(b) $(\mathcal{P}(S), d_A)$ may not be a complete metric space, and
(c) it is not a priori clear that convergence in d_A implies weak convergence.

Problems (b) and (c) are related, and are resolved by the following:

Lemma C5.1 *Suppose S is a closed subset of $\mathbb{R}^\ell$. Then $(\mathcal{P}(S), d_A)$ is a complete metric space, and convergence in d_A implies weak convergence.*

Proof. Let $P_n (n \geq 1)$ be a Cauchy sequence in $(\mathcal{P}(S), d_A)$. Let $\hat{P}_n$ be the extension of P_n to $\mathbb{R}^\ell$, i.e., $\hat{P}_n(B) = P_n(B \cap S)$, where B is a Borel subset. By taking ϕ in (C5.1) to be the identity map, it is seen that the d.f.s F_n of $\hat{P}_n$ are Cauchy in the uniform distance for functions on $\mathbb{R}^\ell$. The limit F on $\mathbb{R}^\ell$ is the d.f. of a probability measure $\hat{P}$ on $\mathbb{R}^\ell$, and $\hat{P}_n$ converges weakly to $\hat{P}$ on $\mathbb{R}^\ell$. On the other hand, $P_n(A)$ converges to some $P_\infty(A)$ uniformly $\forall A \in \mathcal{A}$. It follows that $P_\infty(A) = \hat{P}(A)$ for all $A = (-\infty, \mathbf{x}] \cap S (\forall \mathbf{x} \in \mathbb{R}^\ell)$. To show that $\hat{P}(S) = 1$, recall that since S is closed and $P_n(S) = 1 \;\; \forall n$, by Alexandroff's theorem (Chapter 2, Complements and Details, Theorem C11.1), one has

$$\hat{P}(S) \geq \overline{\lim_{n \to \infty}} \hat{P}_n(S) = \overline{\lim_{n \to \infty}} P_n(S) = 1.$$

Let P be the restriction of $\hat{P}$ to S. Then $P \in \mathcal{P}(S)$, and we will show that P_n converges weakly to P on S. To see this let O be an arbitrary open subset of S. Then $O = U \cap S$, where U is an open subset of $\mathbb{R}^\ell$, and

$$\underline{\lim} P_n(O) = \underline{\lim} \hat{P}_n(U) \geq \hat{P}(U) = P(O),$$

showing, again by Alexandroff's theorem, that P_n converges weakly to P. Now $P_n \circ \phi^{-1}$ converges weakly to $P \circ \phi^{-1}$ for every continuous nondecreasing ϕ on S. This implies $(P_n \circ \phi^{-1})^\wedge$ converges weakly to $(P \circ \phi^{-1})^\wedge$ (use Alexandroff's theorem with open set $U \subset \mathbb{R}^\ell$). But applying the same argument to the d.f.s of $(P_n \circ \phi^{-1})^\wedge$ as used above for the d.f.s. of $\hat{P}_n$ (letting $\mathbf{x}$ in (C5.1) range over all of $\mathbb{R}^\ell$) and, since P_n is Cauchy in $(\mathcal{P}(S), d_A)$, the d.f.s of $(P_n \circ \phi^{-1})^\wedge$ converge to that of $(P \circ \phi^{-1})^\wedge$ uniformly on $\mathbb{R}^\ell$. Thus $P(A) = P_\infty(A) \quad \forall A \in \mathcal{A}$, and $\sup_{A \in \mathcal{A}} |P_n(A) - P(A)| \to 0$. ∎

Problem (a) mentioned above is taken care of by defining a new metric on $\mathcal{P}(S)$. Let $\mathcal{G}_a$ denote the set of all real-valued nondecreasing measurable maps f on S such that $0 \leq f \leq a$ $(a > 0)$. Consider the metric

$$d_a(\mu, \nu) := \sup_{f \in \mathcal{G}_a} \left| \int f \, d\mu - \int f \, d\nu \right| \quad (\mu, \nu \in \mathcal{P}(S)). \quad (C5.4)$$

Clearly,

$$d_a(\mu, \nu) = a d_1(\mu, \nu). \quad (C5.5)$$

Also, the topology under d_1 is stronger than that under d_A,

$$d_1(\mu, \nu) \geq d_A(\mu, \nu). \quad (C5.6)$$

To see this, note that the indicator 1_{A^c} of the complement of every set $A \in \mathcal{A}$ belongs to $\mathcal{G}_1$.

We are now ready to prove our main theorem of this section.

Theorem C5.1 *Let S be a closed subset of $\mathbb{R}^\ell$, and $\alpha_n (n \geq 1)$ an i.i.d. sequence of nondecreasing random maps on S. If $(\mathbf{H}_2)$ holds, then*

$$d_1(T^{*n}\mu, T^{*n}\nu) \leq (1 - \chi^0)^{[n/N]} d_1(\mu, \nu) \quad (\mu, \nu) \in \mathcal{P}(S), \quad (C5.7)$$

where $\chi^0 := \min\{\chi_1^0, \chi_2^0\}$.

Also, there exists a unique invariant probability π on S for the Markov process $X_n := \alpha_n \alpha_{n-1} \cdots \alpha_1 X_0$, and

$$d_1(T^{*n}\mu, \pi) \leq (1 - \chi^0)^{[n/N]}, \quad (\mu \in \mathcal{P}(S)). \quad (C5.8)$$

Proof. Let $\Gamma_1 := \{\gamma \in \Gamma^N : \tilde{\gamma}(S) \subset (-\infty, x_0] \cap S\}$,
$\Gamma_2 := \{\gamma \in \Gamma^N : \tilde{\gamma}(S) \subset [x_0, \infty) \cap S\}$,
where $\tilde{\gamma} = \gamma_N \cdots \gamma_1$ for $\gamma = (\gamma_1, \gamma_2, \cdots, \gamma_N) \in \Gamma^N$. Let $f \in \mathcal{G}_1$. Write

$$\int_S f \, d_1(T^{*N}\mu) - \int_S f \, d_1(T^{*N}\nu)$$

$$= \int_S \left\{ \int_{\Gamma^N} f(\tilde{\gamma}\mathbf{x}) Q^N(d\gamma) \right\} \mu(d\mathbf{x}) - \int_S \left\{ \int_{\Gamma^N} f(\tilde{\gamma}\mathbf{x}) Q^N(d\gamma) \right\} \nu(d\mathbf{x})$$

$$= \sum_{i=1}^{4} \left\{ \int_S h_i(\mathbf{x})\mu(d\mathbf{x}) - \int_S h_i(\mathbf{x})\nu(d\mathbf{x}) \right\}, \quad (C5.9)$$

where

$$h_1(\mathbf{x}) := \int_{F_1 \setminus (F_1 \cap F_2)} f(\tilde{\gamma}\mathbf{x}) Q^N(d\gamma),$$

$$h_2(\mathbf{x}) := \int_{F_2 \setminus (F_1 \cap F_2)} f(\tilde{\gamma}\mathbf{x}) Q^N(d\gamma),$$

$$h_3(\mathbf{x}) := \int_{\Gamma^N \setminus (F_1 \cup F_2)} f(\tilde{\gamma}\mathbf{x}) Q^N(d\gamma),$$

$$h_4(\mathbf{x}) := \int_{(F_1 \cap F_2)} f(\tilde{\gamma}\mathbf{x}) Q^N(d\gamma). \tag{C5.10}$$

Note that, on $F_1 \cap F_2$, $\tilde{\gamma}(\mathbf{x}) = \mathbf{x}_0$, so that the integrals of h_4 in (C5.9) cancel each other. Also,

$$\int_S h_2(\mathbf{x})\mu(d\mathbf{x}) - \int_S h_2(\mathbf{x})v(d\mathbf{x}) = \int_S h_2'(\mathbf{x})v(d\mathbf{x}) - \int_S h_2'(\mathbf{x})\mu(d\mathbf{x}),$$

$$h_2'(\mathbf{x}) := \int_{F_2 \setminus (F_1 \cap F_2)} (1 - f(\tilde{\gamma}\mathbf{x})) Q^N(d\gamma).$$

Now $h_1(\mathbf{x})$ and $h_3(\mathbf{x})$ are nondecreasing, as is $a_2 - h_2'(\mathbf{x})$, where

$$0 \le h_1(\mathbf{x}) \le f(\mathbf{x}_0)(Q^N(F_1) - Q^N(F_1 \cap F_2)) = a_1,$$
$$0 \le h_2'(\mathbf{x}) \le (1 - f(\mathbf{x}_0))(Q^N(F_2) - Q^N(F_1 \cap F_2)) = a_2,$$
$$0 \le h_3(\mathbf{x}) \le 1 - Q^N(F_1) - Q^N(F_2) + Q^N(F_1 \cap F_2) = a_3.$$

Hence, $h_1 \in \mathcal{G}_{a_1}, a_2 - h_2' \in \mathcal{G}_{a_2}, h_3 \in \mathcal{G}_{a_3}$, and, since

$$\left| \int h_i \, d\mu - \int h_i \, dv \right| \le d_{a_i}(\mu, v) = a_i d_1(\mu, v)(i = 1, 3),$$

and

$$\left| \int (a_2 - h_2') \, d\mu - \int (a_2 - h_2') \, dv \right| \le d_{a_2}(\mu, v) = a_2 d_1(\mu, v)$$

it follows that

$$\left| \int_S f \, d(T^{*N}\mu) - \int_S f \, d(T^{*N}v) \right|$$

$$= \left| \int_S h_1 \, d\mu - \int_S h_1 \, dv + \int_S (a_2 - h_2') \, d\mu - \int_S (a_2 - h_2') \, dv \right.$$

$$\left. + \int_3 h_3 \, d\mu - \int_S h_3 \, dv \right| \le (a_1 + a_2 + a_3) d_1(\mu, v), \tag{C5.11}$$

where

$$a_1 + a_2 = f(\mathbf{x}_0)(Q^N(F_1) - Q^N(F_1 \cap F_2))$$
$$+ (1 - f(\mathbf{x}_0)(Q^N(F_2) - Q^N(F_1 \cap F_2))$$
$$\leq \max\{Q^N(F_1), Q^N(F_2)\} - Q^N(F_1 \cap F_2),$$

and $a_3 = 1 - Q^N(F_1) - Q^N(F_2) + Q^N(F_1 \cap F_2)$. It follows that $a_1 + a_2 + a_3 \leq 1 - \chi_2^0$ if $\chi_1 = Q^N(F_1) \geq Q^N(F_2) = \chi_2^0$, and $a_1 + a_2 + a_3 \leq 1 - \chi_1^0$ if $\chi_1^0 \leq \chi_2^0$. Thus $a_1 + a_2 + a_3 \leq 1 - \chi^0$. Hence we have derived

$$d_1(T^{*N}\mu, T^{*N}\nu) \leq (1 - \chi^0) d_1(\mu, \nu), \quad (\mu, \nu \in \mathcal{P}(S)). \quad \text{(C5.12)}$$

Also, $\forall f \in \mathcal{G}_1$,

$$\left| \int_S f d(T^*\mu) - \int_S f d(T^*\nu) \right| = \left| \int_S g(\mathbf{x})\mu(d\mathbf{x}) - \int_S g(\mathbf{x})\nu(d\mathbf{x}) \right|,$$

where $g(\mathbf{x}) := \int_\Gamma f(\tilde{\gamma}\mathbf{x}) Q(d\gamma)$, so that $g \in \mathcal{G}_1$. Thus

$$d_1(T^*\mu, T^*\nu) \leq d_1(\mu, \nu), \quad (\mu, \nu \in \mathcal{P}(S)). \quad \text{(C5.13)}$$

Combining (C5.12) and (C5.13) one arrives at (C5.7). In particular, this implies

$$d_A(T^{*n}\mu, T^{*n}\nu) \leq (1 - \chi^0)^{[n/N]}, \quad (\mu, \nu \in \mathcal{P}(S)). \quad \text{(C5.14)}$$

Fix $\mathbf{x} \in S$, and a positive integer r arbitrarily, and let $\mu = \delta_\mathbf{x}$, $\nu = T^{*r}\delta_\mathbf{x}$ in (C5.14), where $\delta_\mathbf{x}$ is the point mass at $\mathbf{x}$. Then $T^{*n}\mu = p^{(n)}(\mathbf{x}, d\mathbf{y})$, $T^{*n}\nu = p^{(n+r)}(\mathbf{x}, d\mathbf{y})$, and (C5.14) yields

$$d_A(p^{(n)}(\mathbf{x}, d\mathbf{y}), p^{(n+r)}(\mathbf{x}, d\mathbf{y})) \leq (1 - \chi^0)^{[n/N]} \to 0 \quad \text{as } n \to \infty. \quad \text{(C5.15)}$$

That is, $p^{(n)}(\mathbf{x}, d\mathbf{y})$ $(n \geq 1)$ is a Cauchy sequence in $(\mathcal{P}(S), d_A)$. By completeness (see Lemma C5.1), there exists $\pi \in \mathcal{P}(S)$ such that $d_A(p^{(n)}(\mathbf{x}, d\mathbf{y}), \pi(d\mathbf{y})) \to 0$ as $n \to \infty$. Letting $r \to \infty$ in (C5.15) one gets

$$d_A(p^{(n)}(\mathbf{x}, d\mathbf{y}), \pi(d\mathbf{y})) \leq (1 - \chi^0)^{[n/N]}. \quad \text{(C5.16)}$$

Since this is true $\forall \mathbf{x}$, one may average the left-hand side with respect to $\mu(d\mathbf{x})$ to get

$$d_A(T^{*n}\mu, \pi) \leq (1 - \chi1^0)^{[n/N]}. \quad \text{(C5.17)}$$

It remains to show that π is the unique invariant probability for $p(\mathbf{x}, d\mathbf{y})$. This follows from the Lemma C5.2. ∎

Lemma C5.2 *Let S be a Borel subset of $\mathbb{R}^\ell$, and $\alpha_n(n \geq 1)$ i.i.d. monotone nondecreasing maps on the state space S. If $p^{(n)}(\mathbf{x}, d\mathbf{y})$ converges weakly to π as $n \to \infty, \forall x \in S$, then π is the unique invariant probability for $p(\mathbf{x}, d\mathbf{y})$.*

Proof. Suppose first that there exist $\mathbf{a} = (a_1, a_2, \ldots, a_k)$ and $\mathbf{b} = (b_1, b_2, \ldots, b_k)$ in S such that $\mathbf{a} \leq \mathbf{x} \leq \mathbf{b} \ \forall \mathbf{x} \in S$. Let $Y_n(\mathbf{x}) := \alpha_1 \alpha_2 \cdots \alpha_n \mathbf{x}$. Then, as shown in Remark 5.1, $Y_n(\mathbf{a}) \uparrow \underline{Y}$ as $n \uparrow \infty$, and $Y_n(\mathbf{b}) \downarrow \overline{Y}$ as $n \uparrow \infty$. Therefore, $p^{(n)}(\mathbf{a}, d\mathbf{y})$ converges weakly to the distribution of $\underline{Y}$, which must be π. Similarly, $p^{(n)}(\mathbf{b}, d\mathbf{y})$ converges weakly to the distribution of $\overline{Y}$, also π. Since $\underline{Y} \leq \overline{Y}$, it now follows that $\underline{Y} = \overline{Y}$. Let α be independent of $\{\alpha_n: n \geq 1\}$, and have the same distribution Q. Then, $\alpha Y_n(\mathbf{z})$ has the distribution $p^{(n+1)}(\mathbf{z}, d\mathbf{y})$. Hence, $\forall \mathbf{x} \in \mathbb{R}^\ell$,

$$p^{(n+1)}(\mathbf{b}, S \cap [\mathbf{a}, \mathbf{x}]) = \text{Prob}(\alpha Y_n(\mathbf{b}) \leq \mathbf{x}) \leq \text{Prob}(\alpha \overline{Y} \leq \mathbf{x})$$
$$= \int p(z, S \cap [\mathbf{a}, \mathbf{x}])\pi(dz) = \text{Prob}(\alpha \underline{Y} \leq \mathbf{x})$$
$$\leq \text{Prob}(\alpha Y_n(\mathbf{a}) \leq \mathbf{x})$$
$$= p^{(n+1)}\mathbf{a}, S \cap [\mathbf{a}, \mathbf{x}]). \tag{C5.18}$$

By hypothesis, the two extreme sides of (C5.18) both converge to $\pi(S \cap [\mathbf{a}, \mathbf{x}])$ at all points $\mathbf{x}$ of continuity of the d.f. of π. Therefore,

$$\int p(\mathbf{z}, S \cap [\mathbf{a}, \mathbf{x}])\pi(d\mathbf{z}) = \pi(S \cap [\mathbf{a}, \mathbf{x}]) \tag{C5.19}$$

at all such $\mathbf{x}$. From the right-continuity of the d.f. of π (C5.19) holds for all $\mathbf{x}$.

This implies that π is invariant. For, the two sides of (C5.19) are the d.f.'s of $\alpha_1 X_0$ and X_0, respectively, where X_0 has distribution π, and invariance of π just means $\alpha_1 X_0$ and X_0 have the same distribution. If π' is another invariant probability, then $\forall n$

$$\int p^{(n)}(\mathbf{z}, S \cap [\mathbf{a}, \mathbf{x}])\pi'(d\mathbf{z}) = \pi'(S \cap [\mathbf{a}, \mathbf{x}]) \quad \forall \mathbf{x}. \tag{C5.20}$$

But the integrand on the left converges to $\pi(S \cap [\mathbf{a}, \mathbf{x}])$ as $n \to \infty$ by (C5.18). Therefore, the left-hand side converges to $\pi(S \cap [\mathbf{a}, \mathbf{x}])$, showing that $\pi' = \pi$.

Finally, suppose there do not exist $\mathbf{a}$ and/or $\mathbf{b} \in S$, which bound S from below and/or above. In this case, one can find an increasing home-omorphism of S onto a bounded set S_1. It is then enough to consider S_1.

Let $a_i := \inf\{x_i : \mathbf{x} = (x_1, \ldots, x_k) \in S_1\}$, $1 \le i \le k$, and $b_i := \sup\{y_i : \mathbf{y} = y_1, \ldots, y_k) \in S_1\}$. Let $\mathbf{a} = (a_1, a_2, \ldots, a_k)$, $\mathbf{b} = (b_1, b_2, \ldots, b_k)$. For simplicity, assume $N = 1$. Let $\Gamma_1 := \{\gamma \in \Gamma : \gamma \mathbf{x} \le \mathbf{x}_0 \ \forall \mathbf{x} \in S_1\}$, $\Gamma_2 := \{\gamma \in \Gamma : \gamma \mathbf{x} \ge \mathbf{x}_0 \ \forall \mathbf{x} \in S_1\}$. Extend $\gamma \in \Gamma_1 \backslash (\Gamma_1 \cap \Gamma_2)$ by setting $\gamma(\mathbf{a}) = \mathbf{a}$, $\gamma(\mathbf{b}) = \mathbf{x}_0$. Extend $\gamma \in \Gamma_1 \cap \Gamma_2$ to $S_1 \cup \{\mathbf{a}, \mathbf{b}\}$ by setting $\gamma(\mathbf{a}) = \mathbf{x}_0 = \gamma(\mathbf{b})$. Similarly extend $\gamma \in \Gamma_2 \backslash (\Gamma_1 \cap \Gamma_2)$ by setting $\gamma(\mathbf{a}) = \mathbf{x}_0$, $\gamma(\mathbf{b}) = \mathbf{b}$. For $\gamma \notin \Gamma_1 \cup \Gamma_2$, let $\gamma(\mathbf{a}) = \mathbf{a}$, $\gamma(\mathbf{b}) = \mathbf{b}$. On the space $S_1 \cup \{\mathbf{a}, \mathbf{b}\}$ the splitting condition $(\mathbf{H}_2)$ holds, and the proof given above shows that π is the unique invariant probability on $S_1 \cup \{\mathbf{a}, \mathbf{b}\}$. Since $\pi(S_1) = 1$, the proof is now complete. If one of $\mathbf{a}, \mathbf{b}$ belongs to S_1, while the other does not, simply omit the specification at the point that belongs to S_1.   ∎

Remark C5.1 The assumption of continuity of i.i.d. monotone maps in Corollary 5.2 is used only to verify the second hypothesis (5.12) of the *splitting hypothesis* $\mathbf{H}_1$. In the absence of continuity of $\tilde{\gamma}$ (for $\gamma \in \Gamma^N$), $\tilde{\gamma}^{-1} A = \{\mathbf{z} \in S : \tilde{\gamma}.\phi(\mathbf{z}) \le \mathbf{x}\}$ may not belong to $\mathcal{A}$ for A in $\mathcal{A}$ of the form $A = \{\mathbf{y} \in S : \phi(\mathbf{y}) \le \mathbf{x}\}$, where ϕ is continuous and monotone. Hence we are unable to infer (5.12) without the continuity of $\tilde{\gamma}$ for $\gamma \in \Gamma^N$. In the one-dimensional case, a direct proof of the result is given (see the proof of Corollary 5.1), which we are unable to extend to higher dimensions.

Remark C5.2 The assumption (in Corollary 5.2 and Theorem C5.1) that S be a closed subset of $\mathbb{R}^\ell$ may be relaxed somewhat. For example, if there exists a homeomorphism $h : S \to S'$ where h is a strictly increasing map and S' is a closed subset of $\mathbb{R}^k$, then by relabeling S by S', Corollary 5.2 and Theorem C5.1 hold for the state space S. This holds, for example, to sets such as $S = (0, 1)^\ell$, $[0, 1)^\ell$, $(0, 1]^\ell$, or $(0, \infty)^\ell$, which are mapped (by appropriate continuous strictly increasing h) homeomorphically to $S' = \mathbb{R}^\ell$, $[0, \infty)^\ell$, $(-\infty, 0]^\ell$, $\mathbb{R}^\ell$, respectively. However, the following is an example of an open subset S of $\mathbb{R}$ on which $(\mathcal{P}(S), d_A)$ is not complete. Note that in one dimension, d_A is the Kolmogorov distance k_K (see (5.3)).

Example C5.1 (*B.V. Rao*). Let $C \subset [0, 1]$ be the classical Cantor "middle-third" set, and $S = \mathbb{R} \backslash C$. Let X be a random variable having the Cantor distribution μ with support C. Let μ_n be the distribution of $X + 1/3^n$ ($n = 1, 2, \ldots$). It is simple to see that $\mu_n(\mathbb{R} \backslash C) = 1$ and the d.f. F_n of μ_n converges to the d.f. F of μ uniformly, i.e., in the distance $d_A = d_K$ (since, considered as probability measures on R, μ_n converges weakly to μ, and the distribution function F of μ is continuous). On the other hand $\mu(S) = 0$, so that $\mu \notin \mathcal{P}(S)$.

Remark C5.3 The metric d_1 (see (C5.5), (C5.6)) was introduced in Bhattacharya and Lee (1988). If $(P(S), d_1)$ is complete, then one can dispense with Lemmas C5.1, C5.2 for the proof of Theorem C5.1, and (H$_2$) for i.i.d. monotone increasing maps suffices. But the completeness of $(\mathcal{P}(S), d_1)$ is in general a delicate matter as shown in Chakraborty and Rao (1998), where a comprehensive analysis is given for dimensions 1 and 2.

Section 3.6. The transition probability in Example 6.3 does not satisfy the stochastic monotonicity condition assumed in Hopenhayn and Prescott (1992). To see this consider the increasing function $h(x) = x$ on $[a, b]$, and note that

$$Th(x) \equiv \int h(y)p(x, dy) = ph(F(x)) + (1 - p)h(G(x)) \qquad \text{(C6.1)}$$

$$= pF(x) + (1 - p)G(x) = p(3/4)x^{1/2} + (1 - p)\theta x(1 - x).$$

Take $p = 1/2$. Then, writing the right-hand side of (C6.1) as $\phi(x)$, one has

$$\phi(x) = 3/8x^{1/2} + 1.55(x - x^2),$$
$$\phi'(x) = 3/16x^{-1/2} + 1.55(1 - 2x).$$

It is simple to check that $\phi'(r) \equiv \phi'(9/16) > 0$, and $\phi'(0.7) < 0$. Thus ϕ is not monotone. Hence the results of Hopenhayen and Prescott (1992) do not imply Theorem 5.1 or its special cases considered in Subsection 3.6.3.

Subsection 3.6.4. Theorem 6.2 follows from Kifer (1986, Theorem 1.1). An analogous result is contained in Stokey and Lucas (1989, p. 384), but, unlike the latter, we do not require that X_n has a *unique* invariant probability.

Section 3.7. Theorem 7.2 is in Silverstrov and Stenflo (1998). See also Stenflo (2001) and Elton (1990). For additional results on iterations of i.i.d. contractions, including measurability questions, rates of convergence to equilibrium, and a number of interesting applications, we refer to Diaconis and Freedman (1999). Random iterates of linear and distance-diminishing maps have been studied extensively in psychology: the monograph by Norman (1972) is an introduction to the literature with a comprehensive list of references. In the context of economics, the literature on learning has more recently been studied and surveyed by Evans and Honkapohja (1995, 2001). The "reduced form" model of Honkapohja and Mitra (2003) is a random dynamical system with a state space $S \subset \mathbb{R}^\ell$. They use Theorem 7.3 to prove the existence and stability of an invariant distribution.

3.9 Supplementary Exercises

(1) On the state space $S = \{1, 2\}$ consider the maps $\gamma_1, \gamma_2, \gamma_3$ on S defined as $\gamma_1(1) = \gamma_1(2) = 1$, $\gamma_2(1) = \gamma_2(2) = 2$, $\gamma_3(1) = 2$, and $\gamma_3(2) = 1$. Let $\{\alpha_n: n \geq 1\}$ be i.i.d. random maps on S such that $P(\alpha_n = \gamma_j) = \theta_j (j = 1, 2, 3)$, $\theta_1 + \theta_2 + \theta_3 = 1$.

(a) Write down the transition probabilities of the Markov process $\{X_n: n \geq 0\}$ generated by iterations of $\alpha_n (n \geq 1)$ (for an arbitrary initial state X_0 independent of $\{\alpha_n: n \geq 1\}$). [Hint: $p_{12} = \theta_2 + \theta_3$.]

(b) Given arbitrary transition probabilities $p_{ij} (i, j = 1, 2)$ satisfying $p_{11} + p_{22} \leq 1$ (i.e., $p_{11} \leq p_{21}$), find $\theta_1, \theta_2, \theta_3$ to construct the corresponding Markov process. [Hint: $p_{11} = \theta_1$, $p_{22} = \theta_2$.]

(c) In addition to the maps above, let γ_4 be defined by $\gamma_4(1) = 1$, $\gamma_4(2) = 2$. Let $\{\alpha_n: n \geq 1\}$ be i.i.d. with $P(\alpha_n = \gamma_j) = \theta_j (j = 1, 2, 3, 4)$. Given arbitrary transition probabilities $p_{ij} (i, j = 1, 2)$, find $\theta_j (j = 1, 2, 3, 4)$ to construct the corresponding Markov process (by iteration of $\alpha_n (n \geq 1)$). [Hint: $p_{11} = \theta_1 + \theta_4$, $p_{21} = \theta_1 + \theta_3$.]

(2) Consider the model in Example 4.2 on the state space $S = [-2, 2]$ with $\{\varepsilon_n: n \geq 1\}$ i.i.d. uniform on $[-1, 1]$.

(a) Write down the transition probability density $p(x, y)$, i.e., the probability density of $X_1(x) = f(x) + \varepsilon_1$. [Hint: Uniform distribution on $[f(x) - 1, f(x) + 1]$.]

(b) Compute the distribution of $X_2(x) = f(X_1(x)) + \varepsilon_2$. [Hint: Suppose $x \in [-2, 0]$. Then $X_1(x)$ is uniform on $[x, x + 2]$, and $f(X_1(x))$

is uniform on $[-1, 1]$. For f maps $[x, 0]$ onto $[x + 1, 1]$, and it maps $(0, x + 2]$ onto $(-1, x + 1]$, the maps being both linear and of slope 1. Finally, $f(X_1(x)) + \varepsilon_2$ is the sum of two independent uniform random variables on $[-1, 1]$ and, therefore, has the triangular distribution on $[-2, 2]$ with density $(2 - |y|)/4$, $-2 \le y \le 2$ (by direct integration).]

4

Random Dynamical Systems: Special Structures

But a Grand Unified Theory will remain out of the reach of economics, which will keep appealing to a large collection of individual theories. Each one of them deals with a certain range of phenomena that it attempts to understand and to explain.

Gerard Debreu

The general equilibrium models used by macroeconomists are special.

Thomas J. Sargent

4.1 Introduction

In this chapter, we explore in depth the qualitative behavior of several classes of random dynamical systems that have been of long-standing interest in many disciplines. Here the special structures are exploited to derive striking results on the existence and stability of invariant distributions. A key role is played by the central results on splitting and contractions, reviewed in Chapter 3. In Sections 4.2 and 4.3 our focus is on linear time series, especially autoregressive time series models. Next, we consider in Section 4.4 random iterates of quadratic maps. We move on to nonlinear autoregressive (NLAR(k)) and nonlinear autoregressive conditional heteroscedastic (NLARCH(k)) models, which are of considerable significance in modern econometric theory (Section 4.5). Section 4.6 is devoted to random dynamical systems on the state space $\mathbb{R}_{++}$, called random continued fractions, defined by

$$X_{n+1} = Z_{n+1} + \frac{1}{X_n} \quad (n \geq 0). \tag{1.1}$$

Two short final sections concern stability in distribution of processes under additive shocks that satisfy nonnegativity constraints (Section 4.7):

$$X_{n+1} = \max(X_n + \varepsilon_{n+1}, 0) \tag{1.2}$$

and the survival probability of an economic agent in a model with multiplicative shocks (Section 4.8).

4.2 Iterates of Real-Valued Affine Maps (AR(1) Models)

We turn to a random dynamical system where $S = \mathbb{R}$ and $\{\varepsilon_n\}_{n=1}^{\infty}$ is a sequence of i.i.d. random variables (independent of a given initial random variable X_0), and the evolution of the system is described by

$$X_{n+1} = bX_n + \varepsilon_{n+1} \quad (n \geq 0), \tag{2.1}$$

where b is a constant. This is the basic linear autoregressive model of order 1, or the AR(1) model extensively used in time series analysis. It is useful to recall Example 4.2 of Chapter 1, which suggests (2.1) rather naturally. Also, with $b = 1$, the main result applies to models of random walk.

The sequence X_n given by (2.1) is a Markov process with initial distribution given by the distribution of X_0 and the transition probability (of going from x to some Borel set C of $\mathbb{R}$ in one step)

$$p(x, C) = Q(C - bx),$$

where Q is the common distribution of ε_n. To be sure, in terms of our standard notation, the i.i.d. random maps $\alpha_n(n \geq 1)$ are given by

$$\alpha_n(x) = bx + \varepsilon_n. \tag{2.2}$$

From (2.1) we get

$$X_n = b^n X_0 + b^{n-1}\varepsilon_1 + \cdots + b\varepsilon_{n-1} + \varepsilon_n \quad (n \geq 1). \tag{2.3}$$

Then, for each n, X_n has the same distribution as Y_n where

$$Y_n = b^n X_0 + \varepsilon_1 + b\varepsilon_2 + \cdots + b^{n-1}\varepsilon_n. \tag{2.4}$$

Note that the *distributions of the two processes $\{X_n: n \geq 0\}$ and $\{Y_n: n \geq 0\}$ are generally quite different* (Exercise 2.1).

Assume that

$$|b| < 1 \tag{2.5}$$

and $|\varepsilon_n| \le \eta$ with probability 1 for some constant η. Then it follows from (2.4) that

$$Y_n \to \sum_{n=0}^{\infty} b^n \varepsilon_{n+1} \quad \text{a.s.,} \tag{2.6}$$

regardless of X_0. Let π be the distribution of the random variable on the right-hand side of (2.6). Then Y_n converges in distribution to π as $n \to \infty$. As $X_n \overset{\mathcal{L}}{=} Y_n$, X_n converges in distribution to π. Hence π is the unique invariant distribution for the Markov process $\{X_n\}$.

The assumption that the random variable ε_1 is bounded can be relaxed; indeed, it suffices to assume that

$$\sum_{n=1}^{\infty} P(|\varepsilon_1| > m\delta^n) < \infty \quad \text{for some } \delta < \frac{1}{|b|} \text{ and for some } m > 0. \tag{2.7}$$

Since the sum on the left equals $\sum_{n=1}^{\infty} P(|\varepsilon_{n+1}| > m\delta^n)$, one has, by the (*first*) Borel–Cantelli lemma,

$$P(|\varepsilon_{n+1}| \le m\delta^n \text{ for all but finitely many } n) = 1.$$

This implies that, with probability 1, $|b^n \varepsilon_{n+1}| \le m(|b|\delta)^n$ for all but finitely many n. Since $|b|\delta < 1$, the series on the right-hand side of (2.6) converges and is the limit of Y_n.

Exercise 2.1 Show that, although $X_n \overset{\mathcal{L}}{=} Y_n$ separately for each n (see (2.3), (2.4)), the distributions of the processes $\{X_n: n \ge 0\}$ and $\{Y_n: n \ge 0\}$ are, in general, not the same. [Hint: Let $X_0 \equiv 0$. Compare the distributions of $(X_1, X_2) \equiv (\varepsilon_1, b\varepsilon_1 + \varepsilon_2)$ and $(Y_1, Y_2) \equiv (\varepsilon_1, \varepsilon_1 + b\varepsilon_2)$, noting that $X_2 - X_1 \equiv (b-1)\varepsilon_1 + \varepsilon_2$ and $Y_2 - Y_1 \equiv b\varepsilon_2$ have different distributions, unless $b = 1$ or $\varepsilon_n = 0$ a.s. For example, their variances are different (if finite).] ∎

For the sake of completeness, we state a more definitive characterization of the process (2.1). Recall that $x^+ = \max\{x, 0\}$, and we write $\log^+ x = \max\{\log x, 0\} \equiv (\log x)^+ \ (x \ge 0)$.

Theorem 2.1

(a) *Assume that ε_1 is nondegenerate and*

$$E(\log^+ |\varepsilon_1|) < \infty. \tag{2.8}$$

Then the condition

$$|b| < 1 \tag{2.9}$$

is necessary and sufficient for the existence of a unique invariant proba-
bility π for $\{X_n\}_{n=0}^{\infty}$. In addition, if (2.8) and (2.9) hold, then X_n converges
in distribution to π as $n \to \infty$, no matter what X_0 is.

(b) If $b \neq 0$ and

$$E(\log^+ |\varepsilon_1|) = \infty, \tag{2.10}$$

then $\{X_n\}_{n=0}^{\infty}$ does not converge in distribution and does not have any
invariant probability.

Proof. One may express X_n as

$$X_n = b^n X_0 + b^{n-1}\varepsilon_1 + b^{n-2}\varepsilon_2 + \cdots + b\varepsilon_{n-1} + \varepsilon_n$$
$$= b^n X_0 + \sum_{j=1}^{n} b^{n-j}\varepsilon_j. \tag{2.11}$$

Now, recalling the notation $\stackrel{\mathcal{L}}{=}$ for equality in distribution, note that for
each $n \geq 1$,

$$\sum_{j=0}^{n-1} b^j \varepsilon_{j+1} \stackrel{\mathcal{L}}{=} \sum_{j=1}^{n} b^{n-j}\varepsilon_j \quad (n \geq 1). \tag{2.12}$$

(a) To prove *sufficiency* in part (a), assume (2.8) and (2.9) hold. Then
$b^n X_0 \to 0$ as $n \to \infty$. Therefore, we need to prove that

$$\sum_{j=0}^{n-1} b^j \varepsilon_{j+1} \text{ converges in distribution as } n \to \infty. \tag{2.13}$$

Now choose and fix δ such that

$$1 < \delta < \frac{1}{|b|}. \tag{2.14}$$

Then $\delta|b| < 1$, and writing $Z = (\log|\varepsilon_1|)^+ / \log\delta$, one has (see
Lemma 2.1 for the last step)

$$\sum_{j=0}^{\infty} P(|b^j \varepsilon_{j+1}| \geq (\delta|b|)^j) = \sum_{j=0}^{\infty} P(|\varepsilon_{j+1}| \geq \delta^j)$$

$$= \sum_{j=0}^{\infty} P(|\varepsilon_1| \geq \delta^j) = \sum_{j=0}^{\infty} P(\log|\varepsilon_1| \geq j \log\delta)$$

$$= \sum_{j=1}^{\infty} P((\log^+ |\varepsilon_1|) \geq j \log \delta) + P(\log |\varepsilon_1| \geq 0)$$

$$= \sum_{j=1}^{\infty} P(|Z| \geq j) + P(\log |\varepsilon_1| \geq 0)$$

$$= E[Z] + P(\log |\varepsilon_1| \geq 0), \tag{2.15}$$

where $[Z]$ is the integer part of Z. By (2.8), $E[Z] < \infty$. Hence, by the (*first*) Borel–Cantelli lemma,

$$P(|b^j \varepsilon_{j+1}| < (\delta|b|)^j \text{ for all but finitely many } j) = 1. \tag{2.16}$$

Since $\delta|b| < 1$, this implies the existence of a (finite) random variable X such that

$$\sum_{j=0}^{n-1} b^j \varepsilon_{j+1} \to X \text{ a.s.} \quad \text{as } n \to \infty. \tag{2.17}$$

Therefore, $\sum_{j=0}^{n-1} b^j \varepsilon_{j+1}$ converges in distribution to X. Hence, by (2.12) (and the fact $b^n X_0 \to 0$), X_n converges in distribution to X. That is, if π denotes the distribution of the a.s. limit in (2.17), the distribution of X_n converges weakly to π, irrespective of X_0. Since the transition probability $p(x, dy)$ of the Markov process (which is the distribution of $bx + \varepsilon_1$) is a weakly continuous function of x (on $\mathbb{R}$ into $\mathcal{P}(\mathbb{R})$), it follows that π is invariant (see Theorem 11.1, Chapter 2), and it is the unique invariant probability.

To prove the *necessity* of (2.9), first, let $|b| > 1$. Choose $\hat{\theta}$ such that $1 < \hat{\theta} < |b|$. Then

$$\sum_{j=0}^{\infty} P(|b^j \varepsilon_{j+1}| \geq \hat{\theta}^j) = \sum_{j=0}^{\infty} P\left(|\varepsilon_{j+1}| \geq \left(\frac{\hat{\theta}}{|b|}\right)^j \right)$$

$$= \sum_{j=0}^{\infty} P\left(|\varepsilon_1| \geq \left(\frac{\hat{\theta}}{|b|}\right)^j \right) = \infty, \tag{2.18}$$

since $P(|\varepsilon_1| \geq (\frac{\hat{\theta}}{|b|})^j) \uparrow P(|\varepsilon_1| > 0) > 0$ as $j \uparrow \infty$, by the assumption of nondegeneracy of ε_1. But (2.18) implies, by the (*second*) Borel–Cantelli Lemma, that

$$P(|b^j \varepsilon_{j+1}| \geq \hat{\theta}^j \text{ for infinitely many } j) = 1, \tag{2.19}$$

so that $\sum_{j=0}^{n-1} b^j \varepsilon_{j+1}$ diverges a.s. as $n \to \infty$. By a standard result on the series of independent random variables (see Loève 1963, p. 251), it follows that $\sum_{j=0}^{n-1} b^j \varepsilon_{j+1}$ does not converge in distribution. By (2.12), therefore, $\sum_{j=1}^{n} b^{n-j} \varepsilon_j$ does not converge in distribution. If $X_0 = 0$, we see from (2.11) that $X_n \equiv \sum_{j=1}^{n} b^{n-j} \varepsilon_j$ does not converge in distribution.

Now, if $X_0 \neq 0$ then, conditionally given X_0, the same argument applies to the process $Z_n := X_n - X_0$, since $Z_{n+1} = b Z_n + \eta_{n+1}$, with $\{\eta_n : n \geq 1\}$ i.i.d., $\eta_n := \varepsilon_n + (b-1)X_0$. Since $Z_0 = 0$, we see that Z_n does not converge in distribution (see (2.11)). Therefore, $X_n \equiv Z_n + X_0$ does not converge in distribution.

To complete the proof of necessity in part (a), consider the case $|b| = 1$. Let $c > 0$ be such that $P(|\varepsilon_1| > c) > 0$. Then

$$\sum_{j=0}^{\infty} P(|b^j \varepsilon_{j+1}| > c) = \sum_{j=0}^{\infty} P(|\varepsilon_{j+1}| > c)$$

$$= \sum_{j=0}^{\infty} P(|\varepsilon_1| > c) = \infty. \qquad (2.20)$$

By the (*second*) Borel–Cantelli lemma,

$$P\left(\sum_{j=0}^{n-1} b^j \varepsilon_{j+1} \text{ converges a.s.}\right) = 0.$$

Therefore, $\sum_{j=0}^{n-1} b^j \varepsilon_{j+1}$ does not converge in distribution (Loève, 1963, p. 251). Hence, $\sum_{j=1}^{n} b^{n-j} \varepsilon_j$ does not converge in distribution. Since $b^n X_0 = X_0$ in case $b = +1$, and $b^n X_0 = \pm X_0$ in case $b = -1$, $X_n \equiv b^n X_0 + \sum_{j=1}^{n-1} b^{n-j} \varepsilon_j$ does not converge in distribution.

(b) Suppose $b \neq 0$ and (2.10) holds. First, consider the case $|b| < 1$. Then $-\log |b| > 0$ and, using Lemma 2.1,

$$\sum_{j=0}^{\infty} P(|b^j \varepsilon_{j+1}| > 1) = \sum_{j=0}^{\infty} P(|\varepsilon_{j+1}| > |b|^{-j})$$

$$= \sum_{j=0}^{\infty} P(|\varepsilon_1| > |b|^{-j})$$

$$= \sum_{j=1}^{\infty} P(\log |\varepsilon_1| > -j \log |b|) + P(\log |\varepsilon_1| > 0)$$

$$= \sum_{j=1}^{\infty} P\left(\frac{\log |\varepsilon_1|}{-\log |b|} > j\right) + P(\log |\varepsilon_1| > 0)$$

$$= E\left[\frac{\log^+ |\varepsilon_1|}{-\log |b|}\right] + P(\log |\varepsilon_1| > 0)$$

$$= \infty. \tag{2.21}$$

For the last step we have used the fact that $[x] \geq x - 1$. Hence, by the second Borel–Cantelli lemma, $|b^j \varepsilon_{j+1}| > 1$ for infinitely many j outside a set of 0 probability. This implies that the series on the left-hand side in (2.12) diverges almost surely. Since $b^n X_0 \to 0$ as $n \to \infty$, it follows that $b^n X_0 + \sum_{j=0}^{n-1} b^j \varepsilon_{j+1}$ diverges almost surely and, therefore, does not converge in distribution (Loève 1963, p. 251). But the last random variable has the same distribution as X_n. Thus X_n does not converge in distribution.

Next, consider the case $|b| > 1$. Write

$$X_n = b^n \left(X_0 + \sum_{j=1}^{n} b^{-j} \varepsilon_j\right). \tag{2.22}$$

Using the preceding argument, $\sum_{j=1}^{n} b^{-j} \varepsilon_j$ diverges almost surely, since $|b^{-1}| < 1$. Since $|b^n| \to \infty$ as $n \to \infty$, it follows that X_n diverges almost surely and, therefore, does not converge in distribution.

Finally, suppose $|b| = 1$. Then

$$\sum_{j=0}^{\infty} P(|b^j \varepsilon_{j+1}| > 1) = \sum_{j=0}^{\infty} P(|\varepsilon_{j+1}| > 1)$$

$$= \sum P(|\varepsilon_1| > 1) = \infty,$$

since $P(|\varepsilon_1| > 1) > 0$ under the hypothesis (2.10). Therefore, X_n does not converge in distribution. ∎

Example 2.1 An example of a distribution (of ε_1) satisfying (2.10) is given by the probability density function

$$f(x) = \begin{cases} 0 & \text{if } x < 2, \\ \frac{c}{x(\log x)^2} & \text{if } x \geq 2, \end{cases} \tag{2.23}$$

where $c = (\log 2)^{-1}$. The exponent 2 of $\log x$ in the denominator may be replaced by $1 + \delta, 0 < \delta \leq 1$ (with $c = (\log 2)^{-\delta}$); but this exponent cannot be larger than 2. ∎

Remark 2.1 The case $b = 1$ leads to the *general random walk* with arbitrary step size distribution: $X_n = X_0 + \varepsilon_1 + \cdots + \varepsilon_n$ $(n \geq 1)$, which does not converge in distribution unless $\varepsilon_n = 0$ with probability 1.

Remark 2.2 The case $b = 0$ in (2.1) yields $X_n = \varepsilon_n$ $(n \geq 1)$, which is trivially stable in distribution and has as its unique invariant probability the distribution of ε_1.

Remark 2.3 The case $|b| < 1$ is generally referred to as the *stable case* of the model (2.1). Part (b) of Theorem 2.1 shows, however, that stability fails no matter how small a nonzero $|b|$ is, if the innovations ε_n cross a threshold of *heavy-tailed* distributions.

The following lemma has been used in the proof:

Lemma 2.1 *Let Z be a nonnegative random variable. Then*

$$\sum_{j=1}^{\infty} P(Z \geq j) = E[Z], \qquad (2.24)$$

where $[Z]$ is the integer part of Z.

Proof.

$$
\begin{aligned}
E[Z] &= 0P(0 \leq Z < 1) + 1P(1 \leq Z < 2) + 2P(2 \leq Z < 3) \\
&\quad + 3P(3 \leq Z < 4) + \cdots + jP(j \leq Z < j+1) + \cdots \\
&= \{P(Z \geq 1) - P(Z \geq 2)\} + 2\{P(Z \geq 2) - P(Z \geq 3)\} \\
&\quad + 3\{P(Z \geq 3) - P(Z \geq 4)\} + \cdots + j\{P(Z \geq j) \\
&\quad - P(Z \geq j+1)\} + \cdots \\
&= P(Z \geq 1) + P(Z \geq 2) + P(Z \geq 3) \\
&\quad + \cdots + P(Z \geq j) + \cdots .
\end{aligned}
\qquad (2.25)
$$

■

Remark 2.4 The hypothesis "ε_1 is nondegenerate" in the theorem is not quite needed, except for the fact that if $P(\varepsilon_n = c) = 1$ and $b \neq 1$, then $X_n = b^n X_0 + c(b^n - 1)/(b - 1)$, and $-c/(b - 1) = z$, say, is a fixed point, i.e., δ_z is an invariant probability. But even in this last case, if $|b| > 1$ then X_n does not converge if $X_0 \neq z$. If $b = +1$ and $\varepsilon_n = 0$ a.s., then x is a fixed point for every x, so that δ_x is an invariant probability for every $x \in \mathbb{R}$. If $b = -1$ and $\varepsilon_n = c$ a.s., then $x = \frac{c}{2}$ is a fixed point and $\{x, -x + c\}$ is a periodic orbit for every $x \neq \frac{c}{2}$; it follows in this case that $\frac{1}{2}\delta_x + \frac{1}{2}\delta_{-x+c}$ are invariant probabilities for every $x \in \mathbb{R}$. These are the only possible extremal invariant probabilities for the case $|b| \geq 1$. In particular, if (2.8) holds (irrespective of whether ε_1 is a degenerate or not), then (2.9) is necessary and sufficient for stability in distribution of $\{X_n\}$.

4.3 Linear Autoregressive (LAR(*k*)) and Other Linear Time Series Models

Although the results of this section hold with respect to other norms also, for specificity, we take the *norm of a matrix A* to be defined by

$$\|A\| = \sup_{|x|=1} |Ax|, \qquad (3.1)$$

where $|x|$ denotes the Euclidean norm of the vector x.

Theorem 3.1 *Let $S = \mathbb{R}^\ell$, and $(A_i, \varepsilon_i)(i \geq 1)$ be an i.i.d. sequence with $A_i's$ random $(k \times k)$ matrices and $\varepsilon_i's$ random vectors (ℓ-dimensional). If, for some $r \geq 1$,*

$$-\infty \leq E \log \|A_1 \cdots A_r\| < 0, \quad E \log^+ |\varepsilon_i| < \infty, \qquad (3.2)$$

and

$$E \log^+ \|A_1\| < \infty, \qquad (3.3)$$

then the Markov process defined recursively by

$$X_{n+1} = A_{n+1}X_n + \varepsilon_{n+1}(n \geq 0) \qquad (3.4)$$

has a unique invariant probability π and it is stable in distribution.

Proof. To apply Theorem 7.2 of Chapter 3 with $\alpha_i(x) = A_ix + \varepsilon_i$, and Euclidean distance for d, note that

$$d(\alpha_r \cdots \alpha_1 x, \alpha_r \cdots \alpha_1 y) = d(A_r \cdots A_1 x, A_r \cdots A_1 y)$$
$$= |A_r A_{r-1} \cdots A_1(x - y)|$$
$$\leq \|A_r \cdots A_1\| \cdot |x - y|. \qquad (3.5)$$

Thus the first condition in (3.2) implies (7.2) of Chapter 3 in this case. To verify (7.9) of Chapter 3, take $x_0 = 0$. If $r = 1$ then $d(\alpha_1(0), 0) = |\varepsilon_1|$, and the second relation in (3.2) coincides with (7.9) of Chapter 3. In this case (3.3) follows from the first relation in (3.2): $E \log^+ \|A_1\|$ is finite because $E \log \|A_1\|$ exists and is negative. For $r > 1$,

$$d(\alpha_r \cdots \alpha_1 0, 0) = |A_r \cdots A_2 \varepsilon_1 + A_r \cdots A_3 \varepsilon_2 + \cdots$$
$$+ A_r \varepsilon_{r-1} + \varepsilon_r|$$
$$\leq \sum_{j=2}^{r} \|A_r\| \cdots \|A_j\| \cdot |\varepsilon_{j-1}| + |\varepsilon_r|,$$

$$\log^+ d(\alpha_r \cdots \alpha_1(0), 0) \leq \sum_{j=2}^{r} \log(\|A_r\| \cdots \|A_j\| \cdot |\varepsilon_{j-1}| + 1)$$
$$+ \log(|\varepsilon_r| + 1),$$

$$\log(\|A_r\| \cdots \|A_j\| \cdot |\varepsilon_{j-1}| + 1) \leq \log(\|A_r\| + 1) + \cdots + \log(\|A_j\| + 1)$$
$$+ \log(|\varepsilon_{j-1}| + 1). \qquad (3.6)$$

Now $E \log(\|A_1\| + 1) \leq E \log^+ \|A_1\| + 1$, and $E \log(|\varepsilon_1| + 1) \leq E(\log^+ |\varepsilon_1|) + 1)$, and both are finite by (3.2) and (3.3). Hence (3.6) implies $E \log^+ d(\alpha_r \cdots \alpha_1(0), 0) < \infty$, i.e., the hypotheses of Theorem 7.2 of Chapter 3 are satisfied. ∎

Remark 3.1 For the case $r = 1$, Theorem 3.1 is contained in a result of Brandt (1986). For $r > 1$, the above result was obtained by Berger (1992) using a different method.

As an immediate corollary of Theorem 3.1, we have the following multidimensional extension of Theorem 2.1. To state it consider the Markov process defined recursively by

$$X_{n+1} = AX_n + \varepsilon_{n+1} \ (n \geq 0), \qquad (3.7)$$

where A is a (constant) $k \times k$ matrix, $\varepsilon_n (n \geq 1)$ an i.i.d. sequence of k-dimensional random vectors independent of a given k-dimensional random vector X_0.

Corollary 3.1 *Suppose*

$$\|A^{n_0}\| < 1 \text{ for some positive integer } n_0, \qquad (3.8)$$

and

$$E \log^+ |\varepsilon_i| < \infty. \tag{3.9}$$

Then the Markov process (3.7) has a unique invariant probability π and it is stable in distribution.

For applications, it is useful to know that the assumption (3.8) holds if the maximum modulus of eigenvalues of A, also known as the *spectral radius $r(A)$ of A*, is less than 1. This fact is implied by the following result from linear algebra, whose proof is given in Complements and Details.

Lemma 3.1 *Let A be an $m \times m$ matrix. Then the spectral radius $r(A)$ satisfies*

$$r(A) \geqslant \overline{\lim_{n \to \infty}} \|A^n\|^{1/n}. \tag{3.10}$$

Two well-known time series models will now be treated as special cases of Corollary 3.1. These are the *kth-order autoregressive* (or AR(k)) *model* and the *autoregressive moving-average model* (ARMA(p, q)).

Example 3.1 (*AR(k) model*) Let $k > 1$ be an integer and $\beta_0, \beta_1, \ldots, \beta_{k-1}$ real constants. Given a sequence of i.i.d. real-valued random variables $\{\eta_n : n \geq k\}$, and k other random variables $U_0, U_1, \ldots, U_{k-1}$ independent of $\{\eta_n\}$, defined recursively

$$U_{n+k} := \sum_{i=0}^{k-1} \beta_i U_{n+i} + \eta_{n+k} \quad (n \geq 0). \tag{3.11}$$

The sequence $\{U_n\}$ is not, in general, a Markov process, but the sequence of k-dimensional random vectors

$$X_n := (U_n, U_{n+1}, \ldots, U_{n+k-1})' \quad (n \geq 0) \tag{3.12}$$

is Markovian. Here the prime ($'$) denotes transposition, so X_n is to be regarded as a column vector in matrix operations. To prove the Markov property, consider the sequence of k-dimensional i.i.d. random vectors

$$\varepsilon_n := (0, 0, \ldots, 0, \eta_{n+k-1})' \quad (n \geq 1), \tag{3.13}$$

and note that

$$X_{n+1} = AX_n + \varepsilon_{n+1}, \qquad (3.14)$$

where A is the $k \times k$ matrix

$$A := \begin{bmatrix} 0 & 1 & 0 & 0 & \cdots & 0 & 0 \\ 0 & 0 & 1 & 0 & \cdots & 0 & 0 \\ \cdot & \cdot & \cdot & \cdot & \cdots & \cdot & \cdot \\ 0 & 0 & 0 & 0 & \cdots & 0 & 1 \\ \beta_0 & \beta_1 & \beta_2 & \beta_3 & \cdots & \beta_{k-2} & \beta_{k-1} \end{bmatrix}. \qquad (3.15)$$

Hence, as in (3.4), $\{X_n\}$ is a Markov process on the state space $\mathbb{R}^k$. Write

$$A - \lambda I = \begin{bmatrix} -\lambda & 1 & 0 & 0 & \cdots & 0 & 0 \\ 0 & -\lambda & 1 & 0 & \cdots & 0 & 0 \\ \cdot & \cdot & \cdot & \cdot & \cdots & \cdot & \cdot \\ 0 & 0 & 0 & 0 & \cdots & -\lambda & 1 \\ \beta_0 & \beta_1 & \beta_2 & \beta_3 & \cdots & \beta_{k-2} & \beta_{k-1} - \lambda \end{bmatrix}.$$

Expanding $\det(A - \lambda I)$ by its last row, and using the fact that the determinant of a matrix in *triangular form* (i.e., with all off-diagonal elements on one side of the diagonal being zero) is the product of its diagonal elements (see Exercise 3.1), one gets

$$\det(A - \lambda I) = (-1)^{k+1}(\beta_0 + \beta_1\lambda + \cdots + \beta_{k-1}\lambda^{k-1} - \lambda^k). \qquad (3.16)$$

Therefore, the eigenvalues of A are the roots of the equation

$$\beta_0 + \beta_1\lambda + \cdots + \beta_{k-1}\lambda^{k-1} - \lambda^k = 0. \qquad (3.17)$$

Finally, in view of (3.10), the following proposition is a consequence of Theorem 3.1:   ∎

Proposition 3.1 *Suppose that the roots of the polynomial Equation (3.17) are all strictly inside the unit circle in the complex plane, and that the common distribution G of $\{\eta_n\}$ satisfies*

$$E \, \log^+ |\eta_n| < \infty. \qquad (3.18)$$

Then (i) there exists a unique invariant distribution π for the Markov process $\{X_n\}$ defined by (3.12) and (ii) no matter what the initial distribution, X_n converges in distribution to π.

It is simple to check that (3.18) holds if G has a finite absolute moment of some order $r > 0$ (Exercise).

An immediate consequence of Proposition 3.1 is that *the time series* $\{U_n : n \geq 0\}$ *converges in distribution to a steady state* π_U *given, for all Borel sets* $C \subset \mathbb{R}^1$, *by*

$$\pi_U(C) = \pi(\{x \in \mathbb{R}^k : x^{(1)} \in C\}). \qquad (3.19)$$

Here $x^{(1)}$ is the first coordinate of $x = (x^{(1)}, \dots, x^{(k)})$. To see this, simply note that U_n is the first coordinate of X_n, so that X_n converges to π in distribution implies U_n converges to π_U in distribution.

Example 3.2 (*ARMA(k, q) model*) The *autoregressive moving-average model of order* (k, q), in short ARMA(k, q), is defined by

$$U_{n+k} = \sum_{i=0}^{k-1} \beta_i U_{n+i} + \sum_{j=1}^{q} \delta_j \eta_{n+k-j} + \eta_{n+k} \quad (n \geq 0), \qquad (3.20)$$

where k, q are positive integers, $\beta_i (0 \leq i \leq k-1)$ and $\delta_j (1 \leq j \leq q)$ are real constants, $\{\eta_n : n \geq k - q\}$ is an i.i.d. sequence of real-valued random variables, and $U_i (0 \leq i \leq k-1)$ are arbitrary initial random variables independent of $\{\eta_n\}$. Consider the sequences $\{X_n\}$, $\{\varepsilon_n\}$ of $(k + q)$-dimensional vectors

$$X_n = (U_n, \dots, U_{n+k-1}, \eta_{n+k-q}, \dots, \eta_{n+k-1})',$$
$$\varepsilon_n = (0, 0, \dots, 0, \eta_{n+k-1}, 0, \dots, 0, \eta_{n+k-1})' \quad (n \geq 0), \qquad (3.21)$$

where η_{n+k-1} occurs as the kth and $(k + q)$th elements of ε_n. Then

$$X_{n+1} = H X_n + \varepsilon_{n+1} \quad (n \geq 0), \qquad (3.22)$$

where H is the $(k + q) \times (k + q)$ matrix

$$H := \begin{Bmatrix} a_{11} & . & \dots & a_{1k} & 0 & . & . & \dots & 0 & 0 \\ . & . & \dots & . & . & . & . & \dots & . & . \\ a_{k1} & . & \dots & a_{kk} & \delta_q & \delta_{q-1} & . & \dots & \delta_2 & \delta_1 \\ 0 & . & \dots & 0 & 0 & 1 & 0 & \dots & 0 & 0 \\ 0 & . & \dots & 0 & 0 & 0 & 1 & \dots & 0 & 0 \\ . & . & \dots & . & . & . & . & \dots & . & . \\ 0 & 0 & \dots & . & . & 0 & 0 & \dots & 0 & 1 \\ 0 & 0 & \dots & . & . & 0 & 0 & \dots & 0 & 0 \end{Bmatrix},$$

the first k rows and k columns of H being the matrix A in (3.15).

Note that $U_0, \ldots, U_{k-1}, \eta_{k-q}, \ldots, \eta_{k-1}$ determine X_0, so that X_0 is independent of η_k and, therefore, of ε_1. It follows by induction that X_n and ε_{n+1} are independent. Hence $\{X_n\}$ is a Markov process on the state space $\mathbb{R}^{k+q}$.

In order to apply the Lemma 3.1, expand $\det(H - \lambda I)$ in terms of the elements of its last row to get (Exercise)

$$\det(H - \lambda I) = \det(A - \lambda I)(-\lambda)^q. \qquad (3.23)$$

Therefore, the eigenvalues of H are q zeros and the roots of (3.17). Thus, one has the following proposition:

Proposition 3.2 *Under the hypothesis of Proposition 3.1, the ARMA(k, q) process $\{X_n\}$ has a unique invariant distribution π, and X_n converges in distribution to π, no matter what the initial distribution is.*

As a corollary, *the time series $\{U_n\}$ converges in distribution to π_U given, for all Borel sets $C \subset \mathbb{R}^1$, by*

$$\pi_U(C) := \pi(\{x \in \mathbb{R}^{k+q} : x^{(1)} \in C\}), \qquad (3.24)$$

no matter what the distribution of $(U_0, U_1, \ldots, U_{k-1})$ is, provided the hypothesis of Proposition 3.2 is satisfied.

In the case that ε_n is Gaussian, it is simple to check that the random vector X_n in (3.7) is Gaussian, if X_0 is Gaussian. Therefore, under the hypothesis (3.8), π is Gaussian, so that the stationary vector-valued process $\{X_n\}$ with initial distribution π is Gaussian (also see Exercise 3.2). In particular, if η_n are Gaussian in Example 3.1, and the roots of the polynomial equation (3.17) lie inside the unit circle in the complex plane, then the stationary process $\{U_n\}$, obtained when $(U_0, U_1, \ldots, U_{k-1})$ has distribution π in Example 3.1, is Gaussian. A similar assertion holds for Example 3.2. ∎

Exercise 3.1 Prove that the determinant of an $m \times m$ matrix in triangular form equals the product of its diagonal elements. [Hint: Let $(a_{11}, 0, 0, \ldots, 0)$ be the first row of triangular $m \times m$ matrix. Then its determinant is a_{11} times the determinant of an $(m - 1) \times (m - 1)$ triangular matrix. Use induction on m.] ∎

Exercise 3.2 Under the hypothesis of Corollary 3.1, (a) mimic the steps (2.3)–(2.6) to show that the invariant distribution π of $X_n(n \geq 0)$ is given by the distribution of $Y := \sum_{n=0}^{\infty} A^n \varepsilon_{n+1}$. (b) In the special case $\varepsilon_n (n \geq 1)$ are Gaussian, show that the invariant distribution in (a) is Gaussian, or normal, and compute its mean and dispersion matrix (assume those of ε_n are μ and $\sum$, respectively).   ∎

4.4 Iterates of Quadratic Maps

Recall that the family $\{F_\theta : 0 \leq \theta \leq 4\}$ of quadratic, or logistic, maps, is defined by

$$F_\theta(x) = \theta x (1 - x), \quad 0 \leq x \leq 1. \tag{4.1}$$

To study iterations of random (i.i.d.) quadratic maps, we will make use of Proposition 7.1 of Chapter 1, as well as the following lemma:

Lemma 4.1 *Let* $1 < \mu < \lambda < 4$. *Let* $u = \min\{1 - \frac{1}{\mu}, F_\mu(\frac{\lambda}{4})\}$, $v = \frac{\lambda}{4}$. *Then, for every* $\theta \in [\mu, \lambda]$, $[u, v]$ *is invariant under* F_θ.

Proof. We need to prove that (i) $\max\{F_\theta(x) : u \leq x \leq v\} \leq v$ for $\theta \in [\mu, \lambda]$ and (ii) $\min\{F_\theta(x) : u \leq x \leq v\} \geq u$ for $\theta \in [\mu, \lambda]$. The first of these follows from the relations $\max\{F_\theta(x) : u \leq x \leq v\} \leq \max\{F_\lambda(x) : u \leq x \leq v\} \leq \frac{\lambda}{4}$. For (ii) note that by unimodality, $\min\{F_\theta(x) : u \leq x \leq v\} = \min\{F_\theta(u), F_\theta(v)\} \geq \min\{F_\mu(u), F_\mu(v)\}$. If $u = 1 - \frac{1}{\mu} \leq F_\mu(\frac{\lambda}{4})$, then the last minimum is $F_\mu(u) = 1 - \frac{1}{\mu} = u$. If $u = F_\mu(\frac{\lambda}{4}) < 1 - \frac{1}{\mu}$, then $\min\{F_\mu(u), F_\mu(\frac{\lambda}{4})\} = F_\mu(\frac{\lambda}{4}) = u$, since on $(0, 1 - \frac{1}{\mu})$, $F_\mu(x) > x$.   ∎

Let $C_n (n \geq 1)$ be a sequence of i.i.d. random variables with values in $[\mu, \lambda]$, where $1 < \mu < \lambda < 4$. Then

$$\alpha_n := F_{C_n}, \quad n \geq 1 \tag{4.2}$$

is a sequence of random quadratic maps. We consider the Markov process defined recursively by

$$X_0, X_{n+1} = \alpha_{n+1} X_n \quad (n \geq 0), \tag{4.3}$$

where X_0 takes values in $(0, 1)$, and is independent of $\{C_n : n \geq 1\}$. For Theorem 4.1 on the stability in distribution of this Markov process, recall that if ν is a probability measure on $\mathbb{R}$ with compact support, then there is a smallest point and a largest point of its support. As always, $\mathcal{B}(S)$ denotes the Borel sigmafield of a metric space S.

Theorem 4.1 *Let $P(\mu_0 \leq C_n \leq \lambda_0) = 1$, where $1 < \mu_0 < \lambda_0 < 4$. Assume that the distribution Q of C_n has a nonzero absolutely continuous component with density h (with respect to the Lebesgue measure) which is bounded away from zero on an interval $[c, d]$ with $\mu_0 \leq c < d \leq 3, d \leq \lambda_0$. Then the Markov process $X_n(n \geq 0)$ defined by (4.3) on $S = (0, 1)$ has a unique invariant probability π whose support is contained in $[u_0, v_0]$ with u_0, v_0 as in Lemma 4.1 for $\mu = \mu_0$ and $\lambda = \lambda_0$; and for every interval $[u, v]$ with $0 < u \leq u_0 < v_0 < v < 4$, one has*

$$\sup_{x \in [u,v], B \in \mathcal{B}([u,v])} |p^{(n)}(x, B) - \pi(B)| \leq (1 - \chi)^{[n/m]}, \quad n \geq 1, \quad (4.4)$$

where $\chi = \chi(u, v)$ is a positive constant and $m = m(u, v)$ is a positive integer, depending on u, v.

Proof. By Lemma 4.1, if $X_0 \in S_0 \equiv [u_0, v_0]$, $X_n \in S_0$ $n \geq 1$. Hence $X_n(n \geq 0)$ is a Markov process on S_0. We will show that the hypothesis of Corollary 5.3 of Chapter 3 is satisfied. For this first assume, for the sake of simplicity, that the distribution of C_n is *absolutely continuous with a density $h(\theta)$ which is continuous on $[\mu_0, \lambda_0]$ and is positive on $[c, d]$*. By (4.3), the transition probability density of $X_n(n \geq 0)$ is given by

$$p(x, y) = \frac{1}{x(1 - x)} h\left(\frac{y}{x(1 - x)}\right), \quad x, y \in [u_0, v_0]. \quad (4.5)$$

Let $p_\theta = 1 - (1/\theta)$ be the fixed point of F_θ on $(0, 1)$. For $\theta \in [c, d]$, $p(p_\theta, p_\theta) = (p_\theta(1 - p_\theta))^{-1} h(\theta) > 0$. In particular, for $\theta_0 = (c + d)/2$, writing $x_0 = p_{\theta_0}$, one has $p(x_0, x_0) > 0$. Since $(x, y) \to p(x, y)$ is continuous on $S_0 \times S_0$, there exists an interval $[x_1, x_2]$, $x_1 < x_0 < x_2$, such that

$$\delta_1 := \inf_{x,y \in [x_1,x_2]} p(x, y) > 0. \quad (4.6)$$

It follows from Proposition 7.4(b) of Chapter 1 that, for each $y \in [u_0, v_0]$, there exists an integer $n(y)$ such that

$$F_{\theta_0}^n(y) \in (x_1, x_2) \ (y \in [u_0, v_0]), \quad \forall n \geq n(y).$$

The set $0(y) := \{z \in [u_0, v_0] : F_{\theta_0}^{n(y)}(z) \in (x_1, x_2)\}$ is open in $[u_0, v_0]$, and $y \in 0(y)$. By the compactness of $[u_0, v_0]$, there exists a finite set $\{y_1, y_2, \ldots, y_k\}$ such that $\cup_{i=1}^k 0(y_i) = [u_0, v_0]$. Let $N := \max\{n(y_i):$

$1 \leq i \leq k$}. Then $F_{\theta_0}^N(z) \in (x_1, x_2)$ $\forall z \in [u_0, v_0]$. Since the function $g(\theta_1, \theta_2, \ldots, \theta_N, z) := F_{\theta_N} F_{\theta_{N-1}} \cdots F_{\theta_1}(z)$ is (uniformly) continuous on $[c, d]^N \times [u_0, v_0]$, $\underline{g}(\theta_1, \theta_2, \ldots, \theta_N) := \min\{F_{\theta_N} F_{\theta_{N-1}} \cdots F_{\theta_1}(z):$ $z \in [u_0, v_0]\}$ and $\overline{g}(\theta_1, \theta_2, \ldots, \theta_N) := \max\{F_{\theta_N} F_{\theta_{N-1}} F_{\theta_1}(z):$ $z \in [u_0, v_0]\}$ are continuous on $[c, d]^N$. Also, $\underline{g}(\theta_0, \theta_0, \ldots, \theta_0) = \min\{F_{\theta_0}^N(z): z \in [u_0, v_0]\} > x_1$ and $\overline{g}(\theta_0, \theta_0, \ldots, \theta_0) = \max\{F_{\theta_0}^N(z):$ $z \in [u_0, v_0]\} < x_2$. Therefore there exists $\varepsilon_1 > 0$ such that

$$F_{\theta_N} F_{\theta_{N-1}} \cdots F_{\theta_1}(z) \in (x_1, x_2) \ \forall \theta_i \in (\theta_0 - \varepsilon_1, \theta_0 + \varepsilon_1)$$
$$\subset [c, d], \ 1 \leq i \leq N, \quad \text{and} \quad \forall z \in [u_0, v_0]. \tag{4.7}$$

Hence

$$X_N(z) = F_{C_N} F_{C_{N-1}} \cdots F_{C_1}(z) \in (x_1, x_2),$$
$$\text{if} \quad C_i \in (\theta_0 - \varepsilon_1, \theta_0 + \varepsilon_1) \ \forall i, \quad 1 \leq i \leq N. \tag{4.8}$$

Therefore, $\forall x \in [u_0, v_0]$ and for every Borel set $B \subset [u_0, v_0]$, one has by (4.8), and using (4.6) in the second inequality below,

$$p^{(N+1)}(x, B) = P_x(X_{N+1} \in B)$$
$$\geq P_x(X_{N+1} \in B \cap (x_1, x_2), \{C_i \in (\theta_0 - \varepsilon_1, \theta_0 + \varepsilon_1), \forall i, 1 \leq i \leq N\})$$
$$= E_x[P_x(X_{N+1} \in B \cap (x_1, x_2) \mid X_N) \cdot \mathbf{1}_{\{C_i \in (\theta_0 - \varepsilon_1, \theta_0 + \varepsilon_1), \ 1 \leq i \leq N\}}]$$
$$\geq E_x \int_{B \cap (x_1, x_2)} \delta_1 \, dy \cdot \mathbf{1}_{\{C_i \in (\theta_0 - \varepsilon_1, \theta_0 + \varepsilon_1), \ 1 \leq i \leq N\}}$$
$$= \delta_1 \mathbf{m}(B \cap (x_1, x_2))\delta_2^N \left(\delta_2 = \int_{(\theta_0 - \varepsilon_1, \theta_0 + \varepsilon_1)} h(\theta) \, d\theta \right), \tag{4.9}$$

where P_x, E_x denote probability and expectation under $X_0 = x$, and $\mathbf{m}$ denotes the Lebesgue measure. Thus Corollary 5.3 of Chapter 3 applies on the state space $[u_0, v_0]$, with $N + 1$ in place of m, $\lambda(B) = \delta_1 \delta_2^N \mathbf{m}(B \cap (x_1, x_2))$, $\chi = \chi(u_0, v_0) = \delta_1 \delta_2^N (x_2 - x_1)$. Hence there exists a unique invariant probability π on $[u_0, v_0]$, and (4.4) holds for the case $u = u_0, v = v_0$.

Now note that the same proof goes through if we replace $[u_0, v_0]$ by any larger interval $[u_1, v_1]$ corresponding to $\mu = \mu_1, \lambda = \lambda_1$ in Lemma 3.1 with $1 < \mu_1 < \mu_0 < \lambda_0 < \lambda_1 < 4$. By letting μ_1, λ_1 be sufficiently close to 1 and 4, respectively, one can choose u_1, v_1 as close to 0, 1, respectively, as we like, so that (4.4) holds with $u = u_1$, and $v = v_1$ and $\chi = \chi(u_1, v_1)$. If one is given arbitrary $0 < u < v < 1$, one may find

μ_1, λ_1 and u_1, v_1 as above such that $0 < u_1 < u < v < v_1 < 1$, so that
(4.4) holds with $\chi(u, v) = \chi(u_1, v_1)$.

Finally, if h is merely assumed to be a density component that is
bounded away from zero on $[c, d]$, one can find a continuous nonnegative
function on $(1, 4)$, say, h_1 such that $h_1 > 0$ on $[c, d]$ and $h_1 \leq h$ on $(1, 4)$.
The proof then goes through with h_1 in place of h and $p_1(x, y)$ defined
in place of $p(x, y)$ by (4.5) (with h_1 used in place of h). Letting

$$p_1^{(r)}(x, y) = \int_{(0,1)} p_1^{(r-1)}(x, z) p_1(z, y) \, dz \quad (r \geq 2), \qquad (4.10)$$

one obtains $p^{(r)}(x, B) \geq p_1^{(r)}(x, B) \equiv \int_B p_1^{(r)}(x, y) dy \ \forall r \geq 1, \ B \in$
$\mathcal{B}((0, 1), x \in (0, 1))$. Then (4.9) holds with δ_1 defined by (4.6) with
$p_1(x, y)$ in place of $p(x, y)$, and δ_2 defined as in (4.9) but with h replaced by h_1. ∎

Remark 4.1 Theorem 4.1 implies that, under its hypothesis
$\sup\{|p^{(r)}(x, B) - \pi(B)|\colon B \in \mathcal{B}((0, 1))\} \to 0$ as $n \to \infty \forall x \in (0, 1)$,
but not necessarily uniformly $\forall x \in (0, 1)$. To see this convergence
for a given $x \in (0, 1)$, choose $\mu > 1$ sufficiently close to 1 and $\lambda < 4$
sufficiently close to 4 so that the corresponding u, v given by Lemma
4.1 satisfy $0 < u < x < v < 1$ and $[u_0, v_0] \subset [u, v]$, and apply (4.4)
to get $\delta_n(x) := \sup\{|p^{(n)}(x, B) - \pi(B)|\colon B \in \mathcal{B}([u, v])\} \to 0$. Since the
support of π is contained in $[u_0, v_0]$, $\forall B \in \mathcal{B}((0, 1))$, one has $\pi(B) =$
$\pi(B \cap [u_0, v_0]) = \pi(B \cap [u, v])$ so that $|p^{(n)}(x, [u, v]) - \pi([u, v])| =$
$|p^{(n)}(x, [u, v]) - 1| \leq \delta_n(x)$. Hence, $p^{(n)}(x, [u, v]^C) \equiv 1 - p^{(n)}(x,$
$[u, v]) \leq \delta_n(x)$. Thus, $\sup\{|p^{(n)}(x, B) - \pi(B)|\colon B \in \mathcal{B}((0, 1))\} =$
$\sup\{|p^{(n)}(x, B \cap [u, v]) - \pi(B \cap [u, v]) + p^{(n)}(x, B \cap [u, v]^C)|\colon B \in$
$\mathcal{B}((0, 1))\} \leq \delta_n(x) + \delta_n(x) = 2\delta_n(x) \to 0$.

We next consider cases where C_n does not have a density component.
In Examples 4.1–4.4, the distribution of C_n has a two-point support
$\theta_1 < \theta_2$ with $P(C_n = \theta_1) = w$, $P(C_n = \theta_2) = 1 - w$, $0 < w < 1$. The
points θ_1, θ_2 $(1 < \theta_2 < \theta_2 \leq 1 + \sqrt{5})$ are so chosen in Examples 4.1–
4.4 that F_{θ_i} leaves an interval $[c, d]$ invariant $(i = 1, 2)$ where $[c, d]$ is
contained in either (i) $(0, 1/2]$ or (ii) $[1/2, 1)$. In case (i) the maps α_n
(see (4.2)) are monotone increasing while in case (ii) they are monotone
decreasing. We will apply Theorem 5.1 of Chapter 3 by showing that the
following splitting condition (**H**) holds:

(**H**) *There exist* $x_0 \in [c, d]$. $\chi_i > 0$ $(i = 1, 2)$ *and a positive integer* N *such that*

(1) $P(\alpha_N \alpha_{N-1} \cdots \alpha_1 x \leq x_0 \ \forall x \in [c, d]) \geq \chi_1$, *and*
(2) $P(\alpha_N \alpha_{N-1} \cdots \alpha_1 x \geq x_0 \ \forall x \in [c, d]) \geq \chi_2$.

Example 4.1 Take $1 < \theta_1 < \theta_2 \leq 2$. Then F_{θ_i} has an attractive fixed point $p_{\theta_i} = 1 - 1/\theta_i (i = 1, 2)$. Here $[c, d] = [p_{\theta_1}, p_{\theta_2}] \subset (0, 1/2]$, $x_0 \in (c, d)$, and N is a sufficiently large integer such that $F_{\theta_1}^N(p_{\theta_2}) < x_0$, and $F_{\theta_2}^N(p_{\theta_1}) > x_0$. Then the splitting condition (**H**) above holds with $\chi_1 = w^N$ and $\chi_2 = (1 - w)^N$. Note $\alpha_n (n \geq 1)$ are *monotone increasing* on $[c, d]$. Hence, by Theorem 5.1 of Chapter 3, there exists a unique invariant probability π for the Markov process $X_n (n > 0)$ on $[c, d]$ and the distribution of X_n converges to π exponentially fast in the Kolmogorov distance (see (5.3) in Chapter 3). Next, note that if $x \in (0, 1) \backslash [c, d]$, then in a finite number of steps $n(x)$, say, the process $X_n(x) = \alpha_n \cdots \alpha_1 x$ enters $[c, d]$. Hence, the Markov process on $S = (0, 1)$ has a unique invariant probability π and $X_n(x)$ converges in distribution to π geometrically fast in the Kolmogorov distance, as $n \to \infty$, for every $x \in (0, 1)$. ∎

Example 4.2 Take $2 < \theta_1 < \theta_2 \leq 3$. Then F_{θ_i} has an attractive fixed point $p_{\theta_i} = 1 - 1/\theta_i (i = 1, 2)$ and $[c, d] = [p_{\theta_1}, p_{\theta_2}] \subset [1/2, 1)$. The same conclusion as in Example 4.1 holds in this case as well, with an even integer N. The maps $\alpha_n (n \geq 1)$ are monotone decreasing in this case. Since $F_{\theta_i} F_{\theta_j}$ is increasing on $[1/2, 1)$ $(i, j = 0, 1)$, the same kind of argument as in Example 4.1 applies. ∎

Example 4.3 Take $2 < \theta_1 < 3 < \theta_2 \leq 1 + \sqrt{5}$, $\theta_1 \in [8/(\theta_2(4 - \theta_2)), \theta_2]$. Then, by Lemma 4.1, the interval $[u, v]$, with $u = \min\{1 - (1/\theta_1), F_{\theta_1}(\theta_2/4)\}$, $v = \theta_2/4$, is invariant under $F_{\theta_i}(i = 1, 2)$. The condition $\theta_1 \geq 8/[\theta_2(4 - \theta_2)]$ ensures that $u \geq 1/2$ (and is equivalent to $F_{\theta_1}(\theta_2/4) \geq 1/2$), so that $[u, v] \subset [1/2, 1)$ and $\alpha_n (n \leq 1)$ are monotone decreasing on $[u, v]$. It turns out that one may enlarge this interval to $[1/2, v] = [1/2, \theta_2/4]$ as an invariant interval under $F_{\theta_i}(i = 1, 2)$ (Exercise). With $[c, d] = [u, v]$ or $[1/2, \theta_2/4]$, $X_n (n \geq 0)$ is a Markov process on $[c, d]$ generated by monotone decreasing i.i.d. maps $\alpha_n (n \geq 1)$. Then F_{θ_1} has an attractive fixed point $p_{\theta_1} = 1 - 1/\theta_1$, and by Proposition 7.4(c) of Chapter 1, F_{θ_2} has an attractive period-two orbit $\{q_1, q_2\}$, $q_1 < q_2$. Splitting occurs with $x_0 \in (p_{\theta_2}, q_2)$ and N a sufficiently large

even integer. Also with probability 1, the Markov process $\{X_n(x)\}_{n=0}^{\infty}$
reaches $[c, d]$ in finite time, whatever be $x \in (0, 1)$. Thus, there is a
unique invariant probability π on $S = (0, 1)$ and $X_n(x)$ converges in distribution to π exponentially fast in the Kolmogorov distance, as $n \to \infty$,
for every $x \in (0, 1)$.   ∎

Example 4.4 Take $\theta_1 = 3.18$, $\theta_2 = 3.20$. Then F_{θ_i} has an attractive periodic orbit $\{q_{1i}, q_{2i}\}$, $q_{1i} < q_{2i}$ $(i = 1, 2)$. One may let $[u, v]$
be as in Lemma 4.1 $(u = 0.5088, v = 0.80)$ and let $[c, d] = [u, v]$
be an invariant interval for $F_{\theta_i} (i = 1, 2)$. It is simple to check that
$q_{12} < q_{11} < p_{\theta_1} < p_{\theta_2} < q_{21} < q_{22}$ (in our case $p_{\theta_2} = 0.6875$, $q_{12} =$
0.51304551, $q_{22} = 0.79945549$, $p_{\theta_1} = 0.6855349$, $q_{11} = 0.52084749$,
$q_{21} = 0.793617914$). In this example the *splitting condition* (**H**) *does
not hold*. To see this, note that F_{θ_i} maps the interval $[c, q_{11}]$ into $[q_{21}, d)$
and maps $[q_{21}, d]$ into $(c, q_{11}]$ $(i = 1, 2)$. Hence whatever be the se-
quence $\theta^r \in \{\theta_1, \theta_2\}, 1 \leq r \leq N$, either $F_{\theta^1} F_{\theta^2} \cdots F_{\theta^N}$ maps $[c, q_{11}]$ into
$[q_{21}, d]$ and vice versa (which is the case if N is odd) or it maps $[c, q_{11}]$
into $[c, q_{11}]$ and $[q_{21}, d]$ into $[q_{21}, d]$ (if N is even). From this, it follows
that the splitting condition (**H**) does not hold (Exercise). On the other
hand, since $F_{\theta^1 \theta^2}$ leaves each interval $I_1 := [c, q_{11}]$ and $I_2 := [q_{21}, d]$
invariant $\forall \theta^r \in \{\theta_1, \theta_2\}$ $(r = 1, 2)$, it follows that X_{2n} $(n \geq 0)$ is a
Markov process on I_1, generated by monotone increasing maps $\beta_n :=$
$F_{C_n} F_{C_{n+1}}$ $(n \geq 1)$. If $x_0 \in (q_{12}, q_{11})$, then there exists sufficiently large
integer N such that $F_{\theta_1}^{2N}(x) \geq F_{\theta_1}^{2N}(C) \geq x_0 \; \forall x \in I_1$ (since q_{11} is an attractive fixed point of $F_{\theta_1}^2$), and $F_{\theta_1}^{2N}(x) \leq F_{\theta_2}^{2N}(q_{11}) \leq x_0 \; \forall x \in I_1$ (since
q_{12} is an attractive fixed point of $F_{\theta_2}^2$). Hence the splitting condition holds
for $Y_n := X_{2n} (n \geq 0)$ on I_1, so that its distribution converges exponentially fast in the Kolmogorov distance to a probability measure π_1 on $I_1 =$
$[c, q_{11}]$ whatever be the initial $X_0 \in I_1$, and π_1 is the unique invariant
probability for Y_n (i.e., for a Markov process on I_1 with (one-step) transition probability $p^{(2)}(x, dy)$). Similarly, the Markov process $Z_n := X_{2n}$ on
I_2 converges, exponentially fast in Kolmogorov distance, to its unique invariant probability π_2, say, on I_2. For the process $\{X_n : n \geq 0\}$, if $X_0 \in I_1$,
X_{2n} lies in I_1 $\forall n \geq 0$, and $X_1, X_3, \ldots, X_{2n-1} \in I_2$ $(\forall n \geq 1)$. Similarly, if $X_0 \in I_2$, then $X_n \in I_2$ for all even n and $X_n \in I_1$ for all odd n.
    It follows that $\forall x \in I_1 \cup I_2$, as $n \to \infty$,

$$\frac{1}{n} \sum_{m=1}^{n} p^{(m)}(x, dy) \xrightarrow{\mathcal{L}} (\pi_1(dy) + \pi_2(dy))/2. \tag{4.11}$$

This implies that $\pi := (1/2)(\pi_1 + \pi_2)$ is the unique invariant probability for the Markov process $X_n (n \geq 0)$ on the state space $I = I_1 \cup I_2$.
It is not difficult to show that if $x \in (0, 1)\backslash I$, then there exists a finite
(random) integer $n(x)$ such that $X_n(x) \in I \; \forall n \geq n(x)$. From this (or using the fact that $p^{(n)}(x, I) \to 1$ as $n \to \infty$), it follows that (4.11) holds
$\forall x \in S = (0, 1)$, so that π is the unique invariant probability for this
Markov process on $(0, 1)$.   ∎

Remark 4.2 There are a number of interesting features of Example 4.4.
First, the Markov process has two *cyclic classes* I_1 and I_2, providing an
example of such periodicity for an uncountable state space. As a result,
the *stability in distribution is in the average sense*. Second, here one has
a nondegenerate unique invariant probability in the absence of a splitting
condition (**H**) for a Markov process generated by i.i.d. monotone decreasing maps α_n $(n \geq 1)$ on a compact interval $[c, d]$. This is in contrast with
the case of processes generated by i.i.d. monotone nondecreasing maps
on a compact interval, where *splitting is necessary* for the existence of a
unique nondegenerate invariant probability (see Remarks 5.1 and 5.7 in
Chapter 3, Section 3.5).

The next theorem (Theorem 4.2) extends results such as are given in
Examples 4.1–4.4 to distributions of C_n with more general supports, but
without assuming densities. In order to state it, we will recast the splitting
class as $\mathcal{A} = \{[c, x] : c \leq x \leq d\}$ for the case of i.i.d. monotone maps
$\{\alpha_n : n \geq 1\}$ on an interval $[c, d]$ as follows. The Markov process $X_n(x)$,
$(n \geq 1)$, $X_0(x) = x$, $x \in [c, d]$ is said to have the splitting property if
there exist $\delta > 0$, $x_0 \in [c, d]$, and an integer N such that

$$P(X_N(x) \leq x_0, \;\; \forall x \in [c, d]) \geq \delta, \quad P(X_N(x) \geq x_0, \;\; \forall x \in [c, d]) \geq \delta.$$
$$(4.12)$$

We know that for i.i.d. monotone $\alpha_n (n \geq 1)$, (4.12) implies the existence
and uniqueness of an invariant probability π on $[c, d]$; i.e., Theorem
5.1 of Chapter 3 holds. If the inequalities "$\leq x_0$" and "$\geq x_0$," appearing
within parentheses in (4.12) are replaced by strict inequalities "$< x_0$" and
"$> x_0$," respectively, then (4.12) will be referred to as a strict *splitting
property*. Note that (4.12), or its strict version, is a property of the distribution Q of C_n. Denote by Q_N the (product) probability distribution of
$(C_1, \ldots, C_N)$.

Theorem 4.2

(a) *Suppose the distribution Q_0 of C_n has a two-point support $\{\theta_1, \theta_2\}$ $(1 < \theta_1 < \theta_2 < 4)$. If F_{θ_i} leaves an interval $[c, d]$ invariant $(i = 1, 2)$, where $[c, d] \subset (0, 1/2]$ or $[c, d] \subset [1/2, 1)$, and the strict splitting condition holds for $X_n (n \geq 0)$ on $[c, d]$, under Q_0, then (i) under every distribution Q of C_n with θ_1 and θ_2 as the smallest and largest points of support of Q, respectively, the strict splitting condition holds, (ii) the corresponding Markov process $X_n (n \geq 0)$ has a unique invariant probability π, and (iii) X_n converges in distribution to π exponentially fast in the Kolmogorov distance, no matter what the distribution of X_0 is on $S = (0, 1)$.*

(b) *If $3 < \theta_1 < \theta_2 \leq 1 + \sqrt{5}$, are the smallest and largest points of the support of the distribution Q of C_n, and $u \equiv F_{\theta_1}(\theta_2/4) \equiv \theta_1\theta_2(4 - \theta_2)/16 \geq 1/2$, then (i) the Markov process $X_n (n \geq 0)$ on $S = (0, 1)$ is cyclical with cyclical classes $I_1 = [u, q_{11}]$ and $I_2 = [q_{21}, v]$ $(v = \theta_2/4)$, (ii) there exists a unique invariant probability π, and (iii) there is stability in distribution in the average sense.*

Proof. See Complements and Details. ∎

Remark 4.3 The question naturally arises as to whether there always exists a unique invariant probability on $S = (0, 1)$ for every distribution Q of C_n if Q has a two-point support $\{\theta_1, \theta_2\} \subset (1, 4)$. It has been shown by Athreya and Dai (2002) that this is not the case. That is, there exist θ_1, θ_2 such that for Q with support $\{\theta_1, \theta_2\}$ the transition probability admits more than one invariant probability (see Complements and Details).

4.5 NLAR(k) and NLARCH(k) Models

Consider a time series of the form

$$U_{n+1} = f(U_{n-k+1}, U_{n-k+2}, \ldots, U_n) + g(U_{n-k+1}, U_{n-k+2}, \ldots, U_n)\eta_{n+1}$$

$$(n \geq k - 1), \quad (5.1)$$

where $f: \mathbb{R}^k \to \mathbb{R}$, $g: \mathbb{R}^k \to \mathbb{R}$ are measurable functions, and $\{\eta_n : n \geq k\}$ is a sequence of i.i.d. random variables independent of given initial random variables $U_0, U_1, \ldots, U_{k-1}$. Here $k \geq 1$. In the case $g \equiv 1$, this is the so-called *kth-order nonlinear autoregressive model*, NLAR(k). The case NLAR(1) was considered in Chapter 3, Subsection 3.6.1. In its general form (5.1), $\{U_n : n \geq 0\}$ is a *kth-order nonlinear autoregressive conditional heteroscedastic*, or NLARCH(k), time series.

As in the preceding section, $X_n := (U_{n-k+1}, U_{n-k+2}, \ldots, U_n)$, ($n \geq k - 1$) is a Markov process on the state space $\mathbb{R}^k$. Note that we have dropped the *prime* ($'$) from X_n, since no matrix operations are involved in this section.

Theorem 5.1 *Suppose $a_0 \leq g \leq b_0$, $a_1 \leq f \leq b_1$ for some constants $a_i, b_i (i = 0, 1)$, $0 < a_0 \leq b_0$. Also assume that the distribution Q of η_n has a nonzero absolutely continuous component (with respect to the Lebesgue measure on $\mathbb{R}$), with a density $\hat{\phi}$ bounded away from zero on an interval $[c, d]$, where*

$$cb_0 + b_1 < da_0 + a_1. \qquad (5.2)$$

Then the Markov process $X_n (n \geq k - 1)$ has a unique invariant probability π, and its n-step transition probability $p^{(n)}(\mathbf{x}, dy)$ satisfies

$$\sup_{\mathbf{x} \in \mathbb{R}^k, B \in \mathcal{B}(\mathbb{R}^k)} |p^{(n)}(\mathbf{x}, B) - \pi(B)| \leq (1 - \chi)^{[n/k]}, \qquad (5.3)$$

with χ some positive constant.

Proof.

Step 1. First, consider the nonlinear autoregressive case NLAR(k), i.e., take $g \equiv 1$ (or $a_0 = b_0 = 1$). The case $k = 1$ was considered in Section 3.6.1. Then (5.1) becomes

$$U_{n+1} = f(U_{n-k+1}, \ldots, U_n) + \eta_{n+1} \ (n \geq k - 1). \qquad (5.4)$$

Assume also, for the sake of simplicity, that the distribution Q of η_n is absolutely continuous (with respect to the Lebesgue measure on $\mathbb{R}$) with density $\hat{\phi}$, $\hat{\phi} \geq \delta > 0$ on $[c, d]$. Let us take $k = 2$ first: $U_{n+1} = f(U_{n-1}, U_n) + \eta_{n+1} \ (n \geq 1)$. We will compute $p^{(2)}(\mathbf{x}, \mathbf{y})$– the conditional probability density function (p.d.f.) of $X_4 \equiv (U_3, U_4)$ (at $\mathbf{y} = (y_1, y_2)$), given $X_2 \equiv (U_1, U_2) \equiv \mathbf{x} = (x_1, x_2)$. Now the conditional p.d.f. $U_3 \equiv f(U_1, U_2) + \eta_3$ (at $U_3 = y_1$), given $X_2 \equiv (U_1, U_2) = (x_1, x_2)$, is $\hat{\phi}(y_1 - f(u_1, u_2)) = \hat{\phi}(y_1 - f(x_1, x_2))$. Next, the conditional p.d.f. of $U_4 \equiv f(U_2, U_3) + \eta_4$ (at $u_4 = y_2$), given X_2 and U_3 (i.e., given U_1, U_2, U_3), is $\hat{\phi}(u_4 - f(u_2, u_3)) = \hat{\phi}(y_2 - f(x_2, y_1))$. Hence the p.d.f. $\mathbf{y} \to p^{(2)}(\mathbf{x}, \mathbf{y})$ (with respect to the Lebesgue measure on $\mathbb{R}^2$) is

$$p^{(2)}(\mathbf{x}, \mathbf{y}) = \hat{\phi}(y_1 - f(x_1, x_2))\hat{\phi}(y_2 - f(x_2, y_1)),$$
$$\forall \, \mathbf{x} = (x_1, x_2), \; \mathbf{y} = (y_1, y_2) \in \mathbb{R}^2. \tag{5.5}$$

Since, by assumption, $\forall \; y_i \in [c + b_1, d + a_1]$ and $\mathbf{z} \in \mathbb{R}^2$, $c \le y_i - f(\mathbf{z}) \le d$ $(i = 1, 2)$, one has

$$p^{(2)}(\mathbf{x}, \mathbf{y}) \ge \Psi(\mathbf{y}) \; \forall \mathbf{x}, \mathbf{y} \in \mathbb{R}^2,$$
$$\Psi(\mathbf{y}) := \begin{cases} \delta^2 & \forall \mathbf{y} = (y_1, y_2) \in [c + b_1, d + a_1]^2, \\ 0 & \text{otherwise}. \end{cases} \tag{5.6}$$

Hence Corollary 5.3 of Chapter 3 applies with $S = \mathbb{R}^2$, $m = 2$, $\lambda(A) = \int_A \Psi(\mathbf{y}) dy$, and the conclusion of the theorem follows with $\chi = \lambda(\mathbb{R}^2) = \delta^2(d - c - (b_1 - a_1))^2$.

Step 2. Next, consider the case NLAR(k) $\forall \, k \ge 2$, with the assumption on the density of η_n in Step 1 remaining intact. Then $p^{(k)}(\mathbf{x}, \mathbf{y})$ is the conditional p.d.f. of $X_{2k} \equiv (U_{k+1}, \ldots, U_{2k})$ (at a point $\mathbf{y} = (y_1, y_2, \ldots, y_k)$), given $X_k \equiv (U_1, \ldots, U_k) = \mathbf{x} = (x_1, x_2, \ldots, x_k)$. This is given by the products of successive conditional p.d.f.'s:

(1): *p.d.f.* of $U_{k+1} \equiv f(U_1, \ldots, U_k) + \eta_{k+1}$, given $U_1 = x_1, U_2 = x_2, \ldots, U_k = x_k$, namely,

$$\hat{\phi}(y_1 - f(x_1, x_2, \ldots, x_k)), \tag{5.7}$$

(2): *p.d.f.* of $U_{k+2} \equiv f(U_2, \ldots, U_{k+1}) + \eta_{k+2}$, given $U_1 = x_1, \ldots, U_k = x_k, U_{k+1} = y_1$, namely,

$$\hat{\phi}(y_2 - f(x_2, \ldots, x_k, y_1)), \tag{5.8}$$

etc., and, finally,

(k): *p.d.f.* of $U_{2k} \equiv f(U_k, \ldots, U_{2k-1}) + \eta_{2k}$, given $U_1 = x_1, \ldots, U_k = x_k, U_{k+1} = y_1, \ldots, U_{2k-1} = y_{k-1}$, namely,

$$\hat{\phi}(y_k - f(x_k, y_1, y_2 \ldots, y_{k-1})). \tag{5.9}$$

In general, the conditional p.d.f. of $U_{k+j} \equiv f(U_j, U_{j+1}, \ldots, U_{k+j-1}) + \eta_{k+j}$, given $U_1 = x_1, \ldots, U_k = x_k, U_{k+1} = y_1, \ldots, U_{k+j-1} = y_{j-1}$ is

$$\hat{\phi}(y_j - f(x_j, \ldots, x_k, y_1, \ldots, y_{j-1})), \quad 1 \le j \le k, \tag{5.10}$$

where the case $j = 1$ is given by (5.7), i.e., interpret the case $j = 1$ in (5.10) as the expression (5.7). Thus,

$$p^{(k)}(\mathbf{x}, \mathbf{y}) = \prod_{j=1}^{k} \hat{\phi}(y_j - f(x_j, \ldots, x_k, y_1, \ldots, y_{j-1}))$$

$$\forall \; \mathbf{x} = (x_1, \ldots, x_k), \quad \mathbf{y} = (y_1, \ldots, y_k) \in \mathbb{R}^k. \qquad (5.11)$$

As, in the case $k = 2$, $p^{(k)}(\mathbf{x}, \mathbf{y}) \geq \delta^k$ on $[c + b_1, d + a_1]^k$. Define

$$\Psi(\mathbf{y}) = \begin{cases} \delta^k & \text{for } \mathbf{y} \in [c + b_1, d + a_1]^k \\ 0 & \text{otherwise.} \end{cases} \qquad (5.12)$$

Then Corollary 5.3 of Chapter 3 applies with $S = \mathbb{R}^k$, $m = k$, $\lambda(A) = \int_A \Psi(\mathbf{y}) d\mathbf{y}$, $\chi = \delta^k(d - c - (b_1 - a_1))^k$.

Step 3. We now consider the general NLAR(k) case (i.e., $g \equiv 1$), assuming that $\hat{\phi}$ is the density of the nonzero absolutely continuous component of Q, with $\hat{\phi} \geq \delta > 0$ on $[c, d]$. Then the arguments above go through with a density component $p^{(k)}(\mathbf{x}, \mathbf{y})$ given by the right-hand side of (5.11).

Step 4. Next, we consider the case NLARCH(k). Under the assumption that Q is absolutely continuous with density $\hat{\phi}$, the conditional p.d.f. (5.10) is then replaced by the conditional p.d.f. of $U_{k+j} \equiv f(U_j, U_{j+1}, \ldots, U_{k+j-1}) + g(U_j, U_{j+1}, \ldots, U_{k+j-1})\eta_{k+j}$, given $U_1 = x_1, \ldots, U_j = x_j, \ldots, U_k = x_k, U_{k+1} = y_1, \ldots, U_{k+j-1} = y_{j-1}$, namely, the p.d.f. of $f(x_j, x_{j+1}, \ldots, x_k, y_1, \ldots, y_{j-1}) + g(x_j, x_{j+1}, \ldots, x_k, y_1, \ldots, y_{j-1})\eta_{k+j}$, i.e.,

$$\frac{1}{g(x_j, \ldots, x_k, y_1, \ldots, y_{j-1})} \hat{\phi}\left(\frac{y_j - f(x_j, \ldots, x_k, y_1, \ldots, y_{j-1})}{g(x_j, \ldots, x_k, y_1, \ldots, y_{j-1})}\right),$$

$$1 \leq j \leq k. \qquad (5.13)$$

By the hypothesis of the theorem this quantity is no less than $(1/b_0)\delta$ for $cb_0 + b_1 \leq y_j \leq da_0 + a_1$, whatever be $\mathbf{x}$. Hence defining

$$\Psi(\mathbf{y}) := \begin{cases} [(1/b_0)\delta]^k & \text{for } \mathbf{y} \in [cb_0 + b_1, da_0 + a_1]^k \\ 0 & \text{otherwise,} \end{cases} \qquad (5.14)$$

Corollary 5.3 of Chapter 3 applies with $S = \mathbb{R}^k$, $m = k$, $\lambda(A) = \int_A \Psi(\mathbf{y}) d\mathbf{y}$, $A \in \mathcal{B}(\mathbb{R}^k)$, $\chi = (1/b_0)^k \delta^k [da_0 + a_1 - (cb_0 + b_1)]^k$.

Finally, if $\hat{\phi}$ is the density of an absolutely continuous component of Q, the proof follows from the above in exactly the same way as in the NLAR(k) case. ∎

As an immediate consequence of Theorem 5.1, we have the following result (see Section 2.12 for the notion of *asymptotic stationarity*).

Corollary 5.1 *Under the hypothesis of Theorem 5.1, the following hold, irrespective of the (initial) distribution of* $(U_0, U_1, \ldots, U_{k-1})$.

(a) The distribution of U_n converges in total variation distance, exponentially fast as $n \to \infty$, to the probability measure $\bar{\pi}$ induced from π by a coordinate projection, i.e., whatever be j, $1 \leq j \leq k$,

$$\bar{\pi}(B) := \pi(\{\mathbf{x} \equiv (x_1, \ldots, x_k) \in \mathbb{R}^k, x_j \in B\}), \quad B \in \mathcal{B}(\mathbb{R}). \quad (5.15)$$

(b) $\{U_n: n \geq 0\}$ is asymptotically stationary in the strong sense.

Remark 5.1 One may prove the existence and uniqueness of an invariant distribution of $X_n(n \geq 0)$, as well as stability in distribution under broader assumptions than boundedness of f, if the density (component) of the distribution Q of η_n is assumed to be *strictly positive on all of* $\mathbb{R}$ (see Complements and Details).

We next turn to NLAR(k) models which do not require the assumption of having a nonzero absolutely continuous component. Recall that $f: \mathbb{R}^k \to \mathbb{R}$ is nondecreasing (respectively, nonincreasing) if $f(\mathbf{x}) \leq f(\mathbf{y})$ (respectively, $f(\mathbf{y}) \leq f(\mathbf{x})$) if $\mathbf{x} \leq \mathbf{y}$, where $\mathbf{x} \equiv (x_1, \ldots, x_k) \leq (y_1, \ldots, y_k) \equiv \mathbf{y}$ if and only if $x_j \leq y_j$ for all j, $1 \leq j \leq k$.

For the statement of the theorem below, recall from (5.23) of Chapter 3 the class $\mathcal{A}$ of all subsets A of $\mathbb{R}^k$ of the form

$$A = \{\mathbf{y} \in \mathbb{R}^k: \phi(\mathbf{y}) \leq \mathbf{x}\}, \quad (5.16)$$

for some continuous and monotone ϕ on $\mathbb{R}^k$ into $\mathbb{R}^k$. Then the distance d_A on $\mathcal{P}(\mathbb{R}^k)$ is $d_A(\mu, \nu) = \sup\{|\mu(A) - \nu(A)|: A \in \mathcal{A}\}$. As usual, supp($Q$) denotes the support of the probability measure Q.

Theorem 5.2 *Consider the model (5.1), with $g \equiv 1$. Assume $f: \mathbb{R}^k \to [a_1, b_1]$ is continuous and monotone (i.e., monotone nondecreasing or nonincreasing). If* sup(supp(Q)) − inf(supp(Q)) > $b_1 − a_1$, *then*

(a) $X_n (n \geq k - 1)$ has a unique invariant probability π and (b) X_n converges exponentially fast in distribution to π in the distance d_A, uniformly with respect to all distributions of $X_{k-1} \equiv (U_0, \ldots, U_{k-1})$, i.e.,

$$\sup_{\mathbf{x} \in \mathbb{R}^k} d_A(p^{(n)}(\mathbf{x}, .), \pi(.)) \leq (1 - \chi)^{[n/k]} \ (n \geq 1) \tag{5.17}$$

for some constant $\chi > 0$.

Proof. We will apply Corollary 5.2 of Chapter 3. For notational convenience, consider the Markov process $\{Y_n : n \geq 0\}$ by

$$Y_n := X_{n+k-1} \equiv (U_n, U_{n+1}, \ldots, U_{n+k-1}), \quad (n \geq 0), \tag{5.18}$$

defined on the parameter set $\{n \geq 0\}$. Define the i.i.d. monotone maps $(\mathbb{R}^k \rightarrow \mathbb{R}^k)$

$$\alpha_n(\mathbf{x}) := (x_2, \ldots, x_k, f(\mathbf{x}) + \eta_{n+k-1}), \quad n \geq 0,$$
$$\forall \, \mathbf{x} = (x_1, x_2, \ldots, x_k) \in \mathbb{R}^k. \tag{5.19}$$

With initial state $Y_0 \equiv \mathbf{x} = (U_0, U_1, \ldots, U_{k-1})$, write $Y_n = Y_n(\mathbf{x})$. Note that

$$Y_1(\mathbf{x}) = (U_1, U_2, \ldots, U_k) = (x_2, x_3, \ldots, x_k, f(\mathbf{x}) + \eta_k) = \alpha_1(\mathbf{x}),$$
$$U_{k+1} = f(U_1, U_2, \ldots, U_k) + \eta_{k+1} = f(Y_1(\mathbf{x})) + \eta_{k+1} = f(\alpha_1(\mathbf{x})) + \eta_{k+1},$$
$$Y_2(\mathbf{x}) = (U_2, U_3, \ldots, U_{k+1}) = (x_3, x_4, \ldots, x_k, f(\mathbf{x}) + \eta_k, f(\alpha_1(\mathbf{x})) + \eta_{k+1})$$
$$= \alpha_2 \alpha_1(\mathbf{x}),$$
$$Y_k(\mathbf{x}) = (f(\mathbf{x}) + \eta_k, f(\alpha_1(\mathbf{x})) + \eta_{k+1}, f(\alpha_2\alpha_1(\mathbf{x}))$$
$$+ \eta_{k+2}, \ldots, f(\alpha_{k-1} \cdots \alpha_2\alpha_1(\mathbf{x})) + \eta_{2k-1})$$
$$= \alpha_k Y_{k-1}(\mathbf{x}) = \alpha_k \alpha_{k-1} \cdots \alpha_1(\mathbf{x}). \tag{5.20}$$

To verify the splitting condition **(H)** of Corollary 5.2 of Chapter 3, fix $x_0 \in (c + b_1, d + a_1)$, and let $\mathbf{z}_0 = (x_0, x_0, \ldots, x_0) \in \mathbb{R}^k$. Since $a_1 \leq f(\mathbf{x}) \leq b_1 \ \forall \mathbf{x}$, $x_0 - b_1 > c$. Now $f(\mathbf{x}) + \eta_j \leq b_1 + \eta_j \ \forall \mathbf{x}, \ \forall j$. Therefore, $P(Y_k(\mathbf{x}) \leq \mathbf{z}_0 \ \forall \mathbf{x}) \geq P(b_1 + \eta_k \leq x_0, \ b_1 + \eta_{k+1} \leq x_0, \ldots, b_1 + \eta_{2k-1} \leq x_0) = \delta_1^k, \delta_1 := P(\eta_k \leq x_0 - b_1) > 0$, by the assumption on the support of Q. Similarly, $f(\mathbf{x}) + \eta_j \geq a_1 + \eta_j \ \forall \mathbf{x}, \ \forall j$. Also, $x_0 - a_1 < d$. Therefore,

$$P(Y_k(\mathbf{x}) \geq \mathbf{z}_0 \ \forall \mathbf{x})$$
$$\geq P(\eta_k \geq x_0 - a_1, \eta_{k+1} \geq x_0 - a_1, \ldots, \eta_{2k-1} \geq x_0 - a_1)$$
$$= \delta_2^k, \delta_2 := P(\eta_k \geq x_0 - a_1) > 0.$$

We have shown that Corollary 5.2 of Chapter 3 applies with $N = k$, $\tilde{\chi} = \min\{\delta_1^k, \delta_2^k\}$ and $\mathbf{z}_0$ as above. ∎

Remark 5.2 Under the hypothesis of Theorem 5.2, $\{U_n : n \geq 0\}$ converges in distribution to a probability $\bar{\pi}$, irrespective of the initial variables $U_0, U_1, \ldots, U_{k-1}$. Also, it is asymptotically stationary, by Theorem 12.1 of Chapter 2.

Remark 5.3 The assumption of *continuity* of f in Theorem 5.2 may be relaxed to that of *measurability*, provided f is monotone *nondecreasing*. See Complements and Details, Chapter 3, Theorem C5.1.

4.6 Random Continued Fractions

In this section of the chapter we consider an interesting class of random dynamical systems on the state space $S = (0, \infty)$, defined by

$$X_{n+1} = Z_{n+1} + \frac{1}{X_n} \quad (n \geq 0), \tag{6.1}$$

where $\{Z_n : n \geq 1\}$ are i.i.d. nonnegative random variables independent of an initial X_0. One may express X_n as $X_n = \alpha_n \cdots \alpha_1 X_0$, where the i.i.d. monotone decreasing maps on S into S are defined by

$$\alpha_n x = Z_n + \frac{1}{x} \quad x \in (0, \infty) \quad (n \geq 0). \tag{6.2}$$

Thus, writing $[x; y]$ for $x + 1/y$, $[z; x, y] = z + 1/[x; y]$, etc.,

$$X_1 = Z_1 + \frac{1}{X_0} = [Z_1; X_0],$$

$$X_2 = Z_2 + \frac{1}{X_1} = Z_2 + \frac{1}{Z_1 + \frac{1}{X_0}} = [Z_2; Z_1, X_0],$$

$$X_3 = Z_3 + \frac{1}{X_2} = Z_3 + \frac{1}{Z_2 + \frac{1}{Z_1 + \frac{1}{X_0}}} = [Z_3; Z_2, Z_1, X_0], \ldots,$$

$$X_n = [Z_n; Z_{n-1}, Z_{n-2}, \ldots, Z_1, X_0], \quad (n \geq 1). \tag{6.3}$$

In this notation, the backward iterates $Y_n = \alpha_1 \cdots \alpha_n X_0$ may be expressed as

$$Y_0 = X_0,$$

$$Y_1 = \alpha_1 X_0 = Z_1 + \frac{1}{X_0} = [Z_1; X_0],$$

$$Y_2 = \alpha_1 \alpha_2 X_0 = Z_1 + \frac{1}{\alpha_2 X_0} = Z_1 + \frac{1}{Z_2 + \frac{1}{X_0}}$$

$$= [Z_1; Z_2, X_0], \ldots, Y_n = [Z_1; Z_2, \ldots, Z_n, X_0], \quad (n \geq 1). \quad (6.4)$$

These are *general continued fraction expansions* as we will describe in Subsection 4.6.1.

One may think of (6.1) as a model of growth, which reverses when the threshold $x = 1$ is crossed, fluctuating wildly between small and large values.

4.6.1 Continued Fractions: Euclid's Algorithm and the Dynamical System of Gauss

There is a general way of expressing fractions by successively computing the *integer part of the reciprocal of the preceding fractional part*. If the process stops in four steps, for example, then one has

$$x = \frac{1}{a_1 + \frac{1}{a_2 + \frac{1}{a_3 + \frac{1}{a_4}}}}. \quad (6.5)$$

This is referred to as *Euclid's algorithm* for reducing a rational number, say, c/d to its simplest form p/q where p and q have no common factor other than 1. For example, noting that $231/132 = 1 + 99/132, 132/99 = 1 + 33/99 = 1 + 1/3$, we have

$$\frac{132}{231} = \frac{1}{1 + \frac{1}{1 + \frac{1}{3}}} = \frac{1}{1 + \frac{3}{4}} = \frac{4}{7} = [1, 1, 3].$$

In general, if x is rational, then one has a finite continued fraction expansion. If x is irrational, the process does not terminate, and one has the infinite continued fraction expansion

$$x = [a_1, a_2, \ldots, a_n, \ldots] = \frac{1}{a_1 + \frac{1}{a_2 + \frac{1}{\cdots}}}. \quad (6.6)$$

Here $x = \lim_{n \to \infty} p_n/q_n$ where $p_n/q_n = [a_1, a_2, \ldots, a_n]$ is called the *nth convergent*, with p_n and q_n being relatively prime positive integers.

The process of computing the successive convergents may be described in terms of the *dynamical system* on $I = [0, 1)$ given by

$$T(x) := \begin{cases} \frac{1}{x} - [\frac{1}{x}] & \text{if } x \neq 0, \\ 0 & \text{if } x = 0, \end{cases} \quad \begin{array}{l} \text{where } [y] \text{ denotes the integer} \\ \text{part of y.} \end{array} \quad (6.7)$$

For any $x \in I, x \neq 0$, if we put $x_0 = x$ and $x_n = T^n x$ for $n \geq 1$, then the continued fraction expansion (6.6) is given in terms of $x_n \equiv T^n x$ ($n \geq 0$) by

$$a_1 = \left[\frac{1}{x_0}\right], a_2 = \left[\frac{1}{x_1}\right], \ldots, a_n = \left[\frac{1}{x_n}\right], \ldots \quad (6.8)$$

with the convention that the process terminates at the nth step if $x_n = 0$. In a famous discovery sometime around 1812, Gauss showed that the dynamical system (6.7) has an *invariant probability measure* with density g and distribution function G given by

$$g(y) = \frac{1}{(\log 2)(1 + y)}, \quad G(y) = \frac{\log(1 + y)}{\log 2}, \quad 0 < y < 1, \quad (6.9)$$

and that

$$G_n(y) := \mathbf{m}(\{x \in I : T^n x \leq y\}) \to G(y) \quad \text{as } n \to \infty, \quad 0 < y < 1, \quad (6.10)$$

where $\mathbf{m}$ is the Lebesgue measure on $I = [0, 1)$. Since one may think of a dynamical system as a Markov process with a degenerate transition probability, which in the present case is $p(x, \{Tx\}) = 1 \ \forall x \in I$, (6.10) says that if X_0 has distribution $\mathbf{m}$ on I (i.e., the uniform distribution on I) then the (Markov) process $X_n := T^n X_0$ converges in distribution to the invariant distribution (6.9), as $n \to \infty$. If X_0 has the invariant distribution $g(y)dy$, then $X_n = T^n X_0$ ($n \geq 1$) is a stationary process. Proofs of (6.10), and that of the fact that the dynamical system defined by (6.7) has infinitely many periodic orbits, may be found in Khinchin (1964) and Billingsley (1965).

4.6.2 General Continued Fractions and Random Continued Fractions

For our purposes, we will consider *general continued fractions* defined as follows. Given any sequence of numbers $a_0 \geq 0$, and $a_i > 0 (i \geq 1)$, one may formally define the sequence of "convergents" p_n/q_n by

$$r_0 = \frac{a_0}{1} = \frac{p_0}{q_0},$$

$$r_1 = a_0 + \frac{1}{a_1} = \frac{a_0 a_1 + 1}{a_1} = \frac{p_1}{q_1}, \tag{6.11}$$

$$r_2 = a_0 + \frac{1}{a_1 + \frac{1}{a_2}} = a_0 + \frac{a_2}{a_1 a_2 + 1} = \frac{a_0 a_1 a_2 + a_0 + a_2}{a_1 a_2 + 1} = \frac{p_2}{q_2}.$$

In general, one uses the fact $[a_0; a_1, a_2, \ldots, a_{n+1}] = [a_0; a_1, a_2, \ldots, a_n + \frac{1}{a_{n+1}}]$ to derive (by induction) the recursions

$$
\begin{aligned}
p_0 &= a_0, \; q_0 = 1, \quad p_1 = a_0 a_1 + 1, \quad q_1 = a_1; \\
p_n &= a_n p_{n-1} + p_{n-2}, \quad q_n = a_n q_{n-1} + q_{n-2} (n \geq 2),
\end{aligned}
$$

$$[a_0; a_1, \ldots, a_n] := \frac{p_n}{q_n}. \tag{6.12}$$

If the sequence p_n/q_n converges, then we express the limit as the infinite continued fraction expansion $[a_0; a_1, a_2, \ldots]$.

Since from (6.12) we also get

$$p_{n+1} q_n - p_n q_{n+1} = -(p_n q_{n-1} - p_{n-1} q_n) \text{ for all } n, \tag{6.13}$$

one has, on dividing both sides by $q_{n+1} q_n$ and iterating,

$$\frac{p_{n+1}}{q_{n+1}} - \frac{p_n}{q_n} = -\frac{q_{n-1}}{q_{n+1}} \left(\frac{p_n}{q_n} - \frac{p_{n-1}}{q_{n-1}} \right), \quad \frac{p_{n+1}}{q_{n+1}} - \frac{p_n}{q_n} = \frac{(-1)^n}{q_{n+1} q_n}, \tag{6.14}$$

since $p_1/q_1 - p_0/q_0 = 1/a_1$.

Note that the *condition* $a_i > 0$ $(i \geq 1)$ above ensures that $q_n > 0$ for all $n \geq 1$ (as q_0 is defined to be 1). Hence $[a_0; a_1, \ldots, a_n] = p_n/q_n$ is well defined for all $n \geq 0$.

We will need to relax the above condition to

$$a_i \geq 0 \quad \forall i \geq 0, \quad a_m > 0 \text{ for some } m \geq 1. \tag{6.15}$$

If $m \geq 1$ is the smallest integer such that $a_m > 0$, then it is clear from the expressions (6.5), (6.6) that $[a_0; a_1, \ldots, a_n] \equiv a_0 + [a_1, a_2, \ldots, a_n]$ is well defined with $q_n > 0$ for $n \geq m$. The recursions in (6.12) then hold for $n \geq m + 2$. Then (6.13) holds for all $n \geq m + 1$. Hence, in place of (6.14) one gets

$$\frac{p_{n+1}}{q_{n+1}} - \frac{p_n}{q_n} = (-1)^{n-m} \frac{q_{m+1} q_m}{q_{n+1} q_n} \left(\frac{p_{m+1}}{q_{m+1}} - \frac{p_m}{q_m} \right) \quad (n \geq m). \tag{6.16}$$

Now the right-hand side is alternately positive and negative, and the magnitude of the n-th term is nonincreasing in n, as $q_{n+1} \geq q_{n-1}$ and $q_{n+1}q_n \geq q_n q_{n-1}$. Therefore, one has

$$\left| \frac{p_{n+j}}{q_{n+j}} - \frac{p_n}{q_n} \right| \leq \frac{q_{m+1}q_m}{q_{n+1}q_n} \left| \frac{p_{m+1}}{q_{m+1}} - \frac{p_m}{q_m} \right| \quad \forall n \geq m, \; j \geq 1. \quad (6.17)$$

Therefore, $[a_0; a_1, \ldots, a_n] \equiv p_n/q_n$ converges to a finite limit if and only if $q_{n+1}q_n \to \infty$ as $n \to \infty$. We may use this to prove a result on general stability of the Markov process $X_n (n \geq 0)$ defined by (6.1).

Proposition 6.1 *If* $\{Z_n : n \geq 1\}$ *is a nondegenerate nonnegative i.i.d. sequence, then the Markov process* $X_n (n \geq 0)$ *defined by (6.1) on* $S = (0, \infty)$ *converges in distribution to a unique invariant probability, no matter what the (initial) distribution of* X_0 *is.*

Proof. Since Y_n defined by (6.4) has the same distribution as X_n (for every n), it is enough to show that Y_n converges in distribution to the same distribution, say, π, whatever X_0 may be. Since $P(Z_n > 0) > 0$, by the assumption of nondegeneracy, it follows from the (*second*) Borel–Cantelli lemma that, with probability 1, $Z_n > 0$ for infinitely many n. In particular, with probability 1, there exists $m = m(\omega)$ finite such that $Z_m > 0$. Then defining $p_n/q_n = [Z_1; Z_2, \ldots, Z_{n+1}]$ in place of $[a_0; a_1, \ldots, a_n]$ (i.e., take $a_0 = Z_1$, $a_j = Z_{j+1}$, $j \geq 1$), p_n/q_n is well defined for $n \geq m = m(\omega)$ (indeed, for $n \geq m(\omega) - 1$). Also, by recursion (6.12) for q_n, one has, for all $n \geq m + 2$,

$$q_{n+1} = a_{n+1}q_n + q_{n-1} = Z_{n+2}q_n + q_{n-1},$$
$$q_{n+1}q_n = Z_{n+2}q_n^2 + q_n q_{n-1}$$
$$= \cdots = \sum_{j=m+1}^{n} Z_{j+2}q_j^2 + q_{m+1}q_m$$
$$\geq \sum_{i=2\left[\frac{m+2}{2}\right]}^{2[n/2]} Z_{i+2}q_i^2 \geq q_{2\left[\frac{m+2}{2}\right]}^2 \sum_{i=2\left[\frac{m+2}{2}\right]}^{2[n/2]} Z_{i+2}, \quad (6.18)$$

where $[r/2]$ is the integer part of $r/2$. Note that, for the first inequality in (6.18), we picked out terms $Z_{i+2}q_i^2$ with i an even integer. For the second inequality, we have used the fact that q_{2j} is increasing in j (use the recursion $q_{2j} = a_{2j}q_{2j-1} + q_{2j-2}$ in (6.12)). Now there exists

$c > 0$ such that $P(Z_1 > c) > 0$. Applying the (*second*) Borel-Cantelli lemma, it follows that, with probability one, $Z_{2n} > c$ for infinitely many n. This implies $\sum_{j=1}^{N} Z_{2j} \to \infty$ a.s. as $N \to \infty$. Hence $q_{n+1}q_n \to \infty$ a.s., proving that $[Z_1; Z_2, \ldots, Z_{n+1}]$ converges a.s. to some random variable X, say. Finally, note that $Y_{n+1} - [Z_1; Z_2, \ldots, Z_{n+1}] \equiv [Z_1; Z_2, \ldots, Z_{n+1}, X_0] - [Z_1; Z_2, \ldots, Z_{n+1}] \to 0$ a.s., uniformly with respect to X_0 as $n \to \infty$, by (6.17). Hence Y_n converges in distribution to X, irrespective of the initial X_0, and so does X_n. ∎

We will next strengthen Proposition 6.1 by verifying that the splitting hypothesis (**H**) of Theorem 5.1, Chapter 3, holds.

Theorem 6.1 *Under the hypothesis of* Proposition 6.1, *the convergence in distribution of X_n to a unique nonatomic invariant probability is exponentially fast in the Kolmogorov distance, uniformly for all initial distributions.*

Proof. Since the i.i.d. maps in (6.2) are monotone (decreasing) on $S = (0, \infty)$, in order to prove the theorem it is enough to verify that the splitting hypothesis (**H**) of Theorem 5.1, Chapter 3, holds. For this purpose note that there exist $0 < a < b < \infty$ such that $P(Z_n \leq a) > 0$ and $P(Z_n \geq b) > 0$. Check that the formal expressions $[a; b, a, b, a, \ldots]$ and $[b; a, b, a, b, \ldots]$ are both convergent. Indeed, for the first $q_{2n} \geq (ab + 1)^n$, $q_{2n+1} \geq b(ab + 1)^n$ $(n = 0, 1, 2, \ldots)$, as may be checked by induction on n, using (6.12). For the second expression, the corresponding inequalities hold with a and b interchanged. Thus $q_n q_{n+1} \to \infty$ in both cases.

We will show that $x_0 := [a; b, a, b, a, \ldots] < y_0 := [b; a, b, a, b, \ldots]$ and that the splitting hypothesis (**H**) holds with $z_0 = (x_0 + y_0)/2$, if N is appropriately large. For this we first observe that $y_0 - x_0 > b - a$. For, writing $c = [a, b, a, b, \ldots]$ in the notation of (6.6), one has $y_0 = b + c$ and $x_0 = a + 1/(b + 1/c)$, so that $y_0 - x_0 = b - a + c^2 b/(bc + 1) > b - a$. Now for any given $n \geq 1$, the function $f_n(a_0, a_1, \ldots, a_n) := [a_0; a_1, \ldots, a_n]$ is *strictly increasing in the variable a_{2j} and strictly decreasing in the variable a_{2j+1}* $(2j + 1 \leq n)$, *provided $a_1, \ldots, a_n$ are positive.* To see this, one may use induction on n as follows: For $n = 1, 2$, for example, check this by looking at (6.11). Suppose the statement holds for some $n \geq 1$. Consider the function $f_{n+1}(a_0, a_1, \ldots, a_n, a_{n+1})$. Then, by definition of the continued fraction expression, $f_{n+1}(a_0, a_1, \ldots, a_n, a_{n+1}) = f_n(a_0, a_1, \ldots, a_n + 1/a_{n+1})$. If n is even, $n = 2m$, then by the induction hypothesis $f_n(a_0, a_1, \ldots, a_n + 1/a_{n+1}) > f_n(a_0, a_1, \ldots, a_n + 1/b_{n+1}) \equiv f_{n+1}(a_0, a_1, \ldots, a_n, b_{n+1})$ if $a_{n+1} < b_{n+1}$, proving f_{n+1} is strictly decreasing in the $(n + 1)$th coordinate. Similarly, if n is odd, $n = 2m + 1$, then $f_{n+1}(a_0, a_1, \ldots, a_n, a_{n+1}) \equiv f_n(a_0, a_1, \ldots, a_n + 1/a_{n+1}) < f_n(a_0, a_1, \ldots, a_n + 1/b_{n+1})$ if $a_{n+1} < b_{n+1}$, by the induction hypothesis. Since in this representation of f_{n+1} in terms of f_n, the remaining coordinates a_j, $1 \leq j < n$ are unaltered, by the induction hypothesis, f_{n+1} is strictly increasing in a_{2j} and strictly decreasing in a_{2j+1} ($2j + 1 < n, 2j < n$). The argument for the nth coordinate also follows in the above manner. Thus the induction is complete.

Next, let N be such that the Nth convergent p_N/q_N of $x_0 = [a; b, a, b, a, \ldots]$ differs from x_0 by less than $\varepsilon := (z_0 - x_0)/4$ and the Nth convergent p_N'/q_N', say, of $y_0 = [b; a, b, a, b \ldots]$ differs from y_0 by less than $\varepsilon = (y_0 - z_0)/4$. Note that if one writes $p_N/q_N = [a_0; a_1, \ldots, a_N]$ then whatever be x, $[a_0; a_1, \ldots, a_N, x]$ differs from x_0 by less than ε, since the difference between a number u and its $(N + 1)$th convergent is no larger than that between u and its Nth convergent. (Take $m = 0$ in (6.16) and (6.17), and let $j \uparrow \infty$ in (6.17).) The same is true of the $(N + 1)$th convergent of $y_0 = [b; a, b, a, b, \ldots]$. Without loss of generality, let N be *even*. By the argument in the preceding paragraph, on the set $A_1 = \{Z_1 \leq a, Z_2 \geq b, Z_3 \leq a, \ldots, Z_{N+1} \leq a\}$, one has

$$Y_N(x) \equiv [Z_1; Z_2, \ldots, Z_{N+1}, x] \leq [a; \overbrace{b, a, b, \ldots, a}^{N \text{ terms}}, x]$$
$$\leq x_0 + \varepsilon < z_0 \, \forall x \in (0, \infty). \tag{6.19}$$

Similarly, on the set $A_2 = \{Z_1 \geq b, Z_2 \leq a, Z_3 \geq b, \ldots, Z_{N+1} \geq b\}$,

$$Y_N(x) \geq [b; a, b, a, \ldots, b, x] \geq y_0 - \varepsilon > z_0 \quad \forall x \in (0, \infty). \tag{6.20}$$

Letting $\chi_1 = P(A_1) \geq \delta_1^{N+1}$, $\chi_2 = P(A_2) \geq \delta_1^{N+1}$, where $\delta_1 = \min\{P(Z_1 \leq a), P(Z_1 \geq b)\} > 0$, and noting that $\{X_N(x) < z_0 \, \forall x \in (0, \infty)\}$, $\{X_N(x) > z_0 \, \forall x \in (0, \infty)\}$ have the same probabilities as $\{Y_N(x) < z_0 \, \forall x \in (0, \infty)\}$ and $\{Y_N(x) > z_0 \, \forall x \in (0, \infty)\}$, respectively, we have shown that (**H**) holds with $\chi = \delta_1^{N+1}$. Since the i.i.d. maps in (6.2) are strictly decreasing, it follows from Remark 5.4 of Chapter 3 that the unique invariant probability is nonatomic. ∎

Example 6.1 (*Gamma innovation*).   Consider the case where the distribution of Z_n is $\Gamma(\lambda, \beta)$ ($\lambda > 0, \beta > 0$), with density

$$g_{\lambda,\beta}(z) = \frac{\beta^\lambda}{\Gamma(\lambda)} z^{\lambda-1} e^{-\beta z} \mathbf{1}_{(0,\infty)}(z).$$

From (6.1) it follows that if X is a random variable having the invariant distribution π on $(0, \infty)$, then one has the *distributional identity*

$$X \overset{\mathcal{L}}{=} Z + \frac{1}{X}, \tag{6.21}$$

where, on the right-hand side, X and Z are independent and Z has the innovation distribution, which is $\Gamma(\lambda, \beta)$ in this example. By a direct verification, one can check that (6.21) holds if the distribution π of X has the density

$$f_{\lambda,\beta}(x) = c(\lambda, \beta) x^{\lambda-1} e^{-\beta(x+1/x)} \mathbf{1}_{(0,\infty)}(x), \tag{6.22}$$

where $c(\lambda, \beta)$ is the normalizing constant,

$$c(\lambda, \beta) = 1 \Big/ \int_0^\infty x^{\lambda-1} e^{-\beta(x+1/x)} dx.$$

Note that the density of the random variable $1/X = Y$, say, is

$$h(y) = c(\lambda, \beta) y^{-\lambda-1} e^{-\beta(y+1/y)} \mathbf{1}_{(0,\infty)}(y).$$

It is a simple exercise to check that the convolution $h * g_{\lambda,\beta}$ equals $f_{\lambda,\beta}$ (Exercise). This proves (6.21), establishing the invariance of π with density $f_{\lambda,\beta}$. ∎

4.6.3 Bernoulli Innovation

Consider the Markov process on the state space $S = (0, \infty)$ defined by (6.1), where $Z_n (n \geq 1)$ are i.i.d.,

$$P(Z_n = 0) = \alpha, \quad P(Z_n = \theta) = 1 - \alpha, \tag{6.23}$$

with $0 < \alpha < 1$, $\theta > 0$, and X_0 is independent of $\{Z_n : n \geq 1\}$. If X_n converges in distribution to X, where X is some random variable having the unique invariant distribution π on $S = (0, \infty)$, then the probability measure π is characterized by the distributional identity (6.21), where $Z \overset{\mathcal{L}}{=} Z_1$, X has distribution π, and X and Z are independent. As before,

$\stackrel{\mathcal{L}}{=}$ denotes equality in law. As an immediate corollary of Theorem 6.1, we have the following result:

Proposition 6.2

(a) *The Markov process X_n on $S = (0, \infty)$ above has a unique invariant probability π, and the Kolmogorov distance $d_K(T^{*n}\mu, \pi)$ converges to zero exponentially fast in n, uniformly with respect to the initial distribution μ.*

(b) *π is nonatomic.*

The next proposition shows that the cases $0 < \theta \leq 1$ and $\theta > 1$ are qualitatively different.

Proposition 6.3

(a) *If $0 < \theta \leq 1$, then the support of π is full, namely $(0, \infty)$.*
(b) *If $\theta > 1$, then the support of π is a Cantor subset of $(0, \infty)$.*

For the proof of Proposition 6.3 we need two lemmas.

Lemma 6.1 *Let $0 < \theta \leq 1$. Then the set of all continued fractions $[a_0\theta; a_1\theta, a_2\theta, \ldots]$, a_i's are integers, $a_0 \geq 0$, $a_i \geq 1(i \geq 1)$, comprises all of $(0, \infty)$.*

Proof. Let $x \in (0, \infty)$. Define $a_0 = \max\{n \geq 0: n\theta \leq x\}$. If $x = a_0\theta$, then $x = [a_0\theta]$. If not, define r_1 by

$$x = a_0\theta + \frac{1}{r_1}. \tag{6.24}$$

Since $0 < \frac{1}{r_1} < \theta$, $r_1 > \frac{1}{\theta} \geq \theta$. Define $a_1 = \max\{n \geq 0: n\theta \leq r_1\}$. If $r_1 = a_1\theta$, $x = [a_0\theta; a_1\theta]$. If not, we define r_2 by

$$r_1 = a_1\theta + \frac{1}{r_2}, \tag{6.25}$$

$r_2 > \theta$, and let $a_2 = \max\{n \geq 0: n\theta \leq r_2\}$. Proceeding in this way we arrive at either a terminating expansion $[a_0\theta; a_1\theta, \ldots, a_n\theta]$ or an infinite expansion $[a_0\theta; a_1\theta, a_2\theta, \ldots]$. We need to show that, in the latter case, the expansion converges to x. Now use the relation (6.17) with $m = 0$, and $\alpha_i\theta$ in the place of $a_i(i \geq 0)$. One gets

$p_n/q_n = [a_0\theta; a_1\theta, \ldots, a_n\theta]$, and $x = [a_0\theta; a_1\theta, \ldots, a_{n-1}\theta, \ a_n\theta + \frac{1}{r_{n+1}}]$, $1/r_{n+1} < \theta$,

$$\left| x - \frac{p_n}{q_n} \right| < \frac{1}{q_{n+1} q_n}. \tag{6.26}$$

Since $a_i \geq 1 \ \forall i \geq 1$, it is simple to check as before, using (6.12), that

$$q_0 = 1, \quad q_1 = a_1\theta \geq \theta, \quad q_2 \geq q_0 = 1, \quad q_3 \geq q_1 \geq \theta;$$

$$q_{2n} \geq q_0, q_{2n+1} \geq q_1 \ \forall n;$$

$$q_n = a_n\theta q_{n-1} + q_{n-2} \geq \theta^2 + q_{n-2} \ \forall n \geq 2.$$

Hence $q_n \geq [\frac{n}{2}]\theta^2 \to \infty$, and (6.26) $\to 0$ as $n \to \infty$. $\blacksquare$

Remark 6.1 The arguments made in the above proof are analogous to those for the case $\theta = 1$. But the assertion in the lemma does not hold for $\theta > 1$. The reason why the argument breaks down is that if $\theta > 1$, then r_1 defined by (6.25) may be smaller than θ.

Lemma 6.2 *Let $p^{(n)}(x, dy)$ be the n-step transition probability, $n = 1, 2, \ldots$ of a Markov process on $(S, \mathcal{S})$, where S is a separable metric space and $\mathcal{S}$ its Borel sigma field. Assume $x \to p^{(1)}(x, dy)$ is weakly continuous, and π is an invariant probability. If x_0 is in the support of π and $p^{(n)}(x_0, dy)$ converges weakly to π, then the support of π is the closure of $\cup_{n=1}^{\infty} \{support\ of\ p^{(n)}(x_0, dy)\}$.*

Proof. Let F denote the support of π, and $G := \cup_{n=1}^{\infty} \{$support of $p^{(n)}(x_0, dy)\}$. Fix $\varepsilon > 0, n \geq 1$. Let $B(y, \delta)$ denote the open ball in S with center y and radius $\delta > 0$. Suppose y_0 is such that $p^{(n)}(x_0, B(y_0, \varepsilon)) > \delta_1 > 0$. Since $\lim\inf_{x \to x_0} p^{(n)}(x, B(y_0,\varepsilon)) \geq p^{(n)}(x_0, B(y_0, \varepsilon))$, by weak convergence, there exists $\delta_2 > 0$ such that $p^{(n)}(x, B(y_0, \varepsilon)) > \delta_1 \ \forall x \in B(x_0, \delta_2)$. Then

$$\pi(B(y_0, \varepsilon)) = \int_S p^{(n)}(x, B(y_0, \varepsilon))\pi(dx)$$

$$\geq \delta_1 \pi(B(x_0, \delta_2)) > 0,$$

since x_0 is in the support F of π. This proves $G \subset F$. To prove $F \subset \bar{G} \equiv$ closure of G, let $y_0 \in F$. If $y_0 \notin \bar{G}$, then there exists $\delta > 0$ such that $B(y_0, \delta) \cap G = \emptyset$, which implies $p^{(n)}(x_0, B(y_0, \delta)) = 0 \ \forall n \geq 1$. This is impossible since $\lim\inf_{n \to \infty} p^{(n)}(x_0, B(y_0, \delta)) \geq \pi(B(y_0, \delta)) > 0$. $\blacksquare$

Proof of proposition 6.3.

(a) We will make use of Lemma 6.2. Let $F \subset (0, \infty)$ denote the support of π. If $x \in F$, then since $X \overset{\mathcal{L}}{=} Z + \frac{1}{X}$, $\theta + \frac{1}{x} \in F$ and $\frac{1}{x} \in F$, and, in turn, $\theta + x \in F$. Thus F is closed under *inversion*: $x \to \frac{1}{x}$ and under *translation* by θ: $x \to x + n\theta (n = 1, 2, \ldots)$. Let $0 < \theta \leq 1$, and $x_0 \in F$. Then $x_0 + n\theta \in F$ $\forall n = 1, 2, \ldots$, and $(x_0 + n\theta)^{-1} \in F$ $\forall n$, and, therefore, 0 is the infimum of F. Also, $\theta + (x_0 + n\theta)^{-1} \in F$, $\forall n$, so that $\theta \in F$. Let us show that every finite continued fraction $x = [a_0\theta; a_1\theta, \ldots, a_n\theta]$ with $n + 1$ terms belong to F, where $a_0 \geq 0$, $a_i \geq 1$, for $i \geq 1$, are integers. This is true for $n = 0$. Suppose this is true for some integer $n \geq 0$. Take $a_i \geq 1$ $\forall i = 0, 1, \ldots, n$. Then if a is a nonnegative integer, $[a\theta; a_0\theta, a_1\theta, \ldots, a_n\theta] = a\theta + \frac{1}{x} \in F$. Thus all continued fractions with $n + 2$ terms belong to F. Hence, by induction, all continued fractions with finitely many terms belong to F. Thus the closure of this latter set in $(0, \infty)$ is contained in F. By Lemma 6.1, $F = (0, \infty)$.

(b) Let $\theta > 1$. Let $F \subset (0, \infty)$ denote the support of π. As in (a), (i) $\theta \in F$, (ii) F is closed under *inversion* (i.e., if $x \in F$, then $\frac{1}{x} \in F$) and *translation* by θ (i.e., if $x \in F$, $x + \theta \in F$). Consider now the Markov chain (6.1) with $X_0 = \theta$. From Lemma 6.2, and the fact that the distribution $p^{(n)}(\theta, dy)$ of X_n has a finite support (for each $n = 1, 2, \ldots$), it follows that F is the closure (in $(0, \infty)$) of the set $\cup_{n=1}^{\infty}\{y: p^{(n)}(\theta, \{y\}) > 0\}$.

This and (i) and (ii) imply that F is precisely the set of all continued fractions of the form

$$[a_0\theta; a_1\theta, a_2\theta, \ldots] \ (a_i \text{ integers } \forall i \geq 0, a_0 \geq 0, a_i \geq 1 \ \forall i > 0).$$

$$(6.27)$$

Now use the hypothesis $\theta > 1$ to see, using (6.21), that

$$\text{Prob}\left(\frac{1}{\theta} < X < \theta\right) = \text{Prob}\left(\frac{1}{\theta} < Z + \frac{1}{X} < \theta\right)$$

$$= \alpha \ \text{Prob}\left(\frac{1}{\theta} < \frac{1}{X} < \theta\right) + \beta \ \text{Prob}\left(\frac{1}{\theta} < \theta + \frac{1}{X} < \theta\right)$$

$$= \alpha \ \text{Prob}\left(\frac{1}{\theta} < \frac{1}{X} < \theta\right) + \beta \cdot 0 = \alpha \ \text{Prob}\left(\frac{1}{\theta} < X < \theta\right),$$

$$(6.28)$$

which implies $\pi((\frac{1}{\theta}, \theta)) = 0$, i.e., $(\frac{1}{\theta}, \theta) \cap F = \emptyset$. More generally, the only numbers x that cannot be expressed as (6.27) are of the form (see the proof of Lemma 6.1 and Remark 6.1 following it),

$$x = [a_0\theta; a_1\theta, \ldots, a_n\theta, y], \quad \text{with } \frac{1}{\theta} < y < \theta. \tag{6.29}$$

To show that F is nowhere dense in $(0, \infty)$, let $0 < c < d$. We will show that there exists $x \in (c, d)$ such that $x \notin F$. This is obvious if either c or d does not belong to F. So assume $c, d \in F$. This means c and d have continued fraction expansions

$$c = [a_0\theta; a_1\theta, a_2\theta, \ldots], \quad d = [b_0\theta; b_1\theta, b_2\theta, \ldots]. \tag{6.30}$$

Since $c < d$, it is clear that $a_0 \leq b_0$, and if $a_0 = b_0$, then $a_1 \geq b_1$. If $a_0 = b_0$, $a_1 = b_1$, then $a_2 \leq b_2$, and so on. Let $N = \min\{k \geq 0: a_k \neq b_k\}$. First consider the case N is even. Then $a_N < b_N$. Now if $c = [a_0\theta; a_1\theta, \ldots, a_N\theta]$ and $y \in \left(\frac{1}{\theta}, \theta\right)$, then $x := [a_0\theta; a_1\theta, \ldots, a_N\theta, y] > c$, and $x < [a_0\theta; a_1\theta, \ldots, a_{N-1}\theta, (a_N + 1)\theta] \leq d$. Thus $x \in (c, d)$; but, in view of (6.29), $x \notin F$. If c is not equal to $[a_0\theta; a_1\theta, \ldots, a_N\theta]$, then $c = [a_0\theta; a_1\theta, \ldots, a_N\theta, r_{N+1}]$ with $0 < \frac{1}{r_{N+1}} < \theta$, i.e., $r_{N+1} > \frac{1}{\theta}$. If $y \in \left(\frac{1}{\theta}, r_{N+1} \wedge \theta\right)$, then again $c < x := [a_0\theta; a_1\theta, \ldots, a_N\theta, y]$ and $x < [a_0\theta; a_1\theta, \ldots, a_{N-1}\theta, a_N\theta, r_{N+1} \wedge \theta] \leq d$. Thus $x \in (c, d), x \notin F$. The case of N odd is similar.

Finally, note that F has no isolated point, since π is nonatomic. ∎

We now give a computation of the fascinating invariant distribution for the case $\theta = 1$.

Theorem 6.2 *Let π denote the invariant probability of the Markov process (6.1), where $P(Z_1 = 0) = \alpha$, $P(Z_1 = 1) = 1 - \alpha$, $0 < \alpha < 1$.*
(a) The distribution function F of π is given by

$$F(x) = \sum_{i=0}^{\infty} \left(-\frac{1}{\alpha}\right)^i \left(\frac{\alpha}{1+\alpha}\right)^{a_1+a_2+\cdots+a_{i+1}}, \quad 0 < x \leq 1,$$

$$F(x) = 1 - \frac{1}{\alpha} F\left(\frac{1}{x}\right), \quad x > 1. \tag{6.31}$$

Here $x = [a_1, a_2, \ldots]$ is the usual continued fraction expansion of $x \in (0, 1]$. That is, writing $[y]$ for the integer part of y,

$$a_1 = \left[\frac{1}{x}\right], \quad x_1 = \frac{1}{x} - \left[\frac{1}{x}\right], \quad x_2 = \frac{1}{x_1} - \left[\frac{1}{x_1}\right], \ldots,$$

$$x_n = \frac{1}{x_{n-1}} - \left[\frac{1}{x_{n-1}}\right], \ldots, \quad a_{i+1} = \left[\frac{1}{x_i}\right] (i \geq 1). \tag{6.32}$$

with the understanding that the series on the right-hand side in (6.31) is a finite sum that terminates at $i = n$, in the case n is the smallest index such that $x_{n+1} = 0$. (b) π is singular with full support $S = (0, \infty)$.

Proof.

(a) It has been shown that π is nonatomic (Proposition 6.2). From the basic relation (6.21), we get

$$F(x) = \alpha \left(1 - F \left(\frac{1}{x} \right) \right), \quad 0 < x \le 1, \tag{6.33}$$

or writing $\frac{1}{x}$ for x,

$$F(x) = 1 - \frac{1}{\alpha} F \left(\frac{1}{x} \right), \quad 1 \le x < \infty. \tag{6.34}$$

Also, from (6.21),

$$F(x) = 1 - \alpha F \left(\frac{1}{x} \right) - (1 - \alpha) F \left(\frac{1}{x-1} \right), \quad x > 1. \tag{6.35}$$

Using (6.34) in (6.35), one gets

$$F(x) = 1 - \alpha^2 (1 - F(x)) - (1 - \alpha) F \left(\frac{1}{x-1} \right), \quad x > 1, \tag{6.36}$$

or

$$F(x) = 1 - \frac{1}{1+\alpha} F \left(\frac{1}{x-1} \right), \quad x > 1. \tag{6.37}$$

If $x > 2$, then $\frac{1}{x-1} < 1$, so that using (6.33) in (6.32), one has

$$F(x) = 1 - \frac{\alpha}{1+\alpha} (1 - F(x - 1))$$

$$= \frac{1}{1+\alpha} + \frac{\alpha}{1+\alpha} F(x - 1), \quad x > 2. \tag{6.38}$$

Iterating this, one gets

$$F(m + y) = 1 - \left(\frac{\alpha}{1+\alpha} \right)^{m-1} + \left(\frac{\alpha}{1+\alpha} \right)^{m-1} F(1 + y),$$

$$(m = 1, 2, \ldots; 0 < y \le 1). \tag{6.39}$$

Now use (6.37) in (6.39) to get

$$F(m + y) = 1 - \frac{1}{\alpha} \left(\frac{\alpha}{1+\alpha} \right)^m F \left(\frac{1}{y} \right)$$
$$(m = 1, 2, \dots; 0 < y \le 1). \quad (6.40)$$

For $0 < x < 1$, use (6.34) and then (6.40) to get

$$F(x) = \alpha \left(1 - F \left(\frac{1}{x} \right) \right) = \left(\frac{\alpha}{1+\alpha} \right)^{\left[\frac{1}{x} \right]} F \left(\frac{1}{\frac{1}{x} - \left[\frac{1}{x} \right]} \right). \quad (6.41)$$

Finally, use (6.34) in (6.41) to arrive at the important recursive relation

$$F(x) = \left(\frac{\alpha}{1+\alpha} \right)^{a_1} - \frac{1}{\alpha} \left(\frac{\alpha}{1+\alpha} \right)^{a_1} F \left(\frac{1}{x} - \left[\frac{1}{x} \right] \right). \quad (6.42)$$

Iterating (6.42) one arrives at the first relation in (6.31). If x is rational, the iterative scheme terminates when $x_{n+1} = 0$, so that $F(x_{n+1}) = 0$. If x is irrational, one arrives at the infinite series expansion. If $x \ge 1$, (6.35) gives the second relation in (6.31).

(b) By Propositions 6.2 and 6.3, π is nonatomic and has full support $(0, \infty)$. The proof that π is singular (with respect to the Lebesgue measure) is given in Complements and Details.   ∎

Exercise 6.1 Show that, if X has the invariant distribution π of Theorem 6.2, then $E[X] = 1$, where $[X]$ is the integer part of X. In particular, $1 < EX < 2$. [Hint: $P(X \le 1) = \alpha/(1 + \alpha)$, using (6.33), and $P(X \ge m + 1) = (1/\alpha)(\alpha/(1 + \alpha))^m \, \alpha/(1 + \alpha) \, (m = 0, 1, 2, \dots)$ by (6.40). Hence, by Lemma 2.1, $E[X] = \sum_{m=0}^{\infty} P(X \ge m + 1) = 1$.]   ∎

Exercise 6.2 For the general model (6.1), prove that if X has the invariant distribution, then $EX \ge (EZ)/2 + \sqrt{1 + (EZ/2)^2}$. [Hint: $EX = EZ + E(\frac{1}{X}) \ge EZ + 1/EX$, by Jensen's inequality.]   ∎

4.7 Nonnegativity Constraints

We now consider a process in which the relevant state variable is required to be nonnegative. Formally, let $\{\varepsilon_n\}$ be a sequence of i.i.d. random variables, and write

$$X_{n+1} = \max(X_n + \varepsilon_{n+1}, 0). \quad (7.1)$$

Here the state space $S = [0, \infty)$, and the random maps $\alpha_n = f_{\varepsilon_n}$ are defined as

$$\alpha_n x = f_{\varepsilon_n}(x), \quad \text{where } f_\theta(x) = (x + \theta)^+, \quad (\theta \in \mathbb{R}). \quad (7.2)$$

Then

$$X_n(x) = f_{\varepsilon_{n+1}} f_{\varepsilon_n} \cdots f_{\varepsilon_1}(x). \quad (7.3)$$

Theorem 7.1 *Assume* $-\infty \leq E\varepsilon_1 < 0$. *The Markov process* $\{X_n\}$ *has a unique invariant probability* π, *and* X_n *converges in distribution to* π, *no matter what* X_0 *is.*

Proof. The maps f_{ε_n} are monotone increasing and continuous, and S has a smallest point, namely, 0. Writing the backward iteration

$$Y_n(x) = f_{\varepsilon_1} f_{\varepsilon_2} \cdots f_{\varepsilon_n}(x),$$

we see that $Y_n(0)$ increases a.s. to $\underline{Y}$, say. If $\underline{Y}$ is a.s. finite, then the distribution of $\underline{Y}$ is an invariant probability π of the Markov process X_n. One can show that this is the case if $E\varepsilon_n < 0$, and that if $E\varepsilon_n < 0$, the distribution $p^{(n)}(z, dy)$ of $X_n(z)$ converges weakly to π, whatever be the initial state $z \in [0, \infty)$. To prove this, we first derive the following identity:

$$f_{\theta_1} f_{\theta_2} \cdots f_{\theta_n}(0) = \max\{\theta_1^+, (\theta_1 + \theta_2)^+, \ldots, (\theta_1 + \theta_2 + \cdots + \theta_n)^+\}$$
$$= \max\{(\theta_1 + \cdots + \theta_j)^+ : 1 \leq j \leq n\},$$
$$\forall \, \theta_1, \ldots, \theta_n \in \mathbb{R}, \quad n \geq 1. \quad (7.4)$$

For $n = 1$, (7.4) holds since $f_{\theta_1}(0) = \theta_1^+$. As an induction hypothesis, suppose (7.4) holds for $n - 1$, for some $n \geq 2$. Then

$$f_{\theta_1} f_{\theta_2} \cdots f_{\theta_n}(0) = f_{\theta_1}(f_{\theta_2} \cdots f_{\theta_n}(0))$$
$$= f_{\theta_1}(\max\{(\theta_2 + \cdots + \theta_j)^+ : 2 \leq j \leq n\})$$
$$= (\theta_1 + \max\{(\theta_2 + \cdots + \theta_j)^+ : 2 \leq j \leq n\})^+$$
$$= (\max\{\theta_1 + \theta_2^+, \theta_1 + (\theta_2 + \theta_3)^+, \ldots, \theta_1$$
$$\quad + (\theta_2 + \cdots + \theta_n)^+\})^+$$
$$= \max\{(\theta_1 + \theta_2^+)^+, (\theta_1 + (\theta_2 + \theta_3)^+)^+, \ldots,$$
$$\quad (\theta_1 + (\theta_2 + \cdots + \theta_n)^+)^+\}. \quad (7.5)$$

Now check that $(y + z^+)^+ = \max\{0, y, y + z\} = \max\{y^+, (y + z)^+\}$. Using this in (7.5) we get $f_{\theta_1} f_{\theta_2} \cdots f_{\theta_n}(0) = \max\{\theta_1^+, (\theta_1 + \theta_2)^+ (\theta_1 + \theta_2 + \theta_3)^+, \ldots, (\theta_1 + \theta_2 + \cdots + \theta_n)^+\}$, and the proof of (7.4) is complete. Hence

$$Y_n(0) = \max\{(\varepsilon_1 + \cdots + \varepsilon_j)^+, \quad 1 \le j \le n\}. \tag{7.6}$$

By the strong law of large numbers $(\varepsilon_1 + \cdots + \varepsilon_n)/n \to E\varepsilon_1 < 0$ a.s. as $n \to \infty$. This implies $\varepsilon_1 + \cdots + \varepsilon_n \to -\infty$ a.s. Hence $\varepsilon_1 + \cdots + \varepsilon_n < 0$ for all sufficiently large n, so that outside a P-null set, $(\varepsilon_1 + \cdots + \varepsilon_n)^+ = 0$ for all sufficiently large j. This means that $Y_n(0) =$ constant for all sufficiently large n, and therefore $Y_n(0)$ converges to a finite limit $\underline{Y}$ a.s. The distribution π of $\underline{Y}$ is, therefore, an invariant probability.

Finally, note that, by (7.4),

$$Y_n(z) = f_{\varepsilon_1} \ldots f_{\varepsilon_{n-1}} f_{\varepsilon_n}(z) = f_{\varepsilon_1} f_{\varepsilon_2} \ldots f_{\varepsilon_{n-1}} f_{\varepsilon_n + z}(0)$$
$$= \max\{\varepsilon_1^+, (\varepsilon_1 + \varepsilon_2)^+, \ldots, (\varepsilon_1 + \varepsilon_2 + \cdots + \varepsilon_{n-1})^+,$$
$$(\varepsilon_1 + \varepsilon_2 + \cdots + \varepsilon_{n-1} + \varepsilon_n + z)^+\}. \tag{7.7}$$

Since $\varepsilon_1 + \cdots + \varepsilon_n \to -\infty$ a.s., $\varepsilon_1 + \cdots + \varepsilon_{n-1} + z < 0$ for all sufficiently large n. Hence, by (7.6), $Y_n(z) = Y_{n-1}(0) \equiv \max\{\varepsilon_1^+, (\varepsilon_1 + \varepsilon_2)^+, \ldots, (\varepsilon_1 + \varepsilon_2 + \cdots + \varepsilon_{n-1})^+\}$ for all sufficiently large n, so that $Y_n(z)$ converges to $\underline{Y}$ a.s. as $n \to \infty$. Thus $X_n(z)$ converges in distribution to π as $n \to \infty$, for every initial state z. In particular, π is the unique invariant probability. ∎

Remark 7.1 The above proof follows Diaconis and Freedman (1999). The condition "$E\varepsilon_1 < 0$" can be relaxed to "$\sum_{n=1}^{\infty}(1/n)P(\varepsilon_1 + \cdots + \varepsilon_n > 0) < \infty$" as shown in an important paper by Spitzer (1956). Applications to queuing theory may be found in Baccelli and Bremaud (1994).

4.8 A Model with Multiplicative Shocks, and the Survival Probability of an Economic Agent

Suppose that in a one-good economy the agent starts with an initial stock x. A fixed amount $c > 0$ is consumed $(x > c)$, and the remainder $x - c$ is invested for production in the next period. The stock produced in period

1 is $X_1 = \varepsilon_1(X_0 - c) = \varepsilon_1(x - c)$, where ε_1 is a nonnegative random variable. Again, after consumption, $X_1 - c$ is invested in production of a stock of $\varepsilon_2(X_1 - c)$, provided $X_1 > c$. If $X_1 \leq c$, the agent is *ruined*. In general,

$$X_0 = x, \quad X_{n+1} = \varepsilon_{n+1}(X_n - c) \quad (n \geq 0), \quad (8.1)$$

where $\{\varepsilon_n : n \geq 1\}$ is an i.i.d. sequence of nonnegative random variables. The state space may be taken to be $[0, \infty)$ with *absorption* at 0. The *probability of survival* of the economic agent, starting with an initial stock $x > c$, is

$$\rho(x) := P(X_n > c \text{ for all } n \geq 0 \mid X_0 = x). \quad (8.2)$$

If $\delta := P(\varepsilon_1 = 0) > 0$, then it is simple to check that $P(\varepsilon_n = 0$ for some $n \geq 0) = 1$, so that $\rho(x) = 0$ for all x. Therefore, assume

$$P(\varepsilon_1 > 0) = 1. \quad (8.3)$$

From (8.1) one gets, by successive iteration,

$$X_{n+1} > c$$
$$\text{iff } X_n > c + \frac{c}{\varepsilon_{n+1}}$$
$$\text{iff } X_{n-1} > c + \frac{c + \frac{c}{\varepsilon_{n+1}}}{\varepsilon_n} = c + \frac{c}{\varepsilon_n} + \frac{c}{\varepsilon_n \varepsilon_{n+1}} \cdots$$
$$\text{iff } X_0 \equiv x > c + \frac{c}{\varepsilon_1} + \frac{c}{\varepsilon_1 \varepsilon_2} + \cdots + \frac{c}{\varepsilon_1 \varepsilon_2 \ldots \varepsilon_{n+1}}.$$

Hence, on the set $\{\varepsilon_n > 0 \text{ for all } n\}$,

$$\{X_n > c \text{ for all } n\} = \left\{ x > c + \frac{c}{\varepsilon_1} + \frac{c}{\varepsilon_1 \varepsilon_2} + \cdots + \frac{1}{\varepsilon_1 \varepsilon_2 \cdots \varepsilon_n} \text{ for all } n \right\}$$
$$= \left\{ x \geq c + c \sum_{n=1}^{\infty} \frac{1}{\varepsilon_1 \varepsilon_2 \cdots \varepsilon_n} \right\}$$
$$= \left\{ \sum_{n=1}^{\infty} \frac{1}{\varepsilon_1 \varepsilon_2 \cdots \varepsilon_n} \leq \frac{x}{c} - 1 \right\}.$$

In other words,

$$\rho(x) = P \left\{ \sum_{n=1}^{\infty} \frac{1}{\varepsilon_1 \varepsilon_2 \cdots \varepsilon_n} \leq \frac{x}{c} - 1 \right\}. \quad (8.4)$$

This formula will be used to determine conditions on the common distribution of ε_n, under which (1) $\rho(x) = 0$, (2) $\rho(x) = 1$, (3) $\rho(x) < 1$ ($x > c$). Suppose first that $E \log \varepsilon_1$ exists and $E \log \varepsilon_1 < 0$. Then, by the Strong Law of Large Numbers,

$$\frac{1}{n} \sum_{r=1}^{n} \log \varepsilon_n \xrightarrow{\text{a.s.}} E \log \varepsilon_1 < 0,$$

so that $\log \varepsilon_1 \varepsilon_2 \cdots \varepsilon_n \to -\infty$ a.s., or $\varepsilon_1 \varepsilon_2 \cdots \varepsilon_n \to 0$ a.s. This implies that the infinite series in (8.4) diverges a.s., that is,

$$\rho(x) = 0 \quad \text{for all } x, \quad \text{if } E \log \varepsilon_1 < 0. \tag{8.5}$$

Now, by Jensen's inequality, $E \log \varepsilon_1 \leq \log E\varepsilon_1$, with strict inequality unless ε_1 is degenerate. Therefore, if $E\varepsilon_1 < 1$, or $E\varepsilon_1 = 1$ and ε_1 is nondegenerate, then $E \log \varepsilon_1 < 0$. If ε_1 is degenerate and $E\varepsilon_1 = 1$, then $P(\varepsilon_1 = 1) = 1$, and the infinite series in (8.4) diverges. Therefore, (8.5) implies

$$\rho(x) = 0 \quad \text{for all } x, \quad \text{if } E\varepsilon_1 \leq 1. \tag{8.6}$$

It is not true, however, that $E \log \varepsilon_1 > 0$ implies $\rho(x) = 1$ for large x. To see this and for some different criteria, define

$$m := \inf\{z \geq 0 : P(\varepsilon_1 \leq z) > 0\}, \tag{8.7}$$

the smallest point of the support of the distribution of ε_1.
Let us show that

$$\rho(x) < 1 \quad \text{for all } x, \quad \text{if } m \leq 1. \tag{8.8}$$

For this, fix $A > 0$, however large. Find n_0 such that

$$n_0 > A \prod_{r=1}^{\infty} \left(1 + \frac{1}{r^2} \right). \tag{8.9}$$

This is possible, as $\prod(1 + 1/r^2) < \exp\{\sum 1/r^2\} < \infty$. If $m \leq 1$ then $P(\varepsilon_1 \leq 1 + 1/r^2) > 0$ for all $r \geq 1$. Hence,

$$
\begin{aligned}
0 &< P(\varepsilon_r \leq 1 + 1/r^2 \text{ for } 1 \leq r \leq n_0) \\
&\leq P\left(\sum_{r=1}^{n_0} \frac{1}{\varepsilon_1 \cdots \varepsilon_r} \geq \sum_{r=1}^{n_0} \frac{1}{\prod_{j=1}^{r}\left(1 + \frac{1}{j^2}\right)} \right) \\
&\leq P\left(\sum_{r=1}^{n_0} \frac{1}{\varepsilon_1 \cdots \varepsilon_r} \geq \frac{n_0}{\prod_{j=1}^{\infty}\left(1 + \frac{1}{j^2}\right)} \right) \\
&\leq P\left(\sum_{r=1}^{n_0} \frac{1}{\varepsilon_1 \cdots \varepsilon_r} > A \right) \leq P\left(\sum_{r=1}^{\infty} \frac{1}{\varepsilon_1 \cdots \varepsilon_r} > A \right).
\end{aligned}
$$

Because A is arbitrary, (8.4) is less than 1 for all x, proving (8.8).

One may also show that, if $m > 1$, then

$$
\rho(x) = \begin{cases} < 1 & \text{if } x < c\left(\frac{m}{m-1}\right), \\ = 1 & \text{if } x \geq c\left(\frac{m}{m-1}\right), \end{cases} \quad (m > 1). \qquad (8.10)
$$

To prove this, observe that

$$
\sum_{n=1}^{\infty} \frac{1}{\varepsilon_1 \cdots \varepsilon_n} \leq \sum_{n=1}^{\infty} \frac{1}{m^n} = \frac{1}{m-1},
$$

with probability 1 (if $m > 1$). Therefore, (8.4) implies the second relation in (8.10). In order to prove the first relation in (8.10), let $x < cm/(m-1) - c\delta$ for some $\delta > 0$. Then $(x/c) - 1 < 1/(m-1) - \delta$. Choose $n(\delta)$ such that

$$
\sum_{r=n(\delta)}^{\infty} \frac{1}{m^r} < \frac{\delta}{2}, \qquad (8.11)
$$

and then choose $\delta_r > 0$ ($1 \leq r \leq n(\delta) - 1$) such that

$$
\sum_{r=1}^{n(\delta)-1} \frac{1}{(m + \delta_1) \cdots (m + \delta_r)} > \sum_{r=1}^{n(\delta)-1} \frac{1}{m^r} - \frac{\delta}{2}. \qquad (8.12)
$$

Then

$$0 < P(\varepsilon_r < m + \delta_r \quad \text{for } 1 \leq r \leq n(\delta) - 1)$$

$$\leq P\left(\sum_{r=1}^{n(\delta)-1} \frac{1}{\varepsilon_1 \cdots \varepsilon_r} > \sum_{r=1}^{n(\delta)-1} \frac{1}{m^r} - \frac{\delta}{2}\right)$$

$$\leq P\left(\sum_{r=1}^{\infty} \frac{1}{\varepsilon_1 \varepsilon_2 \cdots \varepsilon_r} > \sum_{r=1}^{\infty} \frac{1}{m^r} - \delta\right)$$

$$= P\left(\sum_{r=1}^{\infty} \frac{1}{\varepsilon_1 \cdots \varepsilon_r} > \frac{1}{m - 1} - \delta\right).$$

If $\delta > 0$ is small enough, the last probability is smaller than $P(\sum 1/(\varepsilon_1 \cdots \varepsilon_r) > x/c - 1)$, provided $x/c - 1 < 1/(m - 1)$, i.e., if $x < cm/(m - 1)$. Thus for such x, one has $1 - \rho(x) > 0$, proving the first relation in (8.10). ∎

These results are due to Majumdar and Radner (1991). The proofs follow Bhattacharya and Waymire (1990, pp. 182–184).

4.9 Complements and Details

Section 4.3.
Proof of Lemma 3.1.   Let $\lambda_1, \ldots, \lambda_m$ be the eigenvalues of A. This means $\det(A - \lambda \mathbf{I}) = (\lambda_1 - \lambda)(\lambda_2 - \lambda) \cdots (\lambda_m - \lambda)$, where det is shorthand for determinant and $\mathbf{I}$ is the identity matrix. Let λ_m have the maximum modulus among the λ_i, i.e., $|\lambda_m| = r(A)$. If $|\lambda| > |\lambda_m|$, then $A - \lambda \mathbf{I}$ is invertible, since $\det(A - \lambda I) \neq 0$. Indeed, by the definition of the inverse, each element of the inverse of $A - \lambda \mathbf{I}$ is a polynomial in λ (of degree $m - 1$ or $m - 2$) divided by $\det(A - \lambda \mathbf{I})$. Therefore, one may write

$$(A - \lambda \mathbf{I})^{-1} = (\lambda_1 - \lambda)^{-1} \cdots (\lambda_m - \lambda)^{-1}(A_0 + \lambda A_1 + \cdots + \lambda^{m-1} A_{m-1})$$
$$(|\lambda| > |\lambda_m|), \quad \text{(C3.1)}$$

where $A_j (0 \leq j \leq m - 1)$ are $m \times m$ matrices that do not involve λ. Writing $z = 1/\lambda$, one may express (C3.1) as

$$(A - \lambda \mathbf{I})^{-1} = (-\lambda)^{-m}(1 - \lambda_1/\lambda)^{-1} \cdots (1 - \lambda_m/\lambda)^{-1}\lambda^{m-1} \sum_{j=0}^{m-1}$$
$$(1/\lambda)^{m-1-j} A_j$$

$$= (-1)^m z (1 - \lambda_1 z)^{-1} \cdots (1 - \lambda_m z)^{-1} \sum_{j=0}^{m-1} z^{m-1-j} A_j$$

$$= \left(z \sum_{n=0}^{\infty} a_n z^n \right) \sum_{j=0}^{m-1} z^{m-1-j} A_j, \quad (|z| < |\lambda_m|^{-1}).$$

(C3.2)

On the other hand,

$$(A - \lambda \mathbf{I})^{-1} = -z(\mathbf{I} - zA)^{-1} = -z \sum_{k=0}^{\infty} z^k A^k \left(|z| < \frac{1}{\|A\|} \right). \quad (C3.3)$$

To see this, first note that the series on the right-hand side is convergent in norm for $|z| < 1/\|A\|$, and then check that term-by-term multiplication of the series $\sum z^k A^k$ by $\mathbf{I} - zA$ yields the identity $\mathbf{I}$ after all cancellations. In particular, writing $a_{ij}^{(k)}$ for the (i, j) element of A^k, the series

$$-z \sum_{k=0}^{\infty} z^k a_{ij}^{(k)} \quad (C3.4)$$

converges absolutely for $|z| < 1/\|A\|$. Since (C3.4) is the same as the (i, j) element of the series (C3.2), at least for $|z| < 1/\|A\|$, their coefficients coincide and, therefore, the series in (C3.4) is absolutely convergent for $|z| < |\lambda_m|^{-1}$ (as (C3.2) is).

This implies that, for each $\varepsilon > 0$,

$$\left| a_{ij}^{(k)} \right| < (|\lambda_m| + \varepsilon)^k \quad \text{for all sufficiently large } k. \quad (C3.5)$$

For if (C3.5) is violated, one may choose $|z|$ sufficiently close to (but less than) $1/|\lambda_m|$ such that $|z^{(k')} a_{ij}^{(k')}| \to \infty$ for a subsequence $\{k'\}$, contradicting the requirement that the terms of the convergent series (C3.4) must go to zero for $|z| < 1/|\lambda_m|$.

Now $\|A^k\| \le m \max\{|a_{ij}^{(k)}| : 1 \le i, j \le m\}$ (Exercise). Since $m^{1/k} \to 1$ as $k \to \infty$, (C3.5) implies (3.10). ∎

Section 4.4. The following generalization of Theorem 4.1 has been derived independently by Bhattacharya and Majumdar (2002, 2004) and Athreya (2003).

Theorem C4.1 *Assume that the distribution Q of C_n in (4.2) has a nonzero absolutely continuous component (with respect to the Lebesgue* *measure on (0,4)), whose density is bounded away from zero on some nondegenerate interval in (1,4). Assume*

$$E \log C_1 > 0 \quad and \quad E |\log(4 - C_1)| < \infty. \qquad (C4.1)$$

Then (a) the Markov process $X_n (n \geq 0)$ defined by (4.3) has a unique invariant probability π on $S = (0, 1)$, and

$$\frac{1}{N} \sum_{n=1}^{N} p^{(n)}(x, dy) \to \pi(dy) \text{ in total variation distance,} \quad as\ N \to \infty$$

$$(C4.2)$$

and (b) if, in addition, the density component is bounded away from zero on an interval that includes an attractive periodic point of prime period m, then

$$p^{(mk)}(x, dy) \to \pi(dy) \text{ in total variation distance, as } k \to \infty.$$

$$(C4.3)$$

Remark 4.2 Theorem 4.1, which is due to Dai (2000), corresponds to the case $m = 1$ of part (b) of Theorem C4.1. In general, (C4.1) is a *sufficient condition* for the tightness of $p^{(n)}(x, dy)$ $(n \geq 1)$, and therefore, for the *existence* of an invariant probability on $S = (0, 1)$ (see Athreya and Dai 2002). However, it does not ensure the *uniqueness* of the invariant probability on $(0, 1)$. Indeed, it has been shown by Athreya and Dai (2002) that there are distributions Q (of C_n) with a two-point support $\{\theta_1, \theta_2\}$ for which $p^{(n)}(x, dy)$ admits more than 1 invariant probability on S. This is the case for $1 < \theta_1 < \theta_2 < 4$, such that $1/\theta_1 + 1/\theta_2 = 1$, and $\theta_2 \in (3, z]$, where z is the solution of $x^3(4 - x) = 16$.

A study of the random iteration of two quadratic maps (i.e., Q having a two-point support) was initiated in Bhattacharya and Rao (1993). Examples 4.1–4.3 are taken from this article. Example 4.4 is due to Bhattacharya and Majumdar (1999a). With regard to Examples 4.1 and 4.2, it has been shown by Carlsson (2002) that the existence of a unique invariant probability (and stability in distribution) holds for the case Q with support $\{\theta_1, \theta_2\}$ for all $1 < \theta_1 < \theta_2 \leq 3$. The extension of these results to general distributions, as given in Theorem 4.2, is due to Bhattacharya and Waymire (2002).

Proof of Theorem 4.2. The result is an immediate consequence of the following Lemma: ∎

Lemma C4.1 *Let $\theta_1 < \theta_2$ and $m \geq 1$ be given.*

(a) If, for all $\theta^i \in \{\theta_1, \theta_2\}$, $1 \leq i \leq m$, the range of $F_{\theta^1} \cdots F_{\theta^m}$ on an interval $I_1 = [u_1, v_1]$ is contained in $I_2 = [u_2, v_2]$, then the same is true for all $\theta^i \in [\theta_1, \theta_2]$. In particular, if $F_{\theta^1} \cdots F_{\theta^m}$ leaves an interval $[c, d]$ invariant for all $\theta^i \in \{\theta_1, \theta_2\}$, $1 \leq i \leq m$, then the same is true for all $\theta^i \in [\theta_1, \theta_2]$, $1 \leq i \leq m$.

(b) Suppose, for all $\theta^i \in \{\theta_1, \theta_2\}$, $1 \leq i \leq m$, $F_{\theta^1} \cdots F_{\theta^m}$ leaves invariant an interval $[c, d]$. Assume also that the strict splitting property above holds, with $N = km$, a multiple of m, for a distribution $Q = Q_0$ whose support is $\{\theta_1, \theta_2\}$. Then (i) $F_{\theta^1} \cdots F_{\theta^m}$ leaves $[c, d]$ invariant for all $\theta^i \in [\theta_1, \theta_2]$ and (ii) the strict splitting property holds for an arbitrary $Q = \hat{Q}$ whose support has θ_1 as the smallest point and θ_2 as its largest.

Proof.

(a) The proof is by induction on m. The assertion is true for $m = 1$, since in this case $u_2 \leq \min\{F_{\theta_1}(y): y \in [u_1, v_1]\} \leq F_\theta(x) \leq \max\{F_{\theta_2}(y): y \in [u_1, v_1]\} \leq v_2$, for all $x \in [u_1, v_1]$. Assume the assertion is true for some integer $m \geq 1$. Let $a = \min\{F_{\theta^2} \cdots F_{\theta^{m+1}}(x): \theta^i \in \{\theta_1, \theta_2\}, 2 \leq i \leq m + 1, x \in [u_1, v_1]\}$, and $b = \max\{F_{\theta^2} \cdots F_{\theta^{m+1}}(x): \theta^i \in \{\theta_1, \theta_2\}, 2 \leq i \leq m + 1, x \in [u_1, v_1]\}$. By the induction hypothesis, for arbitrary $\theta^2, \ldots, \theta^{m+1} \in [\theta_1, \theta_2]$, the range $F_{\theta^2} \cdots F_{\theta^{m+1}}$ on $[u_1, v_1]$ is contained in $[a, b]$. On the other hand, $u_2 \leq \min\{F_{\theta_1}(y): y \in [a, b]\}$, $v_2 \geq \max\{F_{\theta_2}(y): y \in [a, b]\}$. Hence, for arbitrary $\theta^1, \ldots, \theta^{m+1} \in [\theta_1, \theta_2]$, the range of $F_{\theta^1} \cdots F_{\theta^{m+1}}$ on $[u_1, v_1]$ is contained in $[u_2, v_2]$, thus completing the induction argument.

(b) (ii) Suppose a strict splitting property holds with $\delta > 0$, x_0, N. Define the continuous functions L, l on $[\theta_1, \theta_2]^N$ by

$$\begin{cases} L(\theta^1, \ldots, \theta^N) := \max_{c \leq x \leq d} F_{\theta^1} \cdots F_{\theta^N}(x), \\ l(\theta^1, \ldots, \theta^N) := \min_{c \leq x \leq d} F_{\theta^1} \cdots F_{\theta^N}(x). \end{cases} \tag{C4.4}$$

By hypothesis, the open subset U of $[\theta_1, \theta_2]^N$, defined by $U := \{(\theta^1, \ldots, \theta^N) \in [\theta_1, \theta_2]^N: L(\theta^1, \ldots, \theta^N) < x_0\}$, includes a point $(\theta_0^1, \ldots, \theta_0^N) \in \{\theta_1, \theta_2\}^N$. Therefore, U contains a rectangle $R = R_1$

$\times \cdots \times R_N$ where R_i is of the form $R_i = [\theta_1, \theta_1 + h_i)$ or $R_i = (\theta_2 - h_i, \theta_2]$ for some $h_i > 0$ $(1 \le i \le N)$, depending on whether $\theta_0^i = \theta_1$ or $\theta_0^i = \theta_2$. By the hypothesis on $\hat{Q}$, $\delta_1 := \hat{Q}^N(R_N) > 0$. Hence the first inequality within parentheses in (4.12) (with "$< x_0$") holds with probability at least δ_1. Similarly, the second inequality in (4.12) holds (with "$> x_0$") with some probability $\delta_2 > 0$, since $V := \{(\theta^1, \dots, \theta^N) \in [\theta_1, \theta_2]^N : l(\theta^1, \dots, \theta^N) > x_0\}$ is an open subset of $[\theta_1, \theta_2]^N$, which includes a point $(\theta_0^1, \dots, \theta_0^N) \in \{\theta_1, \theta_2\}^N$. Now take the minimum of δ_1, δ_2 for δ in (4.12). Finally, the invariance of $[c, d]$ under $F_{\theta^1} \cdots F_{\theta^N}$ for all $\theta^i \in [\theta_1, \theta_2]$, $1 \le i \le N$, follows from (a).   ■

Section 4.5. Theorem 5.2 is essentially contained in Bhattacharya and Lee (1988).

Extensions of Theorem 5.1 to unbounded f, $g \equiv 1$, but assuming η_n has a positive density $(a \cdot e)$ on $\mathbb{R}^k$, may be found in Bhattacharya and Lee (1995) and An and Huang (1996). We state the result in the form given by the latter authors.

Theorem C5.1 *Consider the model (5.1) with $g \equiv 1$. Suppose η_n has a density which is positive a.e (with respect to the Lebesgue measure on $\mathbb{R}$), and that $E|\eta_n| < \infty$. Assume, in addition, that there exist constants $0 < \lambda < 1$ and c such that*

$$|f(x)| \le \lambda \max\{|x_i| : 1 \le i \le k\} + c \qquad (x \in \mathbb{R}^k). \qquad (C5.1)$$

Then the Markov process $X_n := (U_{n-k+1}, U_{n-k+2}, \dots, U_{n-1})'$, $n \ge k$, has a unique invariant probability π, and its transition probability p satisfies

$$\sup_{B \in \mathcal{B}(\mathbb{R}^k)} \left| p^{(n)}(x, B) - \pi(B) \right| \le c(x)\rho^n \qquad (n \ge 1), \qquad (C5.2)$$

for some nonnegative function $c(x)$ and some constant ρ, $0 < \rho < 1$.

The exponential convergence (in total variation distance) toward equilibrium, such as given in (C5.2), is often referred to as *geometric ergodicity*. Theorem C5.1 and other results on geometric ergodicity are generally derived using the so-called *Foster–Tweedie criterion* for irreducible Markov processes, which requires, in addition to the hypothesis of local minorization (see (9.29) in Chapter 2), that there exists a measurable function $V \ge 1$ on $(S, \mathcal{S})$ and constants $\beta > 0$ and c such that

$$TV(x) - V(x) \le -\beta V(x) + c\mathbf{1}_{A_0}(x), \qquad (C5.3)$$

where A_0 is as in the hypothesis (2) of Theorem 9.2, Chapter 2, satisfying (9.29) with $N = 1$ (for some probability measure ν with $\nu(A_0) = 1$). For full details and complete proofs, see Meyn and Tweedie (1993, Chapter 15).

Section 4.6. In order to prove singularity of π in Theorem 6.2 (b), we need two important results on continued fractions (see Billingsley 1965),

$$\frac{a_1 + a_2 + \cdots + a_n}{n} \to \infty$$

$$\text{a.e. with respect to the Lebesgue measure on } [0, 1] \quad \text{(C6.1)}$$

and

$$\lim_{n \to \infty} \frac{1}{n} \log q_n = \frac{\pi^2}{12 \log 2} \quad \text{a.e. on } [0, 1]. \quad \text{(C6.2)}$$

Now let $x \in (0, \infty)$ and let $y = [a_0, a_1, \ldots, a_n]$ be the nth convergent of x, where x is irrational. Then, since the series in (6.31) is alternating with terms of decreasing magnitude,

$$|F(x) - F(y_n)| = \left| \sum_{i=n}^{\infty} \left(-\frac{1}{\alpha} \right)^i \left(\frac{\alpha}{1 + \alpha} \right)^{a_1 + \cdots + a_{i+1}} \right|$$

$$\leq \frac{1}{\alpha^n} \left(\frac{\alpha}{1 + \alpha} \right)^{a_1 + \cdots + a_{n+1}}. \quad \text{(C6.3)}$$

On the other hand, with $y_n = p_n/q_n$, p_n and q_n relatively prime integers, one has (see Kinchin 1964, p. 17)

$$|x - y_n| > \frac{1}{q_n(q_{n+1} + 1)}, \quad \text{(C6.4)}$$

so that

$$\left| \frac{F(x) - F(y_n)}{x - y_n} \right| \leq \frac{1}{\alpha^n} \left(\frac{\alpha}{1 + \alpha} \right)^{a_1 + \cdots + a_{n+1}} q_n(q_{n+1} + 1). \quad \text{(C6.5)}$$

Let M be a large number satisfying

$$\left(\frac{\alpha}{1 + \alpha} \right)^M \frac{\gamma^2}{\alpha} < 1, \quad \text{(C6.6)}$$

where $\gamma = \exp\{\pi^2/12 \log 2)\}$. It follows from (C6.2) that for all x outside a set N_1 of Lebesgue measure zero, one has

$$a_1 + \cdots + a_{n+1} > M(n+1) \quad \forall \text{ sufficiently large } n. \qquad \text{(C6.7)}$$

Now by (C6.2) and using (C6.6), and (C6.7) in (C6.5), one shows that for all x outside a set N of Lebesgue measure zero,

$$\lim_{n\to\infty} \left| \frac{F(x) - F(y_n)}{x - y_n} \right| = 0. \qquad \text{(C6.8)}$$

The proof of singularity is now complete by the Lebesgue differentiation theorem (see Rudin 1966, *Real and Complex Analysis*, pp. 155, 166).

Proposition 6.1 is essentially due to Letac and Seshadri (1983), who also obtained the interesting invariant distribution in Example 6.1. Theorem 6.2 was originally derived in Chassaing, Letac, and Mora (1984) by a different method. The present approach to this theorem is due to Bhattacharya and Goswami (1998). In this section we have followed the latter article and Goswami (2004).

Bibliographical Notes. The study of a linear stochastic difference Equation (3.7) has a long and distinguished history in econometrics and economic theory (see, e.g., Mann and Wald 1943, Haavelmo 1943, and Hurwicz 1944). For an extended discussion of the survival problem, see Majumdar and Radner (1991, 1992). Two important papers appearing in *Economic Theory* (2004, Symposium on Random Dynamical Systems) are Athreya (2004) and Mitra, Montrucchio, and Privileggi (2004). There are several papers motivated by problems of biology that deal with stochastic difference equations and go beyond the issues we have covered: see, for example, Jacquette (1972), Reed (1974), Chesson (1982), Ellner (1984), Kot and Schaffer (1986), and Hardin, Takáč, and Webb (1988).

Invariant Distributions: Estimation and Computation

I have but one lamp by which my feet are guided, and that is the lamp of experience. I know no way of judging the future but by the past.

Patrick Henry

This talk is a collection of stories about trying to predict the future. I have seven stories to tell. One of them is about a prediction, not made by me, that turned out to be right. One is about a prediction made by me that turned out to be wrong. The other five are about predictions that might be right and might be wrong. Most of the predictions are concerned with practical consequences of science and technology. . . . The moral of the stories is, life is a game of chance, and science is too. Most of the time, science cannot tell what is going to happen.

Freeman J. Dyson.

5.1 Introduction

The practical gain in studying the criteria for stability in distribution of a time series like

$$X_{n+1}(x) = \alpha_{n+1}(X_n(x)) \tag{1.1}$$

is to estimate long-run averages of some characteristic $h(X_n)$ of X_n, based on past data. If the process is stable in distribution (on the average) and $\int |h| d\pi < \infty$ (where π is an invariant probability), then, according to the ergodic theorem,

$$\hat{\lambda}_{h,n} = \frac{1}{n+1} \sum_{j=0}^{n} h(X_j)$$

converges to (the long-run average) $\lambda_n = \int h \, d\pi$ with probability 1, for $\pi -$ *almost all initial states* $X_0 = x$. But one does *not* know π and, in general, there is no a priori guarantee that the initial states x belong to this distinguished set. Indeed, the support of an invariant distribution may be quite small (e.g., a Cantor set of Lebesgue measure zero). If π is widely spread out, as in the case, e.g., when π has a strictly positive density on an Euclidean state space S, one may be reasonably assured of convergence.

A second important problem is to determine the *speed of convergence* of $\hat{\lambda}_{h,n}$ to $\int h\,d\pi$. One knows from general considerations based on central limit theorems that this speed is no faster than $O_p(n^{-1/2})$. Is this optimal speed achieved irrespective of the initial state? An approach toward resolving these issues is a primary theme of this chapter.

5.2 Estimating the Invariant Distribution

Let $\{X_n\}_0^\infty$ be a Markov process with state space $(S, \mathcal{S})$, transition probability $p(\cdot, \cdot)$, and a given initial distribution μ (*which may assign mass 1 to a single point*). Recall that the transition probability $p(\cdot, \cdot)$ satisfies (a) for any $A \in \mathcal{S}$, $p(\cdot, A)$ is $\mathcal{S}$-measurable and (b) for any $x \in S$, $p(x, \cdot)$ is a probability distribution over S. Assume that there is an invariant distribution π on $(S, \mathcal{S})$, i.e.,

$$\pi(A) = \int_S p(x, A)\pi(dx) \qquad (2.1)$$

for every A in $\mathcal{S}$. Suppose that we wish to estimate $\pi(A)$ from observing the Markov process $\{X_j\}$ for $0 \le j \le n$, starting from a historically given initial condition. A *natural* estimate of $\pi(A)$ for any A in $\mathcal{S}$ is the *sample proportion of visits* to A, i.e.,

$$\hat{\pi}_n(A) \equiv \frac{1}{n+1} \sum_0^n \mathbf{1}_A(X_j), \qquad (2.2)$$

where $\mathbf{1}_A(x) = 1$ if $x \in A$ and 0 if $x \notin A$. We say that $\hat{\pi}_n(A)$ is a *consistent estimator* of $\pi(A)$ under P_μ if $\hat{\pi}_n(A) \to \pi(A)$ as $n \to \infty$ in probability, i.e.,

$$\forall \varepsilon > 0, \quad P_\mu(|\hat{\pi}_n(A) - \pi(A)| > \varepsilon) \to 0, \qquad (2.3)$$

where P_μ refers to the probability distribution of the process $\{X_n\}_0^\infty$ when X_0 has distribution μ. Assuming that $\hat{\pi}_n$ is such a consistent estimator, it will be useful to know the *accuracy* of the estimate, i.e., the order of the magnitude of the *error* $|\hat{\pi}_n(A) - \pi(A)|$. Under fairly general second moment conditions this turns out to be of the order $O(n^{-1/2})$. The estimator is then called $\sqrt{n}$-*consistent*. Under some further conditions, *asymptotic normality* holds, asserting that

$$\sqrt{n}(\hat{\pi}_n(A) - \pi(A)) \overset{\mathcal{L}}{\to} N(0, \sigma_A^2) \tag{2.4}$$

for $0 < \sigma_A^2 < \infty$ depending on A and possibly the initial distribution μ. This can then be used to provide confidence intervals for $\pi(A)$ based on the data $\{X_j\}_0^n$.

The above issues can be considered in a more general framework where the goal is to estimate

$$\lambda_h = \int h \, d\pi, \tag{2.5}$$

the integral of a *reward function* h that we require to be a real-valued bounded S-measurable function. A natural estimate for λ_h is the *empirical average*:

$$\hat{\lambda}_{h,n} \equiv \frac{1}{(n+1)} \sum_0^n h(X_j). \tag{2.6}$$

As before, it would be useful to find conditions to assess the accuracy of $\hat{\lambda}_{h,n}$, i.e., the order of the magnitude of $|\hat{\lambda}_{h,n} - \lambda_n|$. In particular, it is of interest to know whether this estimate is $\sqrt{n}$ *consistent*, i.e., whether $|\hat{\lambda}_{h,n} - \lambda_n|$ is of the order $O(n^{-1/2})$ and, further, whether *asymptotic normality* holds, i.e., $\sqrt{n}(\hat{\lambda}_{h,n} - \lambda_h)$ converges in distribution to $N(0, \sigma_h^2)$ for $0 < \sigma_h^2 < \infty$ depending on h, *but irrespective of the initial distribution* μ.

5.3 A Sufficient Condition for $\sqrt{n}$-Consistency

In this section we present some sufficient conditions on h and p that ensure $\sqrt{n}$-consistency and asymptotic normality of $\hat{\lambda}_{h,n}$ defined in (2.6), irrespective of the initial distribution.

In what follows, if g is an S-measurable bounded real-valued function on S,

$$Tg(x) \equiv \int g(y)p(x, dy) = E[g(X_1) \mid X_0 = x] \tag{3.1}$$

is the *conditional expectation* of $g(X_1)$, given $X_0 = x$. The *conditional variance* of $g(X_1)$, given $X_0 = x$, is defined by

$$Vg(x) \equiv \text{variance } [g(X_1) \mid X_0 = x] = E[(g(X_1) - Tg(x))^2 \mid X_0 = x]$$
$$= Tg^2(x) - (Tg(x))^2. \qquad (3.2)$$

5.3.1 $\sqrt{n}$-Consistency

The following proposition provides a rather restrictive condition for the estimate $\hat{\lambda}_{h,n}$ of λ_h ((3.1) and (3.2)) to be $\sqrt{n}$-consistent for *any* initial distribution μ and, in particular, for any historically given initial condition.

Proposition 3.1 *Let π be an invariant distribution for $p(\cdot, \cdot)$ and h be a reward function. Suppose there exists a bounded S-measurable function g such that the Poisson equation*

$$h(x) - \lambda_h = g(x) - Tg(x) \quad \text{for all } x \text{ in } S \qquad (3.3)$$

holds.

Then, for any initial distribution μ,

$$E_\mu(\hat{\lambda}_{h,n} - \lambda_h)^2$$
$$\equiv E_\mu \left(\frac{1}{n+1} \sum_{j=0}^{n} h(X_j) - \lambda_h \right)^2$$
$$\leq (8/(n+1)) \max\{g^2(x): x \in S\}. \qquad (3.4)$$

Proof. Since h and g satisfy (3.3), one has

$$h(x) - \lambda_h = g(x) - Tg(x),$$
$$\frac{1}{n+1} \sum_{0}^{n} h(X_j) - \lambda_h = \frac{1}{n+1} \sum_{0}^{n} (g(X_j) - Tg(X_j))$$
$$= \frac{1}{(n+1)} \sum_{1}^{n} [g(X_j) - Tg(X_{j-1})]$$
$$+ \frac{1}{n+1} (g(X_0) - Tg(X_n)).$$

Thus,

$$\frac{1}{n+1} \sum_{0}^{n} h(X_j) - \lambda_h = \frac{1}{(n+1)} \sum_{1}^{n} Y_j + \frac{1}{(n+1)}(g(X_0) - Tg(X_n)),$$

(3.5)

where $Y_j = g(X_j) - Tg(X_{j-1})$ for $1 \leq j \leq n$. By the Markov property of $\{X_n\}$ and the definition of Tg, it follows that $\{Y_j\}_0^\infty$ is a *martingale difference sequence*, i.e.,

$$E(Y_j \mid X_0, X_1, \ldots, X_{j-1}) = 0.$$

For the conditional expectation on the left-hand side equals that of Y_j, given X_{j-1}, namely, $Tg(X_{j-1}) - Tg(X_{j-1}) = 0$. This implies that $\{Y_j: 1 \leq 1 \leq n\}$ are uncorrelated and have mean zero. Thus, for any initial distribution μ,

$$E_\mu \left(\sum_{1}^{n} Y_j \right) = 0 \quad \text{and} \tag{3.6}$$

$$E_\mu \left(\sum_{1}^{n} Y_j \right)^2 = V_\mu \left(\sum_{1}^{n} Y_j \right) = \sum_{1}^{n} V_\mu(Y_j) = \sum_{1}^{n} E_\mu(Y_j^2),$$

where E_μ and V_μ stand for mean and variance under the initial distribution μ.

Using $(a+b)^2 \leq 2(a^2 + b^2)$ repeatedly, we get from (3.5) and (3.6) that

$$E_\mu \left(\frac{1}{n+1} \sum_{0}^{n} h(X_j) - \lambda_h \right)^2$$

$$\leq \frac{2}{(n+1)^2} \sum_{0}^{n} E_\mu(Y_j^2) + \frac{4}{(n+1)^2}(E_\mu(Tg(X_n))^2 + E_\mu(g(X_0))^2)$$

$$\leq \frac{2}{(n+1)^2} \sum_{1}^{n} (2E_\mu g^2(X_j) + 2E_\mu(Tg(X_j))^2)$$

$$+ \frac{4}{(n+1)^2}(E_\mu(Tg(X_n))^2 + E_\mu(g(X_0))^2) \leq \frac{c}{n+1},$$

where $c = 4(\max_x g^2(x) + \max_x (Tg(x))^2) \leq 8 \max_x g^2(x)$. (3.7)   ∎

Definition 3.1 $\hat\lambda_{h,n}$ is said to be a $\sqrt{n}$-*consistent estimate*, or *estimator*, of λ_n if $\sqrt{n}(\hat\lambda_{h,n} - \lambda_h)$ is *stochastically bounded*, i.e., for $\forall \varepsilon > 0$, $\exists K_\varepsilon$

such that for all n,

$$P_\mu\left(\sqrt{n}\left|\frac{1}{n+1}\sum_0^n h(X_j) - \lambda_h\right| > K_\varepsilon\right) \le \varepsilon, \qquad (3.8)$$

whatever be the initial distribution μ. In this case, one also says that $\hat{\lambda}_{h,n} - \lambda_h$ is $O_p(n^{-1/2})$, or *of order* $n^{-1/2}$ *in probability.*

Corollary 3.1 *Under the hypothesis of Proposition 3.1, $\hat{\lambda}_{h,n}$ is a $\sqrt{n}$-consistent estimate of λ_h.*

Proof. By Chebyshev's inequality, $\forall K > 0$,

$$P_\mu\left(\sqrt{n}\left|\frac{1}{n+1}\sum_0^n h(X_j) - \lambda_h\right| > K\right)$$

$$\le \frac{n}{K^2} E_\mu\left(\frac{1}{n+1}\sum_0^n h(X_j) - \lambda_h\right)^2 \le c/K^2,$$

where c is as in (3.7). Hence for $\forall \varepsilon > 0$, there exists a K_ε such that (3.8) holds, namely, $K_\varepsilon = (c/\varepsilon)^{1/2}$. ∎

A natural question is this: Given a reward function h, how does one find a bounded function g such that the Poisson Equation (3.3) holds?
Rewriting (3.3) as

$$g = \tilde{h} + Tg,$$

where $\tilde{h}(x) = h(x) - \lambda_h$, and iterating this we get

$$g = \tilde{h} + T(\tilde{h} + Tg) = \tilde{h} + T\tilde{h} + T^2 g = \cdots$$
$$= \tilde{h} + T\tilde{h} + T^2\tilde{h} + \cdots + T^n\tilde{h} + T^{n+1}g.$$

This suggests that

$$g \equiv \sum_0^\infty T^n\tilde{h} \qquad (3.9)$$

is a candidate for a solution. Then we face the question of the convergence and boundedness of the infinite series $\sum_0^\infty T^n\tilde{h}$. This can often be ensured by the convergence of the distribution of X_n to π at an appropriate rate in a suitable metric.

Corollary 3.2 *Suppose π is an invariant distribution for $p(\cdot, \cdot)$ and h is a bounded measurable function such that the series (3.9) converges uniformly for all $x \in S$. Then the empirical $\hat{\lambda}_{h,n}$ is a $\sqrt{n}$-consistent estimator of λ_h.*

Proof. Check that $g(x)$ defined by (3.9) satisfies the Poisson equation (3.3), and apply Corollary 3.1. ∎

We give two examples involving random dynamical systems of monotone maps, in which the convergence of the infinite series $\sum_{n=0}^{\infty} T^n \tilde{h}$ can be established.

Example 3.1 Assume that the state space S is an interval and Theorem 5.1 (Chapter 3) holds. Then

$$\sup_{x \in S} |p^{(n)}(x, J) - \pi(J)| \le (1 - \tilde{\chi})^{[n/N]},$$

where J is a subinterval of S.

In particular,

$$\sup_{x,y} \sum_{n=1}^{\infty} |p^{(n)}(x, (-\infty, y] \cap S) - \pi((-\infty, y] \cap S)| < \infty.$$

Let $h(z) = \mathbf{1}_{(-\infty, y] \cap S}$ and $\tilde{h}(z) = h(z) - \pi((-\infty, y] \cap S)$. Then

$$T^n \tilde{h}(x) = p^{(n)}(x, (-\infty, y] \cap S) - \pi((-\infty, y] \cap S), \quad n \ge 1.$$

Hence,

$$\sup_{x} \sum_{n=m+1}^{\infty} |T^n \tilde{h}(x)| \le \sum_{n=m+1}^{\infty} (1 - \tilde{\chi})^{[n/N]} \to 0 \quad \text{as } m \to \infty.$$

This implies the uniform convergence over S of

$$\sum_{n=0}^{m} (T^n \tilde{h})(x) \quad \text{as } m \to \infty.$$

Hence, the *empirical distribution function*

$$\frac{1}{n+1} \sum_{j=0}^{n} \mathbf{1}_{(-\infty, y] \cap S}(X_j)$$

is a $\sqrt{n}$-consistent estimator of $\pi((-\infty, y] \cap S)$. ∎

Example 3.2 For this example, we restrict the interval S to be bounded, say, $[c, d]$, and continue to assume that $X_{n+1} = \alpha_{n+1} X_n$ $(n \geq 0)$, with α_n $(n \geq 1)$ i.i.d. monotone, and that the hypothesis of Theorem 5.1 of Chapter 3 holds.

Consider $h(z) = z$, so that $\tilde{h}(z) = z - \int y \pi(dy)$. Note that (see Exercise 3.1)

$$
\begin{aligned}
(T^n \tilde{h})(x) &= E[\tilde{h}(X_n) \mid X_0 = x] \\
&= E[X_n \mid X_0 = x] - \int y \pi(dy) \\
&= \int_{[c,d]} \text{Prob}(X_n > u \mid X_0 = x) \, du - \int_{[c,d]} \pi((u, d]) \, du \\
&= \int_{[c,d]} [1 - p^{(n)}(x, [c, u]) - (1 - \pi([c, u]))] \, du \\
&= \int_{[c,d]} [p^{(n)}(x, [c, u]) - \pi([c, u])] \, du
\end{aligned}
$$

so that

$$
\sup_x |T^n \tilde{h}(x)| \leq \int_{[c,d]} (1 - \tilde{\chi})^{[n/N]} \, du = (d - c)(1 - \tilde{\chi})^{[n/N]} .
$$

Hence, as before,

$$
\sup_x \sum_{n=m+1}^{\infty} |T^n \tilde{h}(x)| \leq (d - c) \sum_{n=m+1}^{\infty} (1 - \tilde{\chi})^{[n/N]} \to 0 \quad \text{as } m \to \infty,
$$

and we have uniform convergence of

$$
\sum_{n=0}^{m} T^n \tilde{h}(x) \quad \text{as } m \to \infty
$$

Thus, by Corollary 3.1, the *empirical mean*

$$
\hat{\lambda}_{h,n} = \frac{1}{n+1} \left[\sum_{j=0}^{n} X_j \right]
$$

is a $\sqrt{n}$-consistent estimate of the equilibrium mean $\int y \pi(dy)$. ∎

Exercise 3.1 Suppose $Z \geq c$ a.s. Prove that

$$
EZ - c = \int_{[c,\infty)} P(Z > u) \, du.
$$

[Hint: First assume $EZ < \infty$. Write $\int_{[c,\infty)} P(Z > u) \, du = \int_{[c,\infty)}$ $\int_{\Omega} \mathbf{1}_{\{Z(\omega)>u\}} \, dP(\omega) \, du = \int_{\Omega} \int_{[c,\infty)} \mathbf{1}_{\{Z(\omega)>u\}} \, du \, dP(\omega)$ (by Fubini) $= \int_{\Omega}(Z(\omega) - c) \, dP(\omega)$. For the case $EZ = \infty$, take $Z \wedge M$, and let $M \uparrow \infty$.]   ∎

Exercise 3.2 Suppose $Z \geq c$ a.s. Then for every $r > 0$,

$$E(Z - c)^r = r \int_{(0,\infty)} u^{r-1} P(Z > c + u) \, du.$$

[Hint: $\int_0^\infty u^{r-1} P(Z > c + u) \, du = \int_\Omega \int_0^\infty u^{r-1} \mathbf{1}_{\{Z(\omega)>c+u\}} \, du \, dP(\omega) = \int_\Omega (\int_0^{Z(\omega)-c} u^{r-1} \, du) \, dP(\omega) = \int_\Omega \frac{1}{r}(Z(\omega) - c)^r \, dP(\omega) = (1/r)E(Z - c)^r$.]   ∎

Example 3.3 Assume the hypothesis of Example 3.2. Consider the problem of estimating the rth moment of the equilibrium distribution: $m_r = \int_{[c,d]} u^r \pi(du) \, (r = 1, 2, \ldots)$. Letting $h(z) = (z - c)^r$, one has, by Exercise 3.2, and Theorem 5.1 of Chapter 3,

$$|T^n \tilde{h}(z)| = \left| E(X_n - c)^r - \int (u - c)^r \pi(du) \right|$$

$$= \left| r \int_{(0,\infty)} u^{r-1} [P(X_n > c + u) - \pi((c + u, d]] \, du \right|$$

$$\leq r \int_{(0,d-c]} u^{r-1} (1 - \tilde{\chi})^{[\frac{n}{N}]} \, du \leq (d - c)^r (1 - \tilde{\chi})^{[\frac{n}{N}]}.$$

Therefore, by Corollary 3.2,

$$\hat{m}_{n,c,r} := \frac{1}{n+1} \sum_{j=0}^n (X_j - c)^r$$

is a $\sqrt{n}$-consistent estimator of

$$m_{c,r} := \int (z - c)^r \pi(du) \quad (r = 1, 2, \ldots).$$

From this, and expanding $(X_j - c)^r$ and $(z - c)^r$, it follows that the rth *empirical* (or *sample*) *moment* $\hat{m}_{n,r} = 1/(n + 1) \sum_{j=0}^n X_j^r$ is a $\sqrt{n}$-consistent estimator of the rth moment m_r of π, for every $r = 1, 2, \ldots$ (Exercise).   ∎

A broad generalization of Example 3.1 is proved by the following theorem:

Theorem 3.1 *Assume the general splitting hypothesis* ($\mathbf{H}_1$) *of Theorem 5.2 (Chapter 3). Then for every A belonging to the splitting class $\mathcal{A}$, $\hat{\pi}_n(A) := 1/(n+1) \sum_0^n \mathbf{1}_A(X_j)$ is a $\sqrt{n}$-consistent estimator of $\pi(A)$.*

Proof. Let $h = \mathbf{1}_A$, $\tilde{h} = \mathbf{1}_A - \pi(A) \equiv h - \lambda_h$. Note that, by Theorem 5.2 (Chapter 3),

$$|T^n \tilde{h}(x)| = |Eh(X_n \mid X_0 = x) - \pi(A)| = |p^{(n)}(x, A) - \pi(A)|$$
$$\leq (1 - \tilde{\chi})^{[n/N]} \,\forall\, n \geq 1, \quad \forall x \in S.$$

Hence the series (3.9) converges uniformly $\forall x \in S$. Now apply Corollary 3.2. ∎

To apply Theorem 3.1 to random dynamical systems generated by i.i.d. monotone maps on a closed set $S \subset \mathbb{R}^l$, recall that (see Chapter 3 (5.23)) the splitting class $\mathcal{A}$ of Corollary 5.2 (Chapter 3) comprises all sets A of the form

$$A = \{y \in S : \phi(y) \leq x\}, \quad \phi \text{ continuous and monotone}$$
$$\text{on } S \text{ into } \mathbb{R}^l, \quad x \in \mathbb{R}^\ell. \tag{3.10}$$

Then Corollary 5.2 of Chapter 3 and Theorem 3.1 above immediatly lead to the following result, where $X_{n+1} = \alpha_{n+1} X_n$ ($n \geq 0$) with $\{\alpha_n : n \geq 1\}$ i.i.d. *monotone* and continuous on S into S.

Corollary 3.3 *Let S be a closed subset $\mathbb{R}^l$ and $\mathcal{A}$ the class of all sets A of the form (3.10). Suppose there exist $z_0 \in \mathbb{R}^l$, a positive integer N, and $\chi > 0$ such that*

$$P(X_N(x) \leq z_0 \,\forall x \in S) \geq \chi,$$
$$P(X_N(x) \geq z_0 \,\forall x \in S) \geq \chi. \tag{3.11}$$

Then, $\hat{\pi}_n(A)$ is a $\sqrt{n}$-consistent estimator of $\pi(A) \,\forall A \in \mathcal{A}$, π being the unique invariant probability for the transition probability of the Markov process X_n ($n \geq 0$).

Remark 3.1 By letting ϕ be the identity map, $\phi(\mathbf{y}) = \mathbf{y}$, it is seen that $\mathcal{A}$ includes all sets of the form $S \cap (X_{j=1}^\ell (-\infty, x_j]) \,\forall \mathbf{x} = (x_1, x_2, \dots, x_l) \in \mathbb{R}^l$. Thus, for each $\mathbf{x} \in \mathbb{R}^l$, the distribution function $\pi(\{\mathbf{y} : \mathbf{y} \leq \mathbf{x}\})$

(extending π to all $\mathbb{R}^l$, taking $\pi(\mathbb{R}^l \setminus S) = 0$) is estimated $\sqrt{n}$-consistently by $[1/(n+1)] \# \{j: 0 \le j \le n, X_j \le \mathbf{x}\}$.

The final result of this section assumes *Doeblin minorization* (Theorem 9.1 (Chapter 2) or Corollary 5.3 (Chapter 3)) and provides $\sqrt{n}$-consistent estimation of λ_h for every bounded measurable h.

Theorem 3.2 *Let $p(x, A)$ be the transition probability of a Markov process on a measurable state space $(S, \mathcal{S})$ such that there exist $N \ge 1$ and a nonzero measure ν for which the N-step transition probability $p^{(N)}$ satisfies*

$$p^{(N)}(x, A) \ge \nu(A) \quad \forall x \in S, \quad \forall A \in \mathcal{S}. \tag{3.12}$$

Then, for every bounded measurable h on S, the series (3.9) converges and $\hat{\lambda}_{h,n}$ is a $\sqrt{n}$-consistent estimator of λ_h.

Proof. Let h be a (real-valued) bounded measurable function on S, $\tilde{h} = h - \int h \, d\pi \equiv h - \lambda_h$ where π is the unique invariant probability for p. Let $\|h\|_\infty := \sup\{|h(x)|: x \in S\}$. From Theorem 9.1 (Chapter 2) and Remark 9.2 (Chapter 2), it follows that

$$|T_n \tilde{h}(x)| \equiv \left| \int h(y) p^{(n)}(x, dy) - \int h(y) \pi(dy) \right|$$
$$\le 2\|h\|_\infty \sup_{A \in \mathcal{S}} |p^{(n)}(x, A) - \pi(A)| \le 2\|h\|_\infty (1 - \chi)^{[n/N]},$$
$$\tag{3.13}$$

where $\chi = \nu(S)$. Hence the series on the right-hand side of (3.9) converges uniformly to a bounded function g, which then satisfies the Poisson equation (3.3). Now apply Corollary 3.2. ∎

Remark 3.2 By Proposition 3.1, if the solution $g = g_A$ of the Poisson equation (3.3), with $h = \mathbf{1}_A (\tilde{h} = \mathbf{1}_A - \pi(A))$, is bounded on S uniformly over a class $\mathcal{A}$ of sets A, then (see (3.4))

$$\sup_{A \in \mathcal{A}} E(\hat{\pi}_n(A) - \pi(A))^2 \le \frac{8}{n+1} \sup\{\|g_A\|_\infty^2: A \in \mathcal{A}\} = \frac{\bar{c}}{n+1}, \text{ say.}$$
$$\tag{3.14}$$

One may in this case say that the estimator $\hat{\pi}_n(A)$ of $\pi(A)$ is *uniformly $\sqrt{n}$-consistent over the class $\mathcal{A}$* (also see Corollary 3.1, and note that K_ε in (3.8) may be taken to be $(\bar{c}/\varepsilon)^{1/2} \ \forall A \in \mathcal{A}$). In Theorem 3.1,

$\hat{\pi}_n(A)$ is uniformly $\sqrt{n}$-consistent over the splitting class $\mathcal{A}$; and in Theorem 3.2, $\hat{\pi}_n(A)$ is uniformly $\sqrt{n}$-consistent over the entire sigmafield $\mathcal{S} = \mathcal{A}$. To see this check that the series on the right-hand side of (3.9) is bounded by $\sum_{n=0}^{\infty}(1 - \chi)^{[n/N]} \leq \sum_{n=0}^{\infty}(1 - \chi)^{(n/N)-1} = (1 - \chi)^{-1}(1 - (1 - \chi)^{1/N})^{-1}$, where $\chi = \tilde{\chi}$ is as in Theorem 5.2 (Chapter 3) under the hypothesis of Theorem 3.1, and $\chi = \bar{\chi}$ as given in Theorem 9.1 (Chapter 2) in the case of Theorem 3.2.

5.4 Central Limit Theorems

Theorems 3.1, 3.2 of the preceding section may be strengthened, or refined, by showing that if h is bounded and if the series (3.9) converges uniformly to g, then whatever the initial distribution μ,

$$\sqrt{n}(\lambda_{h,n} - \lambda_h) \overset{\mathcal{L}}{\to} N(0, \sigma_h^2) \quad \text{as } n \to \infty, \tag{4.1}$$

$$\sigma_h^2 := E_\pi(g(X_1) - Tg(X_0))^2 = \int g^2(x)\pi(dx) - \int (Tg(x))^2\pi(dx),$$

where $\overset{\mathcal{L}}{\to}$ means "*converges in law, or distribution,*" and $N(0, \sigma_h^2)$ is the normal distribution with mean 0 and variance σ_h^2. Note that (4.1) implies

$$P_\mu(|\hat{\lambda}_{h,n} - \lambda_h| > \frac{K}{\sqrt{n}}) \to \text{Prob}(|Z| > K) \text{ as } n \to \infty, \forall K > 0, \tag{4.2}$$

where Z has the distribution $N(0, \sigma_h^2)$. Thus, in the form of (4.2), the central limit theorem CLT (4.1) provides a fairly precise estimate of the error of estimation $\hat{\lambda}_{h,n} - \lambda_h$ as $Z/\sqrt{n}$, asymptotically, where Z is $N(0, \sigma_h^2)$.

In order to prove (4.1) we will make use of the *martingale CLT*, whose proof is given in Complements and Details.

Let $\{X_{k,n}: 1 \leq k \leq k_n\}$ be, for each $n \geq 1$, a square integrable martingale difference sequence, with respect to an increasing family of sigmafields $\{\mathfrak{F}_{k,n}: 0 \leq k \leq k_n\} : E(X_{k,n}|\mathfrak{F}_{k-1,n}) = 0$ a.s. $(1 \leq k \leq k_n)$. Write

$$\sigma_{k,n}^2 := E(X_{k,n}^2 \mid \mathfrak{F}_{k-1,n}); \; s_{k,n}^2 := \sum_{j=1}^{k}\sigma_{j,n}^2; \; s_n^2 := \sum_{j=1}^{k_n}\sigma_{j,n}^2 \equiv s_{k_n,n}^2$$

$$L_n(\varepsilon) := \sum_{j=1}^{k_n}E(X_{j,n}^2 1_{\{|X_{j,n}|>\varepsilon\}} \mid \mathfrak{F}_{j-1,n}); \; S_n := \sum_{j=1}^{k_n}X_{j,n}. \tag{4.3}$$

Proposition 4.1 *(Martingale CLT)* *Assume that, as $n \to \infty$, (i) $s_n^2 \to \sigma^2 \geq 0$ in probability, and (ii) $L_n(\varepsilon) \to 0$ in probability for every $\varepsilon > 0$. Then S_n converges in distribution to a normal distribution with mean 0 and variance σ^2: $S_n \xrightarrow{\mathcal{L}} N(0, \sigma^2)$.*

We are now ready to prove a refinement of Theorem 3.2.

Theorem 4.1 *Assume the Doeblin hypothesis of* Theorem 3.2. *Then (4.1) holds for every bounded measurable $h : S \to \mathbb{R}$, no matter what the initial distribution μ is.*

Proof. Fix an initial distribution μ (of X_0). Write $X_{k,n} := n^{-1/2} (g(X_k) - Tg(X_{k-1}))$, $1 \leq k \leq n = k_n$, and let $\mathfrak{F}_{k,n} \equiv \mathfrak{F}_k = \sigma\{X_j : 0 \leq j \leq k\}$, the sigmafield of all events determined by $X_j, 0 \leq j \leq k$. Since g and Tg are bounded, for every $\varepsilon > 0$, $\mathbf{1}_{\{|X_{j,n}| > \varepsilon\}} = 0 \forall$ sufficiently large n. Hence $L_n(\varepsilon) = 0$ for all sufficiently large n. Also,

$$s_n^2 = n^{-1} \sum_{k=1}^{n} E_\mu[(g(X_k) - Tg(X_{k-1}))^2 \mid \mathfrak{F}_{k-1,n}]$$

$$= n^{-1} \sum_{k=1}^{n} [Tg^2(X_{k-1}) - (Tg)^2(X_{k-1})]. \tag{4.4}$$

Since Tg^2 is bounded, one may use the fact that the series (3.9) converges uniformly with Tg^2 in place of h, to get, by Proposition 3.1,

$$E_\mu \left(n^{-1} \sum_{k=1}^{n} Tg^2(X_{k-1}) - \int Tg^2 \, d\pi \right)^2 = O\left(\frac{1}{n}\right) \to 0, \tag{4.5}$$

which implies, by Chebyshev's inequality,

$$n^{-1} \sum_{k=1}^{n} Tg^2(X_{k-1}) \to \int Tg^2 \, d\pi \text{ in } P_\mu\text{-probability}, \quad \text{as } n \to \infty. \tag{4.6}$$

But, by invariance of π, $\int Tg^2 d\pi = \int g^2 \, d\pi$. Similarly,

$$n^{-1} \sum_{k=1}^{n} (Tg)^2(X_{k-1}) \to \int (Tg)^2 \, d\pi \text{ in probability.} \tag{4.7}$$

Therefore,

$$s_n^2 \to \sigma_h^2 \text{ in probability.} \tag{4.8}$$

If $\sigma_h^2 > 0$, then we have verified the hypothesis of Proposition 4.1 to arrive at the CLT (4.1). If $\sigma_h^2 = 0$, then using the first inequality in (3.7), one shows that

$$E_\mu(\sqrt{n}(\hat{\lambda}_{h,n} - \lambda_h))^2 \leq \frac{2n}{(n+1)^2} \sum_1^n E_\mu Y_j^2$$

$$+ \frac{4n}{(n+1)^2}[E_\mu(Tg(X_n))^2 + E_\mu(g(X_0))^2]. \qquad (4.9)$$

The first term on the right-hand side equals $(2n/(n+1)^2)E_\mu s_n^2 \to \sigma_h^2 = 0$ by (4.8) (and boundedness of s_n^2), while the second term on the right-hand side in (4.9) is $O(1/n)$. Thus the left-hand side of (4.9) goes to zero, implying $\sqrt{n}(\hat{\lambda}_{h,n} - \lambda_h)$ converges in distribution to $\delta_{\{0\}} = N(0,0)$. ∎

For deriving the next CLT, we assume that $X_{n+1} = \alpha_{n+1}X_n$ with $\{\alpha_n : n \geq 1\}$ i.i.d. *continuous and nondecreasing* on a closed subset S of $\mathbb{R}^l$. Note that in Theorem 3.1 we assumed that α_n $(n \geq 1)$ are continuous and monotone. Thus the refinement of $\sqrt{n}$-consistency in the form of the CLT (4.1) is provided under an additional restriction here.

Theorem 4.2 *Assume the general splitting hypothesis* $(\mathbf{H}_1)$ *of Corollary 5.2 of Chapter 3 for i.i.d. continuous monotone nondecreasing maps* $\{\alpha_n : n \geq 1\}$ *on a closed subset S of $\mathbb{R}^l$. Then for every $h : S \to \mathbb{R}$, which may be expressed as the difference between two continuous bounded monotone nondecreasing functions on S, the CLT (4.1) holds.*

Proof. First let h be continuous, bounded, and monotone nondecreasing. If $c \leq h(x) \leq d \forall x \in S$, then $h_1(x) := h(x) - c$ is nonnegative and monotone nondecreasing, and $\tilde{h}_1(x) \equiv h_1(x) - \int h_1 \, d\pi = \tilde{h}(x) = h(x) - \int h \, d\pi$. It then follows, mimicking the calculations in Example 3.2, that

$$T^n\tilde{h}(x) = T^n\tilde{h}_1(x) = E(\tilde{h}_1(X_n) \mid X_0 = x)$$

$$= \int_{[0,d-c]} P_\mu(h_1(X_n) > u \mid X_0 = x) \, du - \int_{[0,d-c]} P_\pi(h_1(X_0) > u) \, du$$

$$= \int_{[c,d]} P_\mu(h(X_n) > u \mid X_0 = x) \, du - \int_{[c,d]} P_\pi(h(X_0) > u) \, du$$

$$= -\int_{[c,d]} [p^{(n)}(x, A(u)) - \pi(A(u))] \, du, \qquad (4.10)$$

where $A(u) := \{y \in S : h(y) \leq u\}$ belongs to the splitting class $\mathcal{A}$ (see Chapter 3 (5.23)). Thus, by Corollary 5.2 (Chapter 3),

$$\sum_{n=m+1}^{\infty} |T^n \tilde{h}(x)| \leq (d-c) \sum_{n=m+1}^{\infty} (1 - \tilde{\chi})^{[n/N]} \to 0 \quad \text{as } m \to \infty,$$

(4.11)

implying that the series (3.9) converges uniformly to a solution g of the Poisson equation (3.3). Note also, that $X_n(x) := \alpha_n \cdots \alpha_1 x$ is continuous and monotone nondecreasing in x for every $n \geq 0$, which implies that $T^n \tilde{h}(x)$ is continuous and monotone nondecreasing in x for every $n \geq 0$. Hence g is bounded, continuous, and monotone nondecreasing. Suppose $\sigma_h^2 > 0$. Then, as in the proof of Theorem 4.1, $X_{k,n} := n^{-1/2}[g(X_k) - Tg(X_{k-1})]$ for $1 \leq k \leq n$ ($n > 1$) is a bounded martingale difference sequence for which $L_n(\varepsilon) = 0$ for all sufficiently large n. Since g is bounded, there exists a constant c_1 such that $g - c_1$ is *nonnegative*, bounded, continuous, and monotone nondecreasing, as are $T(g - c_1), (g - c_1)^2$, and $T(g - c_1)^2$. Hence, with any of these functions as h, (4.10) and (4.11) hold. Therefore, as in the proof of Theorem 4.1, the relations (4.5)–(4.7) hold with g replaced by $g - c_1$. By expanding $(g - c_1)^2$ as $g^2 - 2 c_1 g + c_1^2$, it now follows that (4.5)–(4.7) hold (for g), verifying the hypothesis of the CLT. This proves the theorem for all bounded continuous monotone nondecreasing h. If $h = f_1 - f_2$, where both f_1 and f_2 are bounded, continuous, and monotone nondecreasing, then $\sum_{n=m+1}^{\infty} |T^n \tilde{h}(x)| \leq \sum_{n=m+1}^{\infty} |T^n \tilde{f}_1(x)| + \sum_{n=m+1}^{\infty} |T^n \tilde{f}_2(x)| \to 0$ as $m \to \infty$, uniformly for all x, by the same argument leading to (4.11). The limits $g_i = \sum_{n=0}^{\infty} T^n \tilde{f}_i$ are bounded, continuous, monotone nondecreasing ($i = 1, 2$), and the same is true for $T(g_i g_j)$ and $(Tg_i)(Tg_j)$ ($i, j = 1, 2$). Hence with $g = g_1 - g_2$ as the solution of the Poisson equation (3.3) corresponding to $\tilde{h} = \tilde{f}_1 - \tilde{f}_2$, one has (see (4.4)–(4.8))

$$s_n^2 = n^{-1} \sum_{k=1}^{n} \left[Tg_1^2(X_{k-1}) + Tg_2^2(X_{k-1}) - 2(Tg_1g_2)(X_{k-1}) \right.$$

$$\left. - (Tg_1)^2(X_{k-1}) - (Tg_2)^2(X_{k-1}) + 2Tg_1(X_{k-1}) \cdot Tg_2(X_{k-1}) \right]$$

$$\to \sigma_h^2 \text{ in probability.} \quad (4.12)$$

The hypothesis of the martingale CLT is thus satisfied by the martingale difference sequence $X_{k,n} = [g(X_k) - Tg(X_{k-1})]/n^{1/2}$. ∎

Remark 4.1 In Section 5.3 and in the present section, we have focused on estimating $\lambda_h = \int h \, d\pi$ for bounded h. While this is fine for estimating probabilities $\pi(A) = \int 1_A \, d\pi$ for appropriate A, one is not able to estimate interesting quantities such as moments $\int x^r \, d\pi$ $(r = 1, 2, \ldots)$ on unbounded state spaces. For the latter purpose, one first shows that if, for an $h \in L^2(\pi)$, *the Poisson equation (3.3) has a solution* $g \in L^2(\pi)$, *then the CLT (4.1) holds under* P_π, i.e., under the initial distribution π (see Complements and Details). Although one may then derive the CLT under certain other initial distributions μ (e.g., if μ is absolutely continuous w.r.t. π), one cannot, in general, derive (4.1) under every initial distribution. In addition, one typically does not know enough about π to decide which $h \in L^2(\pi)$ (except for bounded h) and, even if one has a specific $h \in L^2(\pi)$, for which such h the Poisson equation has a solution in $L^2(\pi)$.

Remark 4.2 The hypothesis of *continuity* of the i.i.d. nondecreasing maps $\alpha_n (n \geq 1)$ in Theorem 4.2 may be relaxed to *measurability*, in view of Theorem C5.1 in Chapter 3, Complements and Details. It may be shown that the Poisson equation (3.3) in this case has a solution $g \in L^2(\pi)$, if $h \in L^2(\pi)$ is such that it may be expressed as the difference between two nondecreasing measurable functions in $L^2(\pi)$. It may be also shown that for such h, the CLT (4.1) holds, no matter what the initial distribution μ may be (see Complements and Details).

Remark 4.3 In Section 2.10, the CLT for positive recurrent Markov chains was derived using the so-called *renewal method*, in which the chain $\{X_n : n \geq 0\}$ is divided up into disjoint independent blocks over the random time intervals $[\eta^{(r-1)}, \eta^{(r)}), r \geq 1$, where $\eta^{(r)}$ is the rth return time to a specified state, say, $i(r \geq 1)$, $\eta^{(0)} = 0$. One may then apply the classical CLT for independent summands, together with a general result (Proposition 10.1 (Chapter 2)) on the classical CLT, but with a random number of summands. Such a method may be extended to irreducible processes satisfying (a) the Doeblin minorization (Theorem 9.1 (Chapter 2) or Corollary 5.3 (Chapter 3)) and, more generally, to irreducible Harris recurrent processes satisfying (b) a local Doeblin minorization (Theorem 9.2 (Chapter 2)). In case (a), following the proof of Corollary 5.3 (Chapter 3), if one decides the nature of $\beta_n (n \geq 1)$ by i.i.d. coin tosses $\theta_n (n \geq 1)$, where $P(\theta_n = 1) = \chi$, $P(\theta_n = 0) = 1 - \chi$, and whenever $\theta_n = 1$, one lets $\beta_n = Z_n (Z_n (n \geq 1)$, being i.i.d. with common distribution λ_δ, as defined in Chapter 3 (5.32)), then $\eta^{(r)}$ is replaced by the rth return time with $\theta_n = 1$. For the general case (b), the successive renewal, or regeneration, times are described in Complements and Details in Section 2.9.

5.5 The Nature of the Invariant Distribution

It is in general quite difficult to compute invariant distributions of Markov processes. A notable exception is the class of (positive recurrent) birth–death chains, for which the invariant probabilities are calculated in Propositions 8.3 and 8.4 of Chapter 2. The problem of explicit analytic computation is particuarly hard on state spaces S, which are uncountable as, e.g., in the case S is an interval or a rectangle (finite or infinite). Two particularly interesting (classes of) examples of analytic computations of invariant probabilities on $S = (0, \infty)$ are given in Section 4.6 on random continued fractions, which also illustrate the delicacy of the general problem. Given such difficulties, one may turn to somewhat qualitative descriptions or broad properties of invariant probabilities on continuum state spaces S. As an example, consider Markov processes $X_n = \alpha_n \cdots \alpha_1 X_0$ on an interval S, where $\alpha_n (n \geq 1)$ are i.i.d. monotone and the splitting condition (**H**) holds. The following result is proved in Remark 5.4, Section 3.5.

Proposition 5.1 *In addition to the hypothesis of* Theorem 5.1 (Chapter 3), *assume* α_n *are strictly monotone. Then the unique invariant probability is nonatomic.*

A simple general result on absolute continuity of π is the following.

Proposition 5.2 *Suppose the transition probabilities $p(x, \cdot)$ of a Markov process on a state space $(S, \mathcal{S})$ are, for each $x \in S$, absolutely continuous with respect to a sigmafinite measure ν on $(S, \mathcal{S})$. Then, if π is invariant under p, π is absolutely continuous with respect to ν.*

Proof. Let $B \in \mathcal{S}$ such that $\nu(B) = 0$. Then $p(x, B) = 0$ for all x, so that

$$\pi(B) = \int_S p(x, B)\pi(dx) = \int_S 0\pi(dx) = 0. \qquad (5.1)$$

∎

Beyond such simple results on absolute continuity with respect to Lebesgue measure $\nu = m$, say, proving absolute continuity or singularity turns out to be a formidable mathematical task in general. To illustrate the difficulty, we will consider $S = [0, 1]$ and Γ consisting of two affine linear maps f_0, f_1 given by

$$f_0(x) = \theta x, \quad f_1(x) = \theta x + (1 - \theta), \quad x \in [0, 1], \quad (5.2)$$

picked with probabilities p and $1 - p$: $P(\alpha_n = f_0) = p$, $P(\alpha_n = f_1) = 1 - p$ for some p, $0 < p < 1$. Here the parameter θ lies in $(0, 1)$. One may also represent this process as

$$X_{n+1} = \theta X_n + \varepsilon_{n+1} \ (n \geq 0), \quad (5.3)$$

where $\varepsilon_n (n \geq 1)$ are i.i.d. Bernoulli: $P(\varepsilon_n = 0) = p$, $P(\varepsilon_n = 1 - \theta) = 1 - p \ (0 < p < 1)$, $\theta \in (0, 1)$. If $0 < \theta < 1/2$, then f_0 maps $[0, 1]$ onto $[0, \theta] = I_0$, say, and f_1 maps $[0, 1]$ onto $[1 - \theta, 1] = I_1$, with I_0 and I_1 disjoint, each of length θ. With two iterations, the possible maps are $f_0 \circ f_0$, $f_0 \circ f_1$, $f_1 \circ f_0$, $f_1 \circ f_1$; the first two maps $[0, 1]$ onto two disjoint subintervals of I_0, namely, $I_{00} = [0, \theta^2]$ and $I_{01} = [\theta(1 - \theta), \theta]$ with a gap in the middle of I_0. Similarly, $f_1 \circ f_0$ and $f_1 \circ f_1$ map $[0, 1]$ onto two disjoint subintervals of I_1, namely, $I_{10} = [1 - \theta, \theta^2 + 1 - \theta]$ and $I_{11} = [1 - \theta^2, 1]$ with a gap in the middle of I_1. All these four intervals are of the same length θ^2. In this way the 2^n possible n-stage iterations map $[0, 1]$ onto 2^n disjoint subintervals, dividing each of the 2^{n-1} intervals at the $(n - 1)$th stage into two disjoint intervals with a gap in the middle. Each of the 2^n n-stage intervals is of length θ^n. Thus the Lebesgue measure of the union D_n, say, of all these 2^n intervals is $2^n \theta^n$, converging to 0 as $n \to \infty$, if $0 < \theta < 1/2$. Thus $D \equiv \bigcap_{n=1}^{\infty} D_n$ *is a Cantor set of Lebesgue measure 0, containing the support of the invariant probability* π of the Markov process (5.3) since $\pi(D_n) = 1$ $\forall n$. For the case $p = 1 - p = 1/2$, and $\theta = 1/2$, one may directly check from (5.3) that the *uniform distribution is invariant.* Thus we have proved the following result (use Proposition 5.1 to show π is nonatomic).

Proposition 5.3 *Consider the Markov process* (5.3) *on* $S = [0, 1]$ *with* $\varepsilon_n (n \geq 1)$ *i.i.d.:* $P(\varepsilon_n = 0) = p$, $P(\varepsilon_n = 1 - \theta) = 1 - p$, *where* $0 < \theta < 1$, $0 < p < 1$. (a) *If* $0 < \theta < 1/2$, *then the unique invariant probability* π *is nonatomic and the support of* π *is a Cantor set of Lebesgue measure zero. In particular,* π *is singular w.r.t. Lebesgue measure.*

(b) If $\theta = 1/2$ and $p = 1/2$, then the unique invariant probability π is the uniform distribution, i.e., the Lebesgue measure on $[0, 1]$.

What happens if $1/2 < \theta < 1$? It is simple to check that for each n, the union D_n of the 2^n n-stage intervals (namely, ranges of $f_{i_1} \circ f_{i_2} \circ \cdots \circ f_{i_n}$, $(i_j = 0$ or $1, \forall j))$ is $[0, 1]$ if $\theta \geq 1/2$. It has been known for many years that there are values of $\theta \in (1/2, 1)$ for which π is absolutely continuous with respect to Lebesgue measure, and other $\theta > 1/2$ for which π is singular with respect to Lebesgue measure. A famous conjecture that, for the symmetric case $p = 1/2$, π *is absolutely continuous for all* $\theta \in [1/2, 1)$ *outside a set of Lebesgue measure zero* stood unresolved for nearly 60 years, until Solomyak (1995) proved it. Peres and Solomyak (1996) extended this result to arbitrary $p \in (0, 1)$. Further, it is known that, if $p = 1/2$ then *the set of $\theta \in (1/2, 1)$ for which π is singular is infinite*; but it is still not known if this set is countable. For comprehensive surveys of the known results in this field, along with some new results and open questions, we refer to Peres, Schlag, and Solomyak (1999) and Mitra, Montrucchio, and Privileggi (2004). The latter article is of particular interest to economists as it relates (5.2), (5.3) to a one-sector optimal growth model.

5.5.1 Random Iterations of Two Quadratic Maps

In this subsection we consider the Markov process $X_n := \alpha_n X_{n-1} (n \geq 1)$ as defined in Section 4.4, with $P(\alpha_1 = F_\mu) = \eta$ and $P(\alpha_1 = F_\lambda) = 1 - \eta$ for appropriate pairs $\mu < \lambda$ and $\eta \in (0, 1)$. If $0 \leq \mu, \lambda \leq 1$, then it follows from Proposition 7.4(a) of Chapter 1 that the backward iterates $Y_n(x) \to 0$ a.s. for all $x \in [0, 1]$, so that the Dirac measure δ_0 is the unique invariant probability. Now take $1 < \mu < \lambda \leq 2$, and let $a = p_\mu \equiv 1 - 1/\mu$ and $b = p_\lambda$. Then F_μ, F_λ are both strictly increasing on $[p_\mu, p_\lambda]$ and leave this interval invariant. It follows from Example 4.1 of Chapter 4 that there exists a unique invariant probability π on the state space $(0, 1)$ and that π is *nonatomic* by Proposition 5.1. Since p_μ and p_λ are attractive fixed points of F_μ and F_λ, respectively, they belong to the *support* $S(\pi)$ of π as do the set of all points of the form (see Lemma 6.2, Chapter 4)

$$F_{\varepsilon_1 \varepsilon_2 \cdots \varepsilon_k} p_\mu \equiv f_{\varepsilon_1} f_{\varepsilon_2} \cdots f_{\varepsilon_k} p_\mu \ (f_0 : = F_\mu, f_1 := F_\lambda),$$

$$F_{\varepsilon_1 \varepsilon_2 \cdots \varepsilon_k} p_\lambda \ (k \geq 1), \quad (5.4)$$

for all k-tuples $(\varepsilon_1, \varepsilon_2, \ldots, \varepsilon_k)$ of 0's and 1's and for all $k \geq 1$. Write $\mathrm{Orb}(x; \mu, \lambda) = \{F_{\varepsilon_1 \varepsilon_2 \cdots \varepsilon_k} x \colon k \geq 0, \varepsilon_i = 0 \text{ or } 1 \forall i\}$ ($k = 0$ corresponds to x). By Lemma 6.2 (Chapter 4), if $x \in S(\pi)$ then $S(\pi) = \overline{\mathrm{Orb}(x; \mu, \lambda)}$. This support, however, need not be $[p_\mu, p_\lambda]$. Indeed, if $F_\lambda(p_\mu) > F_\mu(p_\lambda)$, i.e.,

$$\frac{1}{\lambda^2} - \frac{1}{\lambda^3} < \frac{1}{\mu^2} - \frac{1}{\mu^3} \quad (1 < \mu < \lambda \leq 2), \tag{5.5}$$

then $S(\pi)$ is a *Cantor subset* of $[p_\mu, p_\lambda]$. Before proving this assertion we identify pairs μ, λ satisfying (5.5). On the interval $[1, 2]$ the function $g(x) := x^{-2} - x^{-3}$ is strictly increasing on $[1, 3/2]$ and strictly decreasing on $[3/2, 2]$, and $g(1) = 0, g(3/2) = 4/27, g(2) = 1/8$. Therefore, (5.5) holds iff

$$\lambda \in (3/2, 2] \quad \text{and} \quad \mu \in [\hat{\lambda}, \lambda), \tag{5.6}$$

where $\hat{\lambda} \leq 3/2$ is uniquely defined for a given $\lambda \in (3/2, 2]$ by $g(\hat{\lambda}) = g(\lambda)$. Since the smallest value of $\hat{\lambda}$ as λ varies over $(3/2, 2]$ is $\sqrt{5} - 1$, which occurs when $\lambda = 2(g(2) = 1/8)$, it follows that μ can not be smaller than $\sqrt{5} - 1$ if (5.5) or (5.6) holds.

To show that $S(\pi)$ is a Cantor set (i.e., a closed, nowhere dense set having no isolated point), for μ, λ satisfying (5.5), or (5.6), write $I = [p_\mu, p_\lambda]$, $I_0 = F_\mu(I)$, $I_1 = F_\lambda(I)$, $I_{\varepsilon_1 \varepsilon_2 \cdots \varepsilon_k} = F_{\varepsilon_1 \varepsilon_2 \cdots \varepsilon_k}(I)$ for $k \geq 1$ and k-tuples $(\varepsilon_1, \varepsilon_2, \cdots, \varepsilon_k)$ of 0's and 1's. Here $F_{\varepsilon_1 \varepsilon_2 \cdots \varepsilon_k} = f_{\varepsilon_1} f_{\varepsilon_2} \cdots f_{\varepsilon_k}$ as defined in (5.4). Under the present hypothesis $F_\lambda(p_\mu) > F_\mu(p_\lambda)$ (or (5.5)), the 2^k intervals $I_{\varepsilon_1 \varepsilon_2 \cdots \varepsilon_k}$ are disjoint, as may be easily shown by induction, using the fact that F_μ, F_λ are strictly increasing on $[p_\mu, p_\lambda]$. Let $J_k = \cup I_{\varepsilon_1 \varepsilon_2 \cdots \varepsilon_k}$ where the union is over the 2^k k-tuples $(\varepsilon_1, \varepsilon_2, \ldots, \varepsilon_k)$. Since $X_k(x) \in J_k$ for all $x \in I$, and $J_k \downarrow$ as $k \uparrow$, $S(\pi) \subset J_k$ for all k, so that $S(\pi) \subset J := \cap_{k=1}^\infty J_k$. Further, F_μ is a *strict contraction* on $[p_\mu, p_\lambda]$ while $F_\lambda^n(I) \downarrow \{p_\lambda\}$ as $n \uparrow \infty$. Hence the lengths of $I_{\varepsilon_1 \varepsilon_2 \ldots \varepsilon_k}$ go to zero as $k \to \infty$, for a sequence $(\varepsilon_1, \varepsilon_2, \ldots)$ which has only finitely many 0's or finitely many 1's. If there are infinitely many 0's and infinitely many 1's in $(\varepsilon_1, \varepsilon_2, \ldots)$, then for large k with $\varepsilon_k = 0$, one may express $f_{\varepsilon_1} f_{\varepsilon_2} \cdots f_{\varepsilon_k}$ as a large number of compositions of functions of the type F_μ^n or $F_\lambda^n F_\mu (n \geq 1)$. Since for all μ, λ satisfying (5.5) the derivatives of these functions on $[p_\mu, p_\lambda]$ are bounded by $F_\lambda'(p_\mu) F_\mu'(p_\mu) < 1$ (use induction on n and the estimate $F_\lambda'(F_\lambda p_\mu) < 1$), it follows that the lengths of the nested intervals $I_{\varepsilon_1 \varepsilon_2 \cdots \varepsilon_k}$ go to zero as $k \to \infty$ (for every sequence $(\varepsilon_1 \varepsilon_2, \ldots)$). Thus, J does not contain any (nonempty) open interval.

Also, $J \subset \overline{\text{Orb } (p_\mu; \mu, \lambda)} = \overline{\text{Orb } (p_\lambda; \mu, \lambda)}$ by the same reasoning, so that $J = S(\pi)$. Since π is nonatomic, $S(\pi)$ does not include any isolated point, completing the proof that $S(\pi)$ is a Cantor set.

Write $|A|$ for the Lebesgue measure of A. If, in addition to (5.5),

$$\lambda < \left(\frac{\mu - 1}{2 - \mu}\right)\mu, \tag{5.7}$$

then $|J| = 0$. Indeed, for any subinterval I' of I, one has $|F_\lambda(I')| \leq F_\lambda'(p_\mu)|I'|$, $|F_\mu(I')| \leq F_\mu'(p_\mu)|I'|$, from which it follows that $|J_{k+1}| \leq (2 - \mu)(1 + \lambda/\mu)|J_k|$. If (5.7) holds then $c \equiv (2 - \mu)(1 + \lambda/\mu) < 1$, so that $|J| = 0$ if (5.5), (5.7) hold, i.e., $S(\pi)$ *is a Cantor set of Lebesgue measure zero.*

Note that the proof $|I_{\varepsilon_1\varepsilon_2\cdots\varepsilon_k}| \to 0$ depends only on the facts that on $[p_\mu, p_\lambda]$, (i) F_μ is a contraction, (ii) $F_\lambda^n(I) \downarrow \{p_\lambda\}$, and (iii) $F_\lambda'(F_\lambda p_\mu) < 1$, $F_\lambda'(p_\mu)F_\mu'(p_\mu) < 1$. The last condition (iii) may be expressed as

$$\lambda - 2\lambda^2(\mu - 1)/\mu^2 < 1, \quad \lambda < \mu/(2 - \mu)^2 \ (1 < \mu < \lambda \leq 2). \tag{5.8}$$

If (5.8) holds, but (5.5) does not, then the 2^k intervals $I_{\varepsilon_1\varepsilon_2\cdots\varepsilon_k}$ cover $I = [p_\mu, p_\lambda]$. Since the endpoints of $I_{\varepsilon_1\varepsilon_2\cdots\varepsilon_k}$ are in $S(\pi)$, it follows that $S(\pi) = [p_\mu, p_\lambda]$.

There are many open problems in the topic of this subsection. For example, if $1 < \mu < \lambda \leq 2$, but (5.5) (or (5.6)) does not hold, and (5.8) does not hold, then $S(\pi) = [p_\mu, p_\lambda]$. However, we do not know when π is singular or absolutely continuous in this case. Similarly, we do not know for which μ, λ satisfying $2 < \mu < \lambda \leq 3$, π is absolutely continuous (see Example 4.2 in Chapter 4).

An additional point of complexity, as well as interest, here is that there are μ, λ satisfying $1 < \mu < \lambda < 4$ for which the process $X_n(n \geq 0)$ defined above *does not have a unique invariant probability* on $S = (0, 1)$, as has been shown by Athreya and Dai (2002) by an example. One would like to know precisely for which pairs (μ, λ) does the process $X_n(n \geq 0)$ have a unique invariant probability or, equivalently, for which pairs it does not.

5.6 Complements and Details

Section 5.3. This section is mainly based on Bhattacharya and Majumdar (2001). Also see Athreya and Majumdar (2003).

Section 5.4.

Proof of Proposition 4.1 (Martingale CLT).   Assume that, as $n \to \infty$, (i) $s_n^2 \to 1$ in probability and (ii) $L_n(\varepsilon) \to 0$ in probability, for every $\varepsilon > 0$. Then S_n converges in distribution to $N(0, 1)$.

Proof.   Consider the conditional characteristic functions

$$\varphi_{k,n}(\xi) := E(\exp\{i\xi X_{k,n}\} \mid \mathcal{F}_{k-1,n}), \quad (\xi \in \mathbb{R}^1). \qquad (\text{C4.1})$$

Provided $|\prod_1^{k_n}(\varphi_{k,n}(\xi))^{-1}| \leqslant \delta(\xi)$, a constant, it would be enough to show that

(a) $E(\exp\{i\xi S_n\} / \prod_1^{k_n} \varphi_{k,n}(\xi)) = 1$, and
(b) $\prod_1^{k_n} \varphi_{k,n}(\xi) \to \exp\{-\xi^2/2\}$ in probability.

Indeed, if $|(\prod \varphi_{k,n}(\xi))^{-1})| \leqslant \delta(\xi)$, then (a), (b) imply

$$\left| E \exp\{i\xi S_n\} - \exp\{-\xi^2/2\} \right|$$

$$= \exp\{-\xi^2/2\} \left| \frac{E \exp\{i\xi S_n\}}{\exp\{-\xi^2/2\}} - E\left(\frac{\exp\{i\xi S_n\}}{\prod \varphi_{k,n}(\xi)} \right) \right|$$

$$\leqslant \exp\{-\xi^2/2\} E \left| \frac{1}{\exp\{-\xi^2/2\}} - \frac{1}{\prod \varphi_{k,n}(\xi)} \right| \to 0. \quad (\text{C4.2})$$

Now part (a) follows by taking successive conditional expectations given $\mathcal{F}_{k-1,n}$ ($k = k_n, k_n - 1, \ldots, 1$), if $(\Pi \varphi_{k,n}(\xi))^{-1}$ is integrable. Note that the martingale difference is not needed for this. It turns out, however, that in general $(\Pi \varphi_{k,n}(\xi))$ cannot be bounded away from zero. Our first task is then to replace $X_{k,n}$ by new martingale differences $Y_{k,n}$ for which this integrability does hold, and whose sum has the same asymptotic distribution as S_n. To construct $Y_{k,n}$, first use assumption (ii) to check that $M_n \equiv \max\{\sigma_{k,n}^2 : 1 \leqslant k \leqslant k_n\} \to 0$ in probability. Therefore, there exists a *nonrandom* sequence $\delta_n \downarrow 0$ such that

$$P(M_n \geqslant \delta_n) \to 0 \quad \text{as } n \to \infty. \qquad (\text{C4.3})$$

Similarly, there exists, for each $\varepsilon > 0$, a *nonrandom* sequence $\Theta_n(\varepsilon) \downarrow 0$ such that

$$P(L_n(\varepsilon) \geqslant \Theta_n(\varepsilon)) \to 0 \quad \text{as } n \to \infty. \qquad (\text{C4.4})$$

Write $L_{k,n}$ in place of $L_n(\varepsilon)$ for the sum over the indices $j = 1, 2, \ldots, k (1 \leqslant k \leqslant k_n)$. Consider the events

$$A_{k,n}(\varepsilon) := \{\sigma_{k,n}^2 < \delta_n, L_{k,n} < \Theta_n(\varepsilon), s_{k,n}^2 < 2\}, \quad (1 \leqslant k \leqslant k_n). \tag{C4.5}$$

Then $A_{k,n}(\varepsilon)$ is $\mathcal{F}_{k-1,n}$-measurable. Therefore, $Y_{k,n}$ defined by

$$Y_{k,n} := X_{k,n} \mathbf{1}_{A_{k,n}(\varepsilon)} \tag{C4.6}$$

has zero-conditional expectation, given $\mathcal{F}_{k-1,n}$. Although $Y_{k,n}$ depends on ε, we will suppress this dependence for notational convenience. Note also that

$$P(Y_{k,n} = X_{k,n} \text{ for } 1 \leqslant k \leqslant k_n) \geqslant P\left(\bigcap_{k=1}^{k_n} A_{k,n}(\varepsilon)\right)$$

$$= P(M_n < \delta_n, \ L_n(\varepsilon) < \Theta_n(\varepsilon), s_n^2 < 2) \to 1. \tag{C4.7}$$

We will use the notation (4.3) and (C4.1) with a *prime* (') to denote the corresponding quantities for $\{Y_{k,n}\}$. For example, using the fact $E(Y_{k,n} \mid \mathcal{F}_{k-1,n}) = 0$ and a Taylor expansion,

$$\left|\varphi_{k,n}'(\xi) - \left(1 - \frac{\xi^2}{2}\sigma_{k,n}'^2\right)\right|$$

$$= \left|E\left[\exp(i\xi Y_{k,n}) - \left(1 + i\xi Y_{k,n} + \frac{(i\xi)^2}{2}Y_{k,n}^2\right) \mid \mathcal{F}_{k-1,n}\right]\right|$$

$$= \left|E\left[-\xi^2 Y_{k,n}^2 \int_0^1 (1-u)(\exp\{iu\xi Y_{k,n}\} - 1)\,du \mid \mathcal{F}_{k-1,n}\right]\right|$$

$$\leqslant \varepsilon \frac{|\xi|^3}{2}\sigma_{k,n}'^2 + \xi^2 E(Y_{k,n}^2 \mathbf{1}_{\{|Y_{k,n}|>\varepsilon\}} \mid \mathcal{F}_{k-1,n})$$

$$\leqslant \varepsilon \frac{|\xi|^3}{2}\sigma_{k,n}'^2 + \xi^2 E(X_{k,n}^2 \mathbf{1}_{\{|Y_{k,n}|>\varepsilon\}} \mid \mathcal{F}_{k-1,n}). \tag{C4.8}$$

Fix $\xi \in \mathbb{R}^1$. Since $M_n' < \delta_n$, $0 \leqslant 1 - (\xi^2/2)\sigma_{k,n}'^2 \leqslant 1$ $(1 \leqslant k \leqslant k_n)$ for all large n. Therefore, using (C4.8),

$$\left|\prod \varphi_{k,n}'(\xi) - \prod\left(1 - \frac{\xi^2}{2}\sigma_{k,n}'^2\right)\right| \leqslant \sum_{k=1}^{k_n}\left|\varphi_{k,n}'(\xi) - \left(1 - \frac{\xi^2}{2}\sigma_{k,n}'^2\right)\right|$$

$$\leqslant |\xi|^3 \varepsilon + \xi^2 \Theta_n(\varepsilon), \tag{C4.9}$$

and

$$\left| \prod \left(1 - \frac{\xi^2}{2}\sigma'^2_{k,n} \right) - \exp \left\{ -\frac{\xi^2}{2}s'^2_n \right\} \right|$$

$$= \left| \prod \left(1 - \frac{\xi^2}{2}\sigma'^2_{k,n} \right) - \prod \exp \left\{ -\frac{\xi^2}{2}\sigma'^2_{k,n} \right\} \right|$$

$$\leqslant \frac{\xi^4}{8} \sum \sigma'^4_{k,n} \leqslant \frac{\xi^4}{8}\delta_n s'^2_n \leqslant \frac{\xi^4}{4}\delta_n. \qquad \text{(C4.10)}$$

Therefore,

$$\left| \prod \varphi'_{k,n}(\xi) - \exp \left\{ -\frac{\xi^2}{2}s'^2_n \right\} \right| \leqslant |\xi|^3\varepsilon + \xi^2\Theta_n(\varepsilon) + \frac{\xi^4}{4}\delta_n. \qquad \text{(C4.11)}$$

Moreover, (C4.11) implies

$$\left| \prod \varphi'_{k,n}(\xi) \right| \geqslant \exp \left\{ -\frac{\xi^2}{2}s'^2_n \right\} - |\xi|^3\varepsilon + \xi^2\Theta_n(\varepsilon) + \frac{\xi^4}{4}\delta_n$$

$$\geqslant \exp\{-\xi^2\} - |\xi|^3\varepsilon - \left(\xi^2\Theta_n(\varepsilon) + \frac{\xi^4}{4}\delta_n \right). \qquad \text{(C4.12)}$$

By choosing ε sufficiently small, one has for all sufficiently large n (depending on ε), $|\prod\varphi'_{k,n}(\xi)|$ bounded away from zero. Therefore, (a) holds for $\{Y_{k,n}\}$ for all sufficiently small ε (and all sufficiently large n, depending on ε). By using relations as in (C4.2) and the inequalities (C4.11) and (C4.12) and the fact that $s'^2_n \to 1$ in probability, we get (for ε sufficiently small)

$$\overline{\lim_{n\to\infty}} \left| E \left(\exp\{i\xi S'_n\} - \exp \left\{ -\frac{\xi^2}{2} \right\} \right) \right|$$

$$\leqslant \exp \left\{ -\frac{\xi^2}{2} \right\} \overline{\lim_{n\to\infty}} E \left| \left(\exp \left\{ -\frac{\xi^2}{2} \right\} \right)^{-1} - \left(\prod \varphi'_{k,n}(\xi)^{-1} \right) \right|$$

$$\leqslant \exp \left\{ -\frac{\xi^2}{2} \right\} \exp \left\{ \frac{\xi^2}{2} \right\}$$

$$\times \left(\exp \left\{ -\frac{\xi^2}{2} \right\} - |\xi|^3\varepsilon \right)^{-1} \overline{\lim_{n\to\infty}} E \left| \prod \varphi'_{k,n}(\xi) - e^{-\xi^2/2} \right|$$

$$\leqslant \left(\exp \left\{ -\frac{\xi^2}{2} \right\} - |\xi|^3\varepsilon \right)^{-1} |\xi|^3\varepsilon. \qquad \text{(C4.13)}$$

Finally,

$$\overline{\lim_{n\to\infty}} \left| E \exp\{i\xi S_n\} - \exp\left\{-\frac{\xi^2}{2}\right\} \right|$$

$$\leqslant \overline{\lim_{n\to\infty}} \left| E \exp\{i\xi S_n'\} - E \exp\{i\xi S_n'\} \right|$$

$$+ \overline{\lim_{n\to\infty}} \left| E \exp\{i\xi S_n'\} - \exp\left\{-\frac{\xi^2}{2}\right\} \right|$$

$$= 0 + \overline{\lim_{n\to\infty}} \left| E \exp\{i\xi S_n'\} - \exp\left\{-\frac{\xi^2}{2}\right\} \right|$$

$$\leqslant \left(\exp\left\{-\frac{\xi^2}{2}\right\} - |\xi|^3\varepsilon \right)^{-1} |\xi|^3\varepsilon. \qquad \text{(C4.14)}$$

The extreme right side of (C4.14) goes to zero as $\varepsilon \downarrow 0$, while the extreme left does not depend on ε. ∎

Condition (ii) of the martingale CLT is called the *conditional Lindeberg condition*.

The martingale CLT as presented here is due to Brown (1977). Its proof follows Bhattacharya and Waymire (1990).

The following CLT is due to Gordin and Lifsic (1978). Its proof is based on a version of the margingale CLT due to Billingsley (1961) and Ibragimov (1963). The term "ergodic" in its statement is defined in a general context under Remark C4.1. For our purposes, it is enough to know that a *Markov process* $\{X_n: n \geq 0\}$ *with a unique invariant probability* π *is ergodic* (if X_0 has distribution π). The process $\{X_n: n \geq 0\}$ is then said to be ergodic even when X_0 has an arbitrary initial distribution.

As a very special case, the following is obtained: *If* $p(x; dy)$ *admits a unique invariant probability* π, *and* $\{X_n: n \geqslant 0\}$ *is a Markov process with transition probability* $p(x, dy)$ *and initial distribution* π, *then* $\{X_n: n \geqslant 0\}$ *is ergodic*.

Theorem C4.1 (*CLT for discrete-parameter Markov processes*) Assume $p(x, dy)$ admits an invariant probability π and, under the initial distribution π, $\{X_n\}$ is ergodic. Assume also that $\tilde{h} = h - \int h \, d\pi$ is in the range of $I - T$. Then, under initial π,

$$\frac{1}{\sqrt{n}} \sum_{m=0}^{n-1} \left(h(X_m) - \int h \, d\pi \right) \to N(0, \sigma^2) \quad \text{as } n \to \infty, \quad \text{(C4.15)}$$

where $\sigma^2 = \sigma_h^2$ is given by (4.1) with g defined by $(I - T)g = \tilde{h}$.

Remark C4.1 (*Birkhoff's ergodic theorem*)   Let T_1 be a measurable map on a state space $(\tilde{S}, \tilde{\mathcal{S}})$ into itself, which *preserves* a probability measure μ on $(\tilde{S}, \tilde{\mathcal{S}})$. *Birkhoff's ergodic theorem* says: *For every μ-integrable g, $\frac{1}{n} \sum_{m=0}^{n-1} g(T_1^m x)$ converges a.s. (w.r.t. μ) and in $L^1(\mu)$ to a function $\bar{g}(x)$, say, where $\bar{g}$ is invariant*: $\bar{g}(T_1 x) = \bar{g}(x) \; \forall x \in \tilde{S}$, $\int \bar{g} \, d\mu = \int g \, d\mu$. If the dynamical system $(T_1, \tilde{S}, \mu)$ is ergodic, $\bar{g} = \int g \, d\mu$. Here $(T_1, \tilde{S}, \mu)$ is said to be *ergodic* if all invariant functions are constant a.s. with respect to μ. Suppose one takes $\tilde{S} = S^\infty$, $\mathcal{S} = \mathcal{S}^{\otimes \infty}$, and $X_n (n \geq 0)$ a stochastic process with state space $(S, \mathcal{S})$, defined on the canonical probability space $(S^\infty, \mathcal{S}^{\otimes \infty}, Q)$ such that $X_n(\omega)$ is the projection onto the nth coordinate of $\omega \in S^\infty$ $(n \geq 0)$. The *process* $\mathbf{X} = (X_0, X_1, \ldots)$ *is stationary*, if $\tilde{T}\mathbf{X} \equiv (X_1, X_2, \ldots)$ has the same distribution Q as $\mathbf{X}$. Then, by Birkhoff's ergodic theorem, if $E|h(\mathbf{X})| < \infty$, $\frac{1}{n} \sum_{m=0}^{n-1} h(\tilde{T}^m \mathbf{X})$ converges a.s. and in $L^1(Q)$ to an invariant random variable $\bar{h}(\mathbf{X})$, say. If $\{X_n : n \geq 0\}$ is ergodic then $\bar{h}(\mathbf{X}) = Eh(\mathbf{X}) \equiv \int h \, dQ$ a.s. As a special case, with $h(\mathbf{X}) := f(X_0)$ for some measurable $f : S \to R$ such that $E|f(X_0)| < \infty$, one has $\frac{1}{n} \sum_{m=0}^{n-1} f(X_m)$ converges a.s. and in $L^1(Q)$ to an invariant $\bar{h}(\mathbf{X})$, which equals $Ef(X_0)$ if $\{X_n : n \geq 0\}$ is ergodic. Note that the validity of the last italicized statement depends only on the *distribution* of $\mathbf{X}$. Hence it holds for a stationary process $\{X_n : n \geq 0\}$ on an arbitrary probability space $(\Omega, \mathcal{F}, P)$.

For proofs of Birkhoff's ergodic theorem, we refer to Billingsley (1965, pp. 20–29), or Bhattacharya and Waymire (1990, pp. 224–227).

Theorem C4.2 below is due to Bhattacharya and Lee (1988). Theorem 4.2 is a special case of this.

Theorem C4.2   *Let $\{X_n : n \geq 0\}$ be a Markov process defined by $X_{n+1} = \alpha_{n+1} X_n (n \geq 0)$, where $\{\alpha_n : n \geq 1\}$ is a sequence of i.i.d. monotone nondecreasing maps on a closed subset S of $\mathbb{R}^\ell$, and X_0 is independent of $\{\alpha_n : n \geq 1\}$. Suppose the splitting hypothesis* (**H**) *of Section 3.5 holds, as in the case of collary 5.3. Then, no matter what the distribution of X_0 is, the CLT* (C4.15) *holds for every h that may be expressed as the difference $f_1 - f_2$ between two nondecreasing measurable functions $f_i : S \to S$ such that $f_i \in L^2(\pi)(i = 1, 2)$.*

Bhattacharya (1982) contains additional information on the CLT for Markov processes in continuous time, much of which can be adapted to discrete time.

Central limit theorems are truly "central" for statistical inference in time series. Of the great body of literature on the statistical inference for time series, we mention Anderson (1971), Box and Jenkins (1976), Brillinger (1981), Brockwell and Davis (1991), Grenander and Rosenblatt (1957), Granger and Terasvirta (1993), Hannan (1970), Pristley (1981), and Rosenblatt (2000). A comprehensive account of the asymptotic statistical theory for estimation and testing for *parametric* and *semiparametric linear* as well as *nonlinear* models may be found in Tamiguchi and Kakizawa (2000). Hwang (2002) contains some asymptotic results, including bootstrapping of the *nonparametric estimation* of the driving function f in nonlinear autoregressive models (5.1) of Chapter 4, under the hypothesis of Theorem C5.1 in Complements and Details, Chapter 4. For bootstrapping in the first-order model NLAR(1), see Frankel, Kreiss, and Mammen (2002). The results of Hwang on NLAR(k), $k \geq 1$, were obtained independently of those on the first-order model NLAR(1) by Franke et al. (2002).

The Billingsley–Ibragimov CLT (Billingsley 1961, Ibragimov 1963, applied to Markov processes in Gordin and Lifsic (1978), Bhattacharya (1982), is especially useful when the process is not necessarily irreducible.

5.7 Supplementary Exercises

(1) (*Definitions and simple properties of martingales*). A sequence of random variables $\{X_n\}(n = 0, 1, \ldots)$ is said to be $\{\Im_n\}$-adapted with respect to an increasing sequence of sigmafields $\Im_n \subset \Im_{n+1} \cdots (n = 0, 1, \ldots)$ if X_n is $\Im_n$-measurable (for every n). An $\{\Im_n\}$-adapted sequence $\{X_n\}$ with finite expectations is said to be $\{\Im_n\}$-*martingale* if $E(X_{n+1} \mid \Im_n) = X_n$ a.s. ($n \geq 0$). If, in particular, $\Im_n = \sigma\{X_j : 0 \leq j \leq n\} \forall n$, then $\{X_n\}$ is simply said to be a *martingale*. Suppose that $\{X_n\}$ is an $\{\Im_n\}$-adapted martingale. Prove the following statements:

  (a) $E(X_n \mid \Im_n) = X_m$ a.s. for every $m \leq n$. [Hint: Take successive conditional expectations, recalling that $E(X \mid \mathcal{G}) = E[E(X \mid \Im) \mid \mathcal{G}]$ for all sigmafields $\mathcal{G} \subset \Im$.]

  (b) $\{X_n\}$ is a martingale (with respect to the sigmafields $\mathcal{G}_n = \sigma\{X_j : 0 \leq j \leq n\}$. [Hint: $E(X_{n+1} \mid \mathcal{G}_n) = E[E(X_{n+1} \mid \Im_n) \mid \mathcal{G}_n] = E(X_n \mid \mathcal{G}_n) = X_n$ a.s., noting that if $\{X_n\}$ is $\{\Im_n\}$-adapted then $\mathcal{G}_n \subset \Im_n \forall n$.]

(c) The martingale differences $Z_n \equiv X_n - X_{n-1} (n \geq 1)$ satisfy (i) $E(Z_{n+1} \mid \Im_n) = 0 = E(Z_{n+1} \mid \mathcal{G}_n)(n \geq 0)$, (ii) if $EX_n^2 < \infty$ for all n, then $EZ_n Z_m = 0 \; \forall n \neq m$.

(d) $EX_n = EX_0$ for all n (*constancy of expectations*).

(e) $EX_n^2 = EX_0^2 + \sum_{m=1}^{n} EZ_m^2$, if $EX_n^2 < \infty$ for all n, and $Z_m \equiv X_m - X_{m-1} (m \geq 1)$.

(2) (*Examples of martingales*).

(a) Suppose $\{Z_n: n \geq 0\}$ is a sequence of independent random variables with finite means $EZ_n = \mu_n (n \geq 0)$. Prove that (i) $S_n = \sum_0^n (Z_j - \mu_j)$, $n \geq 0$, is an $\{\Im_n\}$-martingale, where $\Im_n = \sigma\{Z_j: 0 \leq j \leq n\}$, (ii) if $EZ_n^2 < \infty$ for all n, then $X_n \equiv S_n^2 - \sum_0^n \sigma_j^2 (n \geq 0)$ is an $\{\Im_n\}$-martingale, where $\sigma_j^2 \equiv var(Z_j)$. [Hint: $E(X_{n+1} \mid \Im_n) = E((S_n + Z_{n+1}\mu_{n+1})^2 - \sum_0^{n+1} \sigma_j^2 \mid \Im_n) = S_n^2 - \sum_0^n \sigma_j^2 + E((Z_{n+1} - \mu_{n+1})^2 \mid \Im_n) - 2S_n E(Z_{n+1} - \mu_{n+1} \mid \Im_n) - \sigma_{n+1}^2 = X_n$.]

(b) Suppose Y_j ($j \geq 1$) are independent random variables with a common finite nonzero mean μ. Show that $X_n = \mu^{-n} \prod_{j=1}^{n} Y_j$ ($n \geq 1$) is an $\{\Im_n\}$-martingale, where $\Im_n = \sigma\{Y_j: 1 \leq j \leq n\}$ ($n \geq 1$).

(c) Suppose X is a random variable on a probability space $(\Omega, \Im, P)$ with a finite expectation. If $\Im_n (n \geq 1)$ is an increasing sequence of sigmafields, $\Im_n \subset \Im$ for all n, show that $X_n \equiv E(X \mid \Im_n), n \geq 1$, is an $\{\Im_n\}$-martingale.

(3) (*Stopping times*). Let $\{X_n\}$ be an $\{\Im_n\}$-adapted sequence of random variables ($n \geq 0$). A *stopping time*, or an $\{\Im_n\}$-*stopping time*, τ is a random variable taking values in $\{0, 1, 2, \ldots\} \cup \{\infty\}$ such that $\{\tau \leq n\} \in \Im_n$ for every $n = 0, 1, 2, \ldots$. Prove the following:

(a) τ is a stopping time if and only if $\{\tau = n\} \in \Im_n$ for all n.

(b) If τ is a stopping time, then so is $\tau \wedge m \equiv \min\{\tau, m\}$ for every nonnegative integer m.

(c) $\tau \equiv m$, a *constant time*, is a stopping time, whatever be the nonnegative integer m.

(d) The *first passage time* $\tau_B = \inf\{n \geq 0: X_n \in B\}$ is a stopping time for every Borel set B. Here "inf" of an empty set is taken to be ∞.

(4) (*Optional stopping of martingales*). Suppose $\{X_n\}$ is a $\{\Im_n\}$-adapted martingale ($n \geq 0$) and τ is a stopping time.

(a) Prove that $EX_\tau = EX_0$ if τ is bounded. [Hint: Suppose $\tau \leq m$. Define $X_{-1} = 0$. Then $EX_\tau = \sum_{n=0}^{m} EX_n \cdot 1_{\{\tau=n\}} = \sum_{n=0}^{m}$

$EX_n[1_{\{\tau \le n\}} - 1_{\{\tau \le n-1\}}.]$ Now use the fact that $E[(X_n - X_{n-1})$
$1_{\{\tau \le n-1\}}] = E[1_{\{\tau \le n-1\}} E(X_n - X_{n-1} \mid \mathfrak{I}_{n-1})] = 0 \forall n \ge 1$, to get $EX_\tau = $
$\sum_{n=0}^{m} EX_n 1_{\{\tau \le n\}} - \sum_{n=0}^{m} E(X_{n-1} + X_n - X_{n-1}) 1_{\{\tau \le n-1\}} = \sum_{n=0}^{m} EX_n 1_{\{\tau \le n\}} - \sum_{n=0}^{m} EX_{n-1} 1_{\{\tau \le n-1\}} = EX_m 1_{\{\tau \le m\}} = EX_m = EX_0.]$

(b) If τ is finite a.s. and $E|X_{\tau \wedge m} - X_\tau| \to 0$ as $m \to \infty$, then prove that $EX_\tau = EX_0$. [Hint: Take $\tau \wedge m$ as the stopping time in (a) to get $EX_{\tau \wedge m} = EX_0$, and then take the limit as $m \to \infty$.]

(5) (*First passage times for the simple symmetric random walk*). Let $S_0 = 0$, $S_n = Z_1 + Z_2 + \cdots + Z_n$ ($n \ge 1$), where $Z_j (j \ge 1)$ are i.i.d., and $P(Z_j = +1) = P(Z_j = -1) = 1/2$. Let a, b be two positive integers and consider the first passage time $\tau = \tau_{\{-a,b\}}$ (to $-a$ or b). Let $\mathfrak{I}_n = \sigma\{Z_j : 1 \le j \le n\}, \mathfrak{I}_0 = \{\phi, \Omega\}$. Let A_1 be the event "$\{S_n\}$ reaches $-a$ before it reaches b," and A_2 be the event "$\{S_n\}$ reaches b before it reaches $-a$."

(a) Prove that $P(A_1) = b/(a+b)$, $P(A_2) = a/(a+b)$. [Hint: $\{S_n : n \ge 0\}$ is a $\{\mathfrak{I}_n\}$ martingale, and $ES_0 = 0 = ES_\tau = bP(A_2) - aP(A_1) = bP(A_2) - a(1 - P(A_2)) = (a+b)P(A_2) - a$. Here we use Exercise 4(b).]

(b) Prove that $E\tau = ab$. [Hint: Consider the $\{\mathfrak{I}_n\}$-martingale $X_0 = 0$, $X_n = S_n^2 - nEZ_1^2 \equiv S_n^2 - n(n \ge 1)$. By Exercise 4(b), $ES_\tau^2 - E\tau = 0$, or $E\tau = (-a)^2 P(A_1) + b^2 P(A_2) = a^2 b/(a+b) + b^2 a/(a+b) = ab$. To apply Exercise 4(b), one needs to show that $E|X_{\tau \wedge m} - X_\tau| \equiv |E(S_{\tau \wedge m} - S_\tau) - (\tau \wedge m - \tau)| \to 0$. Since $S_{\tau \wedge m}^2, S_\tau^2$ are bounded by $\max\{a^2, b^2\}$, and $\tau \wedge m \to \tau$ a.s. as $m \to \infty$, it is enough to prove that $E\tau < \infty$. For then $\tau \wedge m \uparrow \tau$ a.s., and by the monotone convergence theorem, $E|\tau \wedge m - \tau| \to 0$ as $m \to \infty$. To prove that $E\tau < \infty$, write $E\tau = \sum_{m=0}^{\infty} P(\tau \ge m)$. Now the probability that the random walk reaches $\{-a, b\}$ in $n_0 = \max\{a, b\}$ steps or less is at least $\left(\frac{1}{2}\right)^{n_0} \equiv \delta$, say, n_0 matter where the random walk starts in $(-a, b)$ since the distance between the starting point and $\{-a, b\}$ is never more than n_0. Hence $P(\tau \ge m) \le (1 - \delta)^k$ if $m \ge n_0 k$. That is, $P(\tau \ge m) \le (1 - \delta)^{[m/n_0]} \forall m$, where $[r]$ is the integer part of r. This shows that the series for $E\tau$ converges (exponentially fast).]

(6) (*Wald's identity*). Let $Y_j (j \ge 1)$ be a sequence of i.i.d. random variables with finite mean $\mu = EY_j$, $Y_0 \equiv 0$. Let τ be a $\{\mathfrak{I}_n\}$-stopping time, where $\mathfrak{I}_n = \sigma\{Y_j : 0 \le j \le n\}$. Write $S_n = \sum_0^n Y_j$. Suppose $E\tau < \infty$ and $E|S_\tau - S_{\tau \wedge m}| \to 0$ as $m \to \infty$. Then $ES_\tau = \mu E\tau$. [Hint: Consider

the $\{\mathfrak{I}_n\}$-martingale $X_n = S_n - n\mu$, and apply Exercise 4(b) to get $EX_\tau = EX_0 = 0$, i.e., $ES_\tau - E\tau\mu = 0$.]

(7) (*First passage times for the asymmetric simple random walk*). Let $\{Z_j : j \geq 1\}$ be i.i.d., $P(Z_j = +1) = p$, $P(Z_j = -1) = q = 1 - p$, and assume $p > 1/2$. Let $\tau = \tau_{\{-a,b\}}$. Prove that, for the random walk $S_n = \sum_1^n Z_j - n(p - q)$, $n \geq 1$, $S_0 = 0$,

$$E\tau = \frac{a+b}{p-q}\left(\frac{1-\left(\frac{q}{p}\right)^a}{1-\left(\frac{q}{p}\right)^{a+b}}\right) - \frac{a}{p+q}.$$

[Hint: By Exercise (6), $ES_\tau = (p - q)E\tau$. But $ES_\tau = (-a)P(A_1) + bP(A_2)$, where A_1, A_2 are as in Exercise (5). Now use $P(A_1)(= 1 - P(A_2))$ as given in Example 6.3, Chapter 2 (i.e., (6.22), with $x = 0$, $c = -a$, $d = b$).]

6

Discounted Dynamic Programming
Under Uncertainty

The basic need for a special theory to explain behavior under conditions of uncertainty arises from two considerations: (1) subjective feelings of imperfect knowledge when certain types of choices, typically commitments over time, are made; (2) the existence of certain observed phenomena, of which insurance is the most conspicuous example, which cannot be explained on the assumption that individuals act with subjective certainty.

Kenneth J. Arrow

6.1 Introduction

In this chapter we briefly review some results on discounted dynamic programming under uncertainty, and indicate how Markov processes and random dynamical systems are generated by optimal policies. In Section 6.2, following a precise description of the dynamic programming framework, we turn to the special case where the set S of states is countable, and the set A of actions is finite. Here the link between optimal policies and the celebrated functional equation can be established with no measure theoretic complications. In Section 6.3 we study the maximum theorem, which is of independent interest in optimization theory and is a key to the understanding of the basic result in the next section. In Section 6.4 we explore the more general model where S is a Borel subset of a Polish space, and A is a compact action space, and spell out the conditions under which there is a stationary optimal policy.

The dynamic programming technique reviewed here has been particularly powerful in attacking a variety of problems in intertemporal economics. In Section 6.5 we discuss in detail the aggregative model of optimal economic growth under uncertainty. Here, given the specific structure of the model, the process of optimal stocks (or optimal

379

inputs) can be viewed as a random dynamical system. We draw upon our earlier results to analyze the long-run behavior of the optimal input process.

6.2 The Model

A dynamic programming problem is specified by the following objects: $\langle S, A, q, u, \delta \rangle$, where S is a nonempty Borel subset of a Polish (i.e., complete, separable metric) space, interpreted as the set of *states* of some system; A is a nonempty Borel subset of a Polish space, interpreted as the set of *actions* available to the decision maker; q is the *law of motion* of the system – it associates (Borel measurably) with each pair (s, a) a probability measure $q(.|s, a)$ on $\mathcal{S}$, the Borel sigmafield of S: when the system is in state s and an action a is chosen, it moves to the state s' in the next period according to the distribution $q(.|s, a)$; u is a bounded Borel measurable function on $S \times A$, interpreted as the *utility, income,* or *immediate return* – when the system is in state s, and the chosen action is a, the decision maker receives an income $u(s, a)$; δ is a discount factor, $0 < \delta < 1$. A *policy* (or *plan*) $\zeta = (\zeta_t)$ specifies for each $t \geq 1$ which action to choose in the tth period as a Borel measurable function of the history $h = (s_1, a_1, \ldots, a_{t-1}; s_t)$ of the system up to period t, or more generally, ζ specifies for each h a probability distribution $\zeta_t(.|h)$ on the Borel subsets of A.

A Borel function f from S into A defines a policy. When in state s, choose an action $f(s)$ (*independently of when and how the system has arrived at state s*). We denote the corresponding policy by $(f^{(\infty)})$. Such policies are called *stationary*, and f is somewhat informally referred to as an *optimal policy function*.

A policy ζ associates with each initial state s a corresponding tth period expected return $u_t(\zeta)(s)$ and an expected discounted total return

$$I(\zeta)(s) = \sum_{t=1}^{\infty} \delta^{t-1} u_t(\zeta)(s), \tag{2.1}$$

where δ is the discount factor, $0 < \delta < 1$.

A policy ζ^* will be called *optimal* if $I(\zeta^*)(s) \geq I(\zeta)(s)$ for all policies ζ and $s \in S$. The problem, then, is to find an optimal policy.

6.2.1 Optimality and the Functional Equation
of Dynamic Programming

We begin with the case where the state space S is countable and the action space A is finite and introduce the celebrated functional equation that characterizes optimality. Recall from Chapter 2 that the states in a countable S are labeled i, j, k etc. Also, we write $q_{ij}(a)$ to denote the probability with which the state moves from i to j if (after observing i) the action a is chosen. Thus, for any fixed $i \in S$, and $a \in A$, $q_{ij}(a)$ is the probability distribution of the state in the *next* period (i.e., $q_{ij}(a) \geq 0$, $\sum_j q_{ij}(a) = 1$). The immediate return function is bounded, i.e., there is some constant $B > 0$ such that $|u(i, a)| \leq B$ for all $i \in S$, and $a \in A$.

Define

$$V(i) = \sup_\zeta I(\zeta)(i).$$

$V : S \to \mathbb{R}$ is the *value function* of the dynamic programming problem.

Clearly, a policy ζ^* is *optimal* if

$$I(\zeta^*)(i) = V(i) \quad \text{for all } i \in S.$$

Our first result characterizes the value function V by the functional equation (2.2), which will be casually referred to as the "optimality equation" or the "functional equation" of dynamic programming.

Theorem 2.1 *For all $i \in S$,*

$$V(i) = \max_{a \in A} \left[u(i, a) + \delta \sum_j q_{ij}(a) V(j) \right]. \tag{2.2}$$

Proof. Let ζ be *any* policy, and suppose it chooses some action $a \in A$ in the initial period (period 1). Then,

$$I(\zeta)(i) = u(i, a) + \delta \sum_j q_{ij}(a) w_\zeta(j),$$

where $w_\zeta(j)$ represents the total expected return from period 2 onward, given that ζ is used and the state in period 2 is j. But $w_\zeta(j) \le V(j)$. Hence,

$$I(\zeta)(i) \le u(i, a) + \delta \sum_j q_{ij}(a) V(j)$$

$$\le \max_{a \in A} \left[u(i, a) + \delta \sum_j q_{ij}(a) V(j) \right].$$

Since ζ is arbitrary, we have

$$V(i) \equiv \sup_\zeta I(\zeta)(i) \le \max_{a \in A} \left[u(i, a) + \delta \sum_j q_{ij}(a) V(j) \right].$$

Conversely, using the finiteness of A, let a_0 be such that

$$u(i, a_0) + \delta \sum_j q_{ij}(a_0) V(j)$$

$$\equiv \max_{a \in A} \left[u(i, a) + \delta \sum_j q_{ij}(a) V(j) \right].$$

Let ζ be the policy that chooses a_0 when the initial state is i, and if the next state is j, then views the process as originating in j, and uses a policy ζ_j such that $I(\zeta_j)(j) \ge V(j) - \varepsilon$. Hence,

$$I(\zeta)(i) = u(i, a_0) + \delta \sum_j q_{ij}(a_0) I(\zeta_j)(j)$$

$$\ge u(i, a_0) + \delta \sum_j q_{ij}(a_0) V(j) - \delta\varepsilon.$$

Since $V(i) \ge I(\zeta)(i)$, we have

$$V(i) \ge u(i, a_0) + \delta \sum_j q_{ij}(a_0) V(j) - \delta\varepsilon.$$

Hence,

$$V(i) \ge \max_{a \in A} \left[u(i, a) + \delta \sum_j q_{ij}(a) V(j) \right] - \delta\varepsilon.$$

Since ε is arbitrary,

$$V(i) \ge \max_{a \in A} \left[u(i, a) + \delta \sum_j q_{ij}(a) V(j) \right]. \qquad \blacksquare$$

By using the optimality equation (2.2), we are now able to prove the existence of a *stationary* policy $\zeta^* = (\hat{f}^{(\infty)})$ that is optimal.

Theorem 2.2 *Let $\hat{f}$ be a function that selects for each $i \in S$, an action that maximizes the right-hand side of (2.2), i.e.,*

$$u(i, \hat{f}(i)) + \delta \sum_j q_{ij}(\hat{f}(i))V(j)$$

$$\equiv \max_{a \in A} \left[u(i, a) + \delta \sum_j q_{ij}(a)V(j) \right]. \qquad (2.3)$$

Then, writing $\zeta^* = (\hat{f}^{(\infty)})$, we have

$$I(\zeta^*)(i) = V(i) \quad \text{for all } i \in S.$$

Hence, ζ^* is optimal.

Proof. Let $\mathbf{B}(S)$ be the set of real-valued bounded functions on S. For any $g : S \to A$, the mapping

$$L_g : \mathbf{B}(S) \to \mathbf{B}(S), \text{ defined by}$$

$$(L_g u)(i) = u(i, g(i)) + \delta \sum_j q_{ij}(g(i))u(j).$$

($L_g u$ evaluated at i represents the expected discounted return if the initial state is i and we employ g for one period and then are terminated with a final return $u(j)$). One can show (see Section 6.4) that for bounded functions u and v

(i) "$u \geq v$" implies "$L_g u \geq L_g v$".
(ii) $L_g^n u \to I(g^{(\infty)})$ as $n \to \infty$, where $L_g^1 u = L_g u$, $L_g^n u = L_g(L_g^{n-1}u)$.
(iii) $I(g^{(\infty)})$ is the unique fixed point of L_g.

If $\hat{f}$ is a function that satisfies (2.3)

$$L_{\hat{f}} V = V$$
$$L_{\hat{f}}^n V = L_{\hat{f}}^{n-1}(L_{\hat{f}} V) = L_{\hat{f}}^{n-1} V = \cdots = V.$$

Hence, letting $n \to \infty$, using (ii),

$$I(\hat{f}^{(\infty)}) = V. \qquad \blacksquare$$

Exercise 2.1 In this chapter we focus on dynamic programming models with discounting. But "undiscounted" models have also been explored; examples are "negative" dynamic programming (the reward function $u(i, a) \leq 0$) and "positive" dynamic programming (the reward function $u(i, a) \geq 0$) with $\delta = 1$ (see Blackwell 1967 and Ross 1983). Here is an example of a positive dynamic programming problem with an important feature.

The state space S is denumerable $S = \{0, 1, 2, \ldots\}$, and there are two actions $A = \{a_1, a_2\}$. The immediate return function is defined as

$$u(0, a_1) = u(0, a_2) = 0;$$
$$u(i, a_1) = 0 \text{ for } i \geq 1, \quad i \in S;$$
$$u(i, a_2) = 1 - (1/i) \text{ for } i \geq 1, \quad i \in S.$$

The law of motion is described as follows:

$$q_{00}(a_1) = q_{00}(a_2) = 1;$$
$$q_{i,i+1}(a_1) = 1 \text{ for } i \geq 1, \quad i \in S,$$
$$q_{i,0}(a_2) = 1 \text{ for } i \geq 1, \quad i \in S.$$

Note that if the system is in state 0, then it remains there irrespective of the chosen action, and yields no immediate return. If the state is $i \geq 1$, the choice of a_1 takes the system to the state $i + 1$, again yielding no immediate return. If the state is $i \geq 1$, the choice of a_2 yields an immediate return of $1 - (1/i)$, but no future return is generated as the system moves to state 0 from i.

Again, with $\delta = 1$, the problem is to maximize the sum of expected one-period returns. As before write $I(\zeta)(i) = \sum_{t=1}^{\infty} u_t(\zeta)$, where u_t is the expected return in date t from policy ζ (and i is the initial state). Define $V(i) = \sup_\zeta I(\zeta)(i)$ where ζ is any policy. Show that

(a) for $i \geq 1$, $V(i) = 1$, but $I(\zeta)(i) < 1$ for any policy ζ. Hence, *there is no optimal policy.*

(b) Look at the right-hand side of the functional equation (2.2) with $\delta = 1$. What happens to the solution of the maximization problem in this case? Compare your conclusion with Theorem 2.2. ∎

6.3 The Maximum Theorem: A Digression

Consider a static model in which an agent who chooses an action a after observing the state s. Let S be the (nonempty) set of all states and A be the (nonempty) set of all conceivable actions. Suppose that $\varphi : S \to A$ is a correspondence (a set-valued mapping) which associates with each $s \in S$, a (nonempty) subset $\varphi(s)$ of A. One interprets $\varphi(s)$ as the set of actions that are *feasible* for the agent, given the state s. Let u be a real-valued function on $S \times A$, and interpret $u(s, a)$ as the utility or return for the agent when the state is s and his action is a. Given s, one is interested in the elements $\psi(s)$ of $\varphi(s)$ which maximize $u(s, \cdot)$ (now a function of a alone) *on* $\varphi(s)$, and $m(s)$, the *value* of the maximum of u on $\varphi(s)$. The maximum theorem addresses these issues. For a precise statement of this theorem, we first need to develop some definitions involving set-valued maps.

6.3.1 Continuous Correspondences

We shall assume in this subsection and in the next that S is a (nonempty) metric space, with $\mathcal{S}$ its Borel sigmafield. A is a (nonempty) compact metric space. We write $x_n \to x$ to denote a sequence x_n converging to x.

Definition 3.1 A correspondence $\varphi : S \to A$ is a rule which associates with each element $s \in S$, a *nonempty* subset $\varphi(s)$ of A.

We now define some continuity properties of a correspondence.

Definition 3.2 A correspondence $\varphi : S \to A$ *upper semicontinuous* at $s \in S$ iff:

"$s_n \to s, \quad a_n \in \varphi(s_n), \quad a_n \to a$" implies $a \in \varphi(s)$.

φ is upper semicontinuous if φ is upper semicontinuous at *every* $s \in S$.

Definition 3.3 A correspondence $\varphi : S \to A$ is *lower semicontinuous* at $s \in S$ iff:

"$s_n \to s, \quad a \in \varphi(s)$" implies "there is a_n such that $a_n \in \varphi(s_n)$ and $a_n \to a$."

φ is lower semicontinuous if φ is lower semicontinuous at *every* $s \in S$.

Definition 3.4 A correspondence $\varphi : S \rightarrow A$ is *continuous* at s if it is *both* upper and lower semicontinuous at s; φ is continuous if it is continuous at *every* $s \in S$.

Remark 3.1 When for *all* $s \in S$, $\varphi(s)$ consists of a single element (so that φ can be viewed as a function from S into A) the definition of upper semicontinuity at s (or lower semicontinuity at s) is obviously *equivalent* to the definition of *continuity* at s for the function $\varphi(s)$.

Note that, since A is assumed to be compact, if φ is upper semicontinuous at s, $\varphi(s)$ is a nonempty compact subset of A.

6.3.2 The Maximum Theorem and the Existence of a Measurable Selection

We shall now prove the maximum theorem which has wide applications to models of optimization. Recall that S is a (nonempty) metric space (with its Borel sigmafield S) and A a (nonempty) compact metric space. Let φ be a *continuous correspondence* from S into A, and u a real-valued *continuous function* on $S \times A$. For each $s \in S$, define

$$m(s) = \max_{a \in \varphi(s)} u(s, a). \qquad (3.1)$$

Since $\varphi(s)$ is a nonempty compact set, and $u(s, \cdot)$ is continuous on $\varphi(s)$, $m(s)$ is a well-defined real-valued function on S. Next, for $s \in S$ define

$$\psi(s) = \{a \in \varphi(s) : u(s, a) = m(s)\}. \qquad (3.2)$$

Since $\psi(s)$ is nonempty for each $s \in S$, ψ is a correspondence from $S \rightarrow A$.

Theorem 3.1

    (a) The function $m : S \rightarrow \mathbb{R}$ is continuous.

    (b) The correspondence $\psi : S \rightarrow A$ is upper semicontinuous.

    (c) There is a Borel measurable $\hat{f} : S \rightarrow A$ such that $\hat{f}(s) \in \psi(s)$, i.e.,

$$u(s, \hat{f}(s)) = m(s) \equiv \max_{a \in \varphi(s)} u(s, a). \qquad (3.3)$$

Proof.

(a) To prove the continuity of m on S, let $s \in S$, and $s_n \to s$. One has to show that $m(s_n) \to m(s)$. Write $y_n \equiv m(s_n)$. Choose $a_n \in \varphi(s_n)$ such that

$$u(s_n, a_n) = y_n \equiv m(s_n) \quad \text{for all } n. \tag{3.4}$$

Choose *any* subsequence $y_{n'}$ of y_n. Recalling that $y_{n'} = u(s_{n'}, a_{n'})$, and that A is compact, a subsequence $(a_{n''})$ of $(a_{n'})$ converges to some $a \in A$. Hence, $(s_{n''}, a_{n''}) \to (s, a)$. By upper semicontinuity of φ, we can assert that $a \in \varphi(s)$. Also, by continuity of u,

$$y_{n''} = m(s_{n''}) \equiv u(s_{n''}, a_{n''}) \to u(s, a). \tag{3.5}$$

We shall show that $m(s) = u(s, a)$. Clearly, as $a \in \varphi(s)$,

$$u(s, a) \leqq m(s). \tag{3.6}$$

There is some $z \in \varphi(s)$ such that

$$u(s, z) = m(s). \tag{3.7}$$

By lower semicontinuity of φ, there is some $z_{n''}$ such that $z_{n''} \in \varphi(s_{n''})$ and $z_{n''} \to z$. Hence,

$$u(s_{n''}, a_{n''}) \geqq u(s_{n''}, z_{n''}).$$

Taking limit as $n'' \to \infty$,

$$u(s, a) \geqq u(s, z) = m(s). \tag{3.8}$$

By (3.6) and (3.8)

$$u(s, a) = m(s).$$

Hence by (3.5), the subsequence $y_{n''}$ of $y_{n'}$ converges to $m(s)$. Hence, the original sequence $m(s_n)$ must converge to $m(s)$.

(b) The upper semicontinuity of ψ is left as an exercise.

(c) Since ψ is upper semicontinuous, $\psi(s)$ is closed for any $s \in S$. A is compact, hence, it is complete and separable. To prove the existence of a Borel measurable $\hat{f}$ such that $\hat{f}(s) \in \psi(s)$ for every $s \in S$, we appeal to a well-known theorem on measurable selection due to Kuratowski and Ryll-Nardzewski (1965) (stated and proved in Hildenbrand (1974, Lemma 1, p. 55) and reviewed in detail in Wagner (1977, Theorem 4.2)). To use this result we need to show that if F is any closed subset of A, the set $\psi^-(F) \equiv \{s \in S : \psi(s) \cap F \neq \phi\}$ belongs to the Borel sigmafield $\mathcal{S}$ of S. We show that $\psi^-(F)$ is closed. Let $s_n \in \psi^-(F)$ and $s_n \to s$. Then $\psi(s_n) \cap F \neq \phi$ for all n. Choose some $a_n \in \psi(s_n) \cap F$. By compactness of A, there is some subsequence $(a_{n'})$ converging to some a. Since F is closed, $a \in F$. Since $(s_{n'}, a_{n'})$ converges to (s, a) and ψ is upper semicontinuous, $a \in \psi(s)$. Hence $a \in \psi(s) \cap F$, and this implies that $s \in \psi^-(F)$. ∎

We shall refer to $\hat{f}$ as a *measurable selection* from the correspondence φ.

Example 3.1 Let $S = A = [0, 1]$, and $\varphi(s) = A = [0, 1]$ for all s. Then φ is clearly continuous. Define $u(s, a) = sa$. For $s \neq 0$, $\psi(s)$ consists of a single element: $\psi(s) = 1$. For $s = 0$, $\psi(s) = [0, 1]$. Hence, $\psi(s)$ is not lower semicontinuous at $s = 0$. For all $s \in S$, $m(s) = s$.

6.4 Dynamic Programming with a Compact Action Space

In this section, we deal with a more general framework under the following assumptions on S, A, q, and u:

[A.1] *S is a (nonempty) Borel subset of a Polish (complete, separable metric) space;*

[A.2] *A is a compact metric space;*

[A.3] *u is a bounded continuous function on $S \times A$; and*

[A.4] *if $s_n \to s$, $a_n \to a$, then $q(.|s_n, a_n)$ converges weakly to $q(.|s, a)$.*

The main result asserts the existence of a stationary optimal policy $\zeta^* = (\hat{f}^{(\infty)})$; moreover, it is shown that the value function $V (\equiv I(\hat{f}^{(\infty)}))$ which satisfies the functional equation is a bounded continuous function on S.

Let us first note a preliminary result which follows from [A.3].

Lemma 4.1 *Let $w : S \to \mathbb{R}$ be a bounded continuous function. Then $g : S \times A \to \mathbb{R}$ defined by $g(s, a) = \int w(.) \, dq(.|s, a)$ is continuous.*

Recall that $C_b(S)$ is the class of all bounded continuous functions on S. For v, $v' \in C_b(S)$, we use, as before, the metric $d(v, v') = \|v - v'\| = \sup_{s \in S} |v(s) - v'(s)|$. *The metric space $(C_b(S), d)$ is complete.*

For every $w \in C_b(S)$, let $\Im w : S \to \mathbb{R}$ be the function defined by

$$(\Im w)(s) = \max_{a \in A} \left[\int (u(s, a) + \delta w(.)) \, dq(.|s, a) \right]. \tag{4.1}$$

Note that, by virtue of Lemma 4.1, the expression within square brackets on the right-hand side of (4.1) is continuous in s and a, and, consequently, the maximum is assumed for every s. Moreover, Lemma 4.1 implies that $\Im w$ is continuous and, since it is obviously bounded, $\Im w \in C_b(S)$. Thus, $\Im$ maps $C_b(S)$ into $C_b(S)$.

Lemma 4.2 $\Im$ *is a uniformly strict contraction mapping on $C_b(S)$ and, consequently, has a unique fixed point.*

Proof. Let $w_1, w_2 \in C_b(S)$. Clearly, $w_1 \leqslant w_2 + \|w_1 - w_2\|$. Since, as is easy to check, $\Im$ is monotone,

$$\Im w_1 \leqslant \Im(w_2 + \|w_1 - w_2\|) = \Im w_2 + \delta \|w_1 - w_2\|.$$

Consequently, $\Im w_1 - \Im w_2 \leqslant \delta \|w_1 - w_2\|$.
Interchanging w_1 and w_2 we get

$$\Im w_2 - \Im w_1 \leqslant \delta \|w_1 - w_2\|.$$

Hence $\|\Im w_1 - \Im w_2\| \leqslant \delta \|w_1 - w_2\|$,
which proves that $\Im$ is a uniformly strict contraction mapping, since $\delta < 1$.

Since $C_b(S)$ is a complete metric space, it follows that $\Im$ has a unique fixed point V in $C_b(S)$. ∎

The central result on the existence of a stationary optimal policy is now stated.

Theorem 4.1 *There exists a stationary optimal policy $\zeta^* = (\hat{f}^{(\infty)})$ and the value function $V (\equiv \sup_\zeta I(\zeta)) = I(\zeta^*)$ on S is bounded and continuous.*

Proof. As in Section 6.2, with each Borel map $g : S \to A$, associate the operator L_g on $\mathbf{B}(S)$ that sends $u \in \mathbf{B}(S)$ into $L_g u \in \mathbf{B}(S)$, where $L_g u$ is defined by

$$L_g u(s) = \int [u(s, g(s)) + \delta u(.)] \, dq(.|s, g(s)), \quad s \in S. \tag{4.2}$$

It is known that $I(g^{(\infty)})$ is the unique fixed point of the operator L_g (see Theorem 5.1 (f), Strauch (1966)).

Now let $\Im$ be as in (4.1) and let $V \in C_b(S)$ be its unique fixed point, i.e., $\Im V = V$. It follows from Theorem 3.1 that there exists a Borel map $\hat{f}$ from S into A such that $\Im V = L_{\hat{f}} V$. Consequently $L_{\hat{f}} V = V$ so that by the remark made in the preceding paragraph, $V = I(\hat{f}^{(\infty)})$. Hence $\Im V = V$ can be rewritten as

$$I(\hat{f}^{(\infty)}(s)) = \max_{a \in A} \left[\int (u(s, a) + \delta I(\hat{f}^{(\infty)})(.)) \, dq(.|s, a) \right], \quad s \in S. \quad (4.3)$$

Thus $I(\hat{f}^{(\infty)})$ satisfies the optimality equation so that by a theorem of Blackwell (1965, Theorem 6(f)) $\zeta^* = (\hat{f}^{(\infty)})$ is an optimal policy. Moreover, as $V = I(\hat{f}^{(\infty)})$ and $V \in C_b(S)$, it follows that the value function is continuous. ∎

6.5 Applications

6.5.1 The Aggregative Model of Optimal Growth
Under Uncertainty: The Discounted Case

Initially (at "the beginning" of period $t = 0$), the planner observes the stock $y \geq 0$ (the state) and chooses some point a (an action) in $A \equiv [0, 1]$. Interpret a as the fraction of the stock y to be used as an input, and write $x \equiv ay$ to denote the input. Choice of a determines the current consumption $c \equiv (1 - a)y$. This consumption generates an immediate return or utility according to a function $u : \mathbb{R}_+ \to \mathbb{R}_+$ (satisfying the assumptions listed below): $u(c) = u((1 - a)y)$.

To describe the law of motion of the states, consider a fixed set of functions $\Gamma = \{f_1, \dots f_k, \dots, f_N\}$, where each $f_k : \mathbb{R}_+ \to \mathbb{R}_+$ is interpreted as a possible gross output function (we shall introduce assumptions on the structure of f_k below as we proceed). With each f_k we assign a positive probability q_k $(1 \leq k \leq N)$, $\sum_{k=1}^{N} q_k = 1$. Let $Q = \{q_1, \dots, q_N\}$.

After the choice of a (at "the end" of period $t = 0$), Tyche picks an element f at random from Γ: $f = f_k$ with probability q_k. The stock in the next period is given by

$$y_1 = f(x). \quad (5.1)$$

Thus, given the initial y and the chosen a, $y_1 = f_k(ay)$ with probability q_k. This stock y_1 is observed at "the beginning" of the next period, and action a_1 is chosen from A, and, at the end of period $t = 1$, independent of the choice in period 0, Tyche again picks an element f from Γ according to the same distribution $Q = \{q_1, \ldots, q_N\}$. The story is repeated. One may take the initial stock y to be a random variable with a given distribution μ independent of the choices of gross output functions. To ease notation, we write

$$y_{t+1} = f(x_t),$$

remembering that f is randomly picked from Γ at the end of period t *independent of the past choices* and *according to the same distribution Q on Γ.*

We make the following assumptions on the utility function:

[U.1] *u is continuous on* $\mathbb{R}_+$;
[U.2] *u is increasing on* $\mathbb{R}_+$;
[U.3] *u is strictly concave.*

The following assumptions on each f_k are also made:

[T.1] f_k *is continuous on* $\mathbb{R}_+$;
[T.2] $f_k(0) = 0$; f_k *is increasing on* $\mathbb{R}_+$;
[T.3] *There is* $\beta_k > 0$ *such that*

$$f_k(x) > x \quad \text{for } 0 < x < \beta_k,$$
$$f_k(x) < x \quad \text{for } x > \beta_k.$$

We assume that $\beta_1 < \cdots < \beta_k < \cdots < \beta_N$.

Given an initial stock $y > 0$, a policy $\zeta = (\zeta_t)$ generates an *input* process $x^\zeta = (x_t^\zeta)$, a *consumption* process $c^\zeta = (c_t^\zeta)$, and a *stock* process $y^\zeta = (y_t^\zeta)$ according to the description of Section 6.1. Formally,

$$x_0^\zeta \equiv \zeta_0(y) \cdot y, \quad c_0^\zeta = [1 - \zeta_0(y)]y, \quad y_0^\zeta = y \tag{5.2}$$

and for $t \geq 1$, for each history $h_t = (y_0, a_0; \ldots y_{t-1}, a_{t-1}; y_t)$, one has

$$x_t^\zeta(h_t) = \zeta_t(h_t)y_t, \quad c_t^\zeta = [1 - \zeta_t(h_t)]y_t \quad y_t^\zeta = f(x_{t-1}), \tag{5.3}$$

where $f = f_k$ with probability q_k.

Clearly,

$$c_0 + x_0 = y_0,$$
$$c_t + x_t = f(x_{t-1}) \quad \text{for } t \geq 1,$$
$$c_t \geq 0, \quad x_t \geq 0 \quad \text{for } t \geq 0. \tag{5.4}$$

In view of [T.3] we can find $B > 0$ such that $f(x) < x$ for all $x > B$, and for all k $(k = 1, 2, \ldots N)$. Hence, the following boundedness property is not difficult to derive:

Lemma 5.1 *Assume* [T.1]–[T.3] *and let* y *be any initial stock such that* $y \in [0, B]$.

If ζ *is any policy generating a stock process* y^ζ, *then for all* $t \geq 0$

$$0 \leq y_t^\zeta \leq B. \tag{5.5}$$

It follows that the input and consumption process x^ζ *and* c^ζ *generated by* ζ *also satisfy for all* $t \geq 0$

$$0 \leq x_t^\zeta \leq B, \quad 0 \leq c_t^\zeta \leq B. \tag{5.6}$$

In what follows, we choose the state space $S = [0, B]$. If as in Section 6.1, $q(\cdot \mid y, a)$ denotes the law of motion, i.e., the conditional distribution of the stock y_{t+1} in period $t + 1$, given the stock y and action a in period t, then $q(\cdot \mid y, a)$ has the Feller property (see Example 11.2 of Chapter 2). The following existence theorem follows from an application of Theorem 4.1 in the present context:

Theorem 5.1 *Assume* [T.1]–[T.3] *and* [U.1]. *There exists a stationary optimal policy* $\zeta^* = (\hat{\eta}^{(\infty)})$, *where* $\hat{\eta} : S \rightarrow A$ *is a Borel measurable function. The value function* $V = I(\zeta^*)[= \sup_\zeta I(\zeta)]$ *is continuous on* S *and satisfies*

$$V(y) = \max_{a \in A} \left\{ u[(1 - a)y] + \delta \sum_{k=1}^{N} V[f_k(ay)]q_k \right\}$$

$$= u[y - \hat{\eta}(y) \cdot y] + \delta \sum_{i=1}^{n} V[f_k(\hat{\eta}(y) \cdot y)]q_k. \tag{5.7}$$

Moreover, V *is increasing.*

We refer to the function $i(y) \equiv \hat{\eta}(y) \cdot y$ as an *optimal investment policy function* and the function $c(y) \equiv [1 - \hat{\eta}(y)]y$ as an *optimal consumption policy function.*

Given any optimal policy ζ^* (stationary or not) and initial $y \geq 0$ we refer to $(x^{\zeta^*}, c^{\zeta^*}, y^{\zeta^*})$ as an optimal input, consumption, and stock process (from y), respectively. To simplify notation, we often drop the superscript ζ^*.

A general monotonicity property is first noted. The proof uses the strict concavity of u crucially.

Theorem 5.2 *Assume [T.1]–[T.3] and [U.1]–[U.3]. Let $i : S \to S$ be an optimal investment policy function. Then i is non-decreasing, i.e., "$y > y'$" implies "$i(y) \geq i(y')$".*

Proof. We prove the following:
Suppose $y > y'$ and (x_t) and (x_t') are optimal input processes from y, y', respectively. Then $x_0 \geq x_0'$.

The theorem follows by taking $x_0 = i(y)$ and $x_0' = i(y')$. Suppose, to the contrary, that $x_0 < x_0'$. Define new input processes from y and y' as follows. Let $\bar{x}_t = x_t'$ for $t \geq 0$. Then $\bar{x}_0 = x_0' \leq y' < y$, and for $t \geq 1, \bar{x}_t = x_t' \leq f_k(x_{t-1}') = f_k(\bar{x}_{t-1})$. Hence, $(\bar{x}_t)$ is a feasible input process from y. Next define $\bar{x}_t' = x_t$ for $t \geq 0$. Then $\bar{x}_0' = x_0 < x_0' \leq y'$ and for $t \geq 1, \bar{x}_t' = x_t \leq f_k(x_{t-1}) = f_k(\bar{x}_{t-1}')$. Hence $(\bar{x}_t')$ is a feasible input process from y'. Let $(\bar{c}_t)$ and $(\bar{c}_t')$ be consumption processes generated by $(\bar{x}_t)$ and $(\bar{x}_t')$, respectively, (i.e., $\bar{c}_0 = y - \bar{x}_0$, $\bar{c}_t = f_k(\bar{x}_{t-1}) - \bar{x}_t$ for $t \geq 1$; similarly, $\bar{c}_0' = y' - \bar{x}_0'$, $\bar{c}_t' = f_k(\bar{x}_{t-1}') - \bar{x}_t'$ for $t \geq 1$)). Using the functional equation, we obtain

$$u(c_0) + \delta \sum_{k=1}^{N} V[f_k(x_0)]q_k = V(y) \geq u(\bar{c}_0) + \delta \sum_{k=1}^{N} V[f_k(\bar{x}_0)]q_k.$$
(5.8)

$$u(c_0') + \delta \sum_{k=1}^{N} V[f_k(x_0')]q_k = V(y') \geq u(\bar{c}_0') + \delta \sum_{k=1}^{N} V[f_k(\bar{x}_0')]q_k.$$
(5.9)

Adding these two inequalities and noting that $x_0 = \bar{x}_0', x_0' = \bar{x}_0$, we obtain

$$u(c_0) + u(c_0') \geq u(\bar{c}_0) + u(\bar{c}_0').$$
(5.10)

Now, $\bar{c}_0 = y - x_0' > y' - x_0' = c_0'$ and $\bar{c}_0' = y - x_0' < y - x_0 = c_0$. Hence, there is some θ, $0 < \theta < 1$ such that $\bar{c}_0 = \theta c_0 + (1 - \theta)c_0'$.

Then, $\bar{c}_0' = y' - x_0 = (y - x_0) + (y' - x_0') - (y - x_0') = c_0 + c_0' - \bar{c}_0 = (1 - \theta)c_0 + \theta c_0'$. This gives, using the strict concavity of u,

$$u(\bar{c}_0) > \theta u(c_0) + (1 - \theta)u(c_0') \tag{5.11}$$

and

$$u(\bar{c}_0') > (1 - \theta)u(c_0) + \theta u(c_0'). \tag{5.12}$$

So, by addition,

$$u(\bar{c}_0) + u(\bar{c}_0') > u(c_0) + u(c_0'). \tag{5.13}$$

Thus, (5.10) and (5.13) lead to a contradiction. Hence $x_0 \geq x_0'$. ∎

Let us go back to the functional equation (5.7):

$$V(y) = I(\zeta^*)(y) = \max_{a \in A} \left\{ u[(1 - a)y] + \delta \sum_{k=1}^{N} V[f_k(ay)]q_k \right\}$$

$$= u[y - \hat{\eta}(y)y] + \delta \sum_{k=1}^{N} V[f_k(\hat{\eta}(y)y)]q_k. \tag{5.14}$$

For any $y \geq 0$, let $\psi(y)$ be the set of all actions a, where the right-hand side of the first line in (5.14) attains its maximum. Then $\psi(y)$ is an upper semicontinuous correspondence from S into A, and $\hat{\eta}$ is a selection from ψ. Hence, as we remarked earlier (Remark 3.1), *if $\psi(y)$ is unique for each $y \geq 0$, then $\hat{\eta}(y) = \psi(y)$ is a continuous function*.

We now make the following assumption:

[T.4] *Each $f_k \in \Gamma$ is strictly concave on $\mathbb{R}_+$.*

Theorem 5.3 *Assume [T.1]–[T.4] and [U.1]–[U.4]. Then*

(a) *If (x, c, y) and (x', c', y') are optimal (resp. input, consumption, and stock) processes from the initial stock $y \geq 0$, one has for $t \geq 0$,*

$$x_t = x_t' \quad c_t = c_t' \quad y_t = y_t' \quad a.s.$$

(b) *The value function V is strictly concave.*
(c) *$\hat{\eta}(y)$ is continuous on S.*

Proof.

(a) Suppose that (x, y, c) and (x', y', c') are optimal processes from the same initial stock $y > 0$. Then $V(y) = \sum_{t=0}^{\infty} \delta^t Eu(c_t) = \sum_{t=0}^{\infty} \delta^t Eu(c'_t)$. Define a new process $(\bar{x}, \bar{c}, \bar{y})$ as follows:

Fix some $\theta \in (0, 1)$. Let

$$
\begin{aligned}
\bar{x}_t &\equiv \theta x_t + (1 - \theta)x'_t && \text{for } t \geq 0; \\
\bar{y}_0 &\equiv y; \quad \bar{y}_t = f_k(\bar{x}_{t-1}) && \text{for } t \geq 1; \\
\bar{c}_t &= \bar{y}_t - \bar{x}_t && \text{for } t \geq 0.
\end{aligned}
\tag{5.15}
$$

Clearly, $\bar{x}_t \geq 0$, $\bar{y}_t \geq 0$ for $t \geq 0$. Using the strict concavity of f_k,

$$
\begin{aligned}
\bar{c}_t &= f_k(\bar{x}_{t-1}) - \bar{x}_t \\
&= f_k[\theta x_{t-1} + (1 - \theta)x'_{t-1}] - [\theta x_t + (1 - \theta)x'_t] \\
&\geq \theta f_k(x_{t-1}) + (1 - \theta)f_x(x'_{t-1}) - [\theta x_t + (1 - \theta)x'_t] \\
&= \theta[f_k(x_{t-1}) - x_t] + (1 - \theta)[f_k(x'_{t-1}) - x'_t] \\
&= \theta c_t + (1 - \theta)c'_t.
\end{aligned}
\tag{5.16}
$$

Hence $(\bar{x}_t, \bar{c}_t, \bar{y}_t)$ is a well-defined (resp. input, consumption, and stock) process from $\bar{y}_0$. It follows from the monotonicity of u that

$$
Eu(\bar{c}_t) \geq Eu(\theta c_t + (1 - \theta)c'_t).
\tag{5.17}
$$

If for some period $\tau \geq 1$, $x_{\tau-1}$ is different from $x'_{\tau-1}$ on a set of strictly positive probability, using the *strict* concavity of f_k, the inequality in the third line of (5.16) holds with *strict* inequality, and by [U.2], (5.17) holds with strict inequality for $t = \tau$. Hence,

$$
\begin{aligned}
\sum_{t=0}^{\infty} \delta^t Eu(\bar{c}_t) &> \sum_{t=0}^{\infty} \delta^t Eu(\theta c_t + (1 - \theta)c'_t) \\
&\geq \theta \sum_{t=0}^{\infty} \delta^t Eu(c_t) + (1 - \theta) \sum_{t=0}^{\infty} \delta^t Eu(c'_t) \\
&= \theta V(y) + (1 - \theta)V(y) \\
&= V(y).
\end{aligned}
\tag{5.18}
$$

This is a contradiction. Hence $x_t = x'_t$ a.s. for all $t \geq 0$ and it follows that $y_t = y'_t$ and $c_t = c'_t$ a.s. for all $t \geq 0$.

(b) We shall provide a sketch, since some of the details can be easily provided by studying (a) carefully. Fix $y > y' \geq 0$. Let (x, c, y) and (x', c', y') be optimal processes from y and y', respectively. Fix $\theta \in (0, 1)$ and define the process $(\tilde{x}, \tilde{c}, \tilde{y})$ as follows:

$$\tilde{x}_t = \theta x_t + (1 - \theta)x'_t \quad \text{for } t \geq 0;$$
$$\tilde{y} = \theta y + (1 - \theta)y'; \ \tilde{y}_t = f_k(\tilde{x}_{t-1}) \quad \text{for } t \geq 1;$$
$$\tilde{c}_t = \tilde{y}_t - \tilde{x}_t \quad \text{for } t \geq 0.$$

Show that $(\tilde{x}, \tilde{c}, \tilde{y})$ is a well-defined (resp. input, consumption, stock) process from $\tilde{y} = \theta y + (1 - \theta)y'$, and, recalling the arguments leading to (5.17),

$$Eu(\tilde{c}_t) \geq Eu(\theta c_t + (1 - \theta)c'_t) \quad \text{for } t \geq 0. \tag{5.19}$$

Then

$$V[\theta y + (1 - \theta)y'] = V[\tilde{y}] \geq \sum_{t=0}^{\infty} \delta^t Eu(\tilde{c}_t)$$

$$\geq \sum_{t=0}^{\infty} \delta^t Eu[\theta c_t + (1 - \theta)c'_t]$$

$$\geq \theta \sum_{t=0}^{\infty} \delta^t Eu(c_t) + (1 - \theta) \sum_{t=0}^{\infty} \delta^t Eu(c'_t)$$

$$= \theta V(y) + (1 - \theta)V(y').$$

Hence, V is concave. Now if for some $\tau \geq 0$, c_τ is different from c'_τ on a set of positive probability, then using [U.3] we get

$$Eu(\tilde{c}_\tau) \geq Eu(\theta c_\tau + (1 - \theta)c'_\tau)$$
$$> \theta Eu(c_\tau) + (1 - \theta)Eu(c'_\tau). \tag{5.20}$$

If alternatively $c_t = c'_t$ for all $t \geq 0$, since $y > y'$, we have $x_0 > x'_0$. This means, by [T.4] and [U.2], that

$$Eu(\tilde{c}_1) > Eu(\theta c_1 + (1 - \theta)c'_1). \tag{5.21}$$

Hence, we complete the proof of (b) by noting that

$$V(\theta y + (1 - \theta)y') > \theta V(y) + (1 - \theta)V(y'). \tag{5.22}$$

(c) Going back to the right-hand side of (5.7), consider the optimization problem:

$$\max_{a \in A} \left\{ u[(1-a)y] + \delta \sum_{k=1}^{N} V[f_k(ay)]q_k \right\}.$$

The assumed properties of u and f_k and the derived properties of V imply that the maximum is attained at a unique $\bar{a}$. Hence the continuity of $\hat{\eta}$. ∎

A direct consequence of Theorems 5.2 and 5.3 is the following:

Corollary 5.1 *Assume* [T.1]–[T.4] *and* [U.1]–[U.3]. *Then*

(a) the investment policy function $i(y) : S \to S$ *is continuous and nondecreasing and*
(b) the consumption policy function $c(y)$ *is continuous.*

6.5.2 Interior Optimal Processes

It is useful to identify conditions under which optimal processes are interior. To this effect we introduce the following assumptions:

[T.5] $f_k(x)$ *is continuously differentiable at* $x > 0$, *and* $\lim_{x \downarrow 0} f_k'(x) = \infty$ *for* $k = 1, \ldots, N$.
[U.4] $u(c)$ *is continuously differentiable at* $c > 0$, *and* $\lim_{c \downarrow 0} u'(c) = \infty$.

One can derive some strong implications of the differentiability assumptions.

Theorem 5.4 *Under* [T.1]–[T.5] *and* [U.1]–[U.4] *if* (x^*, c^*, y^*) *is an optimal process from some initial stock* $y > 0$, *then for all* $t \geq 0$, $x_t^*(h_t) > 0$, $c_t^*(h_t) > 0$, *and* $y_t^*(h_t) > 0$ *for all histories* h_t.

Proof. First, we show that

$$i(y) > 0 \quad \text{for } y > 0. \tag{5.23}$$

If (5.23) holds, $y_1^* = f_k[i(y_0^*)] > 0$ for $k = 1, 2, \ldots, N$. Hence $x_1^* = i(y_1^*) > 0$. Repeating this argument, we see that $x_t^*(h_t) > 0$, $y_t^*(h_t) > 0$ for all histories h_t.

Fix $\bar{y} > 0$, and suppose that $x_0^* = i(\bar{y}) = 0$. Then $c_0^* = \bar{y}$. Let ε be some positive number, $\varepsilon < \bar{y}$. Define a process (x', c', y') as follows:

$$
\begin{aligned}
y_0' &= \bar{y} > 0, & x_0' &= \varepsilon, & c_0' &= \bar{y} - \varepsilon > 0, \\
y_1' &= f_k(\varepsilon), & x_1' &= 0, & c_1' &= f_k(\varepsilon) = y_1'; \\
c_t' &= x_t' = y_t' = 0 & \text{for } t \geq 2.
\end{aligned}
$$

Clearly, (x', c', y') is a (well-defined) input, consumption, stock process from $\bar{y}$, and let $I(\bar{y})$ be the expected total discounted utility of (x', c', y'). Let (x^*, c^*, y^*) be the optimal process from $\bar{y}$ (with $V(\bar{y})$ being its expected total discounted utility). Then

$$
0 \leq V(\bar{y}) - I(\bar{y}) = \sum_{t=0}^{\infty} \delta^t E[u(c_t^*) - u(c_t')]
$$

$$
= \left[u(\bar{y}) + \delta u(0) \right] - \left[u(\bar{y} - \varepsilon) + \delta \sum_{k=1}^{N} [u(f_k(\varepsilon))]q_k \right].
$$

Hence,

$$
\sum_{k=1}^{N} [u(f_k(\varepsilon))q_k] - u(0) \leq \frac{1}{\delta}[u(\bar{y}) - u(\bar{y} - \varepsilon)]. \tag{5.24}
$$

Let

$$
u(f_{k_1}(\varepsilon)) = \min_k u(f_k(\varepsilon)).
$$

Then,

$$
\sum_{k=1}^{N} [u(f_k(\varepsilon))q_k] \geq u(f_{k_1}(\varepsilon)). \tag{5.25}
$$

From (5.24) and (5.25)

$$
\frac{u(f_{k_1}(\varepsilon)) - u(0)}{f_{k_1}(\varepsilon)} \cdot \frac{f_{k_1}(\varepsilon)}{\varepsilon} \leq \frac{1}{\delta} \frac{[u(\bar{y}) - u(\bar{y} - \varepsilon)]}{\varepsilon}.
$$

Letting $\varepsilon \downarrow 0$ the left side goes to infinity and the right side goes to $\frac{1}{\delta} u'(\bar{y})$, a contradiction. We now show that

$$
c(y) > 0 \quad \text{for } y > 0.
$$

Suppose that $c(\bar{y}) = 0$ for some $\bar{y} > 0$. Then $i(\bar{y}) = \bar{y}$, and

$$V(\bar{y}) = u(0) + \delta \sum_{k=1}^{N} [V(f_k(\bar{y}))q_k]. \qquad (5.26)$$

Consider some ε satisfying $0 < \varepsilon < \bar{y}$. Let $c_0' = \varepsilon$ and $x_0' = \bar{y} - \varepsilon$. With x_0' as the input in period 0, $y_1' = f_k(x_0') = f_k(\bar{y} - \varepsilon)$ with probability $q_k > 0$. Choose the optimal policy ζ^* from $f_k(\bar{y} - \varepsilon)$ yielding $V[f_k(\bar{y} - \varepsilon)]$ as the expected total discounted utility from $f_k(\bar{y} - \varepsilon)$. Hence,

$$u(0) + \delta \sum_{k=1}^{N} V[f_k(\bar{y})]q_k \geq u(\varepsilon) + \delta \sum_{k=1}^{N} [V[f_k(\bar{y} - \varepsilon)]q_k].$$

Simple manipulations lead to

$$\sum_{k=1}^{N} \left[\frac{V[f_k(\bar{y})] - V[f_k(\bar{y} - \varepsilon)]}{f_k(\bar{y}) - f_k(\bar{y} - \varepsilon)} \cdot \frac{f_k(\bar{y}) - f_k(\bar{y} - \varepsilon)}{\varepsilon} \right] q_k \geq \frac{1}{\delta} \frac{u(\varepsilon) - u(0)}{\varepsilon}.$$

Now, letting $\varepsilon \downarrow 0$, we see that the left side goes to $\sum_{k=1}^{N} [V'(f_k(\bar{y})) \cdot f_k'(\bar{y})]q_k$ whereas the right side goes to infinity, a contradiction. ∎

Theorem 5.5 *Under* [T.1]–[T.5] *and* [U.1]–[U.4], *the value function* $V(y)$ *is differentiable at $y > 0$ and*

$$V'(y) = u'(c(y)). \qquad (5.27)$$

Proof. Recall the corresponding result from the deterministic model. ∎

Theorem 5.6 *Under* [T.1]–[T.5] *and* [U.1]–[U.4],

$$u'(c(y)) = \delta \sum_{k=1}^{N} \left[u'[c(f_k(i(y)))] f_k'(i(y)) \right] q_k. \qquad (5.28)$$

Proof. This proof is elementary but long, and is split into several steps. To prove the theorem, first fix $y^* > 0$ and write

$$c^* \equiv c(y^*), \quad x^* \equiv i(y^*) \quad \text{and} \quad x_1^k \equiv i(f_k(x^*)) \text{ for } k = 1, \ldots, N.$$

Let

$$H(c) \equiv u(c) + \delta \sum_{k=1}^{N} \left[u\big(f_k(y^* - c) - x_1^k\big) \right] q_k \qquad (5.29)$$

for all $c > 0$ such that the second expression in the right-hand side is well defined. We begin with

Step 1. The following maximization problem (P) is solved by c^*:

$$\text{maximize } H(c)$$

$$\text{subject to}$$

$$0 \le c \le y^* \quad \text{and} \quad f_k(y^* - c) \ge x_1^k, \quad k = 1, 2 \ldots N.$$

Proof of Step 1. Note that if c^* were not a solution to (P), there is some $\hat{c}$ such that

$$0 \le \hat{c} \le y^*, \quad f_k(y^* - \hat{c}) \geqq x_1^k \quad \text{for } k = 1, \ldots, N;$$

and $H(\hat{c}) > H(c^*)$.

Define a process $(\hat{x}, \hat{c}, \hat{y})$ from y^* as follows:

$$\hat{c}_0 = \hat{c}, \quad \hat{x}_0 = y^* - \hat{c}; \quad \hat{x}_1 = x_1^k, \quad \hat{c}_1 = f_k(\hat{x}_0) - x_1^k, \quad \hat{y}_1 = f_k(\hat{x}_0),$$

and for $t \ge 2$

$$\hat{c}_t = c_t^* \quad \hat{x}_t = x_t^* \quad \hat{y}_t = y_t^*,$$

where (x^*, c^*, y^*) is the optimal process from y^*. The difference between the expected total discounted utility of (x^*, c^*, y^*) and $(\hat{x}, \hat{c}, \hat{y})$ is $(H(c^*) - H(\hat{c})) < 0$, a contradiction to the optimality of (x^*, c^*, y^*) from y^*. ∎

Step 2. There exists $\xi > 0$ such that for all c in $U \equiv (c^* - \xi, c^* + \xi)$ the constraints "$0 \le c \le y^*$", and "$f_k(y^* - c) - x_1^k \ge 0$, for all $k = 1, \ldots, N$" are satisfied.

Proof of Step 2. Use Theorem 5.4 to conclude that $0 < c^* < y^*$; so let $\xi_1 > 0$ be such that $0 < c^* - \xi_1 < c^* < c^* + \xi_1 < y^*$. Then define $U_1 \equiv (c^* - \xi_1, c^* + \xi_1)$. Clearly for all $c \in U_1$ the first constraint in Step 2 holds. Now, $x_1^k \equiv i(f_k(x^*)) < f_k(x^*)$ for each $k = 1, 2, \ldots, N$. Hence, there is some constant $J_1 > 0$ such that

$$f_k(x^*) - x_1^k \geqq J_1 \quad \text{uniformly for all } k = 1, \ldots, N.$$

By continuity of f_k there exists some $\xi \in (0, \xi_1)$ such that uniformly for all $k = 1, 2, \ldots, N$

$$|f_k(x^* + \varepsilon) - f_k(x^*)| < \frac{J_1}{2}$$

for all $\varepsilon \in (-\xi, \xi)$. Define $U \equiv (c^* - \xi, c^* + \xi)$. Then for each $c \in U$, putting $\varepsilon = c^* - c$ we get uniformly for all $k = 1, \ldots, N$

$$
\begin{aligned}
f_k(y^* - c) - x_1^k &\equiv f_k(x^* + \varepsilon) - x_1^k \\
&= \left[f_k(x^*) - x_1^k \right] + \left[f_k(x^* + \varepsilon) - f_k(x^*) \right] \\
&\geq J_1 - \frac{J_1}{2} = \frac{J_1}{2}.
\end{aligned}
$$

In other words, for each $c \in U$

$$f_k(y^* - c) - x_1^* \geq \frac{J_1}{2} \quad \text{for all } k = 1, 2 \ldots N. \tag{5.30}$$

In particular, the second constraint is satisfied for $c \in U$. Since $U \subset U_1$, both constraints are satisfied for $c \in U$. ∎

Thus, we see that $H(c)$ is well defined on the open interval U, and c^* maximizes $H(c)$ on U. The first-order condition leads to $H'(c^*) = 0$. Hence,

$$u'(c^*) = \delta \sum_{k=1}^{N} \left[u'\left(f_k(y^* - c^*) - x_1^k \right) f_k'(y^* - c^*) \right] q_k.$$

In other words,

$$u'(c(y^*)) = \delta \sum_{k} \left[u'(c(f_k(i(y^*)))) f_k'(i(y^*)) \right] q_k. \quad ∎$$

We refer to (5.28) as the *stochastic Ramsey–Euler condition*.

Corollary 5.2 *Under* [T.1]–[T.5] *and* [U.1]–[U.4]

(a) the investment policy function $i(y)$ is continuous and increasing;
(b) the consumption policy function $c(y)$ is continuous and increasing.

Proof.

(a) One needs to prove that, in fact, $i(y)$ is increasing. Note that for $y > 0$

$$V'(y) = u'(c(y)) = \sum_{k=1}^{N} \left[V'(f_k(i(y))) f'_k(i(y)) \right] q_k. \qquad (5.31)$$

Let $y' > y > 0$. Since V is strictly concave, if $y' > \bar{y} > y$

$$V'(y') < V'(\bar{y}) < V'(y).$$

Suppose $i(y') = i(y)$. Then $i(y') = i(\bar{y}) = i(y)$ for $\bar{y} \in [y, y']$. Now,

$$V'(\bar{y}) > V'(y')$$
$$= \sum_{k=1}^{N} \left[V'(f_k(i(y'))) f'_k(i(y')) \right] q_k$$
$$= \sum_{k=1}^{N} \left[V'(f_k(i(\bar{y}))) f'_k(i(\bar{y})) \right] q_k$$
$$= V'(\bar{y}),$$

a contradiction. If $y' > y = 0$, $i(y') > 0 = i(y)$.

(b) To prove that $c(y)$ is increasing, take $y' > y > 0$ and note that

$$u'(c(y)) = V'(y) > V'(y') = u'(c(y')),$$

and use the strict concavity of u. If $y' > y = 0$, $c(y') > 0$ and $c(y) = 0$. ∎

6.5.3 The Random Dynamical System of Optimal Inputs

We shall study the long-run behavior of the optimal process as a random dynamical system. We spell out the details of the optimal input process $\underset{\sim}{x}^* = (x_t^*)$ generated by the optimal investment policy function i. Let $y > 0$ be any initial stock. Then $x_0^* = i(y) > 0$ is the optimal initial input. The optimal input process is given by

$$x_{t+1}^* = i[f(x_t^*)], \qquad (5.32)$$

where $f = f_k$ with probability $q_k > 0$. To simplify exposition we introduce the following assumption:

[T.6] *For all $x > 0$*

$$f_1(x) < \cdots < f_k(x) < \cdots < f_N(x).$$

We should stress that the central result on the convergence of (x_t^*) to the invariant distribution (in the Kolmogorov metric) can be proved without the restrictive assumption [T.6] (see Complements and Details).

Let us write $H_k(x) \equiv i(f_k(x))$ for $k = 1, \ldots, N$. Then the composition map H_k satisfies the following:

(a) H_k is continuous and increasing, i.e., "$x' > x$" implies "$H_k(x') > H_k(x)$."

(b) For all $x > 0$, $H_1(x) < \cdots H_k(x) \cdots < H_N(x)$.

The following lemma is a crucial step in the proof.

Lemma 5.2 *There is some $D > 0$ such that $H_1(x) > x$ for all $x \in (0, D)$.*

Proof. It is convenient to split the long proof into several steps.

Step 1. $H_1(x)$ cannot have a sequence of positive fixed points (x_n) such that $x_n \to 0$ as $n \to \infty$.

Proof of Step 1. Suppose, to the contrary, that there is a sequence x_n such that

$$H_1(x_n) = x_n > 0,$$

and

$$\lim_{n \to \infty} x_n = 0.$$

Now, from the stochastic Ramsey–Euler condition:

$$u'(c(f_1(x_n))) = \delta \sum_{k=1}^{N} \left[u'(c(f_k(H_1(x_n)))) f_k'(H_1(x_n)) \right] q_k$$

$$= \delta \sum_{k=1}^{N} \left[u'[c(f_k(x_n))] f_k'(x_n) \right] q_k$$

$$\geq \delta u' \left[c(f_1(x_n)) \right] [f_1'(x_n)] q_1.$$

Hence,

$$1 \geq \delta f_1'(x_n)q_1.$$

Using [T.5] we get a contradiction as n tends to infinity, since the right-hand side goes to infinity if x_n goes to 0. ∎

Let $D_1 = \inf [x > 0 : H_1(x) = x]$. By Step 1, $D_1 > 0$. There is some $\bar{D} \in (0, D_1)$ such that $H_1(\bar{D}) < \bar{D}$ (otherwise the lemma holds with $D = D_1$). Now if there were some $0 < x < \bar{D}$ such that $H_1(x) > x$, then by the intermediate value theorem, there would be $y \in (x, \bar{D})$ such that $H_1(y) = y < \bar{D} < D_1$, and this would contradict the definition of D_1. Hence,

$$H_1(x) < x \quad \text{for all } x \in (0, \bar{D}). \tag{5.33}$$

We shall see that validity of (5.33) leads to a contradiction.
Step 2. There is some $K \in (0, \bar{D})$ such that

$$c(f_1(i(y))) > c(y) \quad \text{for all } y \in (0, K). \tag{5.34}$$

Proof of Step 2. Pick any $y \in (0, \bar{D})$. Then from the stochastic Ramsey–Euler condition:

$$u'(c(y)) = \delta \sum_{k=1}^{N} \left[u'[c(f_k(i(y)))] f_k'(i(y)) \right] q_k$$

$$\geq \delta u'[c(f_1(i(y)))] f_1'[i(y)] q_1.$$

Hence,

$$\frac{1}{q_1 \delta f_1'(i(y))} \geq \frac{u'[c(f_1(i(y)))]}{u'(c(y))}. \tag{5.35}$$

Since $f_1'(i(y))$ goes to infinity as y goes to 0 (remember $i(y) \leq y$), the left-hand side of the inequity goes to 0 as $y \downarrow 0$. Hence, there is $K \in (0, \bar{D})$ such that for all $y \in (0, K)$

$$\frac{u'[c(f_1(i(y)))]}{u'(c(y))} < 1.$$

By strict concavity of u, $c(f_1(i(y))) > c(y)$ for all $y \in (0, K)$ establishing (5.34). ∎

Next, pick $y_0 \in (0, K)$ and define $x_0 = i(y_0)$, $c_0 = y_0 - i(y_0)$ and for $n \geq 1$, $x_n = H_1(x_{n-1})$, $y_n = f_1(x_{n-1})$, $c_n = y_n - x_n$. Clearly $c_0 > 0$.

Step 3. "$H_1(x) < x$ for all $x \in (0, \bar{D})$" (i.e., validity of (6.33)) implies "$x_n \to 0$, $y_n \to 0$, $c_n \to 0$ as $n \to \infty$."

Proof of Step 3. It is enough to show that $x_n \to 0$. $x_0 < y_0$, and $y_0 \in (0, \bar{D})$ implies $x_0 \in (0, \bar{D})$. If $x_{n-1} \in (0, \bar{D})$ then $x_n = H_1(x_{n-1}) < x_{n-1}$, so $x_n \in (0, \bar{D})$. Thus, (x_n) is decreasing to some x^* (say). Now the continuity of H_1 and $H_1(x_n) = x_{n+1}$ imply in the limit that $H_1(x^*) = x^*$. Clearly $x^* < x_0 < \bar{D} < D_1$. Hence "$x^* > 0$" contradicts the definition D_1. So $x^* = 0$. Convergence of x_n to 0 and continuity of f_1 are invoked to get $y_n \to 0$, and "$c_n = y_n - x_n$ for all n" implies that $c_n \to 0$. ∎

Next, set $y = y_0$ in (5.34) and we get $c_{n+1} > c_n > \cdots > c_0 > 0$ for all $n \geq 1$. So the sequence (c_n) cannot converge to 0, contradicting Step 3. Hence there is some $D > 0$ such that $H_1(x) > x$ for $x \in (0, D)$, and Lemma 5.2 is finally proved. ∎

By using [T.6] and Lemma 5.2 we can assert that

$$x \in (0, D), \quad H_k(x) > x \quad \text{for} \quad k = 1, 2, \ldots, N. \quad (5.36)$$

It is clear (recall [T.3] that for $x > \beta_N$, $H_k(x) \equiv i(f_k(x)) < f_k(x) < x$ for $k = 1, 2, \ldots, N$. Hence, by the intermediate value theorem, for each $k = 1, 2, \ldots, N$ there is some x_k ($D \leq x_k \leq \beta_N$) such that $x_k = H_k(x_k)$. Consider the positive numbers Z_m^1, Z_M^1, Z_m^N, Z_M^N defined as follows:

$$Z_m^1 = \min[x > 0 : H_1(x) = x].$$
$$Z_M^1 = \max[x > 0 : H_1(x) = x].$$
$$Z_m^N = \min[x > 0 : H_N(x) = x].$$
$$Z_M^N = \max[x > 0 : H_N(x) = x]. \quad (5.37)$$

Lemma 5.3

(a) *The points Z_m^1, Z_M^1, Z_m^N, Z_M^N are well defined.*

(b) $Z_m^1 > 0$.

(c) $H_1(x) > x$ *for all $x \in (0, Z_m^1)$; $H_1(x) < x$ for all $x > Z_M^1$.*

(d) $H_N(x) > x$ *for all $x \in (0, Z_m^N)$; $H_N(x) < x$ for all $x > Z_M^N$.*

(e) $Z_m^1 \leq Z_m^N$; $Z_M^1 \leq Z_M^N$.

Proof.

(a) The set $\{x > 0: H_k(x) = x\}$ is nonempty, closed, and bounded for all k.

(b) Since $H_1(x) > x$ for all $x \in (0, D)$, $Z_m^1 \geq D > 0$.

(c) Since $H_1(x) > x$ for $x \in (0, D)$, if $H_1(\bar{x}) < \bar{x}$ for some $\bar{x} < Z_m^1$, continuity of H_1 and the intermediate value theorem will assert the existence of some positive $\tilde{x} < Z_m^1$ such that $H_1(\tilde{x}) = \tilde{x}$, contradicting the definition of Z_m^1. The second assertion is established in a similar manner.

(d) In view of (c), this is left as an exercise.

(e) These inequalities follow from (c) and (d) and the fact that $H_N(x) > H_1(x)$ for all $x \in S$.   ∎

It will now be shown that $Z_M^1 < Z_m^N$. Indeed, for analyzing the long-run behavior of (5.32), we can focus on the interval $[Z_M^1, Z_m^N]$.

Lemma 5.4 $Z_M^1 < Z_m^N$.

Proof.   Let x_1, x_2 be any positive fixed points of H_1 and H_N, respectively. We show that $x_1 < x_2$. This means that $Z_M^1 < Z_m^N$. From the stochastic Ramsey–Euler condition:

$$u'(c(f_1(x_1))) = \delta \sum_{k=1}^{N} \left[u'(c(f_k(H_1(x_1)))) f_k'(H_1(x_1)) \right] q_k$$

$$= \delta \sum_{k=1}^{N} \left[u'(c(f_k(x_1))) f_k'(x_1) \right] q_k.$$

But $f_1(x_1) < f_k(x_1)$ for all k, and, since $c(\cdot)$ is increasing, $c(f_k(x_1)) > c(f_1(x_1))$ for all k. Now strict concavity of u implies that $u'(c(f_1(x_1))) > u'(c(f_k(x_1)))$ for all k. Hence,

$$1 < \delta \sum_{k=1}^{N} f_k'(x_1) q_k.$$

Similarly, we may show that

$$1 > \delta \sum_{k=1}^{N} f_k'(x_2) q_k.$$

So,

$$\sum_{k=1}^{N} f_k'(x_1)q_k > \sum_{k=1}^{N} f_k'(x_2)q_k. \tag{5.38}$$

Strict concavity of f_k implies that $\sum_{k=1}^{N} f_k'(x)q_k$ is decreasing in x, so (5.38) implies that $x_1 < x_2$. ∎

Observe now that

"$x \in [Z_M^1, Z_m^N]$" implies "$H_k(x) \in [Z_M^1, Z_m^N]$ for all k."

Moreover, it is not difficult to show that if the initial $x \notin [Z_M^1, Z_m^N]$, then x_t^* defined by (5.32) will be in $[Z_M^1, Z_m^N]$ in finite time with probability 1. Hence, we have the following result (recall Theorem 5.1 and Example 6.1 of Chapter 3):

Theorem 5.7 *Consider the random dynamical system (5.32) with* $[Z_M^1, Z_m^N]$ *as the state space, and an initial* $x_0 = x \in [Z_M^1, Z_m^N]$. *Then (a) there is a unique invariant distribution* π *of the Markov process* x_t^*; *and (b) the distribution* $\pi_t(x)$ *of* $x_t^*(x)$ *converges to* π *in the Kolmogorov metric* d_K, *i.e.*

$$\lim_{t \to \infty} d_K(\pi_t(x), \pi) = 0,$$

irrespective of the initial $x \in [Z_M^1, Z_m^N]$.

6.5.4 Accumulation of Risky Capital

Consider an investor whose initial *wealth* is $y \geq 0$, and who chooses an action $a \in A \equiv [0, 1]$. Write $x \equiv ay$, interpreted as *investment*, and $c \equiv (1 - a)y$, interpreted as *consumption*. Consumption generates an immediate utility according to a function $u : \mathbb{R}_+ \to \mathbb{R}$: $u(c) = u((1 - a)y)$.

To describe the law of motion, consider a fixed set $\Gamma = \{\rho_1, \ldots, \rho_N\}$, where each $\rho_k > 0$. Let $Q = \{q_1, \ldots, q_k, q_N\}$ ($q_k > 0$, $\sum_{k=1}^{N} q_k = 1$). The investor's wealth in the next period is given by

$$y_1 = \rho x,$$

where $\rho = \rho_k$ with probability q_k. Thus, given the initial y and the chosen a, $y_1 = \rho_k(ay)$ with probability q_k. The investor "observes" y_1 and chooses an action a_1 from A, and an element ρ from Γ is picked according to the same distribution Q, independent of the initial choice, and the story is repeated.

We assume

[A.1] *u is increasing, continuous, and bounded.*

Taking $S = \mathbb{R}_+$, and applying Theorem 4.1 we get

Theorem 5.8 *There exists a stationary optimal policy* $\zeta^* = (\hat{\eta}^{(\infty)})$, *where* $\hat{\eta}: R_+ \to [0, 1]$ *is a Borel measurable function. The value function* $V = I(\zeta^*)[= \sup_\zeta I(\zeta)]$ *is continuous on* $\mathbb{R}_+$ *and satisfies*

$$V(y) = \max_{a \in A} \left\{ u((1 - a)y) + \delta \sum_{k=1}^{N} \left[V(\rho_k ay) \right] q_k \right\}.$$

Moreover, V is increasing.

For a different perspective see Roy (1995).

Example 5.1 An important but difficult problem is to compute or at least approximate an optimal (nearly optimal in some sense) policy function with numerical specifications of the reward function and the law of motion. For the discounted optimal growth problem of Subsection 6.5.1, consider

$$u(c) = \log c$$

and $y_{t+1} = Ax_t^{r_{t+1}}, t \geq 0$; where $A > 0$, and (r_t) is an i.i.d. sequence with $0 < R_1 < r_t < R_2 < 1$ (where R_1, R_2 are given constants). In this case, one can show that

$$c(y) = [1 - \delta(Er)]y$$

is the optimal consumption policy function.

For the accumulation problem of Subsection 6.5.4, consider

$$u(c) = (1 - a)^{-1} c^{1-a}, \quad 0 < a < 1$$

and (ρ_t) to be an i.i.d sequence of random variables (assuming the values $\rho_t = \rho_k > 0$ with probability $q_k > 0$).

If we assume that

$$0 < \delta[E(\rho^{1-a})] < 1$$

then $c(y) = \lambda y$ is the optimal consumption policy function where the positive constant λ is determined from the relation $(1 - \lambda)^a = \delta E \rho^{1-a}$. We omit the algebraic details (readily available from many sources). ∎

6.6 Complements and Details

Section 6.2. The functional equation (2.2) has been ingeniously used to identify qualitative properties of the value and optimal policy functions in many contexts. The texts by Ross (1983) and Bertsekas (1995a,b) contain a number of insightful applications.

The texts by Sargent (1987) and Ljungqvist and Sargent (2000) also provide examples from dynamic macroeconomics, comprehensive lists of references, as well as rich collections of exercises.

Section 6.3. The maximum theorem (see Berge 1963) is a key tool in proving the existence of a Walrasian equilibrium (see Debreu 1959) as well as the Cournot–Nash equilibrium (see Nash 1950 and Debreu 1952).

Section 6.4. We shall indicate some extensions of the basic model.

6.6.1 Upper Semicontinuous Model

A real-valued function f on a metric space (S, d) is *upper semicontinuous* at $x \in S$ if for every $\varepsilon > 0$ there exists a $\delta > 0$ such that $d(x, y) < \delta$ implies that $f(y) < f(x) + \varepsilon$. One can show that f is upper semicontinuous on S if and only if for each real θ, $\{x: f(x) < \theta\}$ is open. One can show that f is continuous on S if and only if both f and $-f$ are upper semicontinuous. Of particular importance is the following:

Proposition C6.1 *Let f be an upper semicontinuous real-valued function on a (nonempty) compact metric space (S, d). Then f is bounded above and assumes its maximum.*

For a useful summary of some of the properties of upper semicontinuous functions, see Royden (1968, pp. 195–196).

Theorem 4.1 was proved by Maitra (1968) under the assumptions [A.1], [A.2], [A.4] and [A.3′]. The reward function u is a bounded, upper semicontinuous function on $S \times A$.

6.6.2 The Controlled Semi-Markov Model

The dynamic programming problem we consider is specified by (1) a *state space S* (a Borel subset of a complete separable metric space), (2) an *action space A* (a compact metric space), (3) a *reward rate* $r(x, a)$, which accrues in state x when action a is taken, (4) a *distribution* $\gamma(du \mid x, a)$ of the *holding time* in state x when an action a is taken, and (5) a *transition probability* $q(dz \mid x, a)$ of the new state z which occurs at the end of this holding period in state x.

Informally, a *policy* is a rule that determines which action to take based on past actions and past and present states. At the kth stage, the policy chooses an action

$$\hat{a}_k = f_k(X_0, \hat{a}_0, \ldots, X_{k-1}, \hat{a}_{k-1}, X_k)$$

based on past actions $\hat{a}_0, \ldots, \hat{a}_{k-1}$ and past and present states $X_0, \ldots, X_k$. The evolution of the state with time, for a given policy $f = \{f_0, f_1, \ldots, f_k, \ldots\}$, may be described as follows. An initial state $X_0 = x_0$ is given, and an action $\hat{a}_0 = f_0(X_0)$ is chosen; the state remains at X_0 for a random time T_0 whose distribution is $\gamma(du \mid X_0, \hat{a}_1)$, given X_0 and $\hat{a}_0$. After time T_0 a new state X_1 occurs with distribution $q(dz \mid X_0, \hat{a}_0)$, and, observing X_1, the action $\hat{a}_1 = f_1(X_0, \hat{a}_0, X_1)$ is chosen; the state remains at X_1 for a period T_1 with distribution $\gamma(du \mid X_1, \hat{a}_1)$, at the end of which (i.e., at time $T_0 + T_1$) a new state X_2 occurs having distribution $q(dz \mid X_1, \hat{a}_1)$, conditionally given $X_1, \hat{a}_1$. This continues indefinitely.

We now give a more formal description.

A *policy* is a sequence of functions $\zeta = \{f_0, f_1, f_2, \ldots\}$: f_0 is a Borel measurable map on S into A, f_1 is a Borel measurable map on $S \times A \times S$ into $A, \ldots, f_k$ is a Borel measurable map on $S \times A \times S \times A \times \cdots \times S \times A \times S = (S \times A)^k \times S$ into A. A policy ζ is *Markovian* if, for each k, f_k depends only on the last coordinate among its arguments, i.e., each f_k is a Borel measurable map on S into A. A policy ζ is *stationary* if it is Markovian and $f_k = f \ \forall k$, where f is some Borel measurable map on S into A; such a policy is denoted by $(f^{(\infty)})$.

Corresponding to each policy ζ and an *initial state* $X_0 = x_0 \in S$, there exist on a suitable probability space three sequences of random variables $X_k, \hat{a}_k, T_k (k = 0, 1, 2, \ldots)$ whose (joint) distribution is specified as follows:

[P.1] $\hat{a}_k$ *is determined by* $X_0, X_1, \ldots, X_k$,

$$\hat{a}_0 = f_0(X_0), \quad \hat{a}_1 = f_1(X_0; \hat{a}_0, X_1), \ldots,$$

$$\hat{a}_k = f_k(X_0, \hat{a}_0, X_1, \hat{a}_1, \ldots, X_{k-1}, \hat{a}_{k-1}, X_k), \ldots.$$

[P.2] $X_0 = x_0$; *conditionally given* $X_0, X_1, \ldots, X_k$, *the distribution of* X_{k+1} *is* $q(dz \mid X_k, \hat{a}_k)$.

[P.3] *Conditionally given all* X_k's $(k = 0, 1, 2, \ldots)$, *the random variables* $T_0, T_1, T_2, \ldots$ *are an independent sequence, with* T_k *having distribution* $\gamma(du \mid X_k, \hat{a}_k)$.

Now define

$$Y_t = X_k \text{ if } T_0 + T_1 + \cdots + T_{k-1} \leq t < T_0 + T_1 + \cdots + T_k \quad (k = 1, 2, \ldots),$$
$$= X_0 \text{ if } 0 \leq t < T_0.$$
$$a_t = \hat{a}_k \text{ if } T_0 + T_1 + \cdots + T_{k-1} \leq t < T_0 + T_1 + \cdots + T_k \quad (k = 1, 2, \ldots),$$
$$= \hat{a}_0 \text{ if } 0 \leq t < T_0. \tag{C6.1}$$

Given a policy ζ and an initial state x, the *total expected discounted reward* (*with discount rate* $\beta > 0$) is given by

$$V(\zeta)(x) = E_x^\zeta \int_0^\infty e^{-\beta t} r(Y_t, a_t) \, dt, \tag{C6.2}$$

where E_x^ζ denotes *expectation* under ζ and x. Write

$$\delta_\beta(x, a) = \int e^{-\beta u} \gamma(du \mid x, a), \quad \tau_\beta(x, a) = \frac{1 - \delta_\beta(x, a)}{\beta}. \tag{C6.3}$$

As we indicate in Remark C6.4.1 following the statements [A.1]–[A.4] below, the right-hand side of (C6.2) is well defined and one has

$$V(\zeta)(x)$$

$$= E_x^\zeta \left[\left(\int_0^{T_0} e^{-\beta t} dt \right) r(X_0, \hat{a}_0) + \sum_{k=1}^\infty \left(\int_{T_0 + \cdots + T_{k-1}}^{T_0 + \cdots + T_k} e^{-\beta t} dt \right) r(X_k, \hat{a}_k) \right]$$

$$= E_x^\zeta \left[\frac{1 - e^{-\beta T_0}}{\beta} r(X_0, \hat{a}_0) + \sum_{k=1}^\infty e^{-\beta(T_0 + \cdots + T_{k-1})} \left(\frac{1 - e^{-\beta T_k}}{\beta} \right) r(X_k, \hat{a}_k) \right]$$

$$= \tau_\beta(x, f_0(x)) r(x, f_0(x))$$

$$+ E_x^\zeta \sum_{k=1}^\infty \delta_\beta(x, f_0(x)) \delta_\beta(X_1, \hat{a}_1) \cdots \delta_\beta(X_{k-1}, \hat{a}_{k-1}) \tau_\beta(X_k, \hat{a}_k) r(X_k, \hat{a}_k).$$

$$\text{(C6.4)}$$

For convenience we take the extreme right-hand side of (C6.4) as the definition of $V(\zeta)(x)$, when it is absolutely convergent. *The optimal discounted reward is*

$$V(x) = \sup_\zeta V(\zeta)(x) \quad (x \in S), \tag{C6.5}$$

where the supremum is over *all* policies ζ. A policy ζ^* is said to be *optimal for discounted reward if*

$$V(\zeta^*)(x) = V(x) \quad \forall x \in S. \tag{C6.6}$$

The model specified by [P.1]–[P.3] is often called a *semi-Markov* or a *Markov renewal model*, especially under stationary policies.

A function F on a metric space (S, d) into $\mathcal{P}(S)$, is *weakly continuous* if the sequence of probability measures $F(y_n)$ converge weakly to $F(y)$ when $y_n \to y$ in (S, d). It is *strongly continuous* if $y_n \to y$ implies

$$\|F(y_n) - F(y)\| \equiv \sup \{|F(y_n)(B) - F(y)(B)| : B \in \mathcal{S}\} \to 0.$$

The following assumptions are made:

[A.1] *S is a nonempty Borel subset of a complete separable metric space; A is a compact metric space.*

[A.2] *$(x, \hat{a}) \to \delta_\beta(x, \hat{a})$ is continuous on $S \times A$. Also,*

$$\theta \equiv \sup_{x,a} \delta_\beta(x, a) < 1. \tag{C6.7}$$

[A.3] $r(x, a)$ *is upper semicontinuous (u.s.c.) on* $S \times A$ *and bounded above. Also,*

$$R(x) \equiv \sum_{k=0}^{\infty} \theta^k r_k(x) < \infty \quad (x \in S), \qquad \text{(C6.8)}$$

where

$$r_0(x) \equiv \sup_a \ | \ r(x, a) \tau_\beta(x, a),$$

$$r_{k+1}(x) = \sup_a \int r_k(z) q(dz \mid x, a) \quad (k = 0, 1, 2, \ldots). \quad \text{(C6.9)}$$

[A.4] $(x, a) \to q(dz \mid x, a)$ *is weakly continuous on* $S \times A$ *into* $\mathcal{P}(S)$.

Remark C6.4.1 By the property [P.3],

$$E_x^\zeta (e^{-\beta \left(\sum\limits_{k=0}^{\infty} T_k \right)} \bigg| \{X_k : 0 \le k < \infty\}) = \prod_{k=0}^{\infty} \int e^{-\beta u} \gamma(du \mid X_k, \hat{a}_k)$$

$$= \prod_{k=0}^{\infty} \delta_\beta(X_k, \widehat{a}_k). \qquad \text{(C6.10)}$$

By assumption [A.2], the last product is zero. Therefore, $\sum_{k=0}^{\infty} T_k = \infty$ almost surely, and the process Y_t in (C6.1) is defined for all $t \ge 0$. Further, by assumption [A.3], and the computation in (C6.4),

$$\left| E_x^\zeta \int_0^\infty e^{-\beta t} r(Y_t, \hat{a}_t) \, dt \right| \le E_x^\zeta \int_0^\infty e^{-\beta t} |r(Y_t, \hat{a}_t)| \, dt$$

$$\le \tau_\beta(x, f_0(x)) |r(x, f_0(x))|$$

$$+ E_x^\zeta \sum_{k=1}^{\infty} \delta_\beta(x, f_0(x)) \delta_\beta(X_1, \hat{a}_1) \cdots$$

$$\delta_\beta(X_{k-1}, \hat{a}_{k-1}) \tau_\beta(X_k, \hat{a}_k) |r(X_k, \hat{a}_k)|$$

$$\le \frac{1}{\beta} \sum_{k=0}^{\infty} \theta^k r_k(x) < \infty. \qquad \text{(C6.11)}$$

Thus (C6.4) is justified.

Before proceeding further let us specify how the model and the assumptions specialize to two important cases.

Special Case 1. (*Discrete time models*). In this case $\gamma(\{1\}|x, a) = 1$ for all x, a, i.e., $T_k \equiv 1$ for all k. Assumption [A.2] is redundant here. Since $\tau_\beta(x, a) \equiv (1 - e^{-\beta})/\beta$, $r_0(x)$ in (C6.8) may be taken to be $\sup |r(x, a)|$ in the statement of [A.3]. The discounted reward for a policy ζ in this case is usually defined by

$$I(\zeta)(x) = \sum_{k=0}^{\infty} \delta^k E_x^\zeta r(X_k, a_k) = (1 - e^{-\beta})^{-1} \beta V(\zeta)(x), \quad \text{(C6.12)}$$

where β and δ are related by

$$\delta = e^{-\beta} = \theta. \quad \text{(C6.13)}$$

Special Case 2. (*Continuous time Markov models*). This arises if the holding time distribution is exponential, i.e.,

$$\gamma(du \mid x, a) = \lambda(x, a) \exp\{-\lambda(x, a)u\}\, du, \quad \text{(C6.14)}$$

where $\lambda(x, a)$ is a continuous function on $S \times A$ into $[0, \infty)$. The process $\{Y_t\}$ in (C6.1) is then Markovian if the policy ζ is Markovian, and is time-homogeneous Markovian if ζ is stationary. For this case

$$\delta_\beta(x, a) = \frac{\lambda(x, a)}{\beta + \lambda(x, a)}, \quad \tau_\beta(x, a) = \frac{1}{\beta + \lambda(x, a)}. \quad \text{(C6.15)}$$

Hence (C6.7) becomes

$$\sup_{x, a} \lambda(x, a) < \infty. \quad \text{(C6.16)}$$

We are now ready to state our main result.

Theorem C6.4.1 *Assume* [A.1]–[A.4].

(a) *The optimal discounted reward is an upper semicontinuous function on S satisfying*

$$V(x) = \max_a \left\{ r(x, a)\tau_\beta(x, a) + \delta_\beta(x, a) \int V(z)q(dz \mid x, a) \right\}. \quad \text{(C6.17)}$$

(b) There exists a Borel measurable function f^* on S into A such that $a = f^*(x)$ maximizes, for each x, the right-hand side of (C6.17). The stationary policy $\zeta^* = (f^{*(\infty)})$ is optimal.

For a complete analysis, see Bhattacharya and Majumdar (1989a). Controlled semi-Markov models in which the optimality criterion is the maximization of "long-run average returns" are studied in Bhattacharya and Majumdar (1989b).

6.6.3 State-Dependent Actions

Consider an optimization problem specified by $\langle S, A, q(\cdot|s, a), u(s, a), \delta \rangle$, where

- S is a (nonempty) Borel subset of a Polish (complete, separable metric) space (the set of all *states*);
- A is a compact metric space (the set of all *conceivable actions*);
- φ is a continuous correspondence from S into A ($\varphi(s)$ is the set of all actions that are *feasible* at s);
- $q(\cdot|s, a)$ for each $(s, a) \in S \times A$ is a probability measure on the Borel sigmafield of S (the law of motion: the distribution of the state s' tomorrow given and state s and the action a today);
- $u(s, a)$ is the (real-valued) immediate return or utility function on $S \times A$;
- $\delta (0 < \delta < 1)$ is the discount factor;
- A policy $\zeta = (\zeta_t)$ is a sequence where ζ_t associates with each $h_t = (s_0, a_0, \ldots, s_t)$ a probability distribution such that $\zeta_t(\varphi(s_t)|s_0, a_0, \ldots, s_t) = 1$ for all $(s_0, a_0, \ldots, s_t)$.
- A Markov policy is a sequence (g_t) where each g_t is a Borel measurable function from S into A such that $g_t(s) \in \varphi(s)$ for all $s \in S$. A stationary policy is a Markov policy in which $g_t \equiv g$ for some Borel measurable function g for all t. A policy ζ associates with each initial state s, the expected total discounted return over the infinite future $I(\zeta)(s)$. A policy ζ^* is *optimal* if $I(\zeta)^*(s) \geq I(\zeta)(s)$ for all policies ζ and all $s \in S$.
- Note that in this description of the optimization problem, one recognizes explicitly that the actions feasible for the decision maker depend on the observed state, and a policy must specify *feasible* actions.

The existence of an optimal stationary policy was proved by Furukawa (1972). We state a typical result. We make the following assumptions:

[A.1] $u(s, a)$ is bounded continuous function on $S \times A$.

[A.2] For each fixed $s \in S$, and any bounded measurable function w, $\int w(\cdot) \, dq(\cdot | s, a)$ is continuous in $a \in A$.

Proposition C6.4.1 *Let A be a (nonempty) compact subset of $\mathbb{R}^l$ and φ a continuous correspondence. Under assumption [A.1] and [A.2], there exists a stationary optimal policy.*

Section 6.5. Majumdar, Mitra, and Nyarko (1989) provide a detailed analysis of the problem of "discounted" optimal growth under uncertainty. In particular, several results are derived under weaker assumptions (e.g. [T.6] was not assumed). In addition to the class of utility functions that satisfy [U.1]–[U.4] they treat the case where u is continuous on $\mathbb{R}_{++}$, and $\lim_{c \downarrow 0} u(c) = -\infty$. Finally, they consider the long-run behavior of optimal processes when [T.4] is not assumed. See Nyarko and Olson (1991, 1994) for models in which the one-period return function depends on the stock. The role of increasing returns in economic growth was emphasized by Young (1928), which has remained a landmark. For a review of the literature on deterministic growth with increasing returns, see Majumdar (2006). Example 5.1 is elaborated in Levhari and Srinivasan (1969) and Mirman and Zilcha (1975). The dynamic programming model of Foley and Hellwig (1975) is of particular interest. The evolution of the optimal process of money balances is described by a random dynamical system with two possible laws of motion (corresponding to the employment status: the agent is employed with probability q and unemployed with probability $1 - q$). The question of convergence to a unique invariant probability was studied by the authors.

The problem of selecting the optimal investment policy under uncertainty and assessing the role of risk on the qualitative properties of the optimal policy is central to the microeconomic theory of finance. A related important issue is portfolio selection: the problem of the investor who has to allocate the total investment among assets with alternative patterns of risk and returns. The literature is too vast for a sketchy survey. For a representative sample of early papers, see Phelps (1962), Hakansson (1970), Sandmo (1970), Cass and Stiglitz (1972), Chipman (1973), and Miller (1974, 1976).

Models of discrete time Markovian decision processes (MDPs) with incomplete information about (or, partial observation of) the state space have been studied by several authors. An interesting theme is to recast or transform such a model into a new one in which the state space is the set of all probability distributions on the states of the original model. See Sawarigi and Yoshikawa (1970) and Rhenius (1974) for formal treatments, and Easley and Kiefer (1988, 1989) for applications. Dynamic programming techniques have been widely used in variations of the "bandit" model in order to treat optimal information acquisition and learning. See the monographs by Berry and Friested (1985) and Gittins (1989) on the mathematical framework and the wideranging applications in many fields. An early application to economics is to be found in Rothschild (1974). Other examples from economics are spelled out in the review article by Sundaram (2004).

This literature on stochastic games is also of interest. See Majumdar and Sundaram (1991) for an economic model, and the lists of references in Maitra and Parthasarathy (1970), Parthasarathy (1973), and Nowak (1985).

Bibliographical Notes. In our exposition we deal with the infinite horizon problem directly. An alternative is to start with a T-period problem (finite) and to study the impact of changing T. Other comparative static/dynamic problems involve studying the impact of (a) variations of the discount factor δ: and (b) variations of the parameters in the production and utility functions. For a sample of results, see Schal (1975), Bhattacharya and Majumdar (1981), Danthene and Donaldson (1981a, b), Dutta (1987), Majumdar and Zilcha (1987), and Dutta, Majumdar, and Sundaram (1994).

Appendix

A1. METRIC SPACES: SEPARABILITY, COMPLETENESS, AND COMPACTNESS

We review a few concepts and results from metric spaces. For detailed accounts see Dieudonné (1960) and Royden (1968). Students from economics are encouraged to review Debreu (1959, Chapter 1), Nikaido (1968, Chapter 1), or Hildenbrand (1974, Part 1). An excellent exposition of some of the material in this and the next section is Billingsley (1968).

Let (S, d) be a metric space. The *closure* of a subset A of S is denoted by $\bar{A}$, its *interior* by A^o, and its *boundary* by $\partial A(= \bar{A} \backslash A^0)$. $\bar{A}$ is the intersection of all closed sets containing A. A^o is the union of all open sets contained in A. We define the distance from x to A as

$$d(x, A) = \inf \{d(x, y): y \in A\},$$

$d(x, A)$ is uniformly continuous in x: verify that

$$|d(x, A) - d(y, A)| \leq d(x, y).$$

Denote by $B(x, \varepsilon)$ the open ball with center x and radius ε: $B(x, \varepsilon) = \{y: d(x, y) < \varepsilon\}$.

Two metrics d_1 and d_2 on S are said to be *equivalent* if (write $B_i(x, \varepsilon) = \{y: d_i(x, y) < \varepsilon\}$) for each x and ε there is a δ with $B_1(x, \delta) \subset B_2(x, \varepsilon)$ and $B_2(x, \delta) \subset B_1(x, \varepsilon)$, so that S with d_1 is homeomorphic to S with d_2.

For any subset A of S, the *complement of* A is denoted by $A^c \equiv S \backslash A$.

419

A1.1. Separability

A metric space (S, d) is *separable* if it contains a countable, dense subset. A *base* for S is a class of open sets such that each open subset of S is the union of some of the members of the class. An *open cover* of A is a class of open sets whose union contains A. A *set A* is *discrete* if around each point of A there is an open "isolated" ball containing no other points of A – in other words, if each point of A is isolated in the relative topology. If S itself is discrete, then taking the distance between distinct points to be 1 defines a metric equivalent to the original one.

Theorem A.1 *These three conditions are equivalent:*

(i) S is separable.
(ii) S has a countable base.
(iii) Each open cover of each subset of S has a countable subcover.

Moreover, separability implies

(iv) S contains no uncountable discrete set, and this in turn implies
(v) S contains no uncountable set A with

$$\inf\{d(x, y): x, y \in A, \ x \neq y\} > 0. \tag{A.1}$$

A1.2. Completeness

A sequence (x_n) in a metric space (S, d) is a *Cauchy sequence* if for every $\varepsilon > 0$ there exists a positive integer $\bar{n}(\varepsilon)$ such that $d(x_p, x_q) < \varepsilon$ for all $p, q \geq \bar{n}(\varepsilon)$.

Clearly, every convergent sequence in a metric space (S, d) is a Cauchy sequence. Every subsequence of a Cauchy sequence is also a Cauchy sequence. One can show that a Cauchy sequence either converges or has no convergent subsequence.

It should be stressed that in the definition of a Cauchy sequence the metric d is used explicitly. The same sequence can be Cauchy in one metric but not Cauchy for an equivalent metric. Let $S = \mathbb{R}$, and consider the usual metric $d(x, y) = |x - y|$.

This metric d is equivalent to

$$\bar{d}(x, y) = \left| \frac{x}{1 + |x|} - \frac{y}{1 + |y|} \right|.$$

The sequence $\{n: n = 1, 2, \ldots\}$ in $\mathbb{R}$ is *not* Cauchy in the d-metric, but is Cauchy in the $\bar{d}$-metric.

A metric space (S, d) is *complete* if every Cauchy sequence in S converges to some point of S.

The *diameter* of A is denoted by $\text{diam}(A) \equiv \sup_{x, y \in A} d(x, y)$. A is *bounded* if $\text{diam}(A)$ is finite.

Theorem A.2 *In a complete metric space, every non-increasing sequence of closed nonempty sets A_n such that the sequence of their diameters $\text{diam}(A_n)$ converges to 0 has a nonempty intersection consisting of one point only.*

Proof. Choose $x_n \in A_n$ ($n \geq 1$). Then $\{x_n\}$ is a Cauchy sequence since $n', n'' \geq n$, $d(x_{n'}, x_{n''}) \leq \text{diam}(A_n)$, which, by completeness of S, must converge to a limit, say, y. We will show that $y \in A_n$ for every n. Suppose, if possible, that $y \notin A_m$. Since A_m is closed, this means there is $\delta > 0$ such that $B(y, \delta) \subset A_m^c$. Then $B(y, \delta) \subset A_n^c$ for all $n \geq m$, which implies $d(x_n, y) \geq \delta \; \forall n \geq m$, contradicting the fact $x_n \to y$. ∎

Theorem A.3 (*Baire category*) *If a complete metric space (S, d) is a countable union of closed sets, at least one of these closed subsets contains a nonempty open set.*

Proof. Suppose that $S = \bigcup_n A_n$, where $\{A_n\}$ is a sequence of closed sets. Suppose, to the contrary, that no A_n contains a nonempty open set. Thus, $A_1 \neq S$, and A_1^c, the complement of A, is open and contains an open ball $B_1 = B(p_1, \varepsilon_1)$ with $0 < \varepsilon_1 < \frac{1}{2}$. By assumption, A_2 does not contain the open set $B(p_1, \varepsilon/2)$. Hence, the nonempty open set $A_2^c \cap B(p_1, \varepsilon/2)$ contains an open ball $B_2 = B(p_2, \varepsilon/2)$ with $0 < \varepsilon_2 < \frac{1}{2^2}$. By induction, a sequence $\{B_n\} = \{B(p_n, \varepsilon_n)\}$ of open balls is obtained with the properties $0 < \varepsilon_n < \frac{1}{2^n}$, $B_{n+1} \subset B(p_n, \varepsilon_n/2)$, $B_n \cap A_n = \phi$, $n = 1, 2$.

Since for $n < m$

$$d(p_n, p_m) \leq d(p_n, p_{n+1}) + \cdots + d(p_{m-1}, p_m)$$
$$< \frac{1}{2^{n+1}} + \cdots + \frac{1}{2^m} < \frac{1}{2^n},$$

the centers (p_n) form a Cauchy sequence, and hence, converge to a point p. Since

$$d(p_n, p) \le d(p_n, p_m) + d(p_m, p)$$
$$< \frac{\varepsilon_n}{2} + d(p_m, p) \to \frac{\varepsilon_n}{2}$$

it is seen that $p \in B_n$ for all n. This implies that p does *not* belong to any A_n, hence, $p \notin \bigcup_{n=1}^{\infty} A_n = S$, a contradiction. ∎

A set E in a metric space (S, d) is *nowhere dense* in S if $(\bar{E})^o = \phi$. This is equivalent to saying that E is nowhere dense if $S\backslash\bar{E}$ is dense in S. (S, d) is of the *first category* if S is the union of a countable family of nowhere dense sets. S is of the *second category* if it is not of the first category. Thus, by Theorem A.2, *a complete metric space is of the second category*. From this theorem we can also conclude that *if (S, d) is a complete metric space, the intersection of a countable family of open dense subsets of S is itself dense in S.*

A1.3. Compactness

A set A is *compact* if each open cover of A contains a finite subcover. An ε-*net* for A is a set of points $\{x_k\}$ with the property that for each x in A there is an x_k such that $d(x, x_k) < \varepsilon$ (the x_k are not required to lie in A). A set is *totally bounded* if, for every positive ε, it has a finite ε-net.

Theorem A.4 *For an arbitrary set A in S, these four conditions are equivalent:*

(i) $\bar{A}$ is compact.

(ii) Each countable open cover of $\bar{A}$ has a finite subcover.

(iii) Each sequence in A has a limit point (has a subsequence converging to a limit, which necessarily lies in $\bar{A}$).

(iv) A is totally bounded and $\bar{A}$ is complete.

A subset of k-dimensional Euclidean space $\mathbb{R}^k$ (with the usual metric) has compact closure if and only if it is bounded (Bolzano–Weirstrass theorem).

Theorem A.5 *If h is a continuous mapping of S into another metric space S' and if K is a compact subset of S, then hK is a compact subset of S'.*

Theorem A.6 *If* $\{K_\theta\}$ *is a collection of compact subsets of a metric space* (S, d) *such that the intersection of every finite subfamily of* $\{K_\theta\}$ *is nonempty, then* $\bigcap_\theta K_\theta$ *is nonempty.*

Proof. For proofs of Theorems A.4–A.6, as well as the Bolzano–Weirstrass theorem, refer to Dieudonné (1960, pp. 56 (3.16.1), 60 (3.17.6), 61 (3.17.9)) or Billingsley (1968, pp. 217, 218). ∎

Corollary A.1 *If* $\{K_n\}$ *is a sequence of nonempty compact sets such that*

$$K_n \supset K_{n+1} \quad (n = 1, 2, \ldots),$$

then $\bigcap_{n=1}^\infty K_n$ *is nonempty.*

A2. INFINITE PRODUCTS OF METRIC SPACES AND THE DIAGONALIZATION ARGUMENT

To begin with, let $\mathbb{R}^\infty$ be the space of sequences $x = (x_1, x_2, \ldots)$ of real numbers. If $d_0(a, b) = |a - b| / (1 + |a - b|)$, then d_0 is a metric on the line $\mathbb{R}^1$ equivalent to the ordinary metric $|a - b|$. The line $(\mathbb{R}^1)$ is complete under d_0. It follows that, if $d(x, y) = \sum_{k=1}^\infty d_0(x_k, y_k)2^{-k}$, then d is a metric on $\mathbb{R}^\infty$. If

$$N_{k,\varepsilon}(x) = \{y : |y_i - x_i| < \varepsilon, \; i = 1, \ldots, k\}, \qquad \text{(A.2)}$$

then $N_{k,\varepsilon}(x)$ is open in the sense of the metric d. Moreover, since $d(x, y) < \varepsilon 2^{-k}/(1 + \varepsilon)$ implies $y \in N_{k,\varepsilon}(x)$, which in turn implies $d(x, y) < \varepsilon + 2^{-k}$, the sets (A.2) form a base for the topology given by d.

This topology is the *product topology*, or the topology of coordinate-wise convergence: for a sequence $x(n)$ in $\mathbb{R}^\infty$, $\lim_n x(n) = x$ if and only if $\lim_n x_k(n) = x_k$, for each k.

The space $\mathbb{R}^\infty$ is *separable*; one countable, dense subset consists of those points with coordinates that are all rationals and that, with only finitely many exceptions, vanish.

Suppose $\{x(n)\}$ is a Cauchy sequence in $\mathbb{R}^\infty$. Since

$$d_0(x_k(m), x_k(n)) \leq 2^k d(x(m), x(n)),$$

it follows easily that, for each k, $\{x_k(1), x_k(2), \ldots\}$ is a Cauchy sequence on the line with the usual metric, so that the limit $x_k = \lim_n x_k(n)$ exists. If $x = (x_1, x_2, \ldots)$, then $x(n)$ converges to x in the sense of $\mathbb{R}^\infty$. Thus $\mathbb{R}^\infty$ is *complete*.

Theorem A.7 *A subset A of $\mathbb{R}^\infty$ has compact closure if and only if the set $\{x_k : x \in A$ is, for each k, a bounded set on the line.*

Proof. It is easy to show that the stated condition is necessary for compactness. We prove sufficiency by the "diagonalization" argument. Given a sequence $\{x(n)\}$ in A, we may choose a sequence of subsequences

$$\begin{cases} x(n_{11}), & x(n_{12}), & x(n_{13}), & \ldots \\ x(n_{21}), & x(n_{22}), & x(n_{23}), & \ldots \\ \quad\cdots\cdots\cdots\cdots\cdots\cdots\cdots\cdots\cdots \\ \quad\cdots\cdots\cdots\cdots\cdots\cdots\cdots\cdots\cdots \end{cases} \tag{A.3}$$

in the following way. The first row of (A.3) is a subsequence of $\{x(n)\}$, so chosen that $x_1 = \lim_i x_1(n_{1i})$ exists; there is such a subsequence because $\{x_1 : x \in A\}$ is a bounded set of real numbers. The second row of (A.3) is a subsequence of the first row, so chosen that $x_2 = \lim_i x_2(n_{2i})$ exists; there is such a subsequence because $\{x_2 : x \in A\}$ is bounded.

We continue in this way; row k is a subsequence of row $k - 1$, and $x_k = \lim_i x_k(n_{ki})$ exists. Let x be the point of $\mathbb{R}^\infty$ with coordinates x_k. If $n_i = n_{ii}$, then $\{x(n_i)\}$ is a subsequence of $\{x(n)\}$. For each k, moreover, $x(n_k), x(n_{k+1}), \ldots$ all lie in the kth row of (A.3), so that $\lim_i x_k(n_i) = x_k$. Thus, $\lim_i x(n_i) = x$, and it follows that $\bar{A}$ is compact. ∎

Theorem A.8 *Every separable metric space S is homeomorphic to a subset of $\mathbb{R}^\infty$.*

Proof. An adaptation of the proof of Lemma C11.1, Chapter 2, is left as an exercise. ∎

Example A.1 (*The space $C[0, 1]$*) Let $C = C[0, 1]$ be the space of continuous, real-valued functions on the unit interval $[0, 1]$ with the uniform metric. The distance between two elements $x = x(t)$ and $y = y(t)$ of C is

$$d(x, y) = \sup_{t \in [0,1]} |x(t) - y(t)| \, ;$$

it is easy to check that d is a metric. Convergence in the topology is uniform pointwise convergence of (continuous) functions.

The space C is separable; one countable, dense set consists of the (polygonal) functions that are linear on each subinterval $[(i - 1)/k,$

i/k], $i = 1, \ldots, k$, for some integer k, and assume rational values at the points i/k, $i = 0, 1, \ldots, k$.

If $\{x_n\}$ is a Cauchy sequence in C, then, for each value of t, $\{x_n(t)\}$ is a Cauchy sequence on the line and hence has a limit $x(t)$. It is easy to show that the convergence $x_n(t) \to x(t)$ is uniform in t, so that x lies in C and is the limit in C of $\{x_n\}$. Thus C is complete. ∎

Example A.2 Let (S, d) be a metric space and $C(S)$ the set of all real-valued bounded continuous functions on S. The distance between two elements $x = x(t)$ and $y = y(t)$ of C is defined as

$$d(x, y) = \sup_{t \in S} |x(t) - y(t)|.$$

The separability of $C[0, 1]$ (Example A.1) is a special case of the following:

Theorem A.9 $C(S)$ *is separable if and only if S is compact.*

Proof. See Dieudonné (1960, p. 134), Taylor (1985, p. 175, Problem 4 with hints), Kuller (1969, p. 120, Theorem 4.6.5), and Kelly (1955, p. 245, Problem $S(d)$). ∎

A3. MEASURABILITY

A sigmafield $\mathcal{F}$ of subsets of a nonempty set Ω is a family of subsets of Ω which contains ϕ (the empty set) and Ω and is closed under the operations of complementation, countable union, and countable intersection. The pair $(\Omega, \mathcal{F})$ consisting of a set Ω and a sigmafield of subsets of Ω is called a *measurable space*. Subsets of Ω that belong to $\mathcal{F}$ are called $\mathcal{F}$-measurable.

Given a class $\mathcal{C}$ of subsets Ω, the smallest sigmafield containing $\mathcal{C}$ (i.e., the intersection of all sigmafields containing $\mathcal{C}$) is called the sigmafield *generated by $\mathcal{C}$*.

Let $(\Omega, \mathcal{F})$ and $(\Omega', \mathcal{F}')$ be measurable spaces; $\mathcal{F}[\mathcal{F}']$ is a sigmafield of subsets of $\Omega[\Omega']$. A *mapping* $h : \Omega \to \Omega'$ from Ω into Ω' *is* said to be *measurable* $(\mathcal{F}, \mathcal{F}')$ if the inverse image $h^{-1}M'$ belongs to $\mathcal{F}$ for each M' in $\mathcal{F}'$. If $h^{-1}\mathcal{F}'$ denotes the family $\{h^{-1}M' : M' \in \mathcal{F}'\}$, this condition is formally stated as $h^{-1}\mathcal{F}' \subset \mathcal{F}$. Since $\{M' : h^{-1}M' \in \mathcal{F}\}$ is a sigmafield, *if $\mathcal{F}_0'$ is contained in $\mathcal{F}'$ and generates it, then $h^{-1}\mathcal{F}_0' \subset \mathcal{F}$ implies*

$h^{-1}\mathcal{F}' \subset \mathcal{F}$. To simplify notation, *measurable* stands for *measurable* $(\mathcal{F}, \mathcal{F}')$.

Let $(\Omega'', \mathcal{F}'')$ be a third measurable space, let $j : \Omega' \to \Omega''$ map Ω' into Ω'', and denote by jh the composition of h and $j : (jh)(\omega) = j(h(\omega))$. It is easy to show that, if $h^{-1}\mathcal{F}' \subset \mathcal{F}$ and $j^{-1}\mathcal{F}'' \subset \mathcal{F}'$, then $(jh)^{-1}\mathcal{F}'' \subset \mathcal{F}$. Hence, the *composition* of two (or more) measurable maps is measurable.

Let S be a metric space. The *Borel sigmafield* of S, denoted by $\mathcal{B}(S)$ (with abandon, by $\mathcal{S}$, to simplify notation), is the sigmafield generated by the open sets of S. Since every closed set is the complement of an open set, $\mathcal{S}$ is also the smallest sigmafield of subsets of S which contains all closed subsets of S.

If $\Omega = S$ and $\Omega' = S'$ are metric spaces, h is continuous when $h^{-1}G'$ is open in S for each open G' in S'. Let $\mathcal{S}$ and $\mathcal{S}'$ be the sigmafield of Borel sets in S and S'. If h is continuous, then, since $h^{-1}G' \in \mathcal{S}$ for G' open in S', and since the open sets in S' generate $\mathcal{S}'$, $h^{-1}\mathcal{S}' \subset \mathcal{S}$, so that h is measurable.

A3.1. Subspaces

Let S be a metric space, and let $\mathcal{S}$ be its sigmafield of Borel sets. A subset S_0 (not necessarily in $\mathcal{S}$) is a metric space in the relative topology. Let $\mathcal{S}_0$ be the sigmafield of Borel sets in S_0. We shall prove

$$\mathcal{S}_0 = \{S_0 \cap A : A \in \mathcal{S}\}. \tag{A.4}$$

Proof. If $h(x) = x$ for $x \in S_0$, then h is a continuous mapping from S_0 to S and hence $h^{-1}\mathcal{S} \subset \mathcal{S}_0$, so that $S_0 \cap A \in \mathcal{S}_0$ if $A \in \mathcal{S}$. Since $\{S_0 \cap A : A \in \mathcal{S}\}$ is a sigmafield in S_0 and contains all sets $S_0 \cap G$ with G open in S, that is, all open sets in S_0, (A.4) follows. ∎

If S_0 lies in $\mathcal{S}$, (A.4) becomes

$$\mathcal{S}_0 = \{A : A \subset S_0, \ A \in \mathcal{S}\}. \tag{A.5}$$

A3.2. Product Spaces: Separability Once Again

Let S' and S'' be metric spaces with metrics d' and d'' and sigmafields $\mathcal{S}'$ and $\mathcal{S}''$ of Borel sets. The rectangles

$$A' \times A'' \tag{A.6}$$

with A' open in S' and A'' open in S'' are a basis for the *product topology* in $S = S' \times S''$. This topology may also be described as the one under which $(x'_n, x''_n) \to (x', x'')$ if and only if $x'_n \to x'$ and $x''_n \to x''$. Finally, the topology may be specified by various metrics, for example,

$$d((x', x''), (y', y'')) = \sqrt{[d'(x', y')]^2 + [d''(x'', y'')]^2} \qquad \text{(A.7)}$$

and

$$d((x', x''), (y', y'')) = \max\{d'(x', y'), d''(x'', y'')\}. \qquad \text{(A.8)}$$

Let the *product sigmafield* $S' \otimes S''$ be the sigmafield generated by the measurable rectangles (sets $(A' \times A'')$ with $A' \in S'$ and $A'' \in S''$), and let S be the sigmafield of Borel sets in S for the product topology. One can show (see Billingsley 1968, pp. 224, 225)

$$S' \otimes S'' \subset S. \qquad \text{(A.9)}$$

Now S is separable if and only if S' and S'' are both separable. Let us show that, if S is separable, then

$$S' \otimes S'' = S. \qquad \text{(A.10)}$$

Proof. In view of (A.9) it remains to prove that if G is open in S, then $G \in S' \otimes S''$. But G is a union of rectangles of the form $A' \times A''$ with A' open in S'. A'' open in S'' (so that $A' \times A'' \in S' \otimes S''$), and if S is separable G is a countable such union. ∎

It should be stressed that without separability (A.10) may fail.

Let (S, d) be a separable metric space, and S its Borel sigmafield. Consider the (Kolmogorov) product sigmafield $S^{\otimes \infty}$. Endow S^∞ with the product topology, and the metric

$$d^\infty(\mathbf{x}, \mathbf{y}) = \sum_{n=0}^{\infty} \frac{1}{2^n}(d(x_m, y_n) \wedge 1) \quad \forall (\mathbf{x}, \mathbf{y}) \in S^\infty.$$

For the next two exercises, see Parthasarathy (1967, pp. 5–6).

Exercise A.1

(a) Show that d^∞ is a metric on S^∞.
(b) Show that (S^∞, d^∞) is separable if (S, d) is separable. ∎

Exercise A.2 Prove that the Borel sigmafield of (S^∞, d^∞) is the same as the product (Kolmogorov) sigmafield $S^{\otimes\infty}$, where S is the Borel sigmafield of S). [Hint: One needs to prove that the class $\mathcal{O}_1$ of open sets of S^∞ under the product topology is the same as the class $\mathcal{O}_2$ of open sets of S^∞ under the metric d^∞.] ∎

A3.3. The Support of a Measure

Let S be a separable metric space and P a probability measure on $\mathcal{B}(S)$.

Theorem A.10 *There is a unique closed set* $\mathrm{supp}(P)$ *satisfying*

(i) $P[\mathrm{supp}(P)] = 1$.

(ii) *If* D *is any closed set such that* $P(D) = 1$, *then* $\mathrm{supp}(P) \subset D$.

(iii) $\mathrm{supp}(P)$ *is the set of all points* $s \in S$ *having the property that* $P(U) > 0$ *for each open set* U *containing* s.

Proof. Let $\mathcal{U} = \{U : U \text{ open}, P(U) = 0\}$. Since S is separable, there are countably many open sets $U_1, U_2, \ldots$ such that $\bigcup_n U_n = \bigcup\{U : U \in \mathcal{U}\}$. Write $U_p \equiv \bigcup_n U_n$, and $\mathrm{supp}(P) = S \backslash U_p$. Since $P(U_p) = P(\bigcup_n U_n) \le \sum_n P(U_n) = 0$, $P[\mathrm{supp}(P)] = 1$.

If D is any closed set with $P(D) = 1$, then $S \backslash D \in \mathcal{U}$; hence, $S \backslash D \subset U_p$, i.e., $\mathrm{supp}(P) \subset D$.

To prove (iii), let $s \in U_p = S \backslash \mathrm{supp}(P)$. U_p is an open set containing s and $P(U_p) = 0$. On the other hand, if $s \in \mathrm{supp}(P)$, and U is an open set containing s, it must be that $P(U) > 0$: otherwise $U \subset U_p$, by the definition of U_p. The uniqueness of $\mathrm{supp}(P)$ is obvious. ∎

We refer to the set $\mathrm{supp}(P)$ as the *support of* P.

A3.4. Change of Variable

If P is a probability measure on $(\Omega, \mathcal{F})$, and if $h^{-1}\mathcal{F}' \subset \mathcal{F}$, Ph^{-1} denotes the probability measure on $(\Omega', \mathcal{F}')$ defined by $(Ph^{-1})(M') = P(h^{-1}M')$ for $M' \in \mathcal{F}'$. If f is a real function on Ω', measurable $\mathcal{F}'$, then the real function fh on Ω is measurable $\mathcal{F}$.

Under these circumstances, f is integrable with respect to Ph^{-1} if and only if fh is intergrable with respect to P, in which case we have

$$\int_{h^{-1}M'} f(h(\omega))P(d\omega) = \int_{M'} f(\omega')Ph^{-1}(d\omega') \qquad (A.11)$$

for each M' in $\mathcal{F}'$.

Note that (A.11) holds for indicator functions $F = \mathbf{1}_{A'}(A' \in \mathcal{F}')$ and, therefore, for simple functions $f = \sum_1^n c_i \mathbf{1}_{A_i}$ ($A_i \in \mathcal{F}'$ are pairwise disjoint). The general equality (A.11) now follows on approximating f by simple functions.

If X takes Ω into a metric space S and $X^{-1}S' \subset S$, so that X is a random element of S, we generally write $E\{f(X)\}$ to denote the *expectation* of $f(X)$ in place of $\int f(X(\omega))P(d\omega)$. If $P = PX^{-1}$ is the distribution of X in S, (A.11) implies

$$E\{f(X)\} = \int_S f(x)P(dx). \qquad (A.12)$$

Let X be a nonnegative and integrable random variable on a probability space $(\Omega, \mathcal{F}, P)$.

Then, as proved under Exercise 3.1, Chapter 5,

$$E\{X\} = \int_0^\infty P\{X \geq t\}\, dt. \qquad (A.13)$$

(The equation also holds in the extended sense that if one side of (A.13) is infinite, so is the other.)

Let X be a real-valued random variable and g a nonnegative Borel function such that $E\{g(X)\}$ is finite. Assume that $g(-x) = g(x)$, and g is nondecreasing on $[0, \infty)$. Then for every $\varepsilon > 0$

$$P[|X| \geq \varepsilon] \leq \frac{E\{g(X)\}}{g(\varepsilon)}. \qquad (A.14)$$

Proof. Let $A = \{|X| \geq \varepsilon\}$. Then

$$\begin{aligned} E\{g(X)\} &= \int_A g(X)\, dP + \int_{A^C} g(X)\, dP \\ &\geq g(\varepsilon)P(A). \end{aligned} \qquad \blacksquare$$

In particular, with $g(x) = |x|^\lambda$ ($\lambda > 0$), one has the *Chebyshev inequality*; *if* $E|X|^\lambda$ *is finite for some* $\lambda > 0$, *then for every* $\varepsilon > 0$,

$$P[|X| \geq \varepsilon] \leq \frac{E|X|^\lambda}{\varepsilon^\lambda}. \tag{A.14'}$$

A4. BOREL-CANTELLI LEMMA

Let $(\Omega, \mathcal{F}, P)$ be a probability space. Here $(\Omega, \mathcal{F})$ is a measurable space. We refer to the elements of $\mathcal{F}$ as *events*, for example, Ω as the "sure event," ϕ as the "impossible event."

Two immediate consequences of the *countable additivity* property of probability measures are the so-called *continuity* properties:

If $A_1 \subset A_2 \subset \cdots$ *is a nondecreasing sequence of events in* $\mathcal{F}$, *then*

$$P\left(\bigcup_{i=1}^\infty A_i\right) = \lim_{n\to\infty} P(A_n). \tag{A.15}$$

To prove this, define $B_n \equiv A_n \setminus A_{n-1}$ for $n \geq 1$, $A_0 \equiv \phi$ and apply countable additivity to $\bigcup_{n=1}^\infty B_n = \bigcup_{n=1}^\infty A_n$.

By considering complements we can also conclude the following:

Let $A_1 \supset A_2 \supset \cdots$ *be a nonincreasing sequence of events in* $\mathcal{F}$, *then*

$$P\left(\bigcap_{i=1}^\infty A_i\right) = \lim_{n\to\infty} P(A_n). \tag{A.16}$$

We now come to the powerful Borel-Cantelli lemma that has been repeatedly invoked.

Lemma A.1 *(Borel–Cantelli)*   *Let* $\{A_n\}$ *be any sequence of events in* $\mathcal{F}$.

Part 1. If $\sum_{n=1}^\infty P(A_n) < \infty$ *then*

$$P(A_n \ i.o.) := P\left(\bigcap_{n=1}^\infty \bigcup_{k=n}^\infty A_k\right) = 0.$$

Part 2. If $A_1, A_2, \ldots$ *are independent events and if* $\sum_{n=1}^\infty P(A_n)$ *diverges, then* $P(A_n \ i.o.) = 1$.

Proof. For Part 1 observe that the sequence of events $B_n = \bigcup_{k=n}^{\infty} A_k$, $n = 1, 2, \ldots$, is a decreasing sequence. Therefore, we have by the continuity property of P

$$P\left(\bigcap_{n=1}^{\infty}\bigcup_{k=n}^{\infty} A_k\right) = \lim_{n\to\infty} P\left(\bigcup_{k=n}^{\infty} A_k\right) \leqslant \lim_{n\to\infty} \sum_{k=n}^{\infty} P(A_k) = 0.$$

For Part 2 note that

$$P(\{A_n \ i.o.\}^c) = P\left(\bigcup_{n=1}^{\infty}\bigcap_{k=n}^{\infty} A_k^c\right) = \lim_{n\to\infty} P\left(\bigcap_{k=n}^{\infty} A_k^c\right) = \lim_{n\to\infty} \prod_{k=n}^{\infty} P(A_k^c).$$

But

$$\prod_{k=n}^{\infty} P(A_k^c) = \lim_{m\to\infty} \prod_{k=n}^{m}(1 - P(A_k))$$

$$\leqslant \lim_{m\to\infty} \exp\left\{-\sum_{k=n}^{m} P(A_k)\right\} = 0. \qquad \blacksquare$$

A5. CONVERGENCE

In what follows, $(\Omega, \mathcal{F}, P)$ is a probability space and $\{X_n, n \geq 1\}$, X are all real-valued random variables on $(\Omega, \mathcal{F}, P)$.

A sequence $\{X_n\}$ is said to *converge almost surely* to a random variable X (written $X_n \xrightarrow{a.s.} X$) if and only if there is a set $E \in \mathcal{F}$ with $P(E) = 0$ such that for every $\omega \in E^c$, $|X_n(\omega) - X(\omega)| \to 0$ as $n \to \infty$.

A weaker notion of convergence is that of convergence in probability. A sequence $\{X_n\}$ is said to *converge in probability* to a random variable X (written $X_n \xrightarrow{P} X$) if for every $\varepsilon > 0$

$$P[\omega \in \Omega : |X_n(\omega) - X(\omega)| \geq \varepsilon] \to 0$$

as $n \to \infty$.

One can show that

$$\text{``}X_n \xrightarrow{a.s} X\text{''} \text{ implies } \text{``}X_n \xrightarrow{P} X.\text{''}$$

Write $S_n \equiv \sum_{k=1}^{n} X_k$. We state the *weak law of large numbers*.

Theorem A.11 *Let X_n be a sequence of independent, identically distributed random variables with mean μ and a finite variance σ^2. Then*

$$\frac{S_n}{n} \xrightarrow{P} \mu.$$

Proof. Use Chebyshev's inequality.   ∎

As an application of the Borel-Cantelli lemma, we first have the following version of the *strong law of large numbers*:

Theorem A.12 *Let $X_1, X_2, \ldots$ be independent random variables having a common distribution with mean μ and finite fourth moment. Then*

$$\frac{S_n}{n} \xrightarrow{a.s.} \mu. \tag{A.17}$$

Proof. (Sketch). By expanding the fourth power, it is not difficult to calculate that

$$E\left(\left[\sum_{i=1}^{n}(X_i - \mu)\right]^4\right) = nE([X_1 - \mu]^4) + 6\binom{n}{2}\sigma^4 \le Cn^2. \tag{A.18}$$

(Here $\sigma^2 = E((X_i - \mu)^2)$ can be shown to be finite.) But applying the Chebyshev inequality we get

$$P\left(\left|\sum_{i=1}^{n}(X_i - \mu)\right| > \varepsilon n\right) \le \frac{Cn^2}{(\varepsilon n)^2}, \tag{A.19}$$

and the sum on n of the right-hand side is finite. From the Borel-Cantelli lemma we therefore can conclude that with probability 1, only finitely many of the events

$$A_n^{(\varepsilon)} = \left\{\omega\colon \left|\frac{S_n}{n} - \mu\right| > \varepsilon\right\}$$

occur; that is, $P(B_\varepsilon) = 0$, where $B_\varepsilon = \limsup A_n^{(\varepsilon)}$. The sets B_ε increase as $\varepsilon \searrow 0$ to the ω set on which $S_n/n \nrightarrow \mu$. Thus, letting $\varepsilon \searrow 0$ through a countable set of values such as k^{-1}, we have

$$P\left(\left\{\omega\colon \frac{X_1 + \cdots + X_n}{n} - \mu \nrightarrow 0\right\}\right) = P\left(\bigcup_k B_{k^{-1}}\right) = 0,$$

which proves Theorem A.11.   ∎

A more general version is now stated. For a nice proof, see Etemadi (1983), also given in Bhattacharya and Waymire (1990, pp. 647–649).

Theorem A.13 *Let $\{X_n\}$ be a sequence of independent, identically distributed real-valued random variables. Let $S_n = \sum_{i=1}^{n} X_i, n = 1, 2, \ldots$. Then the sequence $\{n^{-1} S_n\}$ converges as to a finite limit c if and only if $E|X_n| < \infty$ and in this case $E X_n = c$.*

Corollary A.2 *If $\{X_n\}$ is i.i.d. and $E X_1 = \infty$, then $n^{-1} S_n \to \infty$ a.s.*

Proof. Since $E X_1 = E X_1^+ - E X_1^-, E X_1^- < \infty$. Hence $n^{-1} \sum_1^n X_i^- \to$ a.s. $E X_1^-$. Note that $n^{-1} \sum_1^n X_i = n^{-1} \sum_1^n X_i^- + n^{-1} \sum_1^n X_i^+$. We need to show $n^{-1} \sum_1^n X_i^+ \to \infty$ a.s. Let $Y_{n,M} = \min\{X_n^+, M\}$. Then $E Y_{n,M} \leq M < \infty$. Therefore, $n^{-1} \sum_1^n Y_{i,M} \xrightarrow{a.s.} E Y_{n,M}$. Now $n^{-1} \sum_1^n X_i^+ \geq n^{-1} \sum_1^n Y_{i,M}$ $\forall n$. Therefore, $\liminf_n n^{-1} \sum_1^n X_i^+ \geq E Y_{1,M}$ a.s., $\forall M$. But $E Y_{1,M} \uparrow E X_1^+ = \infty$. ∎

Theorem A.14 *Let X_n be a sequence of i.i.d. random variables with mean 0 and variance $\sigma^2 > 0$. Then $S_n/\sqrt{n} \xrightarrow{\mathcal{L}} N(0, \sigma^2)$.*

Proof. See Proposition 4.1, Chapter 5 for a more general result. ∎

Bibliography

Akerlof, G. and Main, B. (1981): "Pitfalls in Markov Modelling of Labor Market Stocks and Flows," *Journal of Human Resources*, **16**, 141–151.

Allen, R.G.D. (1956): *Mathematical Economics*, St. Martin's Press, New York.

An, H.Z. and Huang, F.C. (1996): "The Geometrical Ergodicity of Non-Linear Autoregressive Models," *Statistica Sinica*, **8**, 943–956.

Anderson, T.W. (1971): *The Statistical Analysis of Time Series*, Wiley, New York.

Arrow, K.J. (1958): "Toward a Theory of Price Adjustment," in *The Allocation of Economic Resources*, (M. Abramovitz, ed.), Stanford University Press, Stanford, CA, pp. 41–51.

Arrow, K.J. (1974): *Essays in the Theory of Risk-Bearing*, North-Holland, Amsterdam.

Arrow, K.J., Block, H.D., and Hurwicz, L. (1959): "On the Stability of Competitive Equilibrium II," *Econometrica*, **27**, 82–109.

Arrow, K.J. and Hurwicz, L. (1958): "On the Stability of Competitive Equilibrium I," *Econometrica*, **26**, 522–552.

Arrow, K.J. and Kurz, M. (1970): *Public Investment: The Rate of Return and Optimal Fiscal Policy*, Johns Hopkins Press, Baltimore, MD.

Athreya, K.B. (2003): "Iterations of IID Random Maps on R^+," in *Probability, Statistics and their Applications: Papers in Honor of Rabi Bhattacharya* (K. Athreya, M. Majumdar, M. Puri, and E. Waymire, eds.), Institute of Mathematical Statistics, Beachwood, OH, pp. 1–14.

Athreya, K.B. (2004): "Stationary Measures for some Markov Chain Models in Ecology and Economics," *Economic Theory*, **23**, 107–122.

Athreya, K.B. and Dai, J.J. (2000): "Random Logistic Maps – I," *Journal of Theoretical Probability*, **13**, 595–608.

Athreya, K.B. and Dai, J.J. (2002): "On the Non-Uniqueness of the Invariant Probability for i.i.d. Random Logistic Maps," *Annals of Probability*, **30**, 437–442.

Athreya, K.B. and Majumdar, M. (2003): "Estimating the Stationary Distribution of a Markov Chain," *Economic Theory*, **27**, 729–742.

Athreya, K.B. and Ney, P. (1978): "A New Approach to the Limit Theory of Recurrent Markov Chains," *Transactions of the American Mathematical Society*, **245**, 493–501.

Azariadis, C. (1993): *Intertemporal Macroeconomics*, Blackwell, Cambridge, MA.

Baccelli, F. and Bremaud, P. (1994): *Elements of Queuing Theory: Palm-Martingale Calculus and Stochastic Recurrences*, Springer Verlag, Berlin.

Bala, V. and Majumdar, M. (1992): "Chaotic Tatonnement," *Economic Theory*, **2**, 437–446.

Bandopadhyay, B. (1999): *Aparajito: The Unvanquished*, Harper Collins, New Delhi (original publication in Bengal in 1932; English translation by Gopa Majumdar).

Barnsley, M. (1993): *Fractals Everywhere*, 2nd edition, Academic Press, New York.

Barnsley, M. and Elton, J. (1988): "A New Class of Markov Processes for Image Encoding," *Advances in Applied Probability*, **20**, 14–32.

Baumol, W.J. (1970): *Economic Dynamics*, The MacMillan Company, New York.

Baumol, W.J. and Benhabib, J. (1989): "Chaos: Significance, Mechanism and Economic Applications," *Journal of Economic Perspectives*, **3**, 77–105.

Becker, R. and Foias, C. (1994): "The Local Bifurcation of Ramsey Equilibrium," *Economic Theory*, **4**, 719–744 (reprinted in Majumdar, M., Mitra, T., and Nishimura, K. (2000), Chapter 5).

Becker, R.A. (1985): "Comparative Dynamics in Aggregate Models of Optimal Capital Accumulation," *Quarterly Journal of Economics*, **100**, 1235–1256.

Benhabib, J. (ed.) (1992): *Cycles and Chaos in Economic Equilibrium*, Princeton University Press, Princeton, NJ.

Benhabib, J. and Day, R. (1981): "Rational Choice and Erratic Behavior," *Review of Economic Studies*, **48**, 459–472.

Benhabib, J. and Day, R. (1982): "A Characterization of Erratic Dynamics in the Overlapping Generations Model," *Journal of Economic Dynamics and Control*, **4**, 37–55.

Benhabib, J. and Nishimura, K. (1979): "The Hopf Bifurcation and the Existence and Stability of Closed Orbits in Multisector Models of Optimal Economic Growth," *Journal of Economic Theory*, **21**, 421–444.

Benhabib, J. and Nishimura, K. (1985): "Competitive Equilibrium Cycles," *Journal of Economic Theory*, **35**, 284–306.

Berge, C. (1963): *Topological Spaces*, Oliver and Boyd, Edinburgh, UK.

Berger, M. (1992): "Random Affine Iterated Function Systems: Mixing and Encoding," in *Diffusion Processes and Related Problems in Analysis*, Vol. II. (Progress in Probability, Vol. 27) Birkhauser, Boston, MA, pp. 315–346.

Berry, D.A. and Friested, B. (1985): *Bandit Problems: Sequential Allocation of Experiments*, Chapman and Hall, London.

Bertsekas, D.P. (1995a): *Dynamic Programming and Optimal Control*, Vol. I, Athena Scientific, Belmont, MA.

Bertsekas, D.P. (1995b): *Dynamic Programming and Optimal Control*, Vol. II, Athena Scientific, Belmont, MA.

Bhaduri, A. and Harris, D.J. (1987): "The Complex Dynamics of the Simple Ricardian Model," *Quarterly Journal of Economics*, **102**, 893–902.

Bhattacharya, R.N. (1978): "Criteria for Recurrence and Existence of Invariant Measures for Multidimensional Diffusions," *Annals of Probability*, **6**, 541–553.

Bhattacharya, R.N. (1982): "On the Functional Central Limit Theorem and the Law of Iterated Logarithm for Markov Processes," *Zeitschrift fur Wahrscheinlichkeitstheorie und Verwandte Gebiete*, **60**, 185–201.

Bhattacharya, R.N. and Goswami, A. (1998): "A Class of Random Continued Fractions with Singular Equilibria," in *Proceedings of the 3rd Calcutta Triennial Symposium in Probability and Statistics*, Oxford University Press, Oxford, pp. 75–85.

Bhattacharya, R.N. and Lee, O. (1988): "Asymptotics of a Class of Markov Processes Which Are Not in General Irreducible," *Annals of Probability*, **16**, 1333–1347 (correction: Bhattacharya, R.N. and Lee, O. (1997), *Annals of Probability*, **25**, 1541–1543).

Bhattacharya, R.N. and Lee, C. (1995): "On Geometric Ergodicity of Non-Linear Autoregressive Models," *Statistics and Probability Letters*, **22**, 311–315 (correction: Bhattacharya, R.N. and Lee, C. (1999), *Statistics and Probability Letters*, **41**, 439–440).

Bhattacharya, R. and Majumdar, M. (1980): "On Global Stability of Some Stochastic Processes: A Synthesis," in *Quantitative Economics and Development* (L. Klein, M. Nerlove, and S.C. Tsiang, eds.), Academic Press, New York, pp. 19–44.

Bhattacharya, R. and Majumdar, M. (1989a): "Controlled Semi-Markov Processes: The Discounted Case," *Journal of Statistical Planning and Inference*, **21**, 365–381.

Bhattacharya, R. and Majumdar, M. (1989b): "Controlled Semi-Markov Processes: The Average Reward Criterion," *Journal of Statistical Planning and Inference*, **22**, 223–242.

Bhattacharya, R. and Majumdar, M. (1999a): "On a Theorem of Dubins and Freedman," *Journal of Theoretical Probability*, **12**, 1067–1087.

Bhattacharya, R.N. and Majumdar, M. (1999b): "Convergence to Equilibrium of Random Dynamical Systems Generated by i.i.d. Monotone Maps with Applications to Economics," *Asymptotics, Nonparametrics and Time Series* (S. Ghosh, ed.), Marcel Dekker, New York, pp. 713–742.

Bhattacharya, R.N. and Majumdar M. (2001): "On a Class of Stable Random Dynamical Systems: Theory and Applications," *Journal of Economic Theory*, **96**, 208–239.

Bhattacharya, R.N. and Majumdar M. (2002): "Stability in Distribution of Randomly Perturbed Quadratic Maps as Markov Processes," Center for Analytical Economics Working Paper 02–03, Cornell University, Ithaca, NY.

Bhattacharya, R.N. and Majumdar M. (2004): "Stability in Distribution of Randomly Perturbed Quadratic Maps as Markov Processes," *Annals of Applied Probability*, **14**, 1802–1809.

Bhattacharya, R.N. and Ranga Rao, R. (1976): *Normal Approximation and Asymptotic Expansion*, Wiley, New York.

Bhattacharya, R.N. and Rao, B.V. (1993): "Random Iterations of Two Quadratic Maps," in *Stochastic Processes: A Festschrift in Honor of Gopinath Kallianpur* (S. Cambanis, J.K. Ghosh, R.L. Karandikar, and P.K. Sen, eds.), Springer-Verlag, New York, pp. 13–21.

Bhattacharya, R.N. and Waymire, E.C. (1990): *Stochastic Processes with Applications*, Wiley, New York.

Bhattacharya, R.N. and Waymire, E.C. (2002): "An Approach to the Existence of Unique Invariant Probabilities for Markov Processes. Limit Theorems," in *Probability and Statistics-I* (I. Berkes, E. Csaki, and M. Csorgo, eds.), Janos Bolyai Soc, Budapest, pp. 181–200.

Bhattacharya, R.N. and Waymire, E.C. (2006a): *A Basic Course in Probability Theory*, Springer-Verlag, Berlin.

Bhattacharya, R.N. and Waymire, E.C. (2006b): *Theory and Applications of Probability and Stochastic Processes*, Graduate texts in Mathematics, Springer-Verlag, Berlin.

Billingsley, P. (1961): "The Lindberg–Levy Theorem for Martingales," *Proceedings of the American Mathematical Society*, **12**, 788–792.

Billingsley, P. (1965): *Ergodic Theory and Information*, Wiley, Huntington, NY.

Billingsley, P. (1968): *Convergence in Probability Measures*, Wiley, New York (2nd edition, 1999).

Billingsley, P. (1986): *Probability and Measure*, 2nd edition. Wiley, New York.

Blackwell, D. (1965): "Discounted Dynamic Programming," *Annals of Mathematical Statistics*, **36**, 226–235.

Blackwell, D. (1967): "Positive Dynamic Programming," in *Proceedings of the 5th Berkeley Symposium on Mathematical Statistics and Probability*, Vol. 1 (L. Le Cam and J. Neyman, eds.), University of California Press, Berkeley, CA, pp. 415–418.

Blumenthal, R.M. and Corson, H. (1972): "On Continuous Collection of Measures," in *Proceedings of the 6th Berkeley Symposium of Mathematical Statistics and Probability*, Vol. 5, University of California Press, Berkeley, CA, pp. 33–40.

Boldrin, M. and Montrucchio, L. (1986): "On the Indeterminacy of Capital Accumulation Paths," *Journal of Economic Theory*, **40**, 26–36.

Box, G.E.P. and Jenkins, G.M. (1976): *Time Series Analysis Forecasting and Control*, revised edition, Holden-Day, San Francisco, CA.

Brandt, A. (1986): "The Stochastic Equation $Y_{n+1} = A_n Y_{n+B} + B_n$ With Stationary Coefficients," *Advances in Applied Probability*, **18**, 211–220.

Brauer, F. and Castillo-Chavez, C. (2001): *Mathematical Models in Population Biology and Epidemiology*, Springer-Verlag, New York.

Brillinger, D.R. (1981): *Time Series Data Analysis and Theory*, expanded edition, Holden-Day, San Francisco, CA.

Brock, W. (1971): "Sensitivity of Optimal Growth Paths With Respect to a Change in Target Stock," *Zeitschrift fur Nationalokonomie Supplement*, **1**, 73–89.

Brock, W. and Gale, D. (1969): "Optimal Growth Under Factor Augmenting Progress," *Journal of Economic Theory*, **1**, 229–243.

Brock, W.A. and Mirman, L.J. (1972): "Optimal Economic Growth and Uncertainty: The Discounted Case," *Journal of Economic Theory*, **4**, 479–513.

Brock, W.A. and Mirman, L.J. (1973): "Optimal Economic Growth and Uncertainty: The No Discounting Case," *International Economic Review*, **4**, 479–513.

Brockwell, P.J. and Davis, R.A. (1991): *Time Series Theory and Methods*, 2nd edition, Springer-Verlag, New York.

Brown, B.M. (1977): "Martingale Central Limit Theorems," *Annals of Mathematical Statistics*, **42**, 59–66.

Carlsson, N. (2002): "A Contractivity Condition for Iterated Function Systems," *Journal of Theoretical Probability*, **15**, 613–630.

Cass, D. and McKenzie, L. (eds.) (1974): *Selected Readings on Macroeconomics and Capital Theory from Econometrica*, MIT Press, Cambridge, MA.

Cass, D. and Stiglitz, J.E. (1972): "Risk Aversion and Wealth Effects on Portfolios With Many Assets," *Review of Economic Studies*, **39**, 331–354.

Chakravarty, S. (1959): *Logic of Investment Planning*, North-Holland, Amsterdam.

Chakraborty, S. and Rao, B.V. (1998): "Completeness of Bhattacharya Metric on the Space of Probabilities," *Statistics and Probability Letters*, **36**, 321–326.

Chassaing, P., Letac, G., and Mora, M. (1984): "Brocot Sequences and Random Walks in SL(2, *R*)," in *Probability Measures on Groups VII, Lecture Notes in Mathematics*, Springer, Berlin, pp. 36–48.

Chesson, P.L. (1982): "The Stabilizing Effect of a Random Environment," *Journal of Mathematical Biology*, **15**, 1–36.

Chipman, J.S. (1973): "The Ordering of Portfolios in Terms of Mean and Variance," *Review of Economic Studies*, **40**, 167–190.

Chung, K.L. (1967): *Markov Chains with Stationary Transition Probabilities*, 2nd edition, Springer, New York.

Clark, C.W. (1971): "Economically Optimal Policies for the Utilization of Biologically Renewable Resources," *Mathematical Biosciences*, **17**, 245–268.

Clark, C.W. (1976): *Mathematical Bioeconomics*, Wiley, New York.

Collet, P. and Eckmann, J.P. (1980): *Iterated Random Maps on the Interval as Dynamical Systems*, Birkhauser, Boston, MA.

Dai, J.J. (2000): "A Result Regarding Convergence of Random Logistic Maps," *Statistics and Probability Letters*, **47**, 11–14.

Dana, R. and Montrucchio, L. (1987): "Dynamic Complexity in Duopoly Games," in *Nonlinear Economic Dynamics* (J.M. Grandmont, ed.), Academic Press, Boston, MA, pp. 40–56.

Danthene, J. and Donaldson, J.B. (1981a): "Stochastic Properties of Fast vs. Slow Growing Economies," *Econometrica*, **49**, 1007–1033.

Danthene, J. and Donaldson, J.B. (1981b): "Certainty Planning in an Uncertain World," *Review of Economic Studies*, **48**, 507–510.

Dasgupta, P.S. (1982): *The Control of Resources*, Blackwell, Oxford, UK.

Dasgupta, P.S. and Heal, G. (1979): *Economic Theory and Exhaustible Resources*, Cambridge University Press, Cambridge, UK.

Day, R.H. (1982): "Irregular Growth Cycles," *American Economic Review*, **78**, 406–414.

Day, R.H. (1994): *Complex Economic Dynamics*, MIT Press, Cambridge, MA.

Day, R.H. and Pianigiani, G. (1991): "Statistical Dynamics and Economics," *Journal of Economic Behavior and Organization*, **16**, 37–83.

Debreu, G. (1952): "A Social Equilibrium Existence Theorem,"*Proceedings of the National Academy of Sciences of the USA*, **38**, 886–893.

Debreu, G. (1959): *Theory of Value*, Wiley, New York.

Debreu, G. and Herstein, I.N. (1953): "Nonnegative Square Matrices," *Econometrica*, **21**, 597–607.

Dechert, W.D. and Nishimura, K. (1983): "A Complete Characterization of Optimal Growth Patterns in an Aggregated Model with a Non-Concave Production Function," *Journal of Economic Theory*, **31**, 332–354.

Demetrius, L. (1971): "Multiplicative Processes," *Mathematical Biosciences*, **12**, 261–272.

Deneckere, R. and Pelikan, S. (1986): "Competitive Chaos," *Journal of Economic Theory*, **40**, 13–25.

Devaney, R.L. (1986): *An Introduction to Chaotic Dynamical Systems*, Benjamin-Cummins, Menlo Park, CA (2nd edition, 1989, Addison-Wesley, New York).

Diaconis, P. (1988): *Group Representations in Probability and Statistics*, Institute of Mathematical Statistics, Hayward, CA.

Diaconis, P. and Freedman, D. (1999): "Iterated Random Functions," *SIAM Review*, **41**, 45–79.

Diaconis, P. and Shashahani, M. (1986): "Product of Random Matrices and Computer Image Generation," *Contemporary Mathematics*, **50**, 173–182.

Diaconis, P. and Strook, D. (1991): "Geometrical Bounds for Eigenvalues of Markov Chains," *Annals of Applied Probability*, **1**, 36–61.

Diamond, P.A. (1965): "National Debt in a Neoclassical Growth Model," *American Economic Review*, **55**, 1126–1150.

Dieudonne, J. (1960): *Foundations of Modern Analysis*, Academic Press, New York.

Diks, C. (1999): *Non-Linear Time Series Analysis*, World Scientific, Singapore.

Doeblin, W. (1937): "Sur les proprietes asymptotique du movement regis par certain types de chaines simples," *Bull. Math. Soc. Rom. Sci.*, **39**, no. 1, 57–115; no. 2, 3–61.

Doob, J.L. (1953): *Stochastic Processes*, Wiley, New York.

Dubins, L.E. and Freedman, D.A. (1966): "Invariant Probabilities for Certain Markov Processes," *Annals of Mathematical Statistics*, **37**, 837–838.

Dudley, R.M. (1989): *Real Analysis and Probability*, Wadsworth and Brooks-Cole, CA.

Dugundji, J. and Granas, A. (1982): *Fixed Point Theory*, Vol. 1, Polish Scientific Publishers, Warszawa.

Durrett, R. (1996): *Probability Theory and Examples*, 2nd edition. Wadsworth, Brooks and Cole, Pacific Grove, CA.

Dutta, P. (1987): "Capital Deepening and Impatience Equivalence in Stochastic Aggregative Growth Models," *Journal of Economic Dynamics and Control*, **18**, 1069–1092.

Dutta, P. (1991): "What Do Discounted Optima Converge To? A Theory of Discount Rate Asymptotics in Economic Models," *Journal of Economic Theory*, **55**, 64–94.

Dutta, P. (1999): *Strategies and Games: Theory and Practice*, MIT Press, Cambridge, MA.

Dutta, P., Majumdar, M., and Sundaram, R.K. (1994): "Parametric Continuity in Dynamic Programming Problem," *Journal of Economic Dynamics and Control*, **19**, 1069–1092.

Dutta, P. and Sundaram, R. (1993): "How Different Can Strategic Models Be?" *Journal of Economic Theory*, **60**, 42–61.

Dynkin, E.B. (1961): *Theory of Markov Processes*, Prentice Hall, Englewood Cliffs, NJ.

Easley, D. and Kiefer, N.M. (1988): "Controlling a Stochastic Process With Unknown Parameters," *Econometrica*, **56**, 1045–1064.

Easley, D. and Kiefer, N.M. (1989): "Optimal Learning with Endogenous Data," *International Economic Review*, **30**, 963–978.

Elaydi, S. (2000): *An Introduction to Difference Equations*, 3rd edition, Springer, New York.

Ellner, S. (1984): "Asymptotic Behavior of Some Stochastic Difference Equation Population Models," *Journal of Mathematical Biology*, **19**, 169–200.

Elton, J.H. (1990): "A Multiplicative Ergodic Theorem for Lipschitz Maps," *Stochastic Processes and Applications*, **34**, 39–47.

Etemadi, N. (1983): "On the Laws of Large Numbers for Nonnegative Random Variables," *Journal of Multivariate Analysis*, **13**, 187–193.

Evans, G.W. and Honkapohja, S. (1995): "Local Convergence of Recursive Learning to Steady States and Cycles in Stochastic Nonlinear Models," *Econometrica*, **63**, 195–206.

Evans, G.W. and Honkapohja, S. (2001): *Learning and Expectations in Macroeconomics*, Princeton University Press, Princeton, NJ.

Feller, W. (1968): *An Introduction to Probability Theory and Its Applications*, Vol. 1, 3rd edition, Wiley, New York.

Fill, J.A. (1991): "Eigenvalue Bounds on Convergence to Stationarity for Non-reversible Markov Chains, With an Application to the Exclusion Processes," *Annals of Applied Probability*, **1**, 62–87.

Foley, D. and Hellwig, M. (1975): "Asset Management With Trading Uncertainty," *Review of Economic Studies*, **42**, 327–246.

Frankel J., Kreiss, J.-P., and Mammen, E. (2002): "Bootstrap of Kernel Smoothing in Non-Linear Time Series," *Bernoulli*, **8**, 1–37.

Frisch, R. (1933): "Propagation Problems and Impulse Problems in Dynamic Economics," in *Economic Essays in Honor of Gustav Cassel*, Allen and Unwin, London, pp. 171–205 (reprinted in *Readings in Business Cycles* (R.A. Gordon and L. Klein, eds., Chapter 9), Richard D. Irwin, Homewood, IL, pp. 155–185).

Frisch, R. and Holme, H. (1935): "The Characteristic Solutions of Mixed Difference and Differential Equations," *Econometrica*, **3**, 225–239.

Furukawa, N. (1972): "Markovian Decision Processes With Compact Action Spaces," *Annals of Mathematical Statistics*, **43**, 1612–1622.

Futia, C.A. (1982): "Invariant Distributions and Limiting Behavior of Markovian Economic Models," *Econometrica*, **50**, 377–408.

Gale, D. (1960): *The Theory of Linear Economic Models*, McGraw-Hill, New York.

Gale, D. (1967): "On the Optimal Development in a Multi-Sector Economy," *Review of Economic Studies*, **34**, 1–18.

Gale, D. (1973): "Pure Exchange Equilibrium of Dynamic Economic Models," *Journal of Economic Theory*, **6**, 12–36.

Galor, O. and Ryder, H.E. (1989): "Existence, Uniqueness and Stability of Equilibrium in an Overlapping Generations Model With Productive Capital," *Journal of Economic Theory*, **49**, 360–375.

Gantmacher, F.R. (1960): *The Theory of Matrices*, Vols. I and II, Chelsea, New York.

Gittins, J.C. (1989): *Multi-Armed Bandit Allocation Indices*, Wiley, New York.

Goldberg, S. (1958): *Introduction to Difference Equations*, Wiley, New York.

Goodwin, R. (1990): *Chaotic Economic Dynamics*, Clarendon Press, Oxford, UK.

Gordin, M.I. and Lifsic, B.A. (1978): "The Central Limit Theorem for Stationary Markov Processes," *Doklady Akademii Nauk SSR*, **19**, 392–393.

Gordon, R.A. and Klein, L. (eds.) (1965): *Readings in Business Cycles*, Richard D. Irwin, Homewood, IL.

Goswami, A. (2004): "Random Continued Fractions: A Markov Chain Approach," *Economic Theory*, **23**, 85–106.

Grandmont, J.M. (1985): "On Endogenous Competitive Business Cycles," *Econometrica*, **53**, 995–1045.

Grandmont, J.M. (1986): "Periodic and Aperiodic Behavior in Discrete One-Dimensional Dynamical Systems," in *Contributions to Mathematical Economics in Honor of Gerard Debreu* (W. Hildenbrand and A. MasColell, eds.), North Holland, New York.

Grandmont, J.M. (ed.) (1987): *Nonlinear Economic Dynamics*, Academic Press, Boston, MA.

Granger, C.W.J. and Terasvirta, T. (1993): *Modelling Non-linear Economic Relationships*, Oxford University Press, Oxford, UK.

Green, J.R. and Majumdar, M. (1975): "The Nature of Stochastic Equilibria," *Econometrica*, **46**, 647–660.

Grenander, U. and Rosenblatt, M. (1957): *Statistical Analysis of Stationary Time Series*, Wiley, New York.

Guckenheimer, J. (1979): "Sensitive Dependence to Initial Conditions for One Dimensional Maps," *Communications in Mathematical Physics*, **70**, 133–160.

Haavelmo, T. (1943): "The Statistical Implications of a System of Simultaneous Equations," *Econometrica*, **11**, 1–112.

Hakansson, H. (1970): "Optimal Investment and Consumption for a Class of Utility Functions," *Econometrica*, **38**, 587–607.

Hannan, E.J. (1970): *Multiple Time Series*, Wiley, New York.

Hardin, D.P., Takáč, P., and Webb, G.F. (1988): "Asymptotic Properties of a Continuous-Space Discrete-Time Population Model in a Random Environment," *Journal of Mathematical Biology*, **26**, 361–374.

Harris, T.E. (1956): The Existence of Stationary Measures for Certain Markov Processes, in *Proceedings of the 3rd Berkeley Symposium of Mathematical Statistics and Probability*, Vol. 2, University of California Press, Berkeley, CA, pp. 113–124.

Hassell M.P. (1975): "Density Dependence in Single-Species Population," *Journal of Animal Ecology*, **44**, 283–295.

Hildenbrand, W. (1974): *Core and Equilibria of a Large Economy*, Princeton University Press, Princeton, NJ.

Honkapohja, S. and Mitra, K. (2003): "Learning with Bounded Memory in Stochastic Models," *Journal of Economic Dynamics and Control*, **27**, 1437–1457.

Hopenhayn, H. and Prescott, E.C. (1992): "Stochastic Monotonicity and Stationary Distributions for Dynamic Economies," *Econometrica*, **60**, 1387–1406.

Hurwicz, L. (1944): "Stochastic Models of Economic Fluctuations," *Econometrica*, **12**, 114–124.

Hwang, E. (2002): *Nonlinear Estimation for Nonlinear Autoregressive Processes*, Ph.D. Dissertation, Department of Mathematics, Indiana University.

Ibragimov, A. (1963): "A Central Limit Theorem for a Class of Dependent Random Variables," *Theoretical Probability and Applications*, **8**, 83–89.

Jacquette, D. (1972): "A Discrete-Time Population Control Model with Set-Up Cost," *Operations Research*, **22**, 298–303.

Jain, N. and Jamison, B. (1967): "Contributions to Doeblin's Theorem of Markov Processes," *Zeitschrift fur Wahrscheinlichkeitstheorie und Verwandte Gebiete*, **8**, 19–40.

Jakobson, M.V. (1981): "Absolutely Continuous Invariant Measures for One-Parameter families of One-Dimensional Maps," *Communications in Mathematical Physics*, **81**, 39–88.

Jones, L.E. and Manuelli, R.E. (1990): "A Convex Model of Equilibrium Growth: Theory and Policy Implications," *Journal of Political Economy*, **98**, 1008–1038.

Kac, M. (1947): "Random Walk and The Theory of Brownian Motion," *American Mathematical Monthly*, **54**, 369–391.

Kamihigashi, T. and Roy, S. (in press): "A Non-Smooth, Non-Convex Model of Economic Growth," *Journal of Economic Theory*.

Kaplan, D. and Glass, L. (1995): *Understanding Nonlinear Dynamics*, Springer-Verlag, New York.

Karlin, S. and Taylor, M. (1975): *A First Course in Stochastic Processes*, 2nd edition. Academic Press, New York.

Karlin, S. and Taylor, M. (1981): *A Second Course in Stochastic Processes*, Academic Press, New York.

Katok, A. and Kifer, Y. (1986). "Random Perturbations of Transformations of an Interval," *Journal D'Analyse Mathematique*, **47**, 193–237.

Kelly, J.L. (1955): *General Topology*, D. van Nostrand, Princeton, NJ.

Khas'minskii, R.Z. (1960): "Ergodic Properties of Recurrent Diffusion Processes and Stabilization of the Solution to the Cauchy Problem for Parabolic Equations," *Theory of Probability and Applications*, **5**, 196–214.

Khinchin, A. Ya. (1964): *Continued Fractions*, English translation of the 3rd Russian edition, Clarendon Press, Oxford, UK.

Kifer, Y. (1986): *Ergodic Theory of Random Transformations*, Birkhauser, Boston, MA.

Kifer, Y. (1988). *Random Perturbations of Dynamical Systems*, Birkhauser, Boston, MA.

Kolmogorov, A.N. (1936): "Anfangsgrunde der Markffschen Ketten mit unendlich vielen moglichen Zustanden," *Matematicheskii Sbornik*, **1**, 607–610.

Kolmogorov, A.N. (1950): *Foundations of the Theory of Probability*, Translation from the Original German edition published by Springer-Verlag in 1933, Chelsea, New York.

Koopmans, T.C. (1957): *Three Essays on the State of Economic Science*, McGraw-Hill, New York.

Koopmans, T.C. (1967): "Objectives, Outcomes and Constraints in Optimal Growth Models," *Econometrica*, **35**, 1–15.

Kot, M. and Schaffer, W.M. (1986): "Discrete-Time Growth-Dispersal Models," *Mathematical Biosciences*, **80**, 109–136.

Kuller, R.G. (1969): *Topics in Modern Analysis*, Prentice-Hall, Englewood Cliff, NJ.

Kuratowski, K. and Ryll-Nardzewski, C. (1965): "A General Theorem on Selectors," *Bulletin of the Polish Academy of Sciences*, **13**, 397–403.

Kurz, M. (1968): "Optimal Growth and Wealth Effects," *International Economic Review*, **9**, 348–357.

Lasota, A. and Mackey, M.C. (1989): "Stochastic Perturbation of Dynamical Systems: The Weak Convergence of Measures," *Journal of Mathematical Analysis and Applications*, **138**, 232–248.

Letac, G. and Seshadri, V. (1983): "A Characterization of the Generalized Inverse Gaussian Distribution by Continued Fractions" *Zentralblatt Wahrscheinlichkeitstheorie Verw. Gebiete*, **62**, 485–489.

Levhari, D. and Srinivasan, T. (1969): "Optimal Savings under Uncertainty," *Review of Economic Studies*, **36**, 153–163 (correction: Levhari, D. and Srinivasan, T. (1977), *Review of Economic Studies*, **44**, p. 197.

Li, T. and Yorke, J. (1975): "Period Three implies Chaos," *American Mathematical Monthly*, **82**, 985–992.

Ljungqvist, L. and Sargent, T.J. (2000): *Recursive Macroeconomic Theory*, MIT Press, Cambridge, MA.

Loeve, M. (1963): *Probability Theory*, Von Nostrand, Princeton, NJ.

Maitra, A. (1968): "Discounted Dynamic Programming on Compact Metric Spaces," *Sankhya Series*, A **27**, 241–248.

Maitra, A. and Parthasarathy, T. (1970): "On Stochastic Games," *Journal of Optimization Theory and Its Applications*, **5**, 289–300.

Majumdar, M. (1974): "Efficient Programs in Infinite Dimensional Spaces: A Complete Characterization," *Journal of Economic Theory*, **7**, 355–369.

Majumdar, M. (1988): "Decentralization in Infinite Horizon Economies: An Introduction," *Journal of Economic Theory*, **45**, 217–227.

Majumdar, M. (ed.) (1992): *Decentralization in Infinite Horizon Economies*, Westview Press, Boulder, CO.

Majumdar, M. (2006): "Intertemporal Allocation with a Non-Convex Technology," in *Handbook of Optimal Growth – Vol. I: The Discrete Time* (C. Le Van, R.A. Dana, T. Mitra, and K. Nishimura, eds.), Springer, Berlin, pp. 171–203.

Majumdar, M. and Mitra, T. (1982): "Intertemporal Allocation with a Non-convex Technology," *Journal of Economic Theory*, **27**, 101–136.

Majumdar, M. and Mitra, T. (1983): "Dynamic Optimization with Non-Convex Technology: The Case of a Linear Objective Function," *Review of Economic Studies*, **50**, 143–151.

Majumdar, M. and Mitra, T. (1994a): "Periodic and Chaotic Programs of Optimal Intertemporal Allocation in an Aggregative Model With Wealth Effects," *Economic Theory*, **4**, 649–676 (reprinted in Majumdar, M., Mitra, T., and Nishimura, K. (2000), Chapter 3).

Majumdar, M. and Mitra, T. (1994b): "Robust Ergodic Chaos in Discounted Dynamic Optimization Model," *Economic Theory*, **4**, 677–688 (reprinted in Majumdar, M., Mitra, T., and Nishimura, K. (2000), Chapter 7).

Majumdar, M., Mitra, T., and Nishimura K. (eds.) (2000): *Optimization and Chaos*, Springer, New York.

Majumdar, M., Mitra, T., and Nyarko, Y. (1989): "Dynamic Optimization Under Uncertainty: Non-Convex Feasible Set," *Joan Robinson and Modern Economic Theory* (G.R. Feiwel, ed.), MacMillian, New York, pp. 545–590.

Majumdar, M. and Nermuth, M. (1982): "Dynamic Optimization in Non-convex Models with Irreversible Investment," *Journal of Economics*, **42**, 339–362.

Majumdar, M. and Radner, R. (1991): "Linear Models of Economic Survival Under Production Uncertainty," *Economic Theory*, **11**, 13–30.

Majumdar, M. and Radner, R. (1992): "Survival Under Production Uncertainty," in *Equilibrium and Dynamics: Essays in Honor of David Gale* (M. Majumdar, ed.), MacMillan, London, pp. 179–200.

Majumdar, M. and Sundaram, R. (1991): "Symmetric Stochastic Games of Resource Extraction: The Existence of Non-Randomized Stationary Equilibrium," in *Stochastic Games and Related Topics: Essays in Honor of Lloyd Shapley* (T.E.S. Raghavan, T.S. Ferguson, T. Parthasarathy, and O.J. Vrieze, eds.), Kluwer Academic Publishers, Norwell, MA, pp. 175–190.

Majumdar, M. and Zilcha, I. (1987): "Optimal Growth in a Stochastic Environment: Some Sensitivity and Turnpike Results," *Journal of Economic Theory*, **43**, 116–134.

Mann, H.B. and Wald, A. (1943): "On the Statistical Treatment of Linear Stochastic Difference Equations," *Econometrica*, **11**, 173–220.

May, R.M. (1976): "Simple Mathematical Models With Very Complicated Dynamics," *Nature*, **261**, 459–467.

McFadden, D. (1967): "The Evaluation of Development Programmes," *Review of Economic Studies*, **34**, 25–51.

Metzler, L.A. (1941): "The Nature and Stability of Inventory Cycles," *Review of Economic Statistics*, **23**, 113–129.

Meyn, S. and Tweedie, R.L. (1993): *Markov Chains and Stochastic Stability*, Springer-Verlag, New York.

Miller, B.L. (1974): "Optimal Consumption With a Stochastic Income Stream," *Econometrica*, **42**, 253–266.

Miller, B.L. (1976): "The Effect on Optimal Consumption of Increased Uncertainty in Labor Income in the Multiperiod Case," *Journal of Economic Theory*, **13**, 154–167.

Mirman, L. and Zilcha, I. (1975): "Optimal Growth Under Uncertainty," *Journal of Economic Theory*, **11**, 329–339.

Misiurewicz, M. (1981): "Absolutely Continuous Measures for Certain Maps of an Interval," *Publication Mathematiques*, **52**, 17–51.

Misiurewicz, M. (1983): "Maps of an Interval," in *Chaotic Behavior of Deterministic Systems* (G. Iooss, R.H.G. Hellerman, and R. Stora, eds.) North-Holland, Amsterdam.

Mitra, K. (1998): "On Capital Accumulation Paths in a Neoclassical Stochastic Growth Model," *Economic Theory*, **11**, 457–464.

Mitra, T. (1983): "Sensitivity of Optimal Programs With Respect to Changes in Target Stocks: The Case of Irreversible Investment," *Journal of Economic Theory*, **29**, 172–184.

Mitra, T. (1998): "On Equilibrium Dynamics Under Externalities in a Model of Economic Development," *Japanese Economic Review*, **49**, 85–107.

Mitra, T. (2000): "Introduction to Dynamic Optimization Theory," in *Optimization and Chaos* (M. Majumdar, T. Mitra, and K. Nishimra, eds.), Springer, New York, Chapter 2.

Mitra, T., Montrucchio, L., and Privileggi P. (2004): "The Nature of Steady State in Models of Optimal Growth Under Uncertainty," *Economic Theory*, **23**, 39–72.

Mitra, T. and Ray, D. (1984): "Dynamic Optimization on a Non-Convex Feasible Set," *Journal of Economics*, **44**, 151–171.

Morishima, M. (1964): *Equilibrium, Stability and Growth*, Oxford University Press, London.

Nash, J.F. (1950): "Equilibrium Points in N-person Games,"*Proceedings of the National Academy of Sciences of the USA*, **36**, 48–49.

Nelson, E. (1959): "Regular Probability Measure on Function Space," *Annals of Mathematics*, **69**, 630–643.

Nermuth, M. (1978): "Sensitivity of Optimal Growth Paths With Respect to a Change in the Target Stocks or in the Length of the Planning Horizon in a Multi-Sector Model," *Journal of Mathematical Economics*, **5**, 289–301.

Neveu, J. (1965): *Mathematical Foundation of the Calculus of Probability*, Holden-Day, San Francisco, CA.

Nikaido, H. (1968): *Convex Structures and Economic Theory*, Academic Press, New York.

Nishimura, K., Sorger, G., and Yano, M. (1994): "Ergodic Chaos in Optimal Growth Models With Low Discount Rate," *Economic Theory*, pp. 705–718 (reprinted in M. Majumdar, T. Mitra, and K. Nishimura, (eds.) (2000), Chapter 9).

Norman, M.F. (1972): *Markov Processes and Learning Models*, Academic Press, New York.

Nowak, A.S. (1985): "Existence of Equilibrium Stationary Strategies in Discounted Non-Cooperative Stochastic Games With Uncountable State Space," *Journal of Optimization Theory and Applications*, **45**, 591–602.

Nummelin, E. (1978a): "A Splitting Technique for Harris Recurrent Chains," *Zeitschrift fur Wahrscheinlichkeitstheorie und Verwandte Gebiete*, **43**, 309–318.

Nyarko, Y. and Olson, L.J. (1991): "Stochastic Dynamic Models With Stock-Dependent Rewards," *Journal of Economic Theory*, **55**, 161–167.

Nyarko, Y. and Olson, L.J. (1994): "Stochastic Growth When Utility Depends on Both Consumption and the Stock Level," *Economic Theory*, **4**, 791–798.

Orey, S. (1971): *Limit Theorems for Markov Chain Transition Probabilities*, Von Nostrand, New York.

Parthasarathy, K.R. (1967): *Probability Measures on Metric Spaces,* Academic Press, New York.

Parthasarathy, T. (1973): "Discounted, Positive and NonCooperative Stochastic Games," *International Journal of Game Theory*, **2**, 25–37.

Peres, Y. and Solomyak, B. (1996): "Absolute Continuity of Bernoulli Convolutions, a Simple Proof," *Mathematical Research Letters*, **3**, 231–239.

Peres, Y., Shlag, W., and Solomyak, B. (1999): "Sixty Years of Bernoulli Convolutions," in Fractal Geometry and Stochastics, vol. 2 (C. Bandt, S. Graf, M. Zahle, eds.), Birkhauser, Basel, 39–65.

Phelps, E. (1962): "Accumulation of Risky Capital," *Econometrica*, **30**, 729–743.

Pollard, D. (1984): *Convergence of Stochastic Processes*, Springer-Verlag, New York.

Polya, G. (1921): "Uber eine Aufgabe der Warschinlichkeitsrechnung betreffend die irrfahrt im Strassennetz," *Mathematische Annalen*, **84**, 149–160.

Pristley, M.B. (1981): *Spectral Analysis and Time Series*, Vol. 1, Academic Press, New York.

Radner, R. (1966): "Optimal Growth in a Linear-Logarithmic Economy," *International Economic Review*, **7**, 1–33.

Radner, R. (1967): "Dynamic Programming of Economic Growth," in *Activity Analysis in the Theory of Growth and Planning* (M. Bacharach and E. Malinvaud, eds.), Macmillan, London, Chapter 2, pp. 111–141.

Radner, R. (1986): "Behavioral Models of Stochastic Control," in *Studies in Mathematical Economics* (S. Reiter, ed.), The Mathematical Association of America, Rhode Island, pp. 351–404.

Radner, R. and Rothschild, M. (1975): "On the Allocation of Effort," *Journal of Economic Theory*, **10**, 358–376.

Ramsey, F. (1928): "A Mathematical Theory of Savings," *Economic Journal*, **38**, 543–549.

Rebelo, S. (1991): "Long-Run Policy Analysis and Long-Run Growth," *Journal of Political Economy*, **99**, 500–521.

Reed, W. (1974): "A Stochastic Model of the Economic Management of a Renewable Animal Resources," *Mathematical Biosciences*, **22**, 313–334.

Rhenius, D. (1974): "Incomplete Information in Markovian Decision Models," *Annals of Statistics*, **2**, 1327–1334.

Ricker, W.E. (1954): "Stock and Recruitment," *Journal of the Fisheries Research Board of Canada*, **11**, 559–623.

Rosenblatt, M. (2000): *Gaussian and Non-Gaussian Linear Time Series and Random Fields*, Springer, New York.

Roskamp, K.W. (1979): "Optimal Economic Growth if Society Derives Benefits From a Capital Stock," *Public Finance*, **34**, 31–34.

Ross, S.M. (1983): *Introduction to Stochastic Dynamic Programming*, Academic Press, New York.

Rothschild, M. (1974): "A Two-Armed Bandit Theory of Market Pricing," *Journal of Economic Theory*, **9**, 175–202.

Roy, S. (1995): "Theory of Dynamic Portfolio Choice for Survival Under Uncertainty, *Mathematical Social Sciences*, **30**, 171–194.

Royden, H. (1968): *Real Analysis*, 2nd edition, Macmillan, New York.

Rudin, W. (1976): *Principles of Mathematical Analysis*, McGraw-Hill, New York.

Ruelle, D. (1991): *Chance and Chaos*, Princeton University Press, Princeton, NJ.

Saari, D.G. (1985): "Iterative Price Mechanisms," *Econometrica*, **53**, 1117–1132.

Samuelson, P.A. (1939): "Interactions Between the Multiplier Analysis and the Principle of Acceleration," *Review of Economic Statistics*, **21**, 75–78.

Samuelson, P.A. (1941): "Conditions That a Root of a Polynomial be Less Than Unity in Absolute Value," *Annals of Mathematical Statistics*, **12**, 360–364.

Samuelson, P.A. (1947): *Foundations of Economic Analysis*, Harvard University Press, MA.

Samuelson, P.A. (1958), "An Exact Consumption Loan Model of Interest With and Without the Social Contrivance of Money," *Journal of Political Economy*, **66**, 467–482.

Sandefur, J.T. (1990): *Discrete Dynamical System*, Clarendon, Oxford, UK.

Sandmo, A. (1970): "The Effect of Uncertainty on Savings Decisions," *Review of Economic Studies*, **37**, 353–360.

Sargent, T.J. (1987): *Dynamic Macroeconomic Theory*, Harvard University Press, Cambridge, MA.

Sawarigi, Y. and Yoshikawa, T. (1970): "Discrete Time Markovian Decision Process With Incomplete State Information," *Annals of Mathematical Statistics*, **41**, 78–86.

Scarf, H. (1960): "Some Examples of Global Instability of Competitive Equilibrium," *International Economic Review*, **1**, 157–172.

Schal, M. (1975): "Conditions for Optimality in Dynamic Programming and for the Limit of n-Stage Optimal Policies to be Optimal," *Zeitschrift fur Wahrscheinlichkeitstheorie und Verwandte Gebiete*, **32**, 179–196.

Silverstrov, D.S. and Stenflo, O. (1998): "Ergodic Theorems for Iterated Function Systems Controlled by Regenerative Sequences," *Journal of Theoretical Probability*, **11**, 589–608.

Simon, H. (1959): "Theories of Decision Making in Economics and Behavioral Science," *American Economic Review*, **49**, 253–283.

Simon, H. (1986): "Theory of Bounded Rationality," in *Decision and Organization*, 2nd edition (C.B. McGuire and R. Radner, eds.), University of Minnesota Press, Inneapolis, MA, Chapter 8.

Singer, D. (1978): "Stable Orbits and Bifurcations of Maps of the Interval," *SIAM Journal on Applied Mathematics*, **35**, 260–266.

Solomyak, B. (1995): "On the Random Series $\sum \pm \lambda^n$ can Erdös Problem," *Annals of Mathematics*, **142**, 611–625.

Solow, R.M. (1956): "A Contribution to the Theory of Economic Growth," *Quarterly Journal of Economics*, **70**, 65–94.

Solow, R.M. and Samuelson, P.A. (1953): "Balanced Growth under Constant Returns to Scale," *Econometrica*, **21**, 412–424.

Sorger, G. (1992): "On the Minimum Rate of Impatience for Complicated Optimal Growth Paths," *Journal of Economic Theory*, **56**, 160–169 (reprinted in Majumdar, M., Mitra, T., and Nishimura, K. (2000), Chapter 10).

Sorger, G. (1994): "On the Structure of Ramsey Equilibrium: Cycles, Indeterminacy and Sunspots," *Economic Theory*, **4**, 745–764 (reprinted in Majumdar, M., Mitra, T., and Nishimura, K. (eds.) (2000), Chapter 6).

Spitzer, F. (1956): "A Combinatorial Lemma and Its Application to Probability Theory," *Transactions of American Mathematical Society*, **82**, 323–339.

Srinivasan, T.N. (1964): "Optimal Growth in a Two-Sector Model of Capital Accumulation," *Econometrica*, **32**, 358–373.

Stenflo, O. (2001): "Ergodic Theorems for Markov Chains Represented by Iterated Function Systems," *Bulletin of Polish Academy of Sciences Mathematics*, **49**, 27–43.

Stokey, N.L. and Lucas, R.E. (1989): *Recursive Methods in Economic Dynamics*, Harvard University Press, Cambridge, MA.

Strauch, R. (1966): "Negative Dynamic Programming," *Annals of Mathematical Statistics*, **37**, 871–890.

Sundaram, R.K. (1989): "Perfect Equilibrium in a Class of Symmetric Dynamic Games," *Journal of Economic Theory*, **47**, 153–177 (corrigendum: Sundaram, R.K. (1989): *Journal of Economic Theory*, **49**, 385–387).

Sundaram, R.K. (2004): "Generalized Bandit Problems," in *Social and Strategic Behavior: Essays in Honor of Jeffrey S. Banks* (D. Austen-Smith and J. Duggan, eds.), Springer-Verlag, Berlin, pp. 1–31.

Tamiguchi, M. and Kakizawa, Y. (2000): *Asymptotic Theory of Statistical Inference for Time Series*, Springer, New York.

Taylor, A.E. (1985): *General Theory of Functions and Integration*, Dover, New York.

Tong, H. (1990): *Nonlinear Time Series: A Dynamical Systems Approach*, Oxford University Press, Oxford, UK.

Topkis, D.M. (1978): "Minimizing a Sub-Modular Function on a Lattice," *Operations Research*, **26**, 305–321.

Torre, J. (1977): "Existence of Limit Cycles and Control in Complete Keynesian System by Theory of Bifurcations," *Econometrica*, **45**, 1457–1466.

Uzawa, H. (1964): "Optimal Growth in a Two-Sector Model of Capital Accumulation," *Review of Economic Studies*, **31**, 1–24.

Verhulst, P.F. (1845): "Recherchers Mathematiques sur la loi d'Accroissement de la Population," *Nouveaux Memoires de l'Academie Royale des Sciences et Belles-Lettres de Bruxelles*, **18**, 1–38.

Wagner, D.H. (1977): "Survey of Measurable Selection Theorems," *SIAM Journal of Control and Optimization*, **15**, 859–903.

Waymire, E. (1982): "Mixing and Cooling From a Probabilistic Point of View," *SIAM Review*, **24**, 73–75.

Wendner, R. (2003): "Existence, Uniqueness and Stability of Equilibrium in an OLG Economy," *Economic Theory*, **23**, 165–174.

Yahav, J.A. (1975): "On a Fixed Point Theorem and Its Stochastic Equivalent," *Journal of Applied Probability*, **12**, 605–611.

Young, A. (1928): "Increasing Returns and Economic Progress," *Economic Journal*, **38**, 527–542.

Zarnowitz, V. (1985): "Recent Work on Business Cycles in Historical Perspective: A Review of Theories and Evidence," *Journal of Economic Literature*, **23**, 523–580.

Author Index

Subject Index

accelerator coefficient, 105
after-n process, 143, 145
aggregative model
  of optimal economic development,
    60–66
  of optimal growth, under uncertainty,
    390–397
Alexandrov's theorem, 219–221, 223, 258,
    287
asymmetric random walk on Z^k, 158
asymptotic productivity, 74
asymptotic stationarity, 121, 196–201
autoregressive moving-average model, 306
  of order (k, q), 308
average productivity function, 24, 29

Baire category theorem, 16, 421
balanced growth and multiplicative
    processes, 53–59
Bernoulli equation, 146, 211, 366
Bernoulli innovation, 330–336
Bernoulli state, 191
Bifurcation theory, 39–46
Billingsley–Ibragimov central limit theorem,
    375
biological reproduction law, 26–28
Birkhoff's ergodic theorem, 374
birth-and-death chains, 120, 164,
    168–170
Blackwell, theorem of, 390
Boltzman's kinetic theory of matter, 170
Bolzano–Weirstrass theorem, 422–423
Borel measurable function, 96, 125,
    230–232, 408, 410, 415
Borel sets, 229, 237, 312, 426
Borel sigmafield, 122, 125, 176, 184, 191,

197, 224, 246, 260, 282, 284, 310, 380,
    386, 426
Borel subset, 235, 260, 291
  of a Polish space, 121–122, 179, 184, 214,
    284, 380
Borel–Cantelli lemma, 280–281, 298,
    300–302, 328, 430–432

canonical construction, 123
Cantor subset, 368
capital, 99–100
Cauchy sequence, 6, 180, 236–237, 254,
    258, 281, 287, 420
central limit theorem, 121, 133, 360–365
  Billingsley–Ibragimov, 375
  classical, 192
  for Markov chains, 187–191
  martingale, 360, 375–378
Chebyshev's inequality, 354, 361, 432
classical optimization models, 28
Cobb–Douglas economies, 101–104
compactness, 422–423
comparative statics and dynamics, 38–46,
    273–275
competitive equilibrium, 115
competitive programs, 66–71
completeness, 420
complex valued function, oscillation of, 229
conditional Lindeberg condition, 373
conditional variance, of $g(X_1)$, 352
conservative thermostat behavior, 175
consistency condition, 122, 202
consumption program, 18, 61, 78
consumption–investment decisions, 93
continuous time Markov models, 414
contraction mapping theorem, 179

457